COLLOIDAL PARTICLES AT LIQUID

Small solid particles adsorbed at liquid interfaces arise in many industrial products and processes, such as anti-foam formulations, crude oil emulsions and flotation. They act in many ways like traditional surfactant molecules, but offer distinct advantages. However, the understanding of how these particles operate in such systems is minimal. This book brings together the diverse topics actively being investigated with contributions from leading experts in the field.

After an introduction to the basic concepts and principles, this book is divided into two sections. The first deals with *particles at planar liquid interfaces*, with chapters of an experimental and theoretical nature. The second concentrates on the behaviour of *particles at curved liquid interfaces*, including particle-stabilised foams and emulsions, and new materials derived from such systems.

This unique collection will be of interest to academic researchers and graduate students in chemistry, physics, chemical engineering, pharmacy, food science and materials science.

BERNARD P. BINKS is Professor of Physical Chemistry at the University of Hull.

TOMMY S. HOROZOV is a Research Fellow in the Department of Chemistry at the University Hull.

Back cover illustrations (from left to right): Spontaneously formed ordered horizontal monolayer of 3 μm diameter very hydrophobic silica particles after their spreading at the silicone oil – water interface. The particle contact angle through water is 170°. Inter-particle distances are 15.4 ± 0.6 μm. Taken from Horozov and Binks, *Colloids Surf. A*, **267** (2005), 64; with permission from Elsevier.

Spontaneous formation of a crystalline disk of 3 μm very hydrophobic silica particles bridging the surfaces of a free-standing octane film in water during its thinning. The particle contact angle through water is 152°. The disk diameter is 117 μm. Taken from Horozov *et al.*, *Langmuir*, **21** (2005), 2330; with permission from the American Chemical Society.

Image of a Janus bubble, diameter ≈430 μm, covered with fluorescent polystyrene particles of diameters 4.9 μm (upper yellow half) and 4.0 μm (lower green half) using a microfluidic focusing device. Taken from Subramaniam *et al.*, *Nat. Mater.* **4** (2005), 553; with permission from the Nature Publishing Group.

Image of a silicone oil droplet of diameter ≈30 μm in water covered with simultaneously self-assembled hydrophobic (green) and hydrophilic (red) fluorescent particles of diameter 1 μm. Taken from Tarimala and Dai, *Langmuir*, **20** (2004), 3492; with permission from the American Chemical Society.

Doughnut-like toroidal assemblies of diameter ≈0.9 μm obtained by drying of aqueous droplets of a 320 nm latex particle suspension placed on the surface of perfluoromethyldecalin in the presence of fluorinated surfactant. Taken from Velev *et al.*, *Science*, **287** (2000), 2240; with permission from the American Institute for the Advancement of Science.

This book is dedicated to the memory of

PROFESSOR JOHN C. EARNSHAW (1944–1999)

John Earnshaw, Professor in the Department of Pure & Applied Physics at Queen's University, Belfast died in a tragic climbing accident in Ireland on 17th January 1999. He was born in Edinburgh and obtained both his B.Sc. and Ph.D. degrees from the University of Durham. His thesis was on the muon content of cosmic ray air showers and he continued this work first as a Research Fellow at Durham and later as a Research Associate at Cornell University. In 1971, he went to Belfast where he spent exactly half of his life. John was extremely energetic and talented and was quickly promoted, becoming a Professor in 1990.

Following his appointment at Queen's, John changed his research field – something he was to do several times in his career. He formed a highly successful group to study the application of laser light scattering to biological systems including human sperm motility. He later became interested in the dynamics of fluid interfaces, phase transitions in surface films and two-dimensional colloid systems. His first paper on the latter (with D.J. Robinson) reported an experimental study of particles spread at air-water surfaces and appeared in 1989. It has been widely quoted since. In later years, he studied the transition to chaos in interfacial systems and also extended his studies on soft condensed matter to encompass the physics of foams. At the time of his death, he was President of the European Colloid and Interface Society, and he was a great supporter of conferences both in the UK and abroad. His contribution to Irish science was acknowledged when he was elected a Member of the Royal Irish Academy. He published over one hundred papers in scientific journals, gave many invited talks and supervised 25 research students.

Outside the scientific world, John had many interests which, apart from his great love of the mountains, lay mainly in the arts. He was widely read and his interest in the works of Beckett was insatiable. John Earnshaw's broad knowledge, his enthusiasm and flair for physics and his great curiosity provided intellectual stimulation to all around him. His enjoyment of life as an academic and physicist was contagious. It was of course much too soon but John died, as he would have liked to die, in the mountains.

I thank Professor C.J. Latimer of Queen's University, Belfast for his help in this.

Bernard P. Binks – Hull, March 2006

COLLOIDAL PARTICLES
AT LIQUID INTERFACES

Edited by

BERNARD P. BINKS

and

TOMMY S. HOROZOV
University of Hull

CAMBRIDGE
UNIVERSITY PRESS

CAMBRIDGE UNIVERSITY PRESS
Cambridge, New York, Melbourne, Madrid, Cape Town, Singapore, São Paulo

Cambridge University Press
The Edinburgh Building, Cambridge CB2 8RU, UK

Published in the United States of America by Cambridge University Press, New York

www.cambridge.org
Information on this title: www.cambridge.org/9780521848466

© Cambridge University Press 2006

First published 2006
This digitally printed version 2008

A catalogue record for this publication is available from the British Library

ISBN 978-0-521-84846-6 hardback
ISBN 978-0-521-07131-4 paperback

Contents

Preface

Solid particles of colloidal dimensions (nm–μm) adsorb at fluid interfaces, either liquid–vapour or liquid–liquid, in many products and processes. Examples include fat crystals around air bubbles in certain foods, particles of sand or clay partially coating water drops in crude oil and the selective attachment of mineral particles to bubbles in froth flotation. The properties of these systems are due in part to the irreversible nature of particle adsorption, and such particles behave in many ways like surfactant molecules. The pioneering work in the area of particle-stabilised foams and emulsions was conducted by Ramsden and Pickering, respectively, early in the 20th century. During the last 10 years or so, there has been a revival of interest in this field, and in the behaviour of particles at planar liquid interfaces, and we felt that it was time to prepare the first book encompassing most of this activity. It is anticipated that this will be the start of a new series in this rapidly evolving field.

Following an introductory chapter to the whole area by the editors, the book is divided into two parts. The first part, dealing with particles at planar interfaces, contains chapters describing simulation and theoretical approaches to the structure, and dynamics of particle monolayers and how particles can assist with the wetting of oils on water. The second part, concerned with particles at curved liquid interfaces, contains chapters on emulsions stabilised solely by particles including mechanisms of stabilisation, various kinds of particle-stabilised foams (aqueous and metal), particle-containing antifoams, and novel materials derived from a range of systems with interfacial particles. The collection will be of interest to chemists, physicists, engineers and materials scientists. It should serve as a reference guide for graduate students and the novice, providing detailed accounts of the current state of research in the various fields.

We would like to thank all of the contributors to the chapters for their patience with us and the various staff members at Cambridge University Press for guiding us through the production stages.

Bernard P. Binks
Tommy S. Horozov
Hull, January 2006

List of Contributors

Norbert Babcsán
Department of Materials Science
Hahn-Meitner-Institut Berlin
Glienicker Straße 100, D-14109
Berlin
Germany

John Banhart
Institute of Materials Science and Technology
Technical University Berlin
Hardenbergstraße 36, D-10632
Berlin
Germany

Lennart Bergström
Department of Physical, Inorganic and Structural Chemistry
University of Stockholm
SE-106 91 Stockholm
Sweden

Bernard P. Binks
Surfactant and Colloid Group
Department of Chemistry
University of Hull
Hull
HU6 7RX
UK

Nikolai D. Denkov
Laboratory of Chemical Physics and Engineering

Faculty of Chemistry
Sofia University
1 J. Bourchier Avenue, 1164 Sofia
Bulgaria

Eric Dickinson
Procter Department of Food Science
University of Leeds
Leeds LS2 9JT
UK

Juan C. Fernández-Toledano
Grupo de Fisica de Fluidos y Biocoloides
Departamento de Fisica Aplicada
Universidad de Granada
Campus de Fuente Nueva, 18071
Granada
Spain

Gerald G. Fuller
Department of Chemical Engineering
Stanford University
Stanford
CA 94305-5025
USA

Werner A. Goedel
Department of Physical Chemistry
Chemnitz University of Technology
Straße der Nationen 6209111
Chemnitz
Germany

Roque Hidalgo-Álvarez
Grupo de Fisica de Fluidos y Biocoloides
Departamento de Fisica Aplicada
Universidad de Granada
Campus de Fuente Nueva, 18071
Granada
Spain

Tommy S. Horozov
Surfactant and Colloid Group
Department of Chemistry
University of Hull
Hull
HU6 7RX
UK

Graeme J. Jameson
Discipline of Chemical Engineering and Centre
for Multiphase Processes
School of Engineering
The University of Newcastle
Callaghan
New South Wales 2308
Australia

Robert J.G. Lopetinsky
Department of Chemical and Materials Engineering
University of Alberta
Edmonton
Alberta T6G 2C7
Canada

Krastanka G. Marinova
Laboratory of Chemical Physics and Engineering
Faculty of Chemistry
Sofia University
1 J. Bourchier Avenue, 1164 Sofia
Bulgaria

Francisco Martínez-López
Grupo de Fisica de Fluidos y Biocoloides
Departamento de Fisica Aplicada
Universidad de Granada
Campus de Fuente Nueva, 18071
Granada
Spain

Jacob H. Masliyah
Department of Chemical and Materials Engineering
University of Alberta

Edmonton
Alberta T6G 2C7
Canada

Sonia Melle
Departamento de Óptica
Facultad de Ciencias Físicas
Universidad Complutense de Madrid
Ciudad Universitaria, 28040
Madrid
Spain

Arturo Moncho-Jordá
Grupo de Fisica de Fluidos y Biocoloides
Departamento de Fisica Aplicada
Universidad de Granada
Campus de Fuente Nueva, 18071
Granada
Spain

Anh V. Nguyen
Discipline of Chemical Engineering and Centre
for Multiphase Processes
School of Engineering
The University of Newcastle
Callaghan
New South Wales 2308
Australia

Robert J. Pugh
Institute for Surface Chemistry
Stockholm, S 11486
Sweden

Edward J. Stancik
Department of Chemical Engineering
Stanford University
Stanford
CA 94305-5025
USA

Orlin D. Velev
Department of Chemical and Biomolecular Engineering
North Carolina State University
Raleigh
NC 27695-7905
USA

Krassimir P. Velikov
FSD, UFHRI
Unilever Research and Development Vlaardingen
Olivier van Noortlaan 120
3133 AT Vlaardingen
The Netherlands

Zhenghe Xu
Department of Chemical and Materials Engineering
University of Alberta
Edmonton
Alberta T6G 2C7
Canada

1

Colloidal Particles at Liquid Interfaces: An Introduction

Bernard P. Binks and Tommy S. Horozov

Surfactant and Colloid Group, Department of Chemistry, University of Hull, Hull, HU6 7RX, UK

1.1 Some Basic Concepts

Colloidal particles are an intrinsic part of systems in which finely divided matter (particles) is dispersed in a liquid or gas. Their size usually ranges from 1 nm to several tens of micrometres, thus covering a broad size domain.[1-3] They are not necessarily solid and examples of "soft" colloidal particles (microgels and bacteria) will be briefly considered later in this chapter. Colloidal particles, similar to surfactant molecules, can spontaneously accumulate at the interface between two immiscible fluids (liquid–gas or liquid–liquid); they are therefore surface active.[4] This fact was realised in the beginning of the last century by Ramsden[5] and Pickering[6] whose merit for instigating the field of particles at liquid interfaces will be discussed later. It is important to emphasise that the surface activity of these particles is not necessarily due to their amphiphilic nature. Solid particles with homogeneous chemical composition and properties everywhere on their surface (Figure 1.1(a)) can strongly attach to liquid interfaces and the reason for their surface activity is made clear below. There is, however, another class of particles with two distinct surface regions

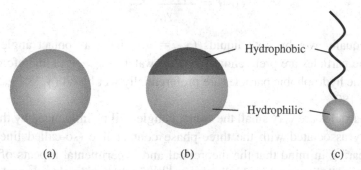

Figure 1.1 Schematic of (a) homogeneous, (b) heterogeneous or amphiphilic (Janus) colloidal particles and (c) a surfactant molecule.

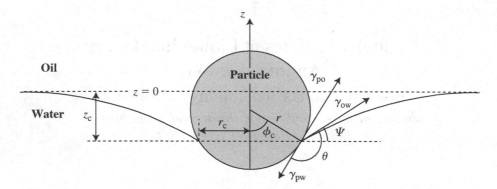

Figure 1.2 Heavy solid spherical particle with radius r and contact angle θ in equilibrium at the oil–water interface levelled at $z = 0$ far from the particle. The three-phase contact line with radius r_c is depressed at depth z_c below the zero level. Other symbols are defined in the text.

with different chemical composition and wetting properties (Figure 1.1(b)). These so-called "Janus" particles are both surface active and amphipilic[7] which makes them very similar to surfactant molecules (Figure 1.1(c)). This class of amphiphilic particles will not be considered here. Information about Janus particle design, synthesis and properties can be found in a recent review.[8]

A key parameter when dealing with solid particles at fluid interfaces is the three-phase contact angle θ. This is the angle between the tangents to the solid surface and the liquid–liquid (or liquid–gas) interface measured through one of the liquids in each point of the three-phase contact line where the solid and two fluids meet. An example for a spherical particle at the oil–water interface is shown in Figure 1.2. We use the convention to measure θ through the more polar liquid (water). The contact angle depends on the surface free energies (interfacial tensions) at the particle–water, γ_{pw}, particle–oil, γ_{po}, and oil–water, γ_{ow}, interface according to Young's equation[9]

$$\cos\theta = \frac{\gamma_{po} - \gamma_{pw}}{\gamma_{ow}} \tag{1.1}$$

Particles equally wet by both liquids ($\gamma_{po} = \gamma_{pw}$) have a contact angle of 90°. Hydrophilic particles are preferentially wet by water ($\gamma_{po} > \gamma_{pw}$), therefore $0° \leqslant \theta < 90°$, while hydrophobic particles are preferentially wet by oil ($\gamma_{po} < \gamma_{pw}$), hence $90° < \theta \leqslant 180°$.

When particles are very small the contact angle will be influenced by the excess free energy associated with the three-phase contact line (so-called line tension effect). Bearing in mind that the theoretical and experimental aspects of the line tension are well discussed in the literature,[10–16] including books[12,13] and a recent review,[14] it will be excluded from our considerations for simplicity. The rest of the chapter is organised as follows. In the next section, without being exhaustive, some key issues about the equilibrium position of a single colloidal particle at a planar fluid interface and the free energy of its detachment to the bulk liquids are presented. The effect of particle shape is considered in the case of rod- and disk-like particles. This is followed by a summary of some very recent developments in the experimental research of particle monolayers at horizontal and vertical fluid interfaces and in thin liquid films. The second section, concentrating on particles adsorbed at curved liquid interfaces, details the important findings on the stabilisation of emulsions and foams by particles alone, and draws examples from a wide range of industrially important products and processes.

1.2 Single Particle at a Fluid Interface

1.2.1 Equilibrium position of a solid particle at a horizontal fluid interface

It is very important in many technological processes (see Chapters 6, 8 and 9) to know the conditions at which a solid particle can stay attached in equilibrium at the liquid–liquid or liquid–gas interface. The problem for the equilibrium of a solid particle at a fluid interface has been extensively treated in the literature[10,12,13,17–21] often in relation to the lateral capillary inter-particle forces caused by the deformation of the fluid interface around two or more floating particles (*e.g.* see Ref. [13] and references therein). This problem can be very difficult in the case of a particle with complex shape and inhomogeneous surface. Solutions have been obtained for particles with simple shape and smooth surface (*e.g.* spheres[12,17–21] and cylinders parallel to the fluid interface[12,17,18]) or sharp edges (*e.g.* disks[21] and long prismatic particles[17] parallel to the fluid interface). In the latter case the three-phase contact line is pinned at the edges and the angle of contact between the fluid and solid interface is not directly defined by equation (1.1). This problem needs a slightly different treatment[17,21] and will not be considered here. The equilibrium position of a particle at a fluid interface can be found either by minimising the free energy of the system[12,18,19,22] or by means of a force analysis[17–21] setting the net force (and the net torque) acting on the particle to zero. The advantage of the first approach is that complex cases (*e.g.* non-uniform particle wetting, line tension effect, *etc.*) can be tackled.[12,18,22] In the case of a smooth homogeneous spherical particle considered below, the force balance approach is equally applicable.[17,19–21]

For clarity we will consider a solid spherical particle with radius r in equilibrium at the oil–water interface when the particle density ρ_p is larger than that of water, ρ_w, and oil ρ_o ($\rho_p > \rho_w > \rho_o$, Figure 1.2). In this situation the oil–water interfacial tension is decisive for keeping the particle attached at the fluid interface. The general case for arbitrary fluids and densities was considered by Princen.[17] Far from the particle the liquid interface is flat and levelled at $z = 0$; the z-axis points upwards (against gravity) normal to the flat liquid interface. The three-phase contact line (a circle with radius r_c) is located at a distance z_c below the zero level, while its position with respect to the particle centre is measured by the angle ϕ_c, hence $r_c = r \sin \phi_c$. The deformed fluid interface (the meniscus) meets the particle surface at angle $\psi = \phi_c + \theta - 180°$ measured to the horizontal level. At equilibrium the net force acting on the particle must be equal to zero. Due to the symmetry the net torque is zero. For the same reason only the vertical force balance (in the z-direction) has to be considered. There are three forces which are involved: the particle weight, mg (m is the particle mass, g is the acceleration due to gravity) acting downwards, the vertical capillary force F_γ due to the vertical component of the oil–water interfacial tension, $\gamma_{ow} \sin(\phi_c + \theta)$ acting upwards at the contact line with length $2\pi r \sin \phi_c$ and the vertical resultant of the hydrostatic pressure distribution around the entire particle, F_p, acting also upwards. The other two interfacial tension forces (γ_{po} and γ_{pw}) depicted in Figure 1.2 must not be included in the force balance because they cannot be considered as external to the particle forces (*e.g.* see Ref. [13, p. 92]). At equilibrium we have

$$F_\gamma + F_p = mg \tag{1.2}$$

The vertical capillary force is

$$F_\gamma = -2\pi r \gamma_{ow} \sin \phi_c \sin(\phi_c + \theta) \tag{1.3}$$

F_p can be obtained by integrating the hydrostatic pressure distribution around the entire particle surface. The result can be written in the form[20]

$$F_p = \rho_w V_{pw} g + \rho_o V_{po} g - (\rho_w - \rho_o) g z_c A_c \tag{1.4}$$

where $V_{pw} = \pi r^3 (2 - 3\cos \phi_c + \cos^3 \phi_c)/3$ and $V_{po} = 4\pi r^3/3 - V_{pw}$ are the particle volumes immersed in water and in oil, respectively and $A_c = \pi (r \sin \phi_c)^2$ is the area of the contact line circle. The first two terms on the right hand side of equation (1.4) are the buoyancy (Archimedes) forces, while the last term accounts for the additional hydrostatic pressure due to depression of the liquid interface below the zero level ($z_c < 0$). The mass of a spherical particle is $m = \rho_p 4\pi r^3/3$. Bearing this in mind, substitution of equations (1.3) and (1.4) in the force balance equation (1.2) after some rearrangement yields[17,19–21]

$$\sin \phi_c \sin(\phi_c + \theta) = -\frac{B}{6}\left[4\frac{\rho_p - \rho_o}{\rho_w - \rho_o} - (1 - \cos \phi_c)^2 (2 - \cos \phi_c) + 3\frac{z_c}{r}\sin^2 \phi_c\right] \tag{1.5}$$

where $B = (\rho_w - \rho_o)r^2 g/\gamma_{ow}$ is a dimensionless parameter (the Bond number). In the considered case of a heavy particle ($\rho_p > \rho_w > \rho_o$), F_γ must always act upwards, therefore $\phi_c + \theta \geqslant 180°$, $\phi_c \leqslant 180°$. Hence, the left hand side of equation (1.5) is restricted in the range $-1 \leqslant \sin\phi_c \sin(\phi_c + \theta) \leqslant 0$. The same should apply to the right hand side of equation (1.5). Obviously, equation (1.5) cannot be solved if B is too large, *i.e.* the particle is too big or too dense. In this case the particle cannot be supported by the fluid interface and will sink in water. To find the critical particle size and density below which the particle can stay attached to the fluid interface at given θ, γ_{ow}, ρ_w and ρ_o is very important for the flotation of minerals considered in Chapter 9. This can be done by solving equation (1.5) if the dependence of z_c on B, ϕ_c and θ is known. The latter can be found by solving the Laplace equation of capillarity

$$\gamma_{ow}\left[z''(1 + z'^2)^{-\frac{3}{2}} + z'(1 + z'^2)^{-\frac{1}{2}}l^{-1}\right] = (\rho_w - \rho_o)gz \qquad (1.6)$$

where $z' \equiv dz/dl$ and $z'' \equiv d^2z/dl^2$ are the first and second derivatives with respect to the radial distance $l \geqslant r_c$ measured from the particle centre in the plane $z = 0$. The two terms in the square brackets are the reciprocals of the principle radii of curvature of the fluid interface, while the right hand side of the equation is the pressure difference across the interface. In the considered case of a circular contact line equation (1.6) has no closed analytical solution and has to be solved numerically[12,13,19] but approximate analytical solutions are available (see Chapter 2 in Ref. [13]). It is worth noting that the deformation of the fluid interface extends to a distance comparable to the capillary length $1/\sqrt{(\rho_w - \rho_o)g/\gamma_{ow}}$ which is usually much larger than the particle size (the capillary length for the pure water–air surface is ~2.7 mm). Therefore when two or more particles are attached to the fluid interface and their menisci overlap a long-range lateral capillary force between particles appears[13,23,24] and can be attractive (when both menisci are depressed or elevated) or repulsive (when depressed and elevated menisci overlap). Directed self assembly of particles at fluid interfaces due to these type of forces is considered in Chapter 7.

When B tends to zero the left hand side of equation (1.5) must also approach zero and $\phi_c + \theta \approx 180°$, thus $\psi \approx 0°$. Hence, for a sufficiently small particle, the deformation of the fluid interface caused by the gravity is very small and can be neglected. In this case the fluid interface can be considered as flat up to the three-phase contact line as shown in Figure 1.3(a). In the case of the air–water surface this is fulfilled for floating particles with radius smaller than ~5 μm when the lateral capillary force is negligible.[13] Deformation of the fluid interface around small spherical particles could exist, however, for reasons different to gravity such as non-uniform wetting of the particle surface.[25,26] The asymmetric electric field around a charged particle at the interface between fluids with very different relative permittivities, ε, *e.g.* water ($\varepsilon \approx 80$) and air ($\varepsilon \approx 1$) or oil ($\varepsilon \approx 2$ for alkanes), could

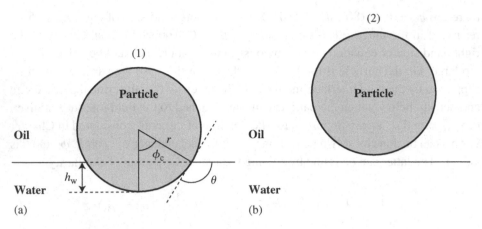

Figure 1.3 Small solid spherical particle with radius r and contact angle θ (a) attached to a planar oil–water interface in its equilibrium state "1" and (b) after its detachment into oil in state "2".

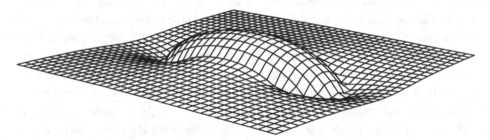

Figure 1.4 Possible interfacial profile around an ellipsoidal particle with contact angle $45°$, aspect ratio $a/b = 4$ and both short axes equal to $2.52 \, \mu m$ oriented with its long axis parallel to the fluid interface. Taken from Ref. [30]; with permission of the American Physical Society.

also generate a deformation of the fluid interface as suggested recently.[27,28] The role of these deformations in particle interactions at fluid interfaces is discussed in Chapter 3. Deformation of the fluid interface may also arise around a non-spherical small particle.[29,30] An example for such a deformation around an ellipsoidal particle at a fluid interface is shown in Figure 1.4.

1.2.2 Free energy of particle detachment from a planar fluid interface

When a small spherical particle at a planar undeformed oil–water interface is in its equilibrium state "1" (Figure 1.3(a)) the surface free energy of the system, $G^{(1)}$ is minimum and is given by the equation

$$G^{(1)} = \gamma_{ow} A^{(1)}_{ow} + \gamma_{pw} A^{(1)}_{pw} + \gamma_{po} A^{(1)}_{po} \tag{1.7}$$

where $A_{ow}^{(1)}$ is the area of the oil–water interface, $A_{pw}^{(1)}$ and $A_{po}^{(1)}$ are the respective areas of the particle–water and particle–oil interfaces whose sum equals the total surface area of the particle A_p

$$A_{pw}^{(1)} + A_{po}^{(1)} = A_p \qquad (1.8)$$

In this state the depth of immersion of the particle in water, h_w (thereby the depth of immersion in oil, $h_o = 2r - h_w$) is directly related to the contact angle by the expression

$$h_w = r(1 + \cos\theta) \qquad (1.9)$$

Therefore this equation can be used for calculating the particle contact angle from the measured value of h_w if r is known.[31-33] If we move the particle from its equilibrium position towards either of the bulk phases by applying some small external force, equation (1.9) will be violated since Young's equation must be satisfied (θ is fixed). Hence, the fluid interface in the new position of the particle after its movement will be deformed. To calculate the change of the surface free energy, ΔG, in this case is not an easy task, since the Laplace equation of capillarity (equation (1.6)) has to be solved, and the approach used by Rapacchietta and Neumann[19] has to be followed. However, if the particle in its final state is detached from the fluid interface and is fully immersed in one of the liquids (Figure 1.3(b)) the calculation of ΔG is straightforward. If the particle after its detachment is in the oil, the surface free energy of the system $G_o^{(2)}$ corresponding to the final state "2" is

$$G_o^{(2)} = \gamma_{ow} A_{ow}^{(2)} + \gamma_{po} A_{po}^{(2)} \qquad (1.10)$$

where $A_{ow}^{(2)}$ is the area of the flat oil–water interface after the detachment and $A_{po}^{(2)} = A_p$. Subtracting equation (1.7) from equation (1.10) and taking into account equations (1.1) and (1.8), the following expression for the free energy of particle detachment into oil, ΔG_{do} is obtained

$$\Delta G_{do} = \gamma_{ow}(A_c + A_{pw}^{(1)} \cos\theta) \qquad (1.11)$$

where $A_c = A_{ow}^{(2)} - A_{ow}^{(1)}$ is the area of the oil–water interface occupied by the particle when it is attached at the fluid interface. A similar derivation leads to the following expression for the free energy of particle detachment into water ΔG_{dw}

$$\Delta G_{dw} = \gamma_{ow}(A_c - A_{po}^{(1)} \cos\theta) \qquad (1.12)$$

By means of equations (1.8) and (1.12), equation (1.11) can be expressed in the form

$$\Delta G_{do} = \Delta G_{dw} + \gamma_{ow} A_p \cos\theta \qquad (1.13)$$

that gives the relation between the two free energies of particle detachment. It is obviously that the detachment of a hydrophilic particle ($\cos\theta > 0$) into oil needs more energy than into water ($\Delta G_{do} > \Delta G_{dw}$), while for the detachment of a hydrophobic particle ($\cos\theta < 0$) the opposite is true ($\Delta G_{do} < \Delta G_{dw}$). This is important for understanding the stabilisation of emulsions by solid particles (see Chapter 6). It is also seen that at $\theta = 90°$ both energies are equal to each other. The minimum energy required for particle detachment irrespective into which of the bulk phases, ΔG_d (called simply the free energy of particle detachment) can be written as

$$\Delta G_d = \begin{cases} \Delta G_{dw} & \text{for} \quad 0 \le \theta \le 90° \\ \Delta G_{do} & \text{for} \quad 90° \le \theta \le 180° \end{cases} \tag{1.14}$$

where ΔG_{dw} and ΔG_{do} are given by equations (1.12) and (1.13) (or (1.11)), respectively. The respective free energies of particle attachment to the fluid interface are given by the same equations taken with the opposite sign. The above equations ((1.8), (1.10)–(1.14)) are written in a rather general form. They depend on the particle shape implicitly through the respective areas. Therefore they are valid for any shape of the particle if the fluid interface can be considered flat up to the particle surface. Some special cases which satisfy the latter requirement are considered below.

1.2.2.1 Spherical particle

In this case the three-phase contact line is a circle with radius $r_c = r\sin\theta$ dividing the particle surface (with area $A_p = 4\pi r^2$) into two spherical caps, so that $A_c = \pi(r\sin\theta)^2$ and $A_{pw}^{(1)} = 2\pi r^2(1 + \cos\theta)$. With these expressions equations (1.12) and (1.13) yield

$$\Delta G_{dw} = \pi r^2 \gamma_{ow}(1 - \cos\theta)^2 \tag{1.15a}$$

$$\Delta G_{do} = \pi r^2 \gamma_{ow}(1 + \cos\theta)^2 \tag{1.15b}$$

These equations were derived by Koretsky and Kruglyakov[34] and later by others.[35,36] In view of equation (1.14) they can be combined to give

$$\Delta G_d = \pi r^2 \gamma_{ow}(1 - |\cos\theta|)^2 \tag{1.16}$$

Therefore, the minimum energy needed to detach a spherical particle from the oil–water interface rapidly increases with particle size (as r^2). The free energies of particle detachment calculated by equations (1.15) and (1.16) with $r = 10\,\text{nm}$ and $\gamma_{ow} = 50\,\text{mN m}^{-1}$ are plotted against the contact angle in Figure 1.5. It is seen that the free energy of particle detachment into water (squares) is smaller than that into oil (circles) for hydrophilic particles ($\theta < 90°$). The opposite is true for hydrophobic particles ($\theta > 90°$). The (minimum) energy of particle detachment, ΔG_d (the line) increases from zero with an increase of the contact angle, reaches its

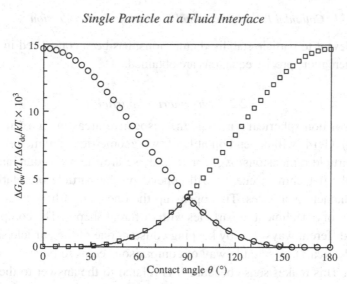

Figure 1.5 Free energy of detachment of a spherical particle into water (squares) and into oil (circles) calculated by equations (1.15) (a) and (b) with $r = 10\,\text{nm}$ and $\gamma_{ow} = 50\,\text{mN m}^{-1}$ *versus* particle contact angle θ. The line is drawn according to equation (1.16).

maximum at $\theta = 90°$ and then decreases to zero at $\theta = 180°$. Except for very small and very large contact angles, ΔG_d is much greater than the thermal energy kT (the Boltzmann constant, k times the temperature, T). At contact angles around $90°$ even nanoparticles can be trapped at the fluid interface with energy which is several orders of magnitude greater than kT and sufficient to make their attachment irreversible. This is in sharp contrast to surfactant molecules, which can adsorb and desorb,[4] and makes certain particles superior stabilisers of emulsions (Chapter 6) and foams (see later in this chapter).

The energy of particle attachment to the fluid interface, $\Delta G_a = -\Delta G_d$, is negative for all contact angles (except for the extremes), hence the particle attachment is thermodynamically favourable (this might not be true in the case of small particles and positive line tension acting to contract the contact line[10,11]). Therefore colloidal particles with chemically homogeneous surfaces can spontaneously attach to fluid interfaces and are surface active. The reason is that part of the fluid interface with area A_c is removed (see equations (1.11) and (1.12)). ΔG_a and ΔG_d for nanoparticles with contact angles close to $0°$ or $180°$ can be comparable to the thermal energy. Such particles can exhibit a reversible attachment–detachment behaviour (similar to surfactants) which has been demonstrated by elegant experiments described later.

The above equations are also applicable for particles at spherically curved oil–water interfaces (drops), if the particle radius is much smaller than the drop radius. This case, which is relevant to particle-stabilised emulsions, has been considered in Refs. [37] and [38] where more precise equations are obtained.

1.2.2.2 Non-spherical particles

In the case of non-spherical particles, the respective areas involved in equations (1.8), (1.10)–(1.14) will depend on at least two geometrical parameters characterising the particle dimensions (*e.g.* for rods these are the rod radius and length). Therefore the detachment energy will depend on the particle orientation and at least two characteristic sizes. This opens up the question of how to compare the free energies of detachment of particles with different shapes. The comparison can be done in different ways (*e.g.* by keeping constant one of the particle sizes or the total particle area[15]). A suitable way of comparison seems to be at constant particle volume. This makes sense because it is related to the answer to the question: *how will the free energy of detachment of a particle with contact angle θ change if we re-shape it keeping its volume constant*? This question is answered below in the case of two smooth bodies: a rod-like particle with rounded hemispherical ends and a rounded disk-like particle (Figure 1.6). They are both shapes of revolution with long and short semi-axes *a* and *b*, respectively. We will assume that the particles

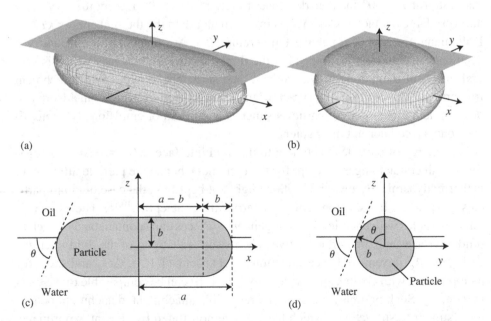

Figure 1.6 Non-spherical particles with contact angle θ = 45° attached to a planar oil–water interface in the case of (a) a rod-like particle with hemispherical ends and (b) a rounded disk-like particle; (c) cross section along their long semi-axis, *a*; (d) cross section of the rod-like particle along its short semi-axis, *b*.

are attached to a planar oil–water interface with their long axis parallel to the fluid interface. In this configuration the fluid interface is flat everywhere, hence equations (1.8), (1.10)–(1.14) can be applied. The additional restriction for the constancy of the particle volume, V_p reads

$$V_p = V_{sphere} = \frac{4\pi r^3}{3} = const \tag{1.17}$$

where V_{sphere} is the volume of a spherical particle with radius r.

The volume of a rod-like particle with hemispherical ends, V_{rod} is

$$V_{rod} = \frac{4}{3}\pi b^3 \left[\frac{2 + 3(a/b - 1)}{2} \right] \tag{1.18}$$

This equation combined with equation (1.17) allows one to express b as a function of the aspect ratio a/b and r as

$$b = r \left[\frac{2 + 3(a/b - 1)}{2} \right]^{-\frac{1}{3}} \tag{1.19}$$

Note that due to the volume constraint b is always smaller than r because $a/b > 1$. At $a/b = 1$ the rod turns into a sphere with radius $b = r$. Simple geometry gives the following expressions for the cross-sectional area, total area and the area of the attached particle in contact with oil, respectively

$$A_c = \pi b^2 \sin^2\theta \left[1 + \frac{4(a/b - 1)}{\pi\sin\theta} \right] \tag{1.20}$$

$$A_p = 4\pi b^2 \frac{a}{b} \tag{1.21}$$

$$A_{po}^{(1)} = 2\pi b^2 (1 - \cos\theta) + 4b^2\theta\left(\frac{a}{b} - 1\right) \tag{1.22}$$

Substitution of equations (1.20)–(1.22) into equations (1.12) and (1.13) yields the following expressions for the free energies of detachment of a rod-like particle into water and into oil, respectively

$$\Delta G_{dw} = \gamma_{ow}\pi b^2 (1 - \cos\theta)^2 \left[1 + \frac{4(a/b - 1)(\sin\theta - \theta\cos\theta)}{\pi(1 - \cos\theta)^2} \right] \tag{1.23}$$

$$\Delta G_{do} = \Delta G_{dw} + 4\pi\gamma_{ow}b^2 \cos\theta\left(\frac{a}{b}\right) \tag{1.24}$$

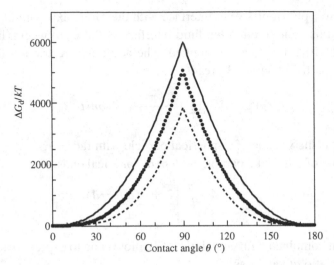

Figure 1.7 Free energy of particle detachment *versus* contact angle θ in the case of spherical (dashed line, equation (1.16)), rod-like (dotted line, equation (1.25)) and disk-like (full line, equation (1.33)) particles with one and the same volume equal to $4.19 \times 10^3 \, \text{nm}^3$. The other parameters are $\gamma_{ow} = 50 \, \text{mN m}^{-1}$, $r = 10 \, \text{nm}$ and aspect ratio of the non-spherical particles $a/b = 2.5$.

where b is given by equation (1.19). The (minimum) free energy of detachment written in the form of equation (1.14) is

$$\Delta G_{d,rod} = \begin{cases} \text{equation (1.23)} & \text{for} \quad 0 \leq \theta \leq 90° \\ \text{equation (1.24)} & \text{for} \quad 90° \leq \theta \leq 180° \end{cases} \qquad (1.25)$$

For a given γ_{ow} and r (*i.e.* at fixed particle volume) equation (1.25) gives the dependence of the free energy of detachment on the particle aspect ratio and contact angle. It is plotted versus the contact angle in Figure 1.7 at $a/b = 2.5$ (dotted line) and compared to that of a spherical particle (dashed line) with the same volume. Both curves have similar shapes with a characteristic maximum at $\theta = 90°$, however $\Delta G_{d,rod}$ is always greater than $\Delta G_{d,sphere}$ except at $\theta = 0°$ or 180°, where both energies are equal to zero. Hence, rod-like particles with rounded ends oriented parallel to the fluid surface are held at the oil–water interface stronger than spherical particles with the same volume.

In the case of a rounded disk-like particle, the particle volume V_{disk} is

$$V_{disk} = \frac{4}{3}\pi b^3 \left[\frac{3(a/b - 1)^2}{2} + \frac{3\pi(a/b - 1)}{4} + 1 \right] \qquad (1.26)$$

The respective expression for $b(a/b, r)$ obtained from equation (1.26) and (1.17) is

$$b = r \left[\frac{3(a/b - 1)^2}{2} + \frac{3\pi(a/b - 1)}{4} + 1 \right]^{-\frac{1}{3}} \qquad (1.27)$$

and the respective areas are

$$A_c = \pi b^2 \left(\frac{a}{b} - 1 + \sin \theta \right)^2 \tag{1.28}$$

$$A_p = 2\pi b^2 \left[\left(\frac{a}{b} - 1 \right)^2 + \pi \left(\frac{a}{b} - 1 \right) + 2 \right] \tag{1.29}$$

$$A_{po}^{(1)} = \pi b^2 \left[\left(\frac{a}{b} - 1 \right)^2 + 2 \left(\frac{a}{b} - 1 \right) \theta - 2 \cos \theta + 2 \right] \tag{1.30}$$

Combining the area equations with equations (1.12) and (1.13) gives expressions for the free energies of detachment of a rounded disk-like particle into water and into oil which are

$$\Delta G_{dw} = \gamma_{ow} \pi b^2 (1 - \cos \theta)^2 \left[1 + \frac{(a/b - 1)^2}{1 - \cos \theta} + \frac{2(a/b - 1)(\sin \theta - \theta \cos \theta)}{(1 - \cos \theta)^2} \right] \tag{1.31}$$

$$\Delta G_{do} = \Delta G_{dw} + 2\pi\gamma_{ow} b^2 \cos \theta \left[\left(\frac{a}{b} - 1 \right)^2 + \pi \left(\frac{a}{b} - 1 \right) + 2 \right] \tag{1.32}$$

where b is given by equation (1.27). Finally, the equation for the free energy of disk-like particle detachment, $\Delta G_{d,disk}$ is

$$\Delta G_{d,disk} = \begin{cases} \text{equation (1.31)} & \text{for} \quad 0 \le \theta \le 90° \\ \text{equation (1.32)} & \text{for} \quad 90° \le \theta \le 180° \end{cases} \tag{1.33}$$

$\Delta G_{d,disk}$ calculated with the same parameters as $\Delta G_{d,rod}$ is also plotted in Figure 1.7 (full line) for comparison. It is seen that its dependence on the contact angle is similar to that for a sphere and rod with rounded ends. However $\Delta G_{d,disk}$ is largest, hence at any contact angle different to 0° or 180° the following relation is valid $\Delta G_{d,disk} > \Delta G_{d,rod} > \Delta G_{d,sphere}$. Since the planar orientation of the non-spherical particles considered here is the most energy favourable,[15,16] the obtained results suggest that the driving force for attachment (*i.e.* the energy gained after attachment) of the particles decreases in the order disk > rod > sphere. Hence, re-shaping spherical particles into rod- or disk-like particles with the same volume can improve their attachment to fluid interfaces. This is even better seen in Figure 1.8 where the detachment energy of rod- and disk-like particles scaled with that of a spherical particle is plotted against the contact angle. Although the deviation of particle shape from that of a sphere is fairly small (see the inset) at the considered aspect ratio (1.25) it has clear impact on the free energy of detachment which is more significant at low and high contact angles. This effect increases at larger aspect ratios (Figure 1.9). For instance,

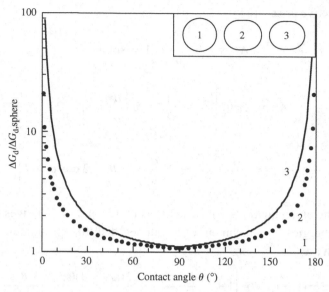

Figure 1.8 Free energy of particle detachment relative to the case of a spherical particle (1, base line) *versus* contact angle θ for rod-like (2, dotted line) and disk-like (3, full line) particles with the same volume equal to $4.19 \times 10^3 \, \text{nm}^3$. The other parameters are $\gamma_{ow} = 50 \, \text{mN m}^{-1}$, $r = 10 \, \text{nm}$ and aspect ratio of the non-spherical particles $a/b = 1.25$. The inset shows the respective cross sections of the particles along their long semi-axis a.

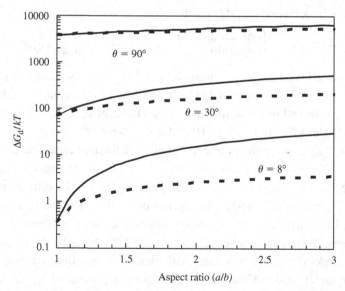

Figure 1.9 Free energy of particle detachment *versus* aspect ratio of rod-like (dashed lines) and disk-like (full lines) particles with the same volume equal to $4.19 \times 10^3 \, \text{nm}^3$. The lines are calculated for three different contact angles by equations (1.25) and (1.33), respectively with $\gamma_{ow} = 50 \, \text{mN m}^{-1}$ and $r = 10 \, \text{nm}$.

spherical particles with radius 10 nm and contact angle of 8° can hardly attach to the oil–water interface since their free energy of detachment is very small ($\sim$0.4 kT). However, rounded disk-like particles of the same volume and material are held at the fluid interface with a rather significant energy of $\sim$10 kT at aspect ratios >2. This might be advantageous when stabilisation of emulsions and foams by solid particles is pursued. Hence, the effect of particle shape on the free energy of particle detachment/attachment demonstrated here might be of significant practical importance.

1.3 Particle Monolayers at Horizontal and Vertical Liquid Interfaces

The structure and stability of particle monolayers at horizontal air–water and oil–water interfaces have been extensively investigated during the last two decades (see Chapters 2 and 3). Such studies can help in understanding the stabilisation of emulsions (Chapters 5, 6 and 8) and foams (Chapters 9–11) by colloidal particles and can be useful for fabricating novel materials (Chapter 7), therefore they are of significant practical importance. Horizontal particle monolayers are often used as model systems for studying colloidal aggregation since particle locations and their evolution can be detected easier in two rather than in three dimensions by optical microscopy. Certain spherical particles, however, self-organise into well-ordered two-dimensional crystal structures with very large separations between particles instead of forming aggregates. This particle behaviour is attributed to long-range electrostatic interactions between dipoles and/or charges (monopoles) at that part of the particle surface which is in contact with oil or air. These and other types of interactions are considered in detail in Chapter 3. Here we will briefly describe some very recent developments in the experimental research of particle monolayers at fluid interfaces relevant to the repulsive monopolar Coulomb interactions through the non-polar fluid (oil or air). The results that follow are also a good illustration of the effect of particle hydrophobicity on the monolayer structure and particle interactions discussed in the following two chapters.

The asymmetric disposition of particles at fluid interfaces (see Figure 1.3(a)) makes their interactions significantly different from those in the bulk. Particles attached to the water–oil (air) interface interact with each other through both fluid phases of remarkably different polarity. This can result in long-range electrostatic repulsion mediated through the non-polar fluid with low relative permittivity, thus causing crystallisation of particle monolayers (see Chapters 2 and 3). Recently, Aveyard *et al.* suggested that strong long-range repulsion between latex particles at the non-polar oil–water interface can arise as a result of small amounts of charge located at the particle–oil surface.[32,39] This was further supported by experiments with monodisperse silica particles at horizontal water–oil interfaces.[40] It was shown that the particle hydrophobicity has a dramatic effect on the particle interactions resulting in an order–disorder transition in the monolayers when the particle contact angle was decreased.[40]

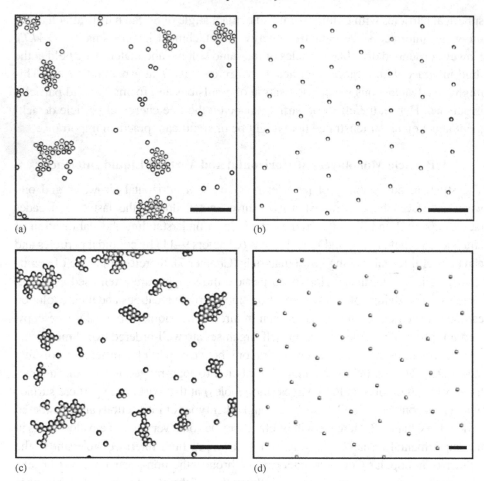

(a)   (b)

(c)   (d)

Figure 1.10 Images of horizontal silica particle monolayers at the octane – water interface: (a) and (b) without added electrolyte, (c) with 10 mM and (d) 100 mM NaCl in water 1 h after their formation. Particle diameter is 3 μm and their contact angles measured through the water are: (a) and (c) 65°, (b) and (d) 152°. All scale bars are 25 μm. Images (a)–(c) are taken from Ref. [41]; with permission of the American Chemical Society and image (d) is taken from Ref. [43]; with permission of Wiley-VCH.

More recently, the effect of pH and electrolyte added to the aqueous phase on the structure of horizontal monolayers of hydrophilic ($\theta \approx 65°$) and very hydrophobic ($\theta \approx 152°$) monodisperse silica particles (3 μm in diameter) was studied.[41–43] The effect of electrolyte on the monolayer structure at the octane–water interface at ambient pH $\approx$ 5.6 has been demonstrated[41] (Figure 1.10). Disordered, aggregated monolayers of hydrophilic silica particles have been observed both in the absence and in the presence of NaCl in the water ((a), (c)).[41] Without electrolyte loose aggregates of particles separated at $\sim$1.5 particle diameters were detected (a). They changed into dense aggregates of close-packed touching particles by adding fairly small amounts of NaCl

(10 mM) to the water (c). It was concluded that the repulsion between hydrophilic particles is weak and mediated mainly through the water phase.[41] In contrast, very hydrophobic particles at the octane–water interface are well separated and ordered in a hexagonal lattice in the absence of NaCl (b).[41] In a separate study[43] the same very hydrophobic particles gave well-ordered monolayers at inter-particle distances larger than 50 μm (>16 particle diameters) even when 100 mM NaCl was present in the water phase (d). This confirmed previous findings[40] that the repulsion between very hydrophobic silica particles is very long ranged and mediated through the oil phase, since the electrostatic repulsion through the water is totally suppressed at such high electrolyte concentration. Similar well-ordered monolayers of very hydrophobic silica particles at large separations have been obtained at the silicone oil–water and air–water interfaces (Figure 1.11).[42] Some aggregates were formed during the formation of the monolayers by spreading but their number and size remained practically unchanged for hours. Lowering the pH to 2 did not affect the ordered monolayer structure at the air–water surface, (b), although the silica surface in contact with water is uncharged at such conditions (the point of zero charge for silica is at pH 2–3).[44] Ordered silica particle monolayers at octane–water[41] and silicone oil–water[42] interfaces were also stable towards aggregation at pH ≈ 2.7 confirming the previous conclusion that the long-range repulsion is mediated through the non-polar fluid.

All reported results for very hydrophobic silica particle monolayers at horizontal oil–water[40,41,43] and air–water[42] interfaces are consistent with Coulomb repulsion through the non-polar fluid due to charges at the particle–oil (air) surface. This is supported by the recent experiments with silica particle monolayers at vertical fluid interfaces.[41,42,45] The experimental setup shown in Figure 1.12 was constructed[45] and used for studying particle monolayers at vertical oil–water[41,42] and air–water[42] interfaces. In these experiments a dilute horizontal monolayer at the fluid interface in the cuvette was formed by spreading silica particles using 2-propanol. Then the monolayer was traversed by a circular frame made of glass or PTFE, poly (tetra fluoroethylene). The glass frame was lifted up from the water into the oil (air) and very thick water films in oil (air) (thicker than 200 μm) with dilute particle monolayers at their surfaces were formed.[42,45] Alternatively,[41,45] the PTFE frame was immersed from oil into water forming a very thick oil film in water. The evolution of the monolayer structure at one of the vertical film surfaces was observed from the side by a microscope. Images were captured by the CCD (charge-coupled device) camera, recorded by a VCR (video cassette recorder) and processed by a computer. A dramatic effect of silica particle hydrophobicity on the monolayer structure was observed[41,45] (Figure 1.13). It was found that the particles with contact angles <99° sediment giving well-ordered arrays of close-packed particles at the bottom and a bare octane–water interface at the top (a). In contrast, very hydrophobic silica particles do not sediment appreciably at vertical octane–water interfaces, (b), and remain very repulsive. The results were analysed with a simple two particle model considering the sedimentation

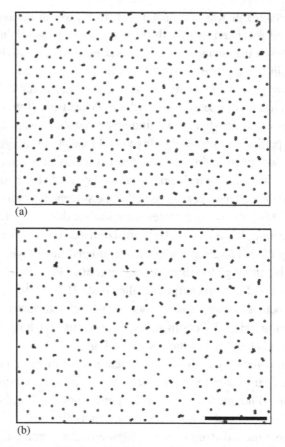

(a)

(b)

Figure 1.11 Images of horizontal silica particle monolayers at the air–water surface at pH equal to (a) 5.6 and (b) 2.0 one h after their formation. Particle diameter is 3 μm and the contact angle measured through the water is 105°. The scale bar is 50 μm. Taken from Ref. [42]; with permission of Elsevier.

equilibrium as a balance between the long-range electrostatic repulsion through the oil, the gravity force and the capillary attraction due to deformation of the fluid interface around particles caused by the non-uniform wetting.[41] A reasonable value of $14 \mu C\,m^{-2}$ for the charge density at the particle–octane surface was found for the most hydrophobic particles with $\theta \approx 152°$. The surface charge density for particles with contact angles $< 99°$ was estimated to be ~ 5 times lower,[41] consistent with the rather sharp order–disorder transition in monolayer structure.[40] The same approach has been used for studying the two-dimensional sedimentation of very hydrophobic silica particles at vertical air–water and silicone oil–water interfaces.[42] It was found that the charge density at the particle surface in contact with air is about two times smaller than that in contact with silicone oil, in accord with the results obtained in Ref. [28] for a similar system using a totally different approach.

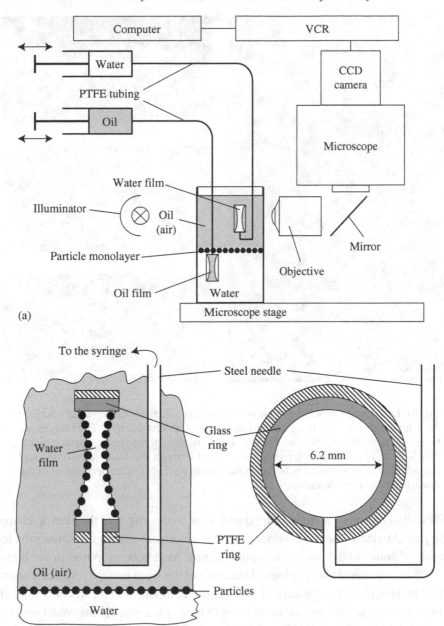

Figure 1.12 Experimental setup for studying particle monolayers at vertical fluid interfaces and in thin liquid films: (a) general scheme, (b) side view and (c) front view cross sections of the frame for water films. The frame for oil films has the same geometry and size, but the glass ring is replaced by a PTFE ring; the steel needle is straight and mounted at the upper wall of the ring. Taken from Ref. [45]; with permission of the American Chemical Society.

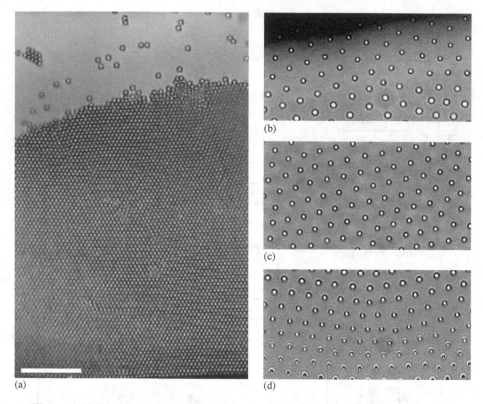

(a)   (d)

Figure 1.13 Images of silica particle monolayers at a vertical octane–water interface 1 h after their formation in a circular frame with diameter 6.2 mm. The images are taken: (a) and (d) at the bottom, (b) top and (c) middle of the frame. Particle diameter is 3 μm and their contact angles measured through the water are (a) 65° and (b)–(d) 152°. The scale bar is 50 μm. Taken from Ref. [41]; with permission of the American Chemical Society.

These findings raise the interesting question as to the origin of the electric charge at the particle–non-polar fluid surface. It is still enigmatic although mechanisms for charging of solids in the bulk of non-polar liquids have been proposed in the literature.[46–48] The chemical and topological structure of the silica surface could be important for charging the solid. Water (which is a major component of the systems studied) might play a very significant role in charging of the particle surface in contact with the non-polar fluid (see Refs. [46,47] and Chapter 3). The above results also demonstrate that studies of vertical monolayers can provide valuable information about the properties of the colloidal particles and their interactions at fluid interfaces.

1.4 Thin Liquid Films with Particle Monolayers at Their Surfaces

The role of the thin liquid film between colliding emulsion droplets in the presence of surfactant is well recognised and extensively studied.[49,50] Much less is known about the properties of films with particle monolayers at their surfaces which are relevant to the stability of particle-stabilised emulsions and foams. Some interesting but rather restricted studies dealing with emulsion films formed between particle-laden drops and planar monolayers were recently reported[51,52] and are discussed in Chapter 5. Very recently, a systematic study of emulsion films with silica particle monolayers at their surfaces was published by Horozov *et al.*[45] where the effect of particle contact angle and surface coverage on the structure and stability of water films in oil (o–w–o) and oil films in water (w–o–w) were investigated. The main findings are very relevant to particle-stabilised emulsions considered in Chapter 6, and will be briefly summarised here.

The experimental setup used in Ref. [45] was already described in the previous subsection (Figure 1.12). The very thick films with particles at their surfaces formed in the glass frame (for water films) or PTFE frame (for oil films) after crossing the octane–water interface were forced to thin by sucking liquid out of the meniscus using the syringes. Monodisperse silica particles (3 μm in diameter) hydrophobised to different extents were used. Their contact angles (through water) at the octane–water interface, measured on glass slides hydrophobised simultaneously with the particles, were in the range 65°–152° (see Table 1.1). The experimental results showed some similarities but also distinct differences in the thinning behaviour of water (o-w-o) and oil (w-o-w) films with dilute particle monolayers at their surfaces. Stable o-w-o films were obtained only when particles with $\theta < 90°$ were used, whereas w-o-w films were stable only in the case of very hydrophobic particles ($\theta \gg 90°$). In both cases some of the particles bridged the film surfaces at the final stages of the film thinning but the particle behaviour during the film thinning and the structure of the resulting thin films were remarkably different (Figure 1.14). Hydrophilic particles were expelled out of the centre of the thinning water film. Some of them were spontaneously attached to both film surfaces forming a ring of bridging particles at the film periphery, (a). As a result the central thinnest region of the water film became unprotected and therefore more vulnerable to rupture. In contrast, very hydrophobic particles were not expelled out of the oil film centre during its thinning. This was attributed mainly to the strong Coulomb repulsion between particles opposing the hydrodynamic drag force. A dense bridging monolayer (crystalline disk) was formed spontaneously in the thinnest central part of the oil film which strongly increased its stability against rupture, (b). The driving force bringing particles together to give a dense crystalline disk is the capillary attraction between adjacent bridging particles caused by the curved menisci of the fluid interface formed when the particles became attached to both film surfaces[13] (see also Chapter 7, Figure 7.18). Emulsion films with close-packed particle monolayers at their surfaces were also studied. It was found that both water films with hydrophilic particles and oil films with very hydrophobic ones were very stable. The results are summarised in Table 1.1. They are in agreement with the findings described in Chapter 6 in that hydrophilic particles can give stable oil-in-water (o/w) emulsions (*i.e.* stable o-w-o

Table 1.1 *Stability of water films in octane and octane films in water stabilised by silica particles. Unstable films break during their formation or several seconds later. Stable films survive for up to ~30 min, while those denoted as "very stable" live more than 1 h. Taken from Ref. [45]; with permission of the American Chemical Society.*

Contact angle/°	Film surfaces with dilute monolayers		Film surfaces with dense monolayers	
	Water films in oil	Oil films in water	Water films in oil	Oil films in water
65 ± 3	Stable (ring formation)	Unstable	Very stable	Unstable
85 ± 2	Stable (ring formation)	Unstable	Stable	Unstable
99 ± 2	Unstable	Unstable	Unstable	Unstable
152 ± 2	Unstable	Very stable (crystallisation)	Unstable	Very stable

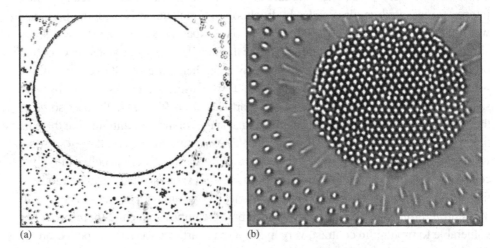

(a)   (b)

Figure 1.14 Images of: (a) a water film in octane with a ring of bridging hydrophilic particles ($\theta = 65°$) and (b) an octane film in water with a disk of bridging hydrophobic particles ($\theta = 152°$) formed spontaneously during film thinning. The hydrophilic particles outside the ring sediment with time (a) but the hydrophobic ones outside the disk do not sediment significantly (b). The scale bar is 50 μm. Taken from Ref. [45]; with permission of the American Chemical Society.

films), whereas hydrophobic ones give water-in-oil (w/o) emulsions (*i.e.* stable w-o-w films). It was hypothesised that the hydrophobic particles could be a more effective stabiliser than the hydrophilic ones because they can make stable films at lower surface coverage. The latter was confirmed experimentally by Horozov and Binks very recently.[43]

1.5 Particle-Stabilised Curved Liquid–Liquid Interfaces Including Emulsions

Since a number of reviews have appeared recently in the areas of emulsions and foams stabilized solely by solid particles,[4,53–56] the purpose of the next two sections is to highlight some very recent findings and to discuss briefly some other fields not commonly thought of in terms of having particles at fluid interfaces. The use of solid particles alone to stabilise mixtures of oil and water in the form of emulsions seems to be credited to Pickering,[6] and the phrase "Pickering emulsions" has been coined for many years. Pickering published his first work in this area in 1907, and in the abstract he writes that "*the subject had already been investigated by Ramsden [in 1903], but his work, unfortunately, did not come under the notice of the writer until that here described had been completed. It is satisfactory to find, however, that Ramsden, pursuing a different line of enquiry, should have arrived at an explanation of emulsification which is essentially the same as that given here*". Given this honesty, it is then strange why such emulsions were not termed "Ramsden emulsions". The writer suspects that some of those later scientists did not actually obtain a copy of Pickering's paper, relying on previous citations to it, and hence were unaware of Ramsden's contribution, as acknowledged by Pickering. It seems appropriate to redress the balance here.

It may be that Ramsden's paper[5] is the first which mentions not only the stabilisation of emulsion drops by solids but also that of bubbles (in foams) too. The paper is entitled "Separation of Solids in the Surface-layers of Solutions and 'Suspensions' (Observations on Surface-membranes, Bubbles, Emulsions, and Mechanical Coagulation).-Preliminary Account". In an obituary of Ramsden, Peters[57] wrote "*the latter work is not as well known as it should be and has a pioneer relationship to some modern protein film work*". In the paper, a summary of the findings of many experiments is given on the spontaneous formation (by agitation) of solid or highly viscous coatings on the free surfaces (*i.e.* air–water) of protein solutions. Similar coatings of solid matter occur on the free surfaces of a large number of non-protein-containing colloidal dispersions, of fine and coarse suspensions and at the interfaces of every pair of liquids (*i.e.* oil–water) capable of forming persistent emulsions. In one case quoted, "*. . .a bubble of air can be seen to pick up the particles in suspension as it passes through the liquid, and to retain them obstinately when it reaches the surface and comes to rest, so that the bubble becomes thickly coated with solid particles, . . .*". It was also mentioned that every solution capable of forming moderately persistent bubbles yielded solid or highly viscous mechanical surface aggregates. It was demonstrated that "*. . .an actual solid membrane forms around the globules of several persistent emulsions, and at the contact interfaces of several pairs of liquids capable of forming persistent emulsions*", and that "*. . .persistently*

Figure 1.15 Photograph of Professor W. Ramsden (1868–1947), M.A., M.D., Fellow of Pembroke College, University of Oxford and Professor of Biochemistry, University of Liverpool, taken in 1933. Thanks to Angela Dell of the School of Biological Sciences, University of Liverpool.

deformed sharply angular and grotesque shapes of the emulsified globules" were produced. Although the paper contains no figures, it is clear from the text that Ramsden reported both foam and emulsion stabilisation by solid particles.

Walter Ramsden, M.A., M.D., (Figure 1.15) was born in 1868 in Saddleworth, Yorkshire (UK) and was the son of a physician. He went to Keble College, Oxford obtaining 1st Class Honours in Physiology in 1892. He was Radcliffe Travelling Fellow from 1893 to 1896, during which time he studied at Zurich and Vienna and finally at Guy's Hospital (London) taking his final M.D. in 1902. Never married and never in medical practice, he devoted himself to teaching and research. In 1899 he was elected at Pembroke College, Oxford to the Sheppard Medical Fellowship which he held to the end of his life in 1947. He became Lecturer in the Department of Physiology and held the post until 1914. He was a founder member of the Biochemical Society. In 1914, Ramsden went to the University of Liverpool to become Johnston Professor of Biochemistry. Colleagues mentioned the "*charm of his old-world courtesy*" and claimed that "*at Liverpool, the University and the medical school felt him to be a rare and distinguished possession*".[58] He retired from there in 1931 to devote his remaining years to research in Oxford. There he worked in the Department of Pathology and, from 1937, in the Department of Biochemistry. He pursued his research on proteins from the silkworm and on liquid surface phenomena. Those who knew him early in his career, and appreciated his brilliant skill as an experimenter in the field of surface chemistry, were disappointed that his early promise did not materialise with sufficient solidity in print. Though he was ever willing to discuss his work and ideas with friends, a certain fastidiousness and a wish to obtain more thorough completion, inhibited the published word.

As a tribute to his work, Clayton's 2nd edition on emulsions in 1928 was dedicated to Ramsden, who wrote Appendix I (in 1927) entitled "*Theory of Emulsions Stabilized by Solid Particles*".[59] In it, characteristically, he stated that with J. Brooks he had devoted 2 years from 1922 to an experimental test of the theory by determining the contact angle made against a variety of solids at benzene–water or paraffin–water interfaces. To the writer's knowledge, this data has never been published, although it was communicated by Ramsden at the meeting of the British Association in Liverpool in 1923.[60] On his death, John Betjeman, the Poet Laureate-to-be, wrote a touching elegy in his poem "*A Few Late Chrysanthemums*" which is highly regarded by critics.[61] It is doubtful if any other professor of biochemistry anywhere will be so honoured. His name will certainly live on for students of English literature, and hopefully now for researchers in colloid science. It reads:

> "I.M. Walter Ramsden ob. March 26, 1947 Pembroke College, Oxford
> Dr. Ramsden cannot read The Times obituary to-day
> He's dead.
> Let monographs on silk worms by other people be
> Thrown away
> Unread
> For he who best could understand and criticize them, he
> Lies clay
> In bed.
>
> The body waits in Pembroke College where the ivy taps the panes
> All night;
> That old head so full of knowledge, that good heart that kept the brains
> All right,
> Those old cheeks that faintly flushed as the port suffused the veins,
> Drain'd white.
>
> Crocus in the Fellows' Garden, winter jasmine up the wall
> Gleam gold.
> Shadows of Victorian chimneys on the sunny grassplot fall
> Long, cold.
> Master, Bursar, Senior Tutor, these, his three survivors, all
> Feel old."

Four years after Ramsden's paper was published saw the appearance of the first paper by Pickering in this area in 1907.[6] Spencer Umfreville Pickering (Figure 1.16) was born in London in 1858 and died in 1920 at the age of 62. He was educated at Eton School and obtained an M.A. in Natural Sciences at Balliol College, Oxford.[62] Even as a school boy he had been devoted to chemistry, and it was while experimenting in the laboratory provided for him by his father at his home in London that, as the result of an explosion, his eye received a serious injury which resulted in its removal. In 1880, he became Lecturer and then Professor in Chemistry at Bedford College, a position which he retained until 1887. He continued to work in the private laboratory in London where he became interested in the nature of solutions.

Figure 1.16 Photograph of Professor S.U. Pickering (1858–1920), M.A., FRS, Professor of Chemistry, Bedford College and Director of Woburn Experimental Fruit Farm, taken in 1893. Thanks to Christine Woollett of The Royal Society, London.

The innumerable determinations of density, freezing point and conductivity were carried out single-handed with a high degree of accuracy, and the results were embodied in more than 70 papers between 1887 and 1896. In his obituary, it was stated that "*...he produced work which, although concerned with widely different branches, was throughout characterised by a disregard of authority and reliance on his own judgement, based on the results of carefully planned and well-executed experiments*".[62] He was elected a Fellow of The Royal Society in 1890. In 1894 he became Director of the Woburn Experimental Fruit Farm, a private venture of Pickering and his college friend, the Duke of Bedford. Consisting of 20 acres, the farm was established for cultural experiments only and all aspects of fruit growing were investigated. In 1905, Pickering began extensive work on insecticides and fungicides, paying particular attention to their composition and use. This study led to work on emulsification, and he succeeded in obtaining remarkable semi-solid emulsions containing as much as 99% of paraffin oil dispersed in only 1% of a 1% soap solution. He also found that insoluble precipitates, such as the basic sulphates of iron and copper, could replace the soap usually employed as emulsifier, yielding extremely stable emulsions, admirably adapted for use as insecticides or fungicides.

His paper in 1907 entitled "Emulsions" had a double objective of obtaining an emulsifying agent which was superior to soap and also of elucidating the nature of emulsification.[6] He found that when copper sulphate (an insecticide) was added to particulate lime (CaO) in water before emulsification with paraffin oil, the emulsion was easier to form and comprised of smaller drops compared with that stabilised by soap molecules. The basic sulphate of copper, precipitated by the action of lime on copper sulphate, was acting as a solid particulate emulsifier. The emulsion behaved as a fungicide and an insecticide, was stable to creaming and showed no signs of coalescence. Other particulate emulsifiers were mentioned including the basic sulphates of iron and nickel, calcium and lead arsenate, ferrous hydroxide, Oxford clay and ferrous hydrosulphide. Pickering suggested that the success of emulsification depended solely on the size of the particles, since "...*when the oil is broken up into small globules by being forced through the syringe, and these globules find themselves in the presence of a number of very much more minute solid particles, the latter will be attracted by the globules, and will form a coating or pellicle over the globules, preventing them from coming in contact and coalescing with their neighbours*".[6] Using microscopy, the aggregation of basic iron sulphate particles around droplets was evident from the brown ring encircling the latter. Pickering also hinted at the importance of the wettability of the particles in oil + water mixtures in obtaining stable emulsions. Thus, in the case of purple of Cassius (colloidal gold/stannic acid mixture) and ferric ferrocyanide (both of which contained finely divided particles), paraffin oil was not emulsified at all but it abstracted the particles from water becoming intensely coloured by them leaving the water colourless. These particles were thus much more wetted by oil than water and incapable of stabilising o/w emulsions.

The work of Ramsden[5] and Pickering[6] initiated a flurry of related work in the 1920s, most notable being the work of Finkle *et al.*[63] on the wettability of solid surfaces and that of Briggs[64] on the effect of particle flocculation on emulsification. Apart from a few papers from then on, including a seminal report by Schulman and Leja[65] linking measured contact angles with emulsion stability, it is surprising why this field lay fallow for nearly 60 years until it was revived again by, *inter alia*, Menon *et al.*[66] in connection with their work on extremely stable water-in-crude oil emulsions and mechanisms of demulsification.

1.5.1 Thermodynamics of emulsification

The formation of particle-stabilised emulsions requires an input of mechanical work. This is needed to break the disperse phase into drops before the solid particles can adsorb on the newly formed oil–water interface, accompanied by a gain in adsorption energy. Such emulsions have a lifetime of years upon storage and, from a thermodynamic viewpoint, correspond to a metastable system. Kralchevsky *et al.*[67] have recently described the thermodynamics of emulsion formation, extending earlier work by Aveyard *et al.*[38] on the influence of line tension and bending energy on this process. Assuming that the particles are spherical, monodisperse and are initially dispersed in phase 2 (*continuous phase*), the work of formation of monodisperse drops of the emulsion 1-in-2, is given by[67]

$$W_1 = \frac{3\gamma_{ow}V}{r} w_1 \qquad (1.34)$$

where γ_{ow} is the bare oil–water interfacial tension, V is the total volume of phase 1 plus phase 2, r is the radius of a particle and w_1 is the dimensionless work written as

$$w_1 = \phi_1\left[\varepsilon(1 - \phi_a b) + \varepsilon^2\phi_a\{f(\theta)(1 - \phi_a b) - 2b\cos\theta\} + O(\varepsilon^3)\right] \quad (1.35)$$

in which ϕ_1 is the volume fraction of phase 1 (volume of phase $1/V$), ϕ_a is the area fraction of interface occupied by particles, ε is the ratio of the particle radius to the drop radius, θ is the contact angle made by particles at the interface measured into phase 2, $b = (1 - \cos\theta)^2$, $f(\theta) = (1 - \cos\theta)^2(2 + \cos\theta)$ and O denotes "of the order of". Since $\varepsilon \ll 1$, the leading term in equation (1.35) is $\phi_1\varepsilon(1 - \phi_a b)$ accounting for the formation of new oil–water interface and for particle adsorption. The curvature effects ($\alpha\ \varepsilon^2$) give a higher order contribution.

The calculated curves of w_1 versus the dimensionless drop curvature, ε, are shown in Figure 1.17(a), for $\phi_1 = 0.3$ (volume fraction of disperse phase) and $\phi_a = 0.9$ (relatively close packed layers).[67] The curves correspond to different values of the contact angle θ. For $\theta \leqslant 90°$, w_1 is positive, indicating that energy is needed to break phase 1 into drops, and increases with decreasing drop size. For $\theta \geqslant 100°$ however, w_1 is negative meaning that, in principle, the emulsion could form spontaneously in the absence of kinetic barriers. This is due to the significant gain in surface energy upon adsorption of particles with $\theta > 90°$. Only for $\theta \approx 93°$ is the first term in equation (1.35) small and comparable to the second term. Then, the dependence of w_1 on ε is parabolic (not linear) having a minimum at a certain value of ε, Figure 1.17(b). Thermodynamically, such a minimum corresponds to the spontaneous formation of drops of the respective size. These minima exist only for a very narrow range of values of θ, and it is unlikely that this situation could be realised in practice.

A similar equation to equation (1.35) exists for the work of formation of the emulsion of 2-in-1, w_2, in which the *disperse phase* is phase 2, again containing the particles. The calculated curves in this case are shown in Figure 1.18(a) for various values of the contact angle and for the same volume fraction of dispersed phase ($\phi_2 = 0.3$).[67] Although the curves are almost identical to those in Figure 1.17(a), around $\theta \approx 93°$ the dependence of w_2 on ε is parabolic displaying in this case a maximum, shown in Figure 1.18(b). Thermodynamically, this maximum corresponds to a critical drop size for the emulsions. For $\varepsilon < \varepsilon_{crit}$, it is favourable for the drops formed to increase in size by coalescence (requiring an input of energy), whereas for $\varepsilon > \varepsilon_{crit}$ it is favourable for the drops to be broken into smaller ones (spontaneous).

When a mixture of oil and water is subjected to homogenisation, it is thought that both o/w and w/o, emulsions can form simultaneously in different spatial domains of the vessel.[68] Only that which is most stable survives however. In the case of particle-stabilised emulsions, the gain in surface free energy upon particle

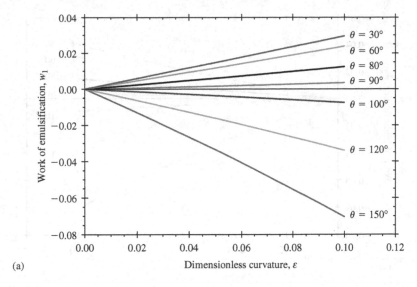

(a)

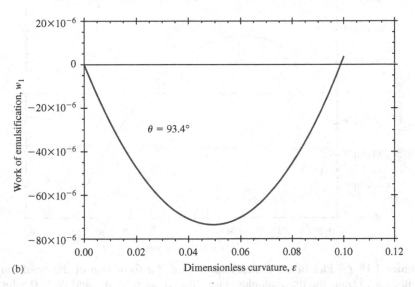

(b)

Figure 1.17 (a) Plot of the dimensionless work for formation of the emulsion 1-in-2, w_1, *versus* the dimensionless curvature, ε, at $\phi_1 = 0.3$ and $\phi_a = 0.9$ for different contact angles θ. (b) The curve for $\theta = 93.4°$ is shown in an enlarged scale revealing a minimum, corresponding to spontaneous formation of drops with the respective size. Taken from Ref. [67]; with permission of the American Chemical Society.

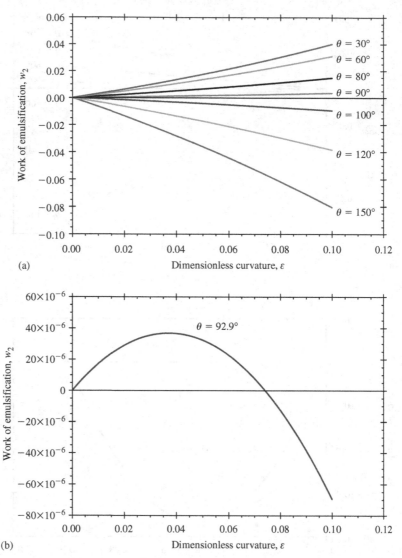

Figure 1.18 (a) Plot of the dimensionless work for formation of the emulsion 2-in-1, w_2, *versus* the dimensionless curvature, ε, at $\phi_2 = 0.3$ and $\phi_a = 0.9$ for different contact angles, θ. (b) The curve for $\theta = 92.9°$ is shown in an enlarged scale to reveal a maximum, corresponding to the critical drop size. Taken from Ref. [67]; with permission of the American Chemical Society.

adsorption is usually much greater than the entropy effects accompanying emulsification such that the difference between the work of formation of emulsion 1-in-2 and 2-in-1, $\Delta w = w_1 - w_2$, provides a thermodynamic criterion implying which emulsion will remain after agitation. For $\Delta w < 0$, emulsion 1-in-2 remains, whereas

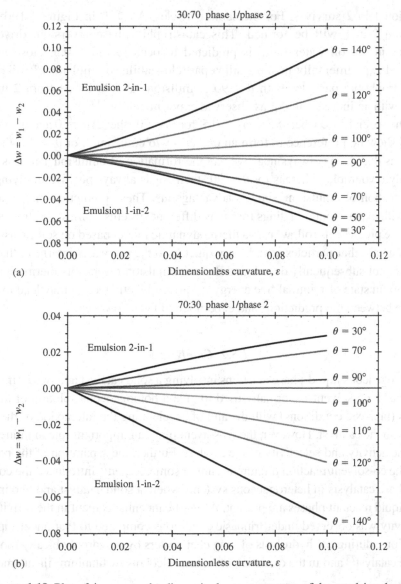

Figure 1.19 Plot of Δw *versus* the dimensionless curvature, ε, of the emulsion drops for different contact angles, θ, at fixed $\phi_a = 0.9$. (a) $\phi_1 = 0.3$, (b) $\phi_1 = 0.7$. $\Delta w < 0$ and $\Delta w > 0$ correspond to the formation of emulsions 1-in-2 and 2-in-1, respectively. Taken from Ref. [67]; with permission of the American Chemical Society.

for $\Delta w > 0$, emulsion 2-in-1 remains. Figure 1.19 shows plots of Δw versus the dimensionless drop curvature, ε, for various values of the contact angle and for two volume fractions of phase 1: $\phi_1 = 0.3$ (Figure 1.19(a)) and $\phi_1 = 0.7$ (Figure 1.19(b)). For hydrophilic particles with $\theta \leqslant 90°$, in Figure 1.19(a) $\Delta w < 0$ and the emulsion 1-in-2 survives. For the same particles, $\Delta w > 0$ in Figure 1.19(b) and emulsion 2-in-1 will be formed. This catastrophic phase inversion, driven by changes in the oil:water ratio, is predicted to occur at $\phi_1 = 0.5$, close to that observed experimentally in some silica particle-stabilised emulsions.[69] Likewise, for hydrophobic particles with $\theta \geqslant 90°$, emulsions should invert from 2-in-1 to 1-in-2 with an increase in ϕ_1, as observed experimentally.[70,71]

In the special case when $\phi_1 = \phi_2 = 0.5$, for $\theta = 90°$ then $\Delta w = 0$ and the system should show no preference to form an o/w or a w/o emulsion. For $\theta < 90°$ the prediction is that Δw is always negative, *i.e.* the formation of emulsion 1-in-2 is energetically favourable, whereas for $\theta > 90°$ then Δw is always positive implying that the formation of emulsion 2-in-1 is advantageous. These predictions are in agreement with experimental findings for many different systems. It must be stressed that the above predictions follow from a thermodynamic model based on several assumptions, namely that particles adsorb very quickly to the oil–water interface, that particles do not subsequently desorb and that the emulsion reaches its thermodynamic equilibrium state of minimal free energy.[67] Various kinetic factors may lead to a difference between the predictions and experimental facts however.

1.5.2 Catalysis

In heterogeneous liquid-phase reactions involving a solid catalyst and two immiscible reactants, a co-solvent is normally used to give a homogeneous solution of the two liquids (biphasic conditions) with the aim of accelerating the interaction of the reactants with the catalyst. However, the co-solvent plays an important role in influencing both the activity and selectivity of the catalyst. Further, the separation of the product from the co-solvent/reaction mixture is cumbersome. Regen[72] introduced the concept of triphase catalysis in heterogeneous systems where a solid catalyst and two immiscible liquid reactant phases are present. A significant enhancement in the activity and selectivity was observed under triphasic conditions compared to that under biphasic conditions, during the hydrolysis of long chain esters by a hydrophobically modified zeolite catalyst[73] and in the oxidation of various alcohols by titanium silicate molecular sieves.[74]

As an example, we describe the recent and more complete work of Ikeda *et al.*[75] on the use of titania impregnated sodium zeolite catalyst particles. Such particles can be placed at the liquid–liquid phase boundary between aqueous hydrogen peroxide and a water-immiscible alkene and act as an efficient catalyst for alkene epoxidation. By modifying the surface of such particles with octadecyltrichlorosilane (OTS), the effect of particle hydrophobicity on catalytic behaviour was established. Three types of particles were studied: as-prepared hydrophilic grade, partially hydrophobic grade (with OTS) and completely hydrophobic grade (with OTS). The particles (50 mg), alkene

Table 1.2 *Effect of hydrophobicity of the solid particle sodium zeolite/titania catalyst, of size 10–40 μm, on the epoxidation of normal alkenes at the alkene–aqueous H_2O_2 interface. Taken from Ref. [75]; re-drawn with permission of the American Chemical Society.*

	1,2-Epoxide yield/μmol					
	Hydrophilic		Partially hydrophobic		Completely hydrophobic	
Alkene	Static	Stirring	Static	Stirring	Static	Stirring
1-hexene	0	4.5	51.6	16.2	7.2	4.5
1-octene	0.2	2.5	27.0	27.4	3.6	5.9
1-dodecene	0	0	17.7	20.0	0	3.2

($4 \, cm^3$) and aqueous H_2O_2 ($1 \, cm^3$) were placed in a glass tube and reacted for 24 h, with or without stirring. As expected, the hydrophilic particles partitioned mostly into the aqueous phase, the most hydrophobic particles likewise into the alkene phase while those of intermediate hydrophobicity adsorbed to the alkene–water interface. Table 1.2 summarises the yield of product obtained during this triphasic catalysis for 3 alkenes.[75] The hydrophilic particles showed the lowest activity, followed by the most hydrophobic ones with those in between exhibiting remarkable activity. The activity of the latter particles was independent of the stirring rate, *i.e.* this catalyst does not require the formation of an emulsion to be active (exception is for 1-hexene where appreciable liberation of by-products occurred when stirred).

In this[75] and subsequent papers,[76,77] the authors insist that the catalytic activity of particles of intermediate hydrophobicity is a result of such particles being truly amphiphilic, *i.e.* during chemical reaction with OTS, one face of the particles becomes coated and rendered hydrophobic while the other face remains uncoated and hence hydrophilic. Such heterogeneously coated particles, frequently called Janus-like after the Roman god of gates and doors represented with a double-faced head each looking in opposite directions, are not a pre-requisite for their adsorption to oil–water interfaces since homogeneously coated particles of the correct wettability will do so also, as mentioned earlier. However, the use of amphiphilic particles may be crucial in relation to their catalytic behaviour. These particles were also shown to act as efficient stabilisers of oil-in-water emulsions.[78]

1.5.3 Emulsions stabilised by stimuli-responsive particles

The majority of studies on particle-stabilised emulsions have employed relatively inert particles like silica or carbon.[4,55] Recently, specific particles have been

deliberately synthesised such that they are responsive to a number of different stimuli, *e.g.* changes in pH, temperature or an applied magnetic field. This sub-section
aims to describe the properties of some of these particles and their ability to act as
stabilisers of emulsions. The ability to influence the behaviour of emulsions
through the effect of these stimuli on the adsorbed particles is clearly of technological interest.

1.5.3.1 Microgel particles

A microgel particle is a cross-linked latex particle which is swollen by a good solvent.[79] Poly(*N*-isopropylacrylamide) or poly(NIPAM) is the most well-studied water-
swellable microgel system, whereas polystyrene (PS) is the best example of an
organic swellable microgel.[80] The latter is swollen by aromatic solvents like toluene.
Ionic microgel particles frequently contain carboxylate groups derived from acrylic or
methacrylic acid (MAA) as co-monomer. This gives scope for variation of the microgel properties by changes in solution conditions like pH.[79]

Novel emulsions stabilised entirely by microgel particles have been prepared and
characterised recently.[81,82] In one, the particles are sensitive to changes in both pH
and temperature.[81] Monodisperse poly(NIPAM)-*co*-MAA microgel particles were
synthesised using surfactant-free precipitation polymerisation based on NIPAM as
monomer, 5 wt.% of MAA as co-monomer and cross-linked with *N,N'*-methylene
bisacrylamide. In water at high pH = 9.4, the particles exhibit a volume phase transition upon changes in temperature, with the hydrodynamic diameter falling progressively from ∼235 nm at 25°C to ∼150 nm at 50°C. This is expected since linear
poly(NIPAM) has a lower critical solution temperature of ∼32°C in water at which
point its conformation changes from a hydrophilic random coil to a more hydrophobic globule. In addition, these particles are also sensitive to pH change due to the
presence of the ionic co-monomer. Thus, at 25°C say, the particle size increases progressively from pH = 3 to pH = 8 after which it levels off. Upon increasing pH,
carboxyl groups in the polymer are ionised so that electrostatic repulsion between
molecules within a particle is enhanced causing it to swell. Thus, by either raising
the temperature or by decreasing the pH, the hydrophobicity of the microgel particles can be increased ultimately affecting their behaviour as an emulsifier.[81]

Emulsions of octanol and water stabilised by 1 wt.% of particles are o/w at all temperatures and pH. At 25°C, batch emulsions are completely stable to coalescence over
4 months between pH = 10 and 6. Their stability to coalescence decreases progressively however from pH = 6 to 3, below which complete oil–water phase separation
ensues. Figure 1.20 shows the appearance of the systems at extremes of pH, before
homogenisation and 2 days after.[81] At pH = 9.4, charged partially hydrophobic particles, residing more in water than in oil when adsorbed, coat octanol drops sufficiently
well to provide stability. The inset in Figure 1.20 shows freeze fracture scanning

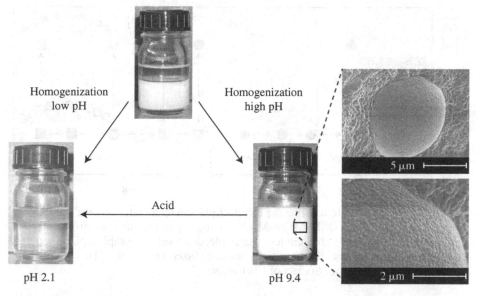

Figure 1.20 Effect of pH on the stability of octanol-in-water emulsions ($\phi_o = 0.3$) stabilised by poly(NIPAM)-*co*-MAA microgel particles (1 wt.%) at 25°C and 48 h after formation. At pH = 2.1, the oil phase is coloured by the dye-labelled particles, whereas the particles cause only a faint tint when distributed in the water phase (middle vessel) or within the stable emulsion (pH 9.4). The inset shows freeze fracture SEM images of particle covered octanol drops at high pH. Taken from Ref. [81]; with permission of the Royal Society of Chemistry.

electron microscopy (SEM) images of such a droplet at two magnifications, where it is clear that particles are densely packed on the interface. Coalescence can also be initiated *in situ* by lowering the pH on addition of HCl to the continuous phase, causing protonation of carboxylate groups in the particles rendering them more hydrophobic. Partial desorption of particles from drop interfaces occurs reducing their surface concentration and promoting coalescence. At pH = 2.1, fast complete phase separation is triggered and particles, originally dispersed in water, partition into the oil phase which becomes coloured. It is relevant to note that these particles do not now stabilise w/o emulsions, either because they are too hydrophobic to remain attached to interfaces or because the volume fraction of water in these systems is too high (0.7).

At fixed pH = 6.1, raising the temperature from 25 to 60°C results in the stable emulsion coarsening and liberating oil via coalescence.[81] The relatively hydrophobic particles are rendered more hydrophobic with increasing temperature, maybe due to dehydration, followed by their desorption from the interface. At pH = 9.4, however, the emulsion remains stable at the higher temperature as highly charged, partially hydrophobic particles remain attached to the interface. A schematic summary of the effects of temperature and pH on microgel-stabilised o/w emulsions is

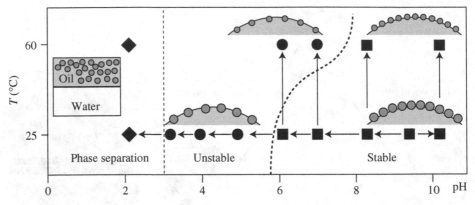

Figure 1.21 Schematic showing the effect of pH and temperature on the stabilising efficiency of poly(NIPAM)-*co*-MAA microgel particles in octanol-in-water emulsions. Squares = stable, circles = unstable, diamonds = complete phase separation. Arrows indicate the transitions studied. Taken from Ref. [81]; with permission of the The Royal Society of Chemistry.

given in Figure 1.21. It remains to understand the influence of these two stimuli on the charge and size of the particles independently.

In work undertaken simultaneously with the above, Fujii *et al.*[82] showed the critical effect of pH at 20°C on o/w emulsions stabilised by another type of microgel particle for different oils. The latter was those of poly(4-vinylpyridine) (P4VP) co-polymerised with nanoparticles of silica (20 nm) and lightly cross-linked with ethylene glycol dimethacrylate. Nanocomposite formation is due to an acid–base interaction between the nitrogen atoms on pyridine and the silanol groups on silica. Their structure is that of "currant bun" morphology in which the silica particles are distributed within the core of the polymer particles and at their surface. The particle surface has dual character, *i.e.* segregated domains of hydrophilic silica and hydrophobic P4VP, promoting adsorption to the oil–water interface. In water, the particles are of diameter 230 nm at pH = 8.8 but swell considerably on lowering the pH to a diameter of 550 nm at pH = 2.5, due to protonation of the 4-vinylpyridine groups imparting cationic character. At high pH, emulsions of equal volumes of methyl myristate (or dodecane) and aqueous dispersion are o/w and stable to coalescence.[82] On lowering the pH *in situ* to 2.0, rapid coalescence is initiated resulting in complete phase separation within 30 s (Figure 1.22(a)). As schematised in Figure 1.22(b), addition of acid causes protonation of the vinylpyridine groups with concomitant swelling of the particles. This increase in their hydrophilicity is sufficient to promote their desorption from drop interfaces allowing coalescence to take place. A comprehensive study of the effects of changes in pH and addition of electrolyte (which enhances ionisation) on the emulsion stabilising properties of these microgel particles has just been completed.[83]

(a)

(b)

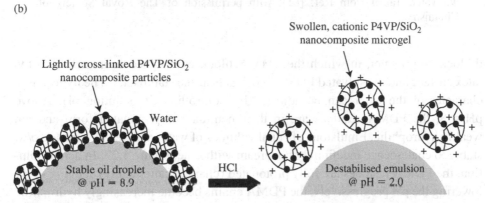

Figure 1.22 (a) Digital photographs illustrating the rapid macroscopic phase separation of a methyl myristate-in-water emulsion prepared at pH = 8.9 using 1 wt.% of P4VP/SiO₂ microgel particles (diameter dried ≈200 nm). Immediately after adjusting the pH of the continuous phase to 2 using HCl (0 s), minimal agitation (5 s) resulted in rapid phase separation (30 s). (b) Schematic of pH-induced demulsification. Taken from Ref. [82]; with permission of Wiley-VCH.

1.5.3.2 Sterically stabilised particles

Charge-stabilised PS latex particles of low charge density, in which the charge originates from persulphate ions used as initiator in their synthesis, prefer to stabilise w/o emulsions.[84] That is, the hydrophobic character of the particles imparted by PS dominates their wettability at an oil–water interface. By grafting a co-polymer onto the PS chains, sterically stabilised latex particles capable of being charged become sufficiently hydrophilic to stabilise o/w emulsions.[85] The steric stabiliser is poly [2-(dimethylamino)ethyl methacrylate-*block*-methyl methacrylate] (PDMA-PMMA)

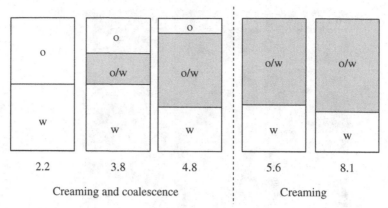

Figure 1.23 Effect of varying the pH (given) on the stability after 24 h of hexadecane-in-water emulsions stabilised by sterically stabilised PDMA-PMMA PS latex particles (diameter = 160 nm). Residual stable emulsion is shaded, o: oil, w: water. Taken from Ref. [85]; with permission of The Royal Society of Chemistry.

diblock co-polymer, in which the PMMA block is designed to adsorb onto the latex surface and the solvated PDMA block acts as the stabiliser.[86] The hydrophilic character of the latter can be adjusted by controlling the solution pH. Above pH = 8, the PDMA residues are in their neutral, non-protonated form and are weakly hydrophilic. Emulsions of equal volumes of water and hexadecane are o/w, stable to coalescence indefinitely but cream with time, Figure 1.23. In acidic solution, the PDMA residues are fully protonated (charged cationically) at pH = 4. By lowering the pH progressively, the PDMA chains become increasingly hydrophilic leading to their gradual desorption from the interface, which in turn destabilises the emulsion.[85] Complete phase separation occurs rapidly by pH = 2.2 (Figure 1.23). This effect is reversible: adjusting the pH from 3.8 to 8.1 allows the original stable emulsion to be re-formed on re-homogenisation. The latex emulsifier can thus be reused.

A similar idea to the one above was reported recently in which silica nanoparticles were modified with an anionic brush of the polyelectrolyte poly(styrenesulphonate).[87] The polymer forms a quenched brush whose charge is independent of pH or ionic strength. These highly charged spherical particles (zeta potential = −77 mV) lower the oil (trichloroethylene)–water interfacial tension significantly and produce stable, non-invertable o/w emulsions at particle concentrations as low as 0.04 wt.%. The vinyl backbone of the polymer is sufficiently non-polar and distant from the charged sulphonate groups to favour particle adsorption.

In addition to being pH sensitive, particles of PS stabilised by PDMA-PMMA chains are also temperature sensitive, since PDMA homopolymer exhibits inverse solubility behaviour between 35°C and 45°C depending on its degree of polymerisation. The particle surface becomes less hydrophilic at higher temperatures and, as shown for the first time, emulsions stabilised by such particles at pH = 8.1 (uncharged) can be inverted from o/w to w/o by a temperature change alone.[88] Inversion takes place between 53°C and 64°C, within which range an emulsion can be of either type which may co-exist. Based on relevant contact angle measurements, it was argued that the dominant influence of temperature is in reducing the degree of hydration of the PDMA chains to such an extent that the particles become preferentially wet by oil. This ability to invert a particle-stabilised emulsion with temperature in this way is akin to emulsions of nonionic surfactant. Inversion is important industrially and is exploited to prepare fine emulsions of increased stability.[89]

1.5.3.3 Ionisable latex particles

Control of the type of emulsion is important as is the ability to invert from one to the other. Until now, obtaining both types of emulsion in a particular oil–water–particle system was only possible by either utilising particles of different wettability[71] or by changing the oil:water ratio[70] or by adding surfactant which adsorbs on particle surfaces and modifies their wettability *in situ*.[53] These routes may not be suitable in many cases. A simple strategy to invert emulsions was devised recently using PS nanoparticles whose surfaces contain carboxylic acid groups.[90] By controlling the degree of charge on particle surfaces, by changes in pH or addition of salt, particles are made sufficiently more or less hydrophilic for them to stabilise both types of emulsion in the same system. It was not possible to achieve this degree of control previously; PS particles containing surface sulphate groups preferred to form w/o emulsions, *i.e.* such particles were relatively hydrophobic.[84] In contrast, PS particles with grafted co-polymer molecules containing amine groups stabilised only o/w emulsions since particles were rendered more hydrophilic.[85]

Spherical, monodisperse, surfactant-free PS particles can be synthesised with different densities of carboxylic acid groups on their surfaces. Under conditions where these groups are predominantly unionised (—COOH), particles are expected to be more hydrophobic stabilising water drops in oil compared to ionising conditions (—COO⁻) in which particles are more hydrophilic and should stabilise oil drops in water. For particles of 200 nm diameter and high surface charge density ($101\ \mu\mathrm{C\,cm}^{-2}$), hexadecane–water emulsions (1:1 by vol.%) were o/w at all pH between 2 and 11 and all salt (NaCl) concentrations between 0 and 2 M.[90] By reducing the charge density however ($12.7\ \mu\mathrm{C\,cm}^{-2}$), the same size particles act as excellent emulsifiers for both emulsion types. Figure 1.24(a) shows the appearance of the emulsions for different pH at high salt concentration, one year after preparation. Emulsions are of low conductivity, disperse in oil and are therefore w/o up to pH = 9.5. Inversion (dotted line) to highly conducting o/w emulsions which

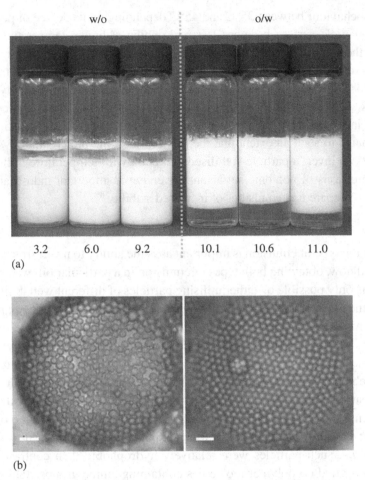

(a)

(b)

Figure 1.24 (a) Appearance of vials at 25°C containing hexadecane and 1 M aqueous NaCl (1:1) and 2 wt.% carboxyl-coated PS latex particles of 200 nm diameter as a function of the pH (given). Emulsions are w/o at pH < 9.5 and undergo sedimentation and are o/w at pH > 10.0 and undergo creaming. (b) Optical micrographs of a w/o (left) and an o/w (right) emulsion drop in the same system but stabilised by 3.2 μm diameter particles. Scale bars = 10 μm. Taken from Ref. [90]; with permission of Wiley-VCH.

disperse in water occurs just below pH = 10.0. All emulsions were completely stable to coalescence since no visible separation of the dispersed phase occurred and the drop size distributions determined by light diffraction were unchanged. Water-in-oil emulsions sediment over time with a clear oil phase separating above (three vessels on left) whereas o/w emulsions cream to the top of the vessel as a clear water phase drains below (three vessels on right). Although the dispersed phase volume fraction in both the sedimented and creamed emulsions increases from 0.5 to over 0.75, emulsion drops remain stable to coalescence due to adsorbed layers of PS particles acting as a mechanical barrier to fusion. In order to visualise particles around drops, emulsions were prepared at low and high pH using larger particles (3.2 μm diameter) of approximately the same surface charge density (11.8 μC cm^{-2}). Optical micrographs of similar sized emulsion drops are seen in Figure 1.24(b).[90] The left image is that of a water drop in oil formed at low pH. The particles cover most of the oil–water interface but arrange themselves in a liquid-like order with parts of the interface devoid of particles. We observed that particles in the aqueous dispersions before emulsification were strongly flocculated at low pH due to the absence of repulsion between uncharged —COOH surface groups. It appears that some of the flocs adsorb intact to drop interfaces leaving voids between them. In contrast, the right image is that of an oil drop in water formed at high pH in which particles appear hexagonally close packed over the entire interface. At these high pH, dissociation of acid groups occurs and repulsion between like-charged particle surfaces in bulk causes breakup of the flocs and a dispersion of discrete particles to be stabilised. The hypothesis that raising the pH results in hydrophobic particles becoming increasingly hydrophilic is given some credence from the contact angle data of water drops under oil on monolayers of a long chain carboxylic acid adsorbed via a terminal thiol group on gold. For a dilute layer of surface —COOH groups, θ_{ow} (through water) decreases from 104° at pH = 2 through 90° to 67° at pH = 13 as acid groups ionise and become more wettable by water.[90] This decrease is in line with emulsion inversion from w/o to o/w with increasing pH.

Charged particles are also sensitive to the presence of electrolyte. A second way to effect emulsion inversion is to start with a hydrophilic system forming an o/w emulsion and reduce the salt concentration rendering the system more hydrophobic.[90] At pH = 10.6, emulsions were shown to invert from o/w to w/o on lowering the salt concentration to 0.6 M as predicted. The reason for this change is that the degree of ionisation of surface acid groups increases not only with pH but also with salt concentration. A relatively high pH value (10) is required for inversion compared with the pH for 50% ionisation (pK_a) of a short chain carboxylic acid in bulk water (4.8). Acid dissociation is more energetically unfavourable when such groups are at an interface (particle or drop) compared with isolated groups in bulk because of the proximity of other similar charged groups, which generate a surface potential ψ. For the particles used here, the area per charged group is ≈1.25 nm^2. Due to the existence of an electrostatic field around a negatively charged particle, protons are held more strongly at the surface and the surface pH is lower than that in bulk. Dissociation of weak acid groups on a surface will therefore only be complete at higher pH in solution compared with that of a weak acid in bulk. Addition of salt causes a reduction in ψ and promotes further dissociation.

1.5.3.4 Paramagnetic particles

An alternative to controlling the stability of particle-stabilised emulsions through changes in solution conditions is to apply an external field to the system. This idea has recently been investigated by employing paramagnetic particles which respond to a magnetic field.[91] The particles of carbonyl iron, of average diameter 1.1 μm, are super-paramagnetic displaying neither hysteresis nor magnetic remanence in the presence of a field. Being relatively hydrophilic, emulsions of decane-in-water stabilised by these particles can be formed either by hand shaking or using a vortex mixer. A magnetic field is applied using an electromagnet made from insulated copper wire wrapped around an iron core. This coil is placed beneath the vial containing the emulsion. Upon application of the field, these emulsions undergo phase separation in two ways. (i) Oil drops coated with particles diffuse through the aqueous phase downwards without affecting the overall emulsion stability. This is clearly seen in Figure 1.25(a); with the field turned off (0 s) the emulsion has already creamed yielding a lower water layer beneath the drops. The drop movement can be controlled externally by the electromagnet. When the applied field strength is not strong enough to induce drop movement, a slight elongation of the drops closest to the magnet along the field direction is observed, Figure 1.25(b). A critical value of the magnetic field strength is identified above which drops begin to move. This value decreases with a decrease in the size of the drops, as larger drops are more deformable. (ii) Complete phase separation (*i.e.* coalescence) occurs in these emulsions when a strong field is applied. The drops as a whole are compressed on reaching the bottom of the vessel and the strong field causes the adsorbed particles to be stripped from the drop interfaces and sediment at the bottom. This releases the oil drops so that they move upwards due to buoyancy and coalesce with bulk oil at the top of the sample (Figure 1.25(c)). A stronger field ($128.9 \, \text{kA m}^{-1}$) destabilises the entire emulsion leading to a completely phase separated system. By applying a square wave magnetic field (on–off), the reversibility of the destruction and re-formation of the emulsion was tested over 100 cycles. Impressively, the original mean drop diameter of the stable emulsion was obtained each time meaning that there are no long-lasting effects following destabilisation.[91]

1.5.3.5 Light-sensitive particles

Directed self-assembly of nanoparticles opens new avenues of technology through the controlled fabrication of nanoscopic materials with unique optical, magnetic and electronic properties.[92] Ligand-stabilised nanoparticles are ideally suited to hierarchical self-assembly since the nanoparticle core dictates the particular property whereas the surface-bound ligand defines the interactions between the particle and its surroundings. At a curved oil–water interface of high area, particles are highly mobile and rapidly achieve an equilibrium assembly. For nanoparticles however, their

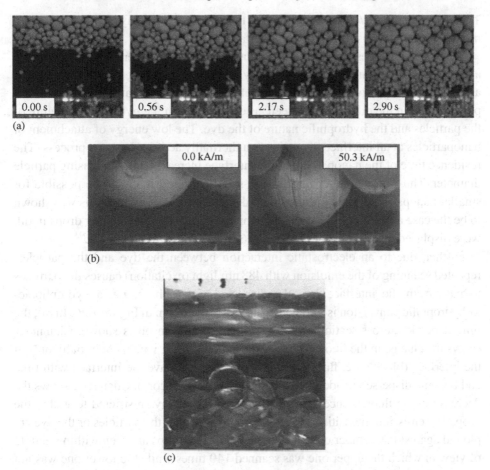

Figure 1.25 (a) Effect of a magnetic field (strength = $29.6\,\mathrm{kA\,m^{-1}}$) on the movement of oil drops towards the bottom of the vial in an o/w emulsion stabilised by paramagnetic particles. At 0 s the drops are at rest in the absence of a field. On applying the field, the drops diffuse downwards without affecting the emulsion stability. (b) Effect of a magnetic field on the shape deformation of a large (diameter 4.5 mm) oil drop. (c) Coalescence of oil drops with the bulk oil phase after moving upwards following application of a high magnetic field (strength = $86\,\mathrm{kA\,m^{-1}}$). Taken from Ref. [91]; with permission of the American Chemical Society.

thermal energy which causes rapid spatial fluctuations is comparable to the adsorption energy.[4] This energy balance can result in a weak segregation of the particles at the interface. The size-dependent adsorption and desorption of nanoparticles from an interface giving rise to two-dimensional phase separation has been investigated by Lin *et al.*[93] The particles, of diameter 2.8 and 4.6 nm, were those of cadmium selenide (CdSe) coated with tri-*n*-octylphosphine oxide (TOPO). They are only dispersible initially in oil (toluene) but adsorb on water drops when shaken with water.

The corresponding w/o emulsion is stable to coalescence for days as a close packed monolayer of particles forms at drop interfaces.

By adding a fluorescent dye, sulforhodamine-B, to the water phase the particles and dye could be examined independently by monitoring their fluorescence at 525 and 585 nm, respectively. No evidence of the presence of particles in the water phase or dye in the oil phase was seen, as expected from the hydrophobic nature of the particles and the hydrophilic nature of the dye. The low energy of attachment of nanoparticles at an interface gives rise to a thermally activated escape process. The residence time of the nanoparticle at the interface increases with increasing particle diameter. Thus, unlike larger micrometre-sized particles, it should be possible for smaller nanoparticles to be preferentially displaced by larger ones. This was shown to be the case as particles of diameter 2.8 nm, initially stabilising water drops in oil, were displaced by particles of diameter 4.6 nm retaining drop stability.[93]

Further, due to an electrostatic interaction between the dye and the particles, repeated scanning of the emulsion with 488 nm light (excitation) causes the particles to pass from the interface into the dispersed water phase, *i.e.* a hydrophobic-to-hydrophilic conversion is initiated. This is clearly shown in Figure 1.26. In (a), the time dependent cross-sectional image of a water drop in oil is shown. Channel 1 shows the change in the fluorescence of the nanoparticles, which are initially only at the interface (black core, fluorescent shell) but which leave the interface with time and become dispersed inside the water drop (fluorescent core). Channel 2 shows the decrease in the fluorescence from the dye which is always restricted to within the drop. The emission intensities *inside* the drop from either the particles or the dye are plotted against the number of scans in (b). In (c), two drops are seen within the field of view in which the upper one was scanned 140 times while the lower one was not irradiated. In the former, in addition to the particles diffusing into the water core, some remained at the interface in the same way that they remained as a shell in the latter. Importantly, this transition in particle wettability did not occur in the absence of the dye. It is suggested that adsorption of dye molecules onto particle surfaces takes place in such a way as to increase the particle hydrophilicity and hence their dispersibility in water. This could be achieved via interaction of the cationic diethyl-amine groups on the dye to anionic groups on the particles exposing hydrophilic sulphonate groups (from the dye) on the particle periphery.

1.5.4 Emulsion/dispersion polymerisation (without surfactant)

In the past few years, organic (latex)–inorganic nanocomposite particles have become the subject of rapidly growing interest. It is thought that combining the properties of the two particle types may be advantageous, *e.g.* merging the thermal stability, mechanical strength or light scattering power of the inorganic particles with the

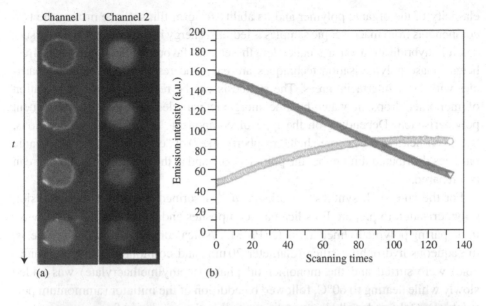

Channel 1   Channel 2

(a)   (b)

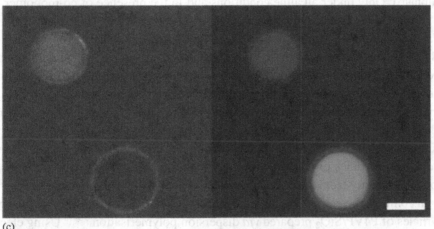

(c)

Figure 1.26 Light-induced hydrophobic-to-hydrophilic conversion of nanoparticles. (a) A 10 μm water drop in oil stabilised by TOPO-coated CdSe particles (2.8 nm) observed with confocal fluorescence microscopy with 488 nm excitation. Channel 1 shows the nanoparticle fluorescence over a period of several minutes during continuous illumination. Channel 2 shows the fluorescence from the sulforhodamine-B dye in water. (b) Average fluorescence intensity inside the water drop *versus* time. Dark points – channel 2 (water), light points – channel 1 (particles). (c) The water drop irradiated with 488 nm light (top left) is compared with the nearby drop without irradiation (bottom left). Images on the right are fluorescence images at 585 nm. Scale bar = 10 μm. Taken from Ref. [93]; with permission of the American Association for the Advancement of Science.

elasticity of the organic polymer and its ability to form films. Since mixing the two components from macroscopic samples is tedious, energy inefficient and not very successful, hybridisation on a smaller length scale is favoured. Such surfactant-free, hetero-phase polymerisation techniques are easy and result in polymer nanoparticles with large interfacial areas. The synthesis can be regarded as the stabilisation of monomer drops in water by fine inorganic particles followed by subsequent polymerisation. Depending on the type of components and prevailing conditions, hybrid particles of either core-shell ("raspberry-like") or "currant bun-like" (inorganic particles distributed throughout the polymer core and at the surface) morphology can be prepared.

For the core-shell synthesis, Tiarks *et al.*[94] described the use of miniemulsion polymerisation to prepare PS/silica nanocomposites and Chen *et al.*[95] did likewise in preparing polymethylmethacrylate (PMMA)/silica composites. In one synthesis, the aqueous hydrophilic silica (diameter 20 nm) and co-monomer (1-vinylimidazole) were stirred and the monomer oil phase (methylmethacrylate) was added slowly while heating to 60°C, followed by addition of the initiator (ammonium persulphate). Only under alkaline conditions and in the presence of co-monomer, *e.g.* 4-vinylpyridine, could stable, nanodispersions be obtained.[95] This is due to the fact that the cationic co-monomer (with amine groups) interacts with silica (with silanol groups) by an acid–base interaction. Without such co-monomers, the pure monomers cannot be stabilised as drops and large polymer particles co-existing with unattached silica particles resulted. As with a surfactant-stabilised process, the final particle size depends on the amount of "emulsifier", *i.e.* silica. The higher the silica content, the smaller the nanocomposite particle size (100–500 nm). As an example, Figure 1.27(a) shows a transmission electron microscope (TEM) image of the hybrid particles of PMMA/SiO$_2$, in which the raspberry-like morphology was obtained directly from the polymerisation of silica-stabilised monomer drops.[95]

In contrast, the currant bun particle morphology has been shown to be present in particles of P4VP/SiO$_2$ prepared *via* dispersion polymerisation.[96–98] Using electron spectroscopy imaging enables the spatial location and concentration of the silica nanoparticles within the composite particles to be determined. As seen in Figure 1.27(b), the spatial distribution of the silica particles is rather uniform. Double-line densitometric profiles indicated high and constant pixel intensities for cross-sections through the two selected nanocomposite particles. The morphology identified is consistent with earlier electrophoresis measurements of these particles which revealed an isoelectric point of pH = 6, *i.e.* the basic 4-vinylpyridine component contributed to the mobility and was therefore located at or very near to the particle surface shear plane along with a fraction of the silica particles. Moreover, X-ray photoelectron studies revealed that the surface Si/N ratio was comparable to the corresponding bulk ratio calculated from gravimetric analysis.

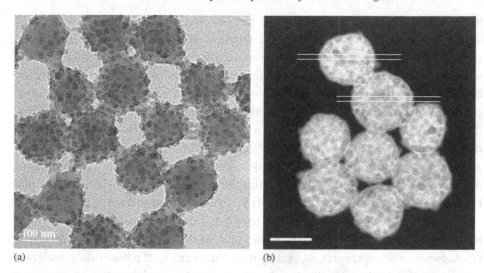

(a)   (b)

Figure 1.27 (a) TEM image of the nanocomposite particles comprising poly(methylmethacrylate), 1-vinylimidazole and silica particles (diameter 20 nm) of "raspberry-like" morphology prepared *via* surfactant-free emulsion polymerisation. Scale bar = 100 nm. Taken from Ref. [95]; with permission of the American Chemical Society. (b) TEM-electron spectroscopy image of nanocomposite particles comprising P4VP and silica particles (diameter 20 nm) of "currant bun-like" morphology prepared *via* dispersion polymerisation. Scale bar = 100 nm. Taken from Ref. [98]; with permission of the American Chemical Society.

The surfactant-free emulsion approach has been utilised very recently by He in the synthesis of both polyaniline/CeO$_2$ composite particles and ZnO microspheres in aqueous solution.[99] In the latter, the solid-stabilised emulsion separated the dimethyl oxalate in oil from the zinc acetate in water, confined the precipitation reaction at the oil–water interface and supplied the drops as templates for the formation of zinc oxalate, the pre-cursor of zinc oxide.

1.5.5 Naturally occurring particles

1.5.5.1 Emulsification using bacterial cells

Emulsions are commonly observed when hydrocarbons and water are mixed during bioremediation or fermentation.[100] While bacteria can produce surfactants that stabilise emulsions, some micro-organisms can emulsify hydrocarbons even in the absence of cell growth. This suggests that emulsification may be associated with the surface properties of the cells, *i.e.* bacterial cells may behave as solid particles at interfaces. The ability of certain intact bacterial cells to stabilise both types of emulsion has been demonstrated recently by Dorobantu *et al.*[101] Using a micropipette technique in which the pipette is filled with oil and placed in a small volume of aqueous cell suspension, the oil–water interfacial tension can be determined. None of the four bacterial types affected the hexadecane–water tension. In order to investigate the mechanism of emulsion stabilisation, two oil drops in water stabilised by bacteria were pushed together. As seen in Figure 1.28(a), despite attempts to induce coalescence through forced contact, the drops remained stable. To demonstrate that the interfacial bacteria interact to form a surface film, a clean oil drop was formed in an aqueous suspension of *Rhodococcus erythropolis* and then slowly withdrawn after several minutes thereby reducing the interfacial area. Figures 1.28(b) and (c) clearly show the formation of a rigid film due to the bacteria at the interface that resisted the area reduction and eventually form a crumpled surface.

Four different strains of bacteria (size $\approx 2 \times 1$ μm) displayed varying abilities to stabilise emulsions of hexadecane and water.[101] The most hydrophilic (*Pseudomonas fluorescents*), judged by the contact angle of a water drop under oil on a lawn of bacterial cells, produced no stable emulsion, while the most hydrophobic (*Rhodococcus erythropolis*) gave more stable emulsions apparently of both types. This system separated within 24 h to yield an upper oil phase, a middle stable emulsion phase and a lower aqueous phase. In Figure 1.29(a), drops of oil in water are visible from a sample of the middle phase with cells adhering to their interfaces and dispersed in the aqueous phase. Figure 1.29(b) shows a drop of water in oil (left) taken from the upper phase. Cells are seen at the interface and cluster in the oil phase (right). The most stable emulsions of o/w type are those with *Acinetobacter venetianus* where contact angles are intermediate between those for the two bacteria mentioned above. Here deformed oil drops surrounded by bacteria (dark) are separated by thin aqueous films (light), Figure 1.29(c). Such a concentrated emulsion did not flow unless shaken, indicating a yield stress like that of a gel. This change in emulsion type and stability with the inherent wettability of the bacterial cells at the oil–water interface is reminiscent of that seen for systems stabilised by fine silica particles.[71,102]

1.5.5.2 Demulsification using bacteria particles

In addition to bacteria acting as emulsion stabilisers, they have also been reported to act as demulsifiers of otherwise stable emulsions of surfactant, sometimes more effectively than conventional surfactant demulsifiers.[103–105] Thus, hydrophobic bacteria induce demulsification of petroleum or alkane o/w emulsions[104] whereas hydrophilic bacteria can demulsify water-in-petroleum oil emulsions.[103] In both cases involving petroleum oil, the hydrophobicity of the bacterial cells of *Nocardia amarae* is controlled by the age of the bacterial culture – the more aged the culture, the more hydrophobic and less charged were the cell surfaces. This particular bacterium can withstand organic chemical extractions, pH changes, high levels of alkalinity and high

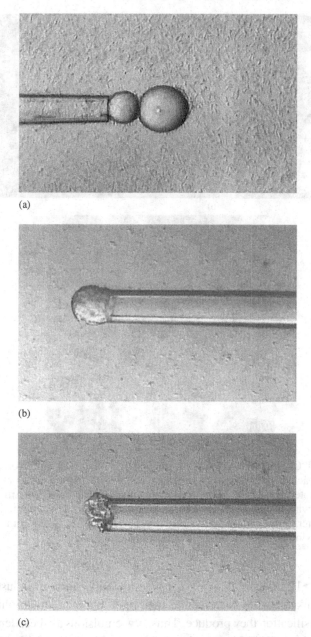

(a)

(b)

(c)

Figure 1.28 Oil drops in water stabilised by bacteria particles. (a) Two drops of hexadecane extruded into an aqueous suspension of *Acinetobacter venetianus* RAG-1 (tip diameter = 14 μm). (b) Deformation of the oil–water interface, after partial withdrawal of the oil drop, caused by *Rhodococcus erythropolis* 20S-E1-c adhering to it. (c) Appearance after complete withdrawal of the oil drop in (b). Taken from Ref. [101]; with permission of the American Society for Microbiology.

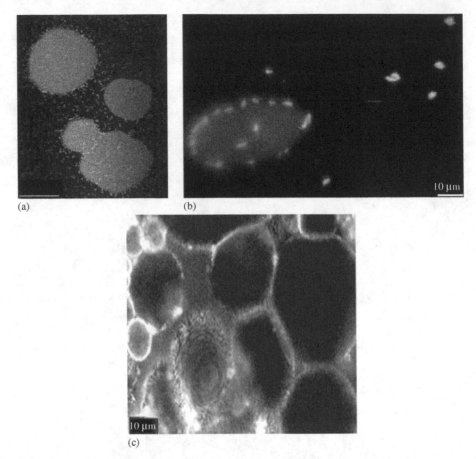

(a)   (b)

(c)

Figure 1.29 Confocal laser scanning microscope images of emulsion drops stabilised by bacterial cells. (a) Oil drops in water with *Rhodococcus erythropolis*
20S-E1-c; note excess cells in the continuous phase, scale bar = 10 μm. (b) Water
drop in oil with *Rhodococcus erythropolis* 20S-E1-c. (c) Concentrated oil drops
(dark) in water (light) with *Acinetobacter venetianus* RAG-1. Taken from Ref. [101];
with permission of the American Society for Microbiology.

temperatures ($>120°C$), so allowing for possible application in industry. Table 1.3
illustrates the links between bacteria age, their hydrophobicity/hydrophilicity and the
extent of demulsification they produce. Thus, o/w emulsions are best demulsified with
older, hydrophobic cells, whereas w/o emulsions are best demulsified with younger,
hydrophilic cells.

It is likely that the mechanism by which bacterial cells cause coalescence and even-
tual phase separation of emulsions is similar to that put forward by Aveyard *et al.* for
the demulsifying ability of commercial or pure surfactant demulsifiers of brine-
in-crude oil (w/o) emulsions.[106,107] The resolution can be explained in terms of simple

Table 1.3 *Correlation between age of bacterial culture of Nocardia amarae, contact angle of bacteria at air–water surface and the extent of demulsification (after 24 hrs.) of either o/w or w/o emulsions involving a different petroleum on adding 500 ppm of cells. Taken from Refs. [103] and [104]; re-drawn with permission of Springer-Verlag.*

		Demulsification (%)	
Culture (age/days)	$\theta_{aw}/°$	o/w	w/o
1.0	20	15	97
2.0	18	44	97
2.5	27[a]	–	97
3.5	46	30	–
5.0	52[a]	–	97
7.5	60	41	96
12.0	62	89	90
18.5	63[a]	–	80
20.0	63	94	–

[a]Estimated.

hydrophilie–lipophile balance (HLB) concepts, if the demulsifier can be thought of as a traditional surfactant. At low concentrations, the demulsifier is thought to first displace the indigeneous stabilising layers around water drops. As the concentration increases, the adsorbed demulsifier layer becomes more concentrated until the layer becomes close packed. The system HLB, and hence preferred emulsion type, is now dictated by the particular properties of the demulsifier. In the case of an initial o/w emulsion, hydrophobic particles once adsorbed prefer to stabilise w/o emulsions and hence, when added, will break the o/w emulsion. Likewise, hydrophilic particles tend to stabilise o/w emulsions and so will break the original w/o emulsion.

1.5.5.3 Spore particles

Similarities between the structural organisation within spore walls and the flocculation of synthetic colloidal particle systems was recognised by Hemsley *et al.*[108] The potential use of naturally occurring spore particles in stabilising emulsions, in addition to investigating their behaviour in spread monolayers at fluid–fluid planar interfaces, has been explored by Binks *et al.*[109] The spores were those of *Lycopodium clavatum* from the evergreen club moss which are harvested in September and have been used for centuries for many different things. These include pyrotechnics, herbal remedies and the more familiar school demonstration of employing it as a very hydrophobic powder which floats on the surface of water such that one's hand remains dry when immersed in it! Architecturally, spores (and pollen grains) are made up of an internal sac containing cytoplasm and an external wall.[110] The wall consists of two main layers: an inner layer known as the intine, made of cellulose and polysaccharides, and an outer, exine layer composed of a substance known as sporopollenin. Spore and pollen grains are the most ubiquitous of fossils and more widely distributed in time and space than any other representative of living matter. During the processes which occur after deposition, many spore grains uniquely preserve their characteristic morphological structure by virtue of the great resistance of their exine to both biological decay and chemical attack; intact spores have been found in sedimentary rock up to 500 million years old.[110]

The spores of *Lycopodium clavatum* shown in Figure 1.30(a) are approximately 30 μm in diameter, monodisperse and have rough surfaces.[109] The 3-fold marking (Y shape) originates from cell division in which the spore is in contact with three other cells in a tetrahedral arrangement. Starting from dispersions of the particles in oil, o/w emulsions are preferred on mixing with water for a range of oils of different polarity. Increasing the concentration of particles results in a decrease in the average size of oil drops and an increase in their stability to coalescence. As seen in Figure 1.30(b) showing the arrangement of spore particles adsorbed to interfaces of millimetre-sized emulsion drops, it is remarkable that only a fraction of the interface is covered by particles, despite such emulsions being stable to coalescence for more than a year.[109] The apparent coverage by particles is no more than 20%, as mentioned earlier for other systems.[111,112] In obtaining this image, drops were in motion and it was noticed that upon approach, interfacial particles diffused toward the area of contact increasing the particle density. It was difficult to reliably distinguish between a bilayer of particles or a bridging monolayer however. Particle re-distribution may play an important role in stabilising sparsely coated drops against coalescence.[112]

As discussed earlier and published recently, the structure of particle layers at a planar oil–water interface is preserved at a curved interface surrounding drops impacting on the long-term stability of emulsions.[43] On either the free surface of water or at planar oil–water interface, the spore particles tend to be predominantly aggregated forming clusters and chains, Figure 1.30(c). This suggests the presence of attractive forces between particles. The fact that particles mainly attach at the end of a chain rather than forming branches along it signifies repulsion between them also. The attraction is due to the presence of capillary forces caused by deformation of the fluid interface due to the effect of gravity (particle weight and buoyancy force). In comparison with silica particle monolayers, the aggregation of the spore particles is consistent with them being relatively hydrophilic and thus preferentially stabilising o/w emulsions. The ability of nanoparticles of cowpea mosaic virus to stabilise o/w emulsions has also been recently shown.[113]

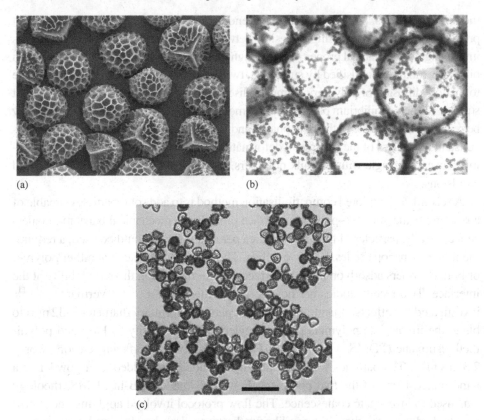

(a)

(b)

(c)

Figure 1.30 (a) SEM image of dry spore particles of *Lycopodium clavatum* of diameter ≈30 μm. Note the rough surface structure and the Y-shaped marking. (b) Optical microscopy image of *iso*propyl myristate-in-water emulsion drops stabilised by the spore particles above; scale bar = 200 μm. (c) Microscopy image of planar monolayer of spore particles at the toluene–water interface; scale bar = 100 μm. Taken from Ref. [109]; with permission of the American Chemical Society.

1.5.6 Immiscible polymer blends

Most chemically different polymers are immiscible and their blending is an economically attractive route to develop new materials that combine the desirable properties of more than one polymer.[114] The conversion of the immiscible blend to a useful polymeric product requires some manipulation of the interface. The final properties of the blend are influenced by the size scale of the microstructure. One of the classical routes to ensure adhesion between the phases is the use of a third component, or compatibiliser, which is miscible with both phases.[114] Such a compatibiliser may be a homopolymer or a block, graft or star co-polymer. The compatibilisation engenders the desired blend morphology by controlling the interfacial properties and the size of the dispersed drops. It also stabilises the blend against coalescence during subsequent processing as well as ensuring adhesion between the phases in the solid state, thus improving the mechanical properties. Although these systems share several features with oil–water mixtures emulsified by surfactant, two important differences exist. Firstly, due to the low diffusion of polymeric compatibilisers compared with low molecular weight surfactants, non-equilibrium phenomena are more likely to occur; the compatibiliser may be regarded as insoluble in the bulk phases and hence its total amount at the interface is constant regardless of the interfacial deformation. Secondly, macromolecular compatibilisers can entangle with the bulk polymers, a feature absent in surfactant-stabilised emulsions.

A second, less explored compatibilisation method is to add solid particles capable of adsorbing at the polymer–polymer interface providing a mechanical barrier to coalescence. Ideally, particles of high surface area per unit weight should be used, a requirement met by nanoparticles of clay or silica. The possibility exists that either polymer or both polymers adsorb on particle surfaces *in situ* modifying their wettability at the interface. Two recent studies are noteworthy in this context.[115,116] Vermant *et al.*[115] investigated the effects of using fumed silica particles (primary diameter ≈32 nm) to blend the immiscible polymers poly*iso*butylene (PIB, viscosity 91 Pa s) and polydimethylsiloxane (PDMS, viscosity 206 Pa s), whose bare interfacial tension is only 2.3 mN m^{-1}. The particles were partially hydrophobic and added as a powder to a hand-mixed blend of the two polymers. For emulsions of PIB-in-PDMS, rheology was used to investigate coalescence. The flow protocol involved applying shear flow at a high shear rate (dynamic equilibrium between drop breakup and coalescence) followed by a decrease in shear rate to induce coalescence. In the absence of particles, the drop radius increased continuously with time due to coarsening. On addition of 0.5 wt.% of particles however, this growth was drastically slowed down, and suppressed completely for the blend containing 1 wt.% of particles. Cryo-freeze fracture SEM images of emulsion samples are given in Figure 1.31 for the no particle case (a) and the 1 wt.% particle case (b). Without particles, a deformed microstructure is seen in which PIB drops are pulled out of the continuous PDMS phase. By contrast, clear evidence is seen for the accumulation of a rigid layer of particles at the interface of PIB drops, and as observed in oil–water emulsions, excess particles are present in the continuous phase.[115] The thickness of the stabilising particle layer (200 nm) implies that the nanoparticles are aggregated at the interface. By varying the volume ratio of the two polymers, the same particles were shown to stabilise the opposite emulsion type, *i.e.* PDMS-in-PIB, the stability of which was predictably much less.

In a similar study, Ray *et al.*[116] have used hydrophobically modified clay particles (diameter < 100 nm) to prepare stable emulsions of PS and polypropylene (PP). The initially hydrophilic layered silicate particles, incompatible with the polymers, were rendered hydrophobic by cation exchange between the silicate ions and a di-chain

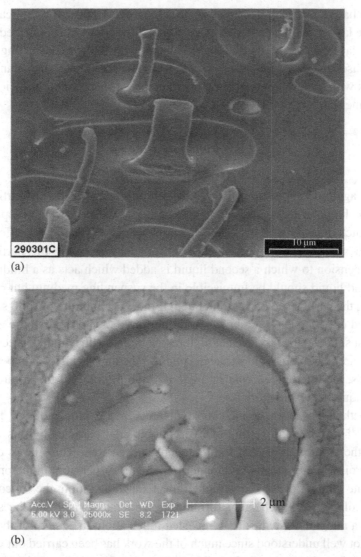

(a)

(b)

Figure 1.31 Cryo-freeze fracture SEM images of a blend of 30 vol.% PIB-in-PDMS with (a) no particles: scale bar = 10 μm, (b) 1 wt.% partially hydrophobic nanoparticles of silica: scale bar = 2 μm. Taken from Ref. [115]; with permission of Springer-Verlag.

dimethylammonium salt. Drops of PS-in-PP decreased in size upon increasing the concentration of particles. The authors attributed this to a concomitant decrease in interfacial tension from 5.1 mN m^{-1} for the virgin polymer pair to 3.4 mN m^{-1} with 0.5 wt.% particles. Bright field TEM images of these emulsions revealed the silicate layers of the clay particles in the interfacial region. Although both polymers separately have no strong interaction with such particles, both were intercalated into the silicate layers when blended together. X-ray diffraction data confirmed the sharing of the layered silicates by the two polymers with the particles acting as a true compatibiliser. It is likely that research of this kind will continue using a range of particle types with a view to establishing the benefits of incorporating particles in the final polymer product.

1.5.7 Flotation using oil

1.5.7.1 Oil agglomeration

Spherical agglomeration, the selective formation of aggregates of particles held together by liquid bridges, is a promising method for the recovery and separation of particles including minerals.[117] The separation method is based on differences in surface chemistry. The agglomerates are formed by agitating the particles in a liquid suspension to which a second liquid is added which acts as a binding agent. This second liquid should be immiscible in the suspending medium but capable of cementing the particles to be agglomerated. If there is a mixture of solid substances present and only one is wetted by the bridging liquid then virtually pure agglomerates of that solid can be produced at high recovery and with a high degree of selectivity. The strength of an agglomerate depends on the interfacial tension between the two liquids, the three-phase contact angle θ_{ow} and the ratio of the volumes of the bridging liquid and solid particles.

The method has been applied most notably in the processing of fine coals (<5 μm).[118] Preferential wetting of hydrophobic coal particles by oils forms the basis for the separation of coal from aqueous suspensions. This allows coal to be wetted by oil while oxide mineral matter remains in suspension. In the presence of an adequate amount of oil and sufficient mechanical agitation, the oil-coated coal particles collide with each other forming agglomerates which can be separated. Although a number of studies have revealed much about the nature of the process, it is still not well understood since much of the work has been carried out with complex materials such as coal and fuel oil. A simpler, model system was investigated by Drzymala *et al.*[119] in which agglomeration experiments were conducted with relatively pure graphite, hand-picked coal of low ash content, iron pyrite (FeS_2), quartz and kaolin (clay). Graphite was chosen because its surface properties are similar to those of some types of coal, while the inorganic minerals were selected because they represent some of the most common impurities in coal. The materials, ground to ultrafine size, were suspended in water individually and agglomerated with heptane. Information about the structure of the resulting agglomerates was obtained by microscopic examination and by measuring the stability to sedimentation of the suspension. The agglomerated solids were recovered by screening and the weight yield

Table 1.4 *Correlation between the recovery of various solids from water using heptane and the contact angle of the solid at the oil–water interface (measured through water). Taken from Ref. [119]; re-drawn with permission of Pergamon Press.*

Solid	$\theta_{ow}/°$	Specific recovery (g solid/g oil)
Kaolin	<5	0
Quartz	20	0
Iron pyrite	85	0.6
Coal	105	7.2
Graphite	113	14.5

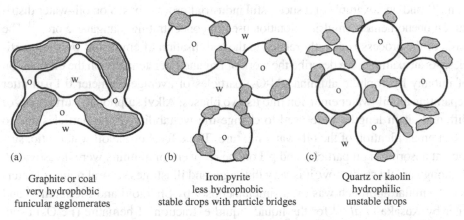

(a)
Graphite or coal
very hydrophobic
funicular agglomerates

(b)
Iron pyrite
less hydrophobic
stable drops with particle bridges

(c)
Quartz or kaolin
hydrophilic
unstable drops

Figure 1.32 Schematic structures following agglomeration of particles (shaded) in water (w) on addition of oil (o) for particles of different hydrophobicity. Taken from Ref. [119]; re-drawn with permission of Pergamon Press.

was measured. In addition, the three-phase contact angle was determined on polished surfaces of the mineral samples. As seen in Table 1.4, a clear correlation exists between the hydrophobicity of the particles and their recovery by oil.

Using relatively hydrophobic materials such as graphite or coal, the agglomerate structure depends on the amount of oil added. At low volumes, loose flocs form which are held together by pendular bridges of oil between the solid particles. With larger amounts, more compact aggregates are formed which involve funicular bridging, Figure 1.32(a). Some water can be trapped within the agglomerates. The specific recovery was high in both mineral cases. For iron pyrite which is less hydrophobic, the particles tended to adsorb on oil drops preventing their coalescence. Some particles even formed bridges between drops, Figure 1.32(b). The recovery is reduced. Finally, very hydrophilic quartz or kaolin particles remained entirely within the water phase and did not attach themselves to oil drops. Instead, they collected around the drops,

Figure 1.32(c), serving as links between them. Such agglomerates were very weak and easily broken so that they were not recovered by screening. These examples serve to illustrate the importance of particle wettability on particle recovery and, as seen in this case, particles of intermediate wettability producing stable o/w emulsions are detrimental to recovery.[119]

1.5.7.2 Emulsion flotation

There is ongoing interest in developing methods for the recovery of particles of size less than 1 μm from disseminated ores. One such method is the extraction of fine particulate solids from aqueous suspensions into an oil phase (liquid–liquid extraction),[120] and one example of a successful industrial process based on oil–water distribution phenomena is emulsion flotation used in concentrating manganese ore.[121] The basis of the process is directly related to the stabilisation of emulsions by solid particles. As an example, we describe the work of Lai and Fuerstenau[122] on the oil flotation of initially hydrophilic alumina (Al_2O_3) particles of average diameter 0.1 μm after separating an oil–water emulsion into its two phases. Alkyl sulphonate surfactants of different chain lengths were used to change the wettability of the particles and to effect emulsification of the oil–water mixture. The effects of the oil:water ratio, surfactant adsorption on particles and pH on the recovery of alumina were investigated. Although an old study now, it is very thorough and illustrates several points of interest. A similar approach was used simultaneously by Shergold and Mellgren[123] and later by Kusaka *et al.*[124] for the liquid–liquid extraction of hematite (Fe_2O_3) using sodium dodecyl sulphate (SDS), and silica ($-$ve) using hexadecyltrimethylammonium chloride ($+$ve), respectively.

Oxide particles are hydrophilic but can be made hydrophobic through the adsorption of a suitable collector. In the case of alumina, which is positively charged below pH = 9, such a collector is an anionic surfactant, *e.g.* sodium alkyl sulphonate, and adsorption proceeds via electrostatic attraction between surfactant head groups and particle surfaces.[122] In order to ascertain how the degree of hydrophobicity affected the extraction of such particles into oil, sulphonates of different alkyl chain length were used. Since the wettability of alumina in aqueous surfactant solution is determined by the adsorption of sulphonate onto its surface, adsorption isotherms were obtained at pH = 7.4. The isotherms are shown in Figure 1.33(a) for the octyl and tetradecyl surfactants (left ordinate). They show that the adsorbed amount increases with both an increase in surfactant concentration and in the surfactant chain length. This adsorption modifies the wettability of the particles at the oil–water interface, as judged from the contact angle of an *iso*octane drop under water (measured through water) on a single crystal of sapphire also shown in Figure 1.33(a) (right ordinate). The angles increase, *i.e.* particles become more hydrophobic, with increasing surfactant concentration and

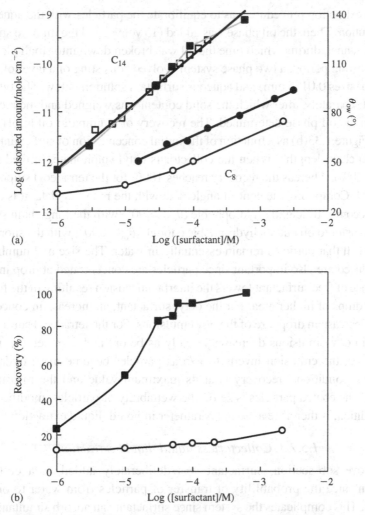

Figure 1.33 (a) Adsorbed amount of surfactant on alumina particles (diameter = 0.1 μm) at pH = 7.4 *versus* equilibrium [surfactant] in water (left ordinate) for C_8 (●) and C_{14} (□) sodium alkyl sulphonate, and *iso*octane-water-solid contact angle (through water, right ordinate) *versus* initial [surfactant] in water for C_8 (○) and C_{14} (■) surfactant. (b) Recovery of alumina in *iso*octane *versus* initial [surfactant] in water at pH = 7.4 for C_8 (○) and C_{14} (■) sodium alkyl sulphonate. Taken from Ref. [122]; re-drawn with permission of American Institute of Mining Engineers.

increasing chain length. The value of θ_{ow} never exceeds 80° with the octyl sulphonate whereas it approaches 140° with the tetradecyl sulphonate.[122]

In order to attain high recovery of alumina, it is necessary to make the solid particles as hydrophobic as possible and to disperse the oil phase as much as possible so as to provide a large oil–water interfacial area for collection of the fine particles. The first step in the extraction procedure was to equilibrate the particles with the aqueous surfactant solution. Then, the oil phase was added (15 vol.%) and the mixture stirred in a separatory funnel, during which time the oil was broken down into small globules.[122] After the stirring period, a two phase system evolved consisting of a layer of oil globules on top (sizes 0.01–5 mm) and aqueous surfactant solution below. The two phases were dried separately, after which the solid content was weighed and the recovery of particles in the oil phase determined. The recovery of alumina in oil at pH = 7.4 is shown in Figure 1.33(b) as a function of the initial concentration of surfactant in water for the two chain lengths. When the collector is octyl sulphonate, limited recovery occurs ($<$ 20%), whereas the recovery reaches 100% for the tetradecyl sulphonate by 6×10^{-4} M. Comparing the contact angle data with the recovery data, it is clear that particles become sufficiently hydrophobic ($\theta_{ow} > 90°$) with the long chain surfactant enabling transfer to oil but not hydrophobic enough ($\theta_{ow} < 90°$) with the shorter chain analogue such that particles remain essentially in water. The size and number of oil drops produced are also important since particles are concentrated at drop interfaces. The presence of free surfactant lowers the interfacial tension resulting in the formation of smaller drops of higher area. For the octyl surfactant, an increase in concentration leads to a decrease in drop size of the o/w emulsions. For the tetradecyl surfactant, the drop size in o/w emulsions decreases initially as before and, at a certain surfactant concentration, the emulsion inverts to w/o as particles become more hydrophobic. Under these conditions, recovery is at its maximum value and the measured zeta potential of the coated particles is zero. The wettability of particles, modified by surfactant addition, is therefore a crucial parameter in liquid–liquid extraction.[122]

1.5.7.3 Collectorless liquid–liquid extraction

In the above sub-section, surfactant was deliberately added as a collector in order to enhance the probablility of transfer of particles from water to oil during extraction. This complicates the system since surfactant can adsorb simultaneously at the solid–liquid and liquid–liquid interfaces. A simpler scenario is extraction in the absence of surfactant, so-called collectorless extraction, allowing a better understanding of the mechanism of adhesion between oil drops and solid particles. Kusaka et al.[125,126] have worked on many aspects of this process recently, using electrolyte addition or changes in pH to modify the wettability of particles by influencing their charge properties. Hydrophilic quartz particles in water, of average diameter 0.24 μm, were equilibrated gently with *iso*octane and the mixture allowed to stand. A dense emulsion phase containing some of the particles separated from the aqueous phase. After drying and weighing, the recovery of particles into the emulsion phase was determined. In addition, separate electrokinetic measurements were conducted on both the oil drops and particles in water under the same conditions as those used in the extraction experiments. In the case of $CaCl_2$ as electrolyte and at pH = 5.5, the effect of salt

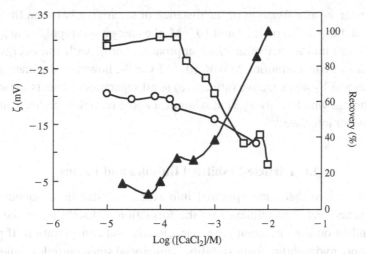

Figure 1.34 Zeta potential of quartz particles (□) and *iso*octane drops (○) in water at pH = 5.5 (left ordinate) as a function of the concentration of CaCl$_2$. Also shown is the recovery (▲) of the particles into the emulsion phase during liquid–liquid extraction (right ordinate). Taken from Ref. [126]; re-drawn with permission of Elsevier.

concentration on the zeta potential of oil drops and quartz particles is shown in Figure 1.34 (left ordinate). Oil drops in water at this pH are negatively charged (probably due to preferential adsorption of OH$^-$ from water) as are the quartz particles (due to ionisation of surface silanol groups). The magnitude of the zeta potential decreases in both cases with salt concentration, reaching values around -10 mV at 10^{-2} M CaCl$_2$, as Ca^{2+} ions adsorb on surfaces. Also shown is the recovery of particles into the emulsion phase (right ordinate), which increases from 10% to 100% upon increasing the salt concentration to 10^{-2} M. Comparison of the three data sets indicates that the increase in recovery of particles correlates with the decrease in the zeta potentials, and therefore with the decrease in the electrostatic repulsion between oil drops and solid particles. The same correlation was recorded for NaCl and LaCl$_3$ as electrolytes, where the concentration of salt required to achieve maximum recovery decreased with an increase in the valency of the cation. Similarly, a reduction in pH towards 2 (isoelectric point for silica) increased the recovery markedly at low salt concentration.[126]

The particle–oil coagulation occurring during extraction is an example of heterocoagulation in which the total potential energy of interaction V_T can be calculated using DLVO (Derjaguin–Landau–Verwy–Overbeek) theory, *i.e.* by summing up the energy of interaction due to the overlap of electrical double layers and that due to van der Waals forces. Since the quartz particles (<1 μm) are much smaller than the oil drops (few mm), the energy can be approximated by an expression for the interaction between a sphere and an infinite flat plate (in water). The results of the calculations of V_T as a function of the distance of separation between the surfaces revealed that, for [CaCl$_2$] of 10^{-4} and 10^{-3} M, the energy was repulsive at large distances (>5 nm) and attractive at close approach (1 nm), with the energy barrier falling with salt concentration. At 6×10^{-3} M CaCl$_2$, however, the energy barrier disappeared and V_T was negative (attractive) at all separations. This is the same salt concentration at which recovery is maximum, *i.e.* the particles readily concentrate at the oil–water interface.[126]

1.6 Particle-Stabilised Bubbles and Foams

Solid particles have been incorporated into surfactant-stabilised aqueous foams for many years, and their influence on the formation and stability of the foam is very dependent on the surfactant type, particle size and concentration. If particles are reasonably hydrophilic, foam stability is enhanced since particles collect in the Plateau borders slowing down film drainage. Hydrophobic particles however can enter the air–water surfaces of the foam and cause destabilisation via the bridging–dewetting or bridging–stretching mechanisms. This is discussed in more detail in Chapter 10.

The literature concerned with the ability of particles to act as foam stabilisers in the absence of any other surface-active material is very sparse,[4,127–133] although much is known in the area of flotation,[120] reviewed in Chapter 9, in which particles attach to air bubbles in the presence of surfactant. Very recently, particles alone have been shown to be effective in stabilising metal foams, *i.e.* air bubbles in molten metal.[133] These attempts and the properties of the porous solid formed after cooling are discussed in Chapter 11. In aqueous systems, different particle types, sizes and protocols have been reported with varying degrees of success in stabilising large volumes of foam. Related to foam formation is the ability to transport particles from water to air–water surfaces. Unlike oil–water interfaces, Okubo[134] and Dong and Johnson[135] have shown that the surface tension of colloidal dispersions of purified charged particles (PS, silica, titania) can be lowered from the value for pure water upon increasing the particle concentration in bulk. The lowering, by as much as 20 mN m^{-1}, depends on particle type and size and whether the particles form crystal-like or liquid-like structures in bulk. As pointed out by Paunov *et al.*,[136] an energy barrier exists between charged particles in bulk water and the charged air–water surface which must be surmounted if particles are to adsorb. The height of this barrier can be reduced by addition of salt or by changing the pH, such that when the kinetic energy of the particles exceeds the barrier height, they adsorb to the surface and become trapped. This has been demonstrated very clearly by Wan and Tokunaga[137] for sub-micrometre-sized clay particles (by tuning pH) and by Hu *et al.*[138] for silver nanoparticles (by adding KCl).

An alternative method of overcoming the energy barrier to the growth of surface crystals of particles involves targeted delivery of the particles through hydrodynamic flows.[139] The method allows an unprecedented degree of control over the surface composition, bubble size and stability. Using a three-channel hydrodynamic focusing device, an aqueous dispersion of colloidal particles is driven through the outer channels with the inner channel carrying the dispersed gas phase. The surface between the continuous and dispersed phases thus serves as the substrate for particle assembly.[139] The curved bubble surface is held stationary relative to the motion of the particles in the continuous phase in order to allow sufficient time for the particles to adsorb. For a particle concentration of 0.1%, armoured bubbles can be ejected at a rate of $10\,s^{-1}$. The shear-driven ejection of the jammed particle shells makes the geometry of the outlet channel the main determinant of the size of the bubbles and the continuous one-step assembly produces monodisperse-coated bubbles. Jammed particle shells are only reproducibly formed when particles approach the surface with high velocity ($10\,cm\,s^{-1}$), suggesting that each capture event resulting in a particle becoming adsorbed to the surface occurs in a timescale of tens of microseconds. The shear experienced by the particles from the continuous phase liquid is insufficient to detach the adsorbed particles. Instead, the particles are transported by the flow to the anterior of the curved air–water surface. Repeated capture and transport events result in a rapid buildup of particles in a close-packed crystallite on bubble surfaces. The thermal motion of the particles is arrested in the tightly packed shell, so that bubbles resist not only the possibility of spontaneous coalescence due to surface area minimisation but also that of shear-induced coalescence. In going a step further, new kinds of particle shells around bubbles can be tailored. Thus, by loading PS particles labelled with rhodamine (yellow) in one channel and PS particles of a different size labelled with fluorescein (green) in the other channel, bubbles coated with hemi-shells of the two particle types could be produced. These Janus-like bubbles may have use for targeting or sorting purposes.[139]

In addition to coalescence, the other major mechanism by which foams collapse is disproportionation, in which gas diffuses from smaller to larger bubbles due to the higher Laplace pressure within the former. Although theory suggests that films around bubbles with high mechanical rigidity should be able to prevent this and bubble shrinkage, experimental measurements reveal that even the most viscoelastic protein films cannot halt disproportionation. It was therefore of interest to see if surface-active particles, associated with high adsorption energies, could generate a sufficiently rigid shell to prevent bubble shrinkage by this route. Du *et al.*[140] showed that this was possible using fumed silica nanoparticles which were partially hydrophobic (40% residual SiOH). Various methods were employed to generate air bubbles in an aqueous particle dispersion beneath a planar air–water surface, including direct injection of bubbles and a sudden reduction of pressure within the cell containing the dispersion causing bubble nucleation. Although some large bubbles (radius 250 μm) were unstable to disproportionation, smaller bubbles (radius 50–100 μm) were stable from their formation over a period of several days. Bubbles of intermediate size shrank to a certain extent and then remained stable subsequently. This is in contrast to bubbles stabilised by either gelatin or β-lactoglobulin proteins which shrank rapidly and disappeared completely in 1–2 h.

In a follow-up paper, Dickinson *et al.*[141] demonstrated that stable bubbles could be formed from an aqueous dispersion of fumed silica nanoparticles of lower inherent hydrophobicity (67% and 80% SiOH) by adding salt (NaCl) to the water. Their stability increased with increasing particle concentration and increasing salt concentration. Since it is known that the hydrophobicity of silica surfaces increases on addition of electrolyte as their charge is neutralised, this method is an easy way to encourage otherwise too hydrophilic particles to adsorb at bubble surfaces. At the highest particle concentration of 1 wt.%, the aqueous dispersion was gel-like due to a particle network obtained via silanol–silanol bonds and it was proven that the adsorbed particle layer on bubbles was contiguous with the excess particle network in the bulk aqueous phase (see Chapter 8 for more discussion). Some of these findings equally apply to the stability of emulsion drops as discussed earlier.

Theoretically, Kam and Rossen[142] offer an explanation for the great stability to disproportionation of particle-coated bubbles using a two-dimensional model. The stability of small bubbles to mass transfer of gas requires a mechanism whereby, as a bubble shrinks, its Laplace pressure falls (rather than rises) closing down gas dissolution. Suppose a single particle is attached to a bubble surface, with its position dictated by the contact angle, θ_{aw}. As more particles become adsorbed, the bubble radius increases by an amount of the submerged portion of the particles because the bubble volume is held constant. This increase in radius continues until the bubble surface is completely coated with particles. If the bubble then shrinks, the reduction in bubble volume requires that the bare air–water surfaces (between particles), originally convex, move along particle surfaces inwards toward the bubble centre. The shape of the surfaces and their curvature change while maintaining the same contact angle on the particle surfaces. These surfaces can be flattened, *i.e.* become planar, or even concave depending on the extent of bubble volume reduction. During this process, the Laplace pressure no longer balances the surface tension and there is a net inward force which may be balanced by the stress between particles in the adsorbed layer. Shrinkage reaches a limit when the air–water surfaces on either side of a particle meet on the solid surface, representing the minimum bubble volume possible before the bubble detaches from at least one solid particle. It is shown that, for typical values of θ_{aw} (>20°), a small decrease in bubble volume from the unstressed state causes the Laplace pressure to decrease to zero and then to negative values. When it is zero, the bubble is in equilibrium with liquid saturated with gas. Thus, when negative, gas in the bubble can be in equilibrium with liquid unsaturated with gas, and this equilibrium is stable to mass transfer. An estimate of the stress on the adsorbed particles, *i.e.* the solid–solid force, was made and, for particles of density $2.5\,\mathrm{g\,cm^{-3}}$ of radius $0.31\,\mu\mathrm{m}$ around bubbles of radius $10\,\mu\mathrm{m}$ with a surface tension equal to $70\,\mathrm{mN\,m^{-1}}$, this force is 970,000 times the force of gravity on the particles! These stresses might drive dissolution and precipitation of solid around the points of particle contact, sintering the solids into a rigid continuous framework[142] and halt completely disproportionation as observed experimentally.[140]

Two recent studies on bulk foams stabilised entirely by solid particles are worthy of comment. In the first, Alargova *et al.*[129] describe the use of rod-shaped microparticles of the photoresist polymer SU-8 in hand-shaken systems. The particles were polydisperse of average length $23.5\,\mu\mathrm{m}$ and average diameter $0.6\,\mu\mathrm{m}$, shown in Figure 1.35(a). The polymer is an epoxy-containing material with methyl side groups and the likely reason it is an efficient foam former is that it is relatively hydrophobic exhibiting a contact angle at the air–water surface, θ_{aw}, of $\approx 80°$. Aqueous dispersions of the particles foamed readily, with air bubbles initially distributed throughout the entire volume of the samples making them milky in appearance. After several minutes, most of the bubbles rose to the top of the sample leaving a less turbid phase of particles dispersed in water below. The initial foam volume increased with particle concentration (0.2–2.2 wt.%), and for all samples decreased slightly in the first few minutes and then remained constant for more than 3 weeks. This is to be compared

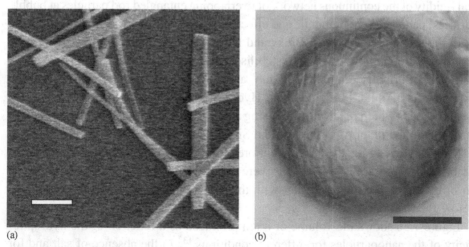

(a)   (b)

Figure 1.35 (a) SEM image of the SU-8 polymer rod particles used to stabilise aqueous foams; scale bar = $5\,\mu\mathrm{m}$. (b) Optical microscope image of a single air bubble in water coated with a hairy layer of SU-8 polymer rod particles; scale bar = $50\,\mu\mathrm{m}$. Taken from Ref. [129]; with permission of the American Chemical Society.

with a popular foam-forming surfactant like SDS just above its critical micelle concentration for which the foam collapses completely within 24 h.

A more stringent test of foam stability is its resistance to drying which normally destroys any surfactant-stabilised foam.[129] In the particle-stabilised foam left open to air, its volume remained constant for more than 1 week even though the water below it slowly evaporated through the foam layer. Attempts to destroy the remaining foam by fast drying and expansion in a vacuum resulted in an increase in foam volume by a factor of two and continued stability. The micro-rods adsorbed on bubble surfaces result in this extraordinary "super-stabilisation" effect. Interestingly, mixtures of particles and SDS surfactant produced foams which were as unstable as pure surfactant ones, with SDS thus acting as a defoamer and suppressing the stabilising effect of the particles. The authors hypothesised that the likely cause of this is due to adsorption of surfactant monomer onto particle surfaces rendering them more hydrophilic so that the coated particles lose their affinity for air–water surfaces.

Microscopic examination of the particle-stabilised foam revealed that it was composed of small (10–100 µm) spherical bubbles. As seen in Figure 1.35(b), each bubble is covered with a dense shell of adsorbed entangled particles which extend into the aqueous phase. The particles appear to be flexible and adsorb parallel to the curved liquid surface. The thickness of a single foam film (air–water–air) corresponds to at least two layers of rods, $\approx$1–2 µm, which is significantly much thicker than that of a surfactant, 10–100 nm. The steric repulsion between the adsorbed particle layers thus keeps the films very thick, reducing their probability of rupture and retarding gas diffusion. The other contributor to the remarkable stability of these foams is the mechanical rigidity of the continuous network of overlapping entangled rod particles at bubble surfaces; particles were immobile and effectively jammed.

In the second recent study, Binks and Horozov[132] investigated the influence of particle hydrophobicity on foams stabilised by fumed silica nanoparticles. These particles, of primary diameter $\approx$30 nm, are the same as those used in the bubble experiments of Refs. [140 and 141]. Hydrophilic silica particles possess surface silanol groups which react with silanising agents to impart hydrophobicity. The latter is quantified in terms of the residual SiOH content, varying here from 100% to 14%. Acknowledging the fact that more hydrophobic particles are required to adsorb for foam stabilisation, foams were prepared either with particles initially in air or by dispersing them in water with the aid of ethanol, which is then removed before the foaming tests.

Figure 1.36 shows the volume of foam produced as a function of the hydrophobicity of the nanoparticles for different conditions.[132] In the absence of salt and for hand-shaken systems (3 w/v% particles initially in air, curve 1), no foam is formed for hydrophilic ($\geq$70% SiOH) or very hydrophobic particles (14% SiOH, dry powder remained on water surface) but the foamability increases progressively for

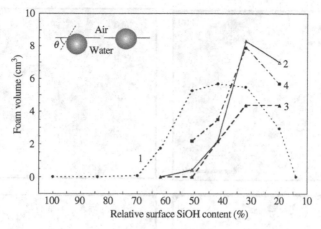

Figure 1.36 Volume of foam produced at room temperature from 7 cm³ of water in the presence of fumed silica nanoparticles of different hydrophobicity (given as % SiOH on their surfaces) at 3 w/v% (hand shaken, 1) and 0.86 w/v% (homogenised, 2–4). The systems and times since foam formation are: (1) no salt, 0 and 1 h, (2) no salt, 10 s, (3) no salt, 27 h, (4) 8.5 mM NaCl, 13.5 h. The inset shows schematically the position of a particle at an air–water surface if hydrophilic (high % SiOH, left) or more hydrophobic (low % SiOH, right). Taken from Ref. [132]; with permission of Wiley-VCH.

particles of intermediate hydrophobicity. These foams contain ∼30% water when formed and are subsequently very stable to collapse. An alternative method is to disperse the particles in a water–ethanol solution, allowing the more hydrophobic ones to become wetted, remove the ethanol by repeated sedimentation-redispersion cycles in pure water and aerate the dispersion using an Ultra Turrax homogeniser. The initial volume of foam in this case (0.86 w/v% particles, curve 2) shows a more pronounced maximum with respect to particle hydrophobicity, with those possessing 32% SiOH being the most effective. The foams obtained with 32% and 20% SiOH are wet and even at long time contain ∼60% water. They are very stable to collapse. The reduction in their volume is due to water drainage and bubble compaction, but not to loss of air (curve 3, 27 h.). Particles with surface SiOH ⩾ 42% reside mainly in the water phase (cloudy) with little foam whereas particles with 32% and 20% SiOH are located entirely within the white, creamy foam enveloping air bubbles. The bubble size falls progressively with an increase in particle hydrophobicity.

Using the recently developed Dispersion Stability Analyser 24,[143] the destructive processes occurring in the foam can be detected and monitored well before it is possible by naked eye observations. The vertical profiles of the scattered light intensity were determined for the silica-stabilised foam and an SDS-stabilised one at different times since foam formation. In both cases, the light intensity close to the sample bottom decreases with time, while the light intensity maximum shifts

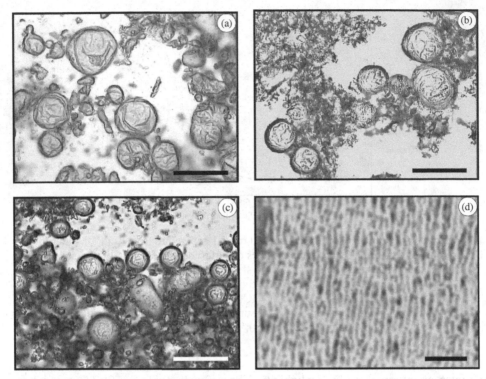

Figure 1.37 Optical microscopy images of aqueous foams stabilised by silica nanoparticles possessing 32% SiOH on their surfaces. Systems and time after formation by homogenisation are: (a) no salt, 3 w/v% particles, 40 h, (b) 8.5 mM NaCl, 0.86 w/v% particles, 5 min, (c) 1 M NaCl, 0.37 w/v% particles, 5 min. Also shown in (d) is the image of a planar monolayer of silica nanoparticles (with 50% SiOH) at the air–water surface after compression to a surface pressure of 70 mN m^{-1}. The corrugations are parallel to the trough barriers. All scale bars = 50 μm. Taken from Ref. [132]; with permission of Wiley-VCH.

upwards. These changes are a result of the drainage of water from between bubbles in the foam, with a clear serum appearing at the bottom in contact with a concentrated foam at the top. In the particle case, this takes place relatively slowly and is over after several hours. By contrast, the fast drainage of water out of the foam in the SDS case (few minutes) leads to a significant shift of the intensity maximum upwards and to a drastic decrease in the intensity in the upper part of the profile. The foam collapse is accompanied by the appearance of local maxima and minima. Such changes in the light intensity are a direct consequence of foam coarsening due to both bubble coalescence and disproportionation. Adsorbed particles clearly prevent both of these occurring.

Optical microscope images of bubbles in foams stabilised by particles with 32% SiOH are given in Figure 1.37(a)–(c), with and without salt.[132] Distinct non-spherical bubbles, of size 5–50 μm, are a feature of these systems. Their surfaces are rough due to ripples.[144] Similar ripples have been observed in the case of a planar air–water silica particle monolayer after compression (d), suggesting that the bubbles are covered with dense particle layers compressed to a high surface pressure close to the surface tension of water. Stable bubbles are probably formed by coalescence between smaller bubbles covered with dilute particle layers during homogenisation. As the bubble area decreases, excess particles cannot be released since irreversibly adsorbed and so the surface corrugates to increase in area. Since silica particles in water and bare air–water surfaces are negatively charged, it is anticipated that addition of salt to water should enhance the transfer of particles to the surface by reducing the energy barrier to adsorption. It may also increase their contact angle as hydrophobicity increases, with both leading to improved foam stabilisation. That this is the case is seen in the data in Figure 1.36 (curve 4) and particles initially relatively hydrophilic (42% and 51% SiOH) can now stabilise foams to a higher extent than in the absence of salt.

Acknowledgements

BPB wishes to thank Dr. G.A.J. Pitt, formerly of the Department of Biochemistry, University of Liverpool, and Mrs. Angela Dell currently at The School of Biological Sciences, University of Liverpool for their help in researching the work of W. Ramsden. He also thanks Mrs. Christine Woollett of The Royal Society, London for her help in tracing the career of S.U. Pickering.

References

1. R.J. Hunter, *Foundations of Colloid Science*, Vol. 1, Clarendon Press, Oxford, 1987, Chapter 3.
2. W.B. Russel, D.A. Saville and W.R. Schowalter, *Colloidal Dispersions*, Cambridge University Press, Cambridge, 1989.
3. D.F. Evans and H. Wennerstrom, *The Colloidal Domain: Where Physics, Chemistry, Biology and Technology Meet*, VCH Publishers, New York, 1994.
4. B.P. Binks, *Curr. Opin. Colloid Interf. Sci.*, **7** (2002), 21.
5. W. Ramsden, *Proc. Roy. Soc.*, **72** (1903), 156.
6. S.U. Pickering, *J. Chem. Soc.*, **91** (1907), 2001.
7. B.P. Binks and P.D.I. Fletcher, *Langmuir*, **17** (2001), 4708.
8. A. Perro, S. Reculusa, S. Ravaine, E. Bourgeat-Lamic and E. Duguet, *J. Mater. Chem.*, **15** (2005), 3745.
9. T. Young, *Phil. Trans.*, **95** (1805), 84.
10. A. Scheludko, B.V. Toshev and D.T. Boyadjiev, *J. Chem. Soc. Faraday Trans. 1*, **72** (1976), 2815.
11. R. Aveyard and J.H. Clint, *J. Chem. Soc. Faraday Trans.*, **92** (1996), 85.
12. A.W. Neumann and J.K. Spelt (eds.), *Applied Surface Thermodynamics*, Marcel Dekker, New York, 1996.

13. P.A. Kralchevsky and K. Nagayama, *Particles at Fluid Interfaces and Membranes*; Elsevier Science, Amsterdam, 2001.
14. A. Amirfazli and A.W. Neumann, *Adv. Colloid Interf. Sci.*, **110** (2004), 121.
15. J. Faraudo and F. Bresme, *J. Chem. Phys.*, **118** (2003), 6518.
16. L. Dong and D.T. Johnson, *Langmuir*, **21** (2005), 3838.
17. H.M. Princen, in *Surface and Colloid Science*, Vol. 2, ed. E. Matijevic, Wiley-Interscience, New York, 1969, p. 1.
18. A.V. Rapacchietta, A.W. Neumann and S.N. Omenyi, *J. Colloid Interf. Sci.*, **59** (1977), 541.
19. A.V. Rapacchietta and A.W. Neumann, *J. Colloid Interf. Sci.*, **59** (1977), 555.
20. D.D. Joseph, J. Wang, R. Bai, B.H. Yang and H.H. Hu, *J. Fluid Mech.*, **469** (2003), 139.
21. P. Singh and D.D. Joseph, *J. Fluid Mech.*, **530** (2005), 31.
22. I.B. Ivanov, P.A. Kralchevsky and A.D. Nikolov, *J. Colloid Interf. Sci.*, **112** (1986), 97.
23. M.M. Nicolson, *Proc. Camb. Phil. Soc.*, **45** (1949), 288.
24. D.Y.C. Chan, J.D. Henry Jr. and L.R. White, *J. Colloid Interf. Sci.*, **79** (1981), 410.
25. D. Stamou, C. Duschl and D. Johannsmann, *Phys. Rev. E*, **62** (2000), 5263.
26. P.A. Kralchevsky, N.D. Denkov and K.D. Danov, *Langmuir*, **17** (2001), 7694.
27. M.G. Nikolaides, A.R. Bausch, M.F. Hsu, A.D. Dinsmore, M.P. Brenner, C. Gay and D.A. Weitz, *Nature*, **420** (2002), 299.
28. K.D. Danov, P.A. Kralchevsky and M.P. Boneva, *Langmuir*, **20** (2004), 6139.
29. A.B.D. Brown, C.G. Smith and A.R. Rennie, *Phys. Rev. E*, **62** (2000), 951.
30. J.C. Loudet, A.M. Alsayed, J. Zhang and A.G. Yodh, *Phys. Rev. Lett.*, **94** (2005), 018301.
31. Z. Horvolgyi, S. Nemeth and J.H. Fendler, *Colloids Surf. A*, **71** (1993), 207.
32. R. Aveyard, J.H. Clint, D. Nees and V.N. Paunov, *Langmuir*, **16** (2000), 1969.
33. V.N. Paunov, *Langmuir*, **19** (2003), 7970.
34. A.F. Koretsky and P.M. Kruglyakov, *Izv. Sib. Otd. Akad. Nauk USSR*, **2** (1971), 139.
35. T.F. Tadros and B. Vincent, in *Encyclopedia of Emulsion Technology*, Vol. 1, ed. P. Becher, Marcel Dekker, New York, 1983, p. 129.
36. S. Levine, B.D. Bowen and S.J. Partridge, *Colloids Surf.*, **38** (1989), 325.
37. S. Levine and B.D. Bowen, *Colloids Surf. A*, **59** (1991), 377.
38. R. Aveyard, J.H. Clint and T.S. Horozov, *Phys. Chem. Chem. Phys.*, **5** (2003), 2398.
39. R. Aveyard, B.P. Binks, J.H. Clint, P.D.I. Fletcher, T.S. Horozov, B. Neumann, V.N. Paunov, J. Annesley, S.W. Botchway, D. Nees, A.W. Parker, A.D. Ward and A.N. Burgess, *Phys. Rev. Lett.*, **88** (2002), 246102.
40. T.S. Horozov, R. Aveyard, J.H. Clint and B.P. Binks, *Langmuir*, **19** (2003), 2822.
41. T.S. Horozov, R. Aveyard, B.P. Binks and J.H. Clint, *Langmuir*, **21** (2005), 7405.
42. T.S. Horozov and B.P. Binks, *Colloids Surf. A*, **267** (2005), 64.
43. T.S. Horozov and B.P. Binks, *Angew. Chem. Int. Ed.*, **45** (2006), 773.
44. J. Persello, in *Adsorption on Silica Surfaces*, ed. E. Papier, Marcel Dekker, New York, 2000, Chapter 10.
45. T.S. Horozov, R. Aveyard, J.H. Clint and B. Neumann, *Langmuir*, **21** (2005), 2330.
46. J. Lyklema, *Adv. Colloid Interf. Sci.*, **2** (1968), 65.
47. M.E. Labib and R. Williams, *J. Colloid Interf. Sci.*, **115** (1987), 330.
48. M.F. Hsu, E.R. Dufresne and D.A. Weitz, *Langmuir*, **21** (2005), 4881.
49. I.B. Ivanov (ed.), *Thin Liquid Films*, Surfactant Science Series, Vol, 29, Marcel Dekker, New York, 1988.
50. D.N. Petsev (ed.), *Emulsions: Structure, Stability and Interactions*, Elsevier, Amsterdam, 2004.

51. N.P. Ashby, B.P. Binks and V.N. Paunov, *Chem. Commun.*, (2004), 436.
52. E.J. Stancik, M. Kouhkan and G.G. Fuller, *Langmuir*, **20** (2004), 90.
53. D.E. Tambe and M.M. Sharma, *Adv. Colloid Interf. Sci.*, **52** (1994), 1.
54. D. Rousseau, *Food Res. Int.*, **33** (2000), 3.
55. R. Aveyard, B.P. Binks and J.H. Clint, *Adv. Colloid Interf. Sci.*, **100–102** (2003), 503.
56. B.S. Murray and R. Ettelaie, *Curr. Opinion Colloid Interf. Sci.*, **9** (2004), 314.
57. R.A. Peters, *Biochem.*, **42** (1948), 321.
58. G.A.J. Pitt, *Biochem. Soc. Trans.*, **31** (2003), 16.
59. W. Ramsden, Appendix I entitled "Theory of Emulsions Stabilized by Solid Particles", in *The Theory of Emulsions and Their Technical Treatment*, ed. W. Clayton, 2nd edn., Churchill, London, 1928, pp. 229–242.
60. *Nature*, **112** (1923), 671.
61. J. Betjeman, *A Few Late Chrysanthemums*, John Murray, London, 1954.
62. A.D. Hall, *Proc. Roy. Soc. A*, **111** (1926), viii.
63. P. Finkle, H.D. Draper and J.H. Hildebrand, *J. Am. Chem. Soc.*, **45** (1923), 2780.
64. T.R. Briggs, *J. Ind. Eng. Chem.*, **13** (1921), 1008.
65. J.H. Schulman and J. Leja, *Trans. Faraday Soc.*, **50** (1954), 598.
66. V.B. Menon, R. Nagarajan and D.T. Wasan, *Sep. Sci. Technol.*, **22** (1987), 2295.
67. P.A. Kralchevsky, I.B. Ivanov, K.P. Ananthapadmanabhan and A. Lips, *Langmuir*, **21** (2005), 50.
68. J.T. Davies and E.K. Rideal, *Interfacial Phenomena*, Academic Press, New York, 1963.
69. B.P. Binks and S.O. Lumsdon, *Phys. Chem. Chem. Phys.*, **2** (2000), 2959.
70. B.P. Binks and S.O. Lumsdon, *Langmuir*, **16** (2000), 2539.
71. B.P. Binks and S.O. Lumsdon, *Langmuir*, **16** (2000), 8622.
72. S.L. Regen, *J. Am. Chem. Soc.*, **97** (1975), 5956.
73. H. Ogawa, K. Tensai, K. Taya and T. Chihara, *J. Chem. Soc., Chem. Commun.*, (1990), 1246.
74. A. Bhaumik and R. Kumar, *J. Chem. Soc., Chem. Commun.*, (1995), 349.
75. S. Ikeda, H. Nur, T. Sawadaishi, K. Ijiro, M. Shimomura and B. Ohtani, *Langmuir*, **17** (2001), 7976.
76. K. Ikeue, S. Ikeda, A. Watanabe and B. Ohtani, *Phys. Chem. Chem. Phys.*, **6** (2004), 2523.
77. K.-M. Choi, S. Ikeda, S. Ishino, K. Ikeue, M. Matsumara and B. Ohtani, *Appl. Cat. A*, **278** (2005), 269.
78. Y. Takahara, S. Ikeda, S. Ishino, K. Tachi, K. Ikeue, T. Sakata, T. Hasegawa, H. Mori, M. Matsumara and B. Ohtani, *J. Am. Chem. Soc.*, **127** (2005), 6271.
79. B.R. Saunders and B. Vincent, *Adv. Colloid Interf. Sci.*, **80** (1999), 1.
80. R. Pelton, *Adv. Colloid Interf. Sci.*, **85** (2000), 1.
81. T. Ngai, S.H. Behrens and H. Auweter, *Chem. Commun.*, (2005), 331.
82. S. Fujii, E.S. Read, B.P. Binks and S.P. Armes, *Adv. Mater.*, **17** (2005), 1014.
83. B.P. Binks, R. Murakami, S.P. Armes and S. Fujii, *Langmuir*, **22** (2006), 2050.
84. B.P. Binks and S.O. Lumsdon, *Langmuir*, **17** (2001), 4540.
85. J.I. Amalvy, S.P. Armes, B.P. Binks, J.A. Rodrigues and G.-F. Unali, *Chem. Commun.*, (2003), 1826.
86. J.I. Amalvy, G.-F. Unali, Y. Li, S. Granger-Bevan, S.P. Armes, B.P. Binks, J.A. Rodrigues and C.P. Whitby, *Langmuir*, **20** (2004), 4345.
87. N. Saleh, T. Sarbu, K. Sirk, G.V. Lowry, K. Matyjaszewski and R.D. Tilton, *Langmuir*, **21** (2005), 9873.
88. B.P. Binks, R. Murakami, S.P. Armes and S. Fujii, *Angew. Chem. Int. Ed.*, **44** (2005), 4795.

89. B.P. Binks (ed.), *Modern Aspects of Emulsion Science*, The Royal Society of Chemistry, Cambridge, 1998.
90. B.P. Binks and J.A. Rodrigues, *Angew. Chem. Int. Ed.*, **44** (2005), 441.
91. S. Melle, M. Lask and G.G. Fuller, *Langmuir*, **21** (2005), 2158.
92. F. Caruso (ed.), *Colloids and Colloid Assemblies*, Wiley-VCH, Weinheim, 2004.
93. Y. Lin, H. Skaff, T. Emrick, A.D. Dinsmore and T.P. Russell, *Science*, **299** (2003), 226.
94. F. Tiarks, K. Landfester and M. Antonietti, *Langmuir*, **17** (2001), 5775.
95. M. Chen, L. Wu, S. Zhou and B. You, *Macromolecules*, **37** (2004), 9613.
96. C. Barthet, A.J. Hickey, D.B. Cairns and S.P. Armes, *Adv. Mater.*, **11** (1999), 408.
97. M.J. Percy, J.I. Amalvy, C. Barthet, S.P. Armes, S.J. Greaves, J.F. Watts and H. Wiese, *J. Mater. Chem.*, **12** (2002), 697.
98. J.I. Amalvy, M.J. Percy, S.P. Armes, C.A.P. Leite and F. Galembeck, *Langmuir*, **21** (2005), 1175.
99. Y. He, *Materials Chem. & Phys.*, **92** (2005), 134 and 609.
100. M. Rosenberg and E. Rosenberg, *Oil Petrochem. Pollut.*, **2** (1985), 155.
101. L.S. Dorobantu, A.K.C. Yeung, J.M. Foght and M.R. Gray, *Appl. Env. Microbiol.*, **70** (2004), 6333.
102. N. Yan, M.R. Gray and J.H. Masliyah, *Colloids Surf. A*, **193** (2001), 97.
103. N.C.C. Gray, A.L. Stewart, W.L. Cairns and N. Kosaric, *Biotechnol. Lett.*, **6** (1984), 419.
104. A.L. Stewart, N.C.C. Gray, W.L. Cairns and N. Kosaric, *Biotechnol. Lett.*, **5** (1983), 725.
105. S.H. Park, J.-H. Lee, S.-H. Ko, D.-S. Lee and H.K. Lee, *Biotechnol. Lett.*, **22** (2000), 1389.
106. R. Aveyard, B.P. Binks, P.D.I. Fletcher and J.R. Lu, *J. Colloid Interf. Sci.*, **139** (1990), 128.
107. R. Aveyard, B.P. Binks, P.D.I. Fletcher, X. Ye and J.R. Lu, in *Emulsions-A Fundamental and Practical Approach*, ed. J. Sjöblom, Kluwer Academic Publishers, Amsterdam, 1992, p. 97.
108. A.R. Hemsley, B. Vincent, M.E. Collinson and P.C. Griffiths, *Ann. Bot.*, **82** (1998), 105, 108.
109. B.P. Binks, J.H. Clint, G. Mackenzie, C. Simcock and C.P. Whitby, *Langmuir*, **21** (2005), 8161.
110. J. Brooks, P.R. Grant, M.D. Muir, P. van Gijzel and G. Shaw (eds.), *Sporopollenin*, Academic Press, London, 1971.
111. B.R. Midmore, *Colloids Surf. A*, **132** (1998), 257.
112. E. Vignati, R. Piazza and T.P. Lockhart, *Langmuir*, **19** (2003), 6650.
113. J.T. Russell, Y. Lin, A. Böker, L. Su, P. Carl, H. Zettl, J. He, K. Sill, R. Tangirala, T. Emrick, K. Littrell, P. Thiyagarajan, D. Cookson, A. Fery, Q. Wang and T.P. Russell, *Angew. Chem. Int. Ed.*, **44** (2005), 2420.
114. P. van Puyvelde, S. Velankar and P. Moldenaers, *Curr. Opin. Colloid Interf Sci.*, **6** (2001), 457.
115. J. Vermant, G. Cioccolo, K. Golapan Nair and P. Moldenaers, *Rheol. Acta*, **43** (2004), 529.
116. S.S. Ray, S. Pouliot, M. Bousmina and L.A. Utracki, *Polymer*, **45** (2004), 8403 and references therein.
117. C.I. House and C.J. Veal in *Colloid and Surface Engineering: Applications in the Process Industries*, ed. R.A. Williams, Butterworth-Heinemann, Oxford, 1992, p.188.
118. V.P. Mehrotra, K.V.S. Sastry and B.W. Morey, *Int. J. Min. Proc.*, **11** (1983), 175.

119. J. Drzymala, R. Markuszewski and T.D. Wheelock, *Min. Eng.*, **1** (1988), 351.
120. K.L. Sutherland and I.W. Wark, *Principles of Flotation*, Australasian Institute of Mining and Metallurgy, Melbourne, 1955.
121. S.J. McCarroll, *AIME Trans.*, **199** (1954), 289.
122. R.W.M. Lai and D.W. Fuerstenau, *AIME Trans.*, **241** (1968), 549.
123. H.L. Shergold and O. Mellgren, *Trans. Inst. Min. Metall.*, **78** (1969), 121.
124. E. Kusaka, Y. Kamata, Y. Fukunaka and Y. Nakahiro, *Colloids Surf. A*, **139** (1998), 155.
125. E. Kusaka, H. Tamai, Y. Nakahiro and T. Wakamatsu, *Min. Eng.*, **6** (1993), 455.
126. E. Kusaka, Y. Nakahiro and T. Wakamatsu, *Int. J. Miner. Process.*, **41** (1994), 257.
127. J.C. Wilson, *Ph.D. Thesis*, University of Bristol, UK, 1980.
128. Y.Q. Sun and T. Gao, *Metall. Mater. Trans. A*, **33** (2002), 3285.
129. R.G. Alargova, D.S. Warhadpande, V.N. Paunov and O.D. Velev, *Langmuir*, **20** (2004), 10371.
130. D.T. Wasan, A.D. Nikolov and A. Shah, *Ind. Eng. Chem. Res.*, **43** (2004), 3812.
131. K. Vijayaraghavan, A. Nikolov, D. Wasan, B. Calloway, M. Stone and D. Lambert, *J. Chin. Inst. Chem. Engrs.*, **36** (2005), 37.
132. B.P. Binks and T.S. Horozov, *Angew. Chem. Int. Ed.*, **44** (2005), 3722.
133. Th. Wübben and S. Odenbach, *Colloids Surf. A*, **266**, (2005), 207.
134. T. Okubo, *J. Colloid Interf. Sci.*, **171** (1995), 55.
135. L. Dong and D.T. Johnson, *J. Disp. Sci. Technol.*, **25** (2004), 575.
136. V.N. Paunov, B.P. Binks and N.P. Ashby, *Langmuir*, **18** (2002), 6946.
137. J. Wan and T.K. Tokunaga, *J. Colloid Interf. Sci.*, **247** (2002), 54.
138. J-.W. Hu, G.-B. Han, B. Ren, S.-G. Sun and Z.-Q. Tian, *Langmuir*, **20** (2004), 8831.
139. A.B. Subramaniam, M. Abkarian and H.A. Stone, *Nature Mat.*, **4** (2005), 553.
140. Z. Du, M.P. Bilbao-Montoya, B.P. Binks, E. Dickinson, R. Ettelaie and B.S. Murray, *Langmuir*, **19** (2003), 3106.
141. E. Dickinson, R. Ettelaie, T. Kostakis and B.S. Murray, *Langmuir*, **20** (2004), 8517.
142. S.I. Kam and W.R. Rossen, *J. Colloid Interf. Sci.*, **213** (1999), 329.
143. T.S. Horozov and B.P. Binks, *Langmuir*, **20** (2004), 9007.
144. D. Vella, P. Aussillous and L. Mahadevan, *Europhys. Lett.*, **68** (2004), 212.

Bernie Binks (left) obtained his B.Sc. (1983) and Ph.D. in surface chemistry (1986) at the University of Hull. After postdoctoral work in Paris (ellipsometry of low tension surfactant monolayers) and Hull (Langmuir-Blodgett films), he became Lecturer in the Dept. of Chemistry, University of Hull in 1991. He was promoted to Professor of Physical Chemistry in 2003. He is Leader of the Surfactant & Colloid Group, 20 strong, with interests in particles at interfaces, foams, emulsions and wetting. He has published over 170 papers and edited 2 books on Emulsion Science and Methods for Characterising Surfactant Systems. He was awarded (with Prof. P.D.I. Fletcher) the Colloid & Interface Science Group Medal for 2004 from The RSC, UK.

Tommy Horozov (right) obtained a B.Sc. in Chemistry and Physics at the University of Shumen, Bulgaria (1985). After some years as Lecturer at the College of Mathematics, Varna he undertook his Ph.D. at the University of Sofia on the dynamics of surfactant solutions (1988–1991). He was a Visiting Researcher at the University of Antwerp for 1 year. After 4 years in research at Chimatech Corp., Sofia, he came to the University of Hull in 2000 as a research fellow and has worked on particles at interfaces since then.

Tel.: +44-01482-465450; *Fax*: +44-01482-466410; *E-mail*: b.p.binks@hull.ac.uk

Section 1

Particles at Planar Liquid Interfaces

2

Structure and Formation of Particle Monolayers at Liquid Interfaces

Lennart Bergström

Department of Physical, Inorganic and Structural Chemistry, Stockholm University, SE-106 91 Stockholm, Sweden

2.1 Introduction

The structure displayed by assemblies of colloidal particles, whether in three dimensions (3-D) or in two dimensions (2-D), is an important aspect in many industrial processes and products, *e.g.* waste-water treatment, paints and ceramics, but also for the assembly of new materials.[1-4] Colloidal particles can be made to organise into ordered arrays or to attain heterogeneous structures with different degrees of disorder. The control of colloidal structure formation starts with the particle interactions (attractive or repulsive) and colloidal dynamics. These interactions balance against thermal forces and external influences such as gravity and applied force fields to determine what configurations the particles will adopt, *e.g.* network-like, random or ordered configurations (see Figure 2.1).

These colloidal structures acquire interesting and useful properties not only from their constituent materials, but also from the spontaneous emergence of mesoscopic order that characterises their internal structure.[5] Ordered arrays of colloidal particles with lattice constants ranging from a few nanometres to a few microns have potential applications as optical computing elements and chemical sensors, and templates for fabricating quantum electronic systems. Restricting the colloidal array to 2-D has particular relevance to sensor and membrane applications. Disordered systems can be used as ceramic membranes.

The construction of photonic crystals emphasises the importance of the correlation between structural features and the material properties. Photonic crystals are assemblies of colloidal particles, packed into mesoscopic, periodic arrays. Their ability to diffract light in the ultraviolet, visible and near-infrared part of the spectrum[6] requires a well-ordered structure without structural defects. This property makes them the ideal material for optical applications, such as rejection filters. Carefully designed photonic crystals may be used as sensors being responsive to light, temperature and chemicals.[7,8] Photonic crystals, in both 2-D and 3-D, with a controlled defect formation have shown promise as optical band gap materials and wave guides.[9]

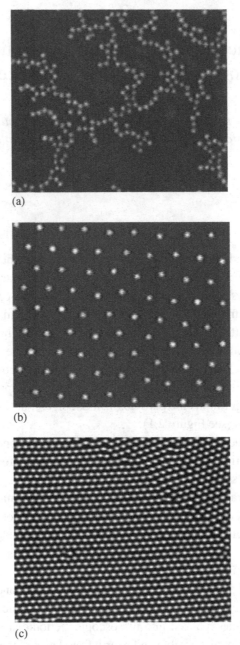

(a)

(b)

(c)

Figure 2.1 Examples of colloidal structures produced by spherical silica particles with a diameter of 1.97 μm trapped at an air–liquid surface; (a) disordered network of octyl-coated silica that aggregate at the air–toluene surface; (b) low density ordered structure caused by a temporal long-range repulsion between the silica particles; (c) dense particle monolayer of octadecyl-coated particles that have been forced to close pack by an increasing surface fraction. (a) and (b) taken from Ref. [45]; with permission of Academic Press; (c) taken from Ref. [22]; with permission of the American Chemical Society.

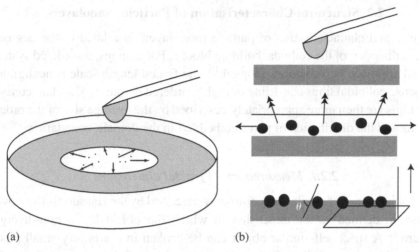

(a)   (b)

Figure 2.2 A schematic view of the spreading of a dispersion droplet on a liquid surface. (a) illustrates the lateral spreading that is facilitated by the very low interfacial tension between the dispersion medium and the liquid; (b) illustrates the evaporation of the dispersion medium and the trapping of the particles at the air–liquid surface. The assembled particle network can be transferred onto a substrate that is slowly raised through the liquid phase, schematically illustrated at the bottom of (b).

Different processes have been proposed to produce 2-D colloidal films with varying degrees of homogeneity. The most common method is controlled drying of colloidal suspensions. This method is based on the convective assembling technique, controlled by the evaporation rate of the solvent.[3] Another possibility is to form colloidal films by the self-assembly of particles attached at liquid interfaces. Figure 2.2 gives an example of how particles can be spread and assembled at the liquid–air surface. The particles are well dispersed in a spreading solvent (commonly methanol or ethanol) and a droplet is applied to the liquid surface with a syringe. The droplet spreads over the liquid surface, forming a thin film. As the alcohol evaporates, the particles are trapped at the liquid–air surface, and the particle assembly is mainly controlled by the range and magnitude of the inter-particle interactions.

This chapter deals mainly with the formation and characterisation of particle monolayers at planar liquid interfaces and intends to give an introduction to the relation between the structure of colloidal particles assembled at interfaces and their interactions. The chapter starts with a brief description of how the structure of both disordered and ordered particle monolayers can be characterised. This is followed by a presentation of investigations of colloidal aggregation, formation of ordered colloidal films and various types of structures with a mesoscopic order at air–liquid and liquid–liquid interfaces.

2.2 Structural Characterisation of Particle Monolayers

The structural characterisation of particle monolayers is related to the degree of order, or disorder, of the colloidal building blocks. For tenuous, disordered systems, a fractal approach is commonly adopted.[10] The fractal length scale is negligible in very dense colloidal films consisting of highly ordered domains. The characteristics of the films are then more appropriately described by the average size of the ordered domains and the distribution of pores embedded in the domain boundaries.

2.2.1 Measurement of fractal dimensions

Fractal objects are self-similar and commonly described by the Hausdorff–Besicovitch dimension, defined as the dimension in which the object has a non-diverging measure.[11] A strict self-similar object can be broken into arbitrary small pieces, each of which is a small replica of the entire object. The structure of colloidal aggregates, as well as gels and percolated structures, may display a statistical self-similar nature. That is, the average structure of an aggregate is self-similar with the average structure of a smaller part. Aggregates and gels are mass fractals in the sense that the fractal description indicates how the mass of n colloidal particles fills the space occupied by the aggregate. Mass fractals display a power law scaling between the mass $n(R)$ within a certain length scale R, and the length scale itself, according to

$$n(R) \propto R^{D_f} \qquad (2.1)$$

where D_f is the fractal dimension of the cluster.

The statistical self-similarity of colloidal aggregates applies to a limited length scale. The lower cut-off is determined by the size of the basic building unit, which can be a sub-cluster or the particle itself. A cluster of s particles can have any dimension between one (a line) and the value for a closed packed structure. It is possible to express how the number of particles in a cluster $n(r)$, (cluster mass), scales with the radius, r, of the cluster [12]

$$n(r) = n_o \left(\frac{r}{r_o} \right)^{D_f} \qquad (2.2)$$

with the basic building unit consisting of n_o particles with a mean radius of r_o. This equation may be re-written to include the volume fraction of the basic building unit, ϕ_o, and the radius of a single particle, a, which is the smallest possible building unit. The resulting expression reads

$$n(r) = \phi_0 \left(\frac{r_0}{a}\right)^D \left(\frac{r}{r_0}\right)^{D_f} \tag{2.3}$$

which shows that it is possible to define two fractal dimensions if the basic building unit itself is a self-similar cluster with a fractal dimension, D.

The fractal dimension is often measured as an exponent in a power law expression relating cluster mass and a measure of length scale. Hence, the resulting fractal dimension depends upon the choice of length scale and on the choice of method used to relate mass to length scale. A complete fractal description of the structure requires knowledge of the pre-factors r_0 and n_0.[12] When the pre-factors are known it is possible to relate the fractal dimensions obtained with different methods. However, without the pre-factors, the fractal dimension is still a good discriminator of structural changes. Comparison between different methods gives a qualitative description of structural changes in a system of aggregating colloids. Below, two of the most common methods used to determine the fractal dimensions are described.

2.2.1.1 Radius of gyration

This method exploits the scaling relation between the mass and one length scale in a set of clusters.[11] The number of primary particles in an aggregate, s, represents the mass of the aggregate if the particles are monodisperse. Furthermore, the radius of gyration, R_g, is the characteristic length scale with

$$s \propto R_g^{D_{fs}} \tag{2.4}$$

This method results in a fractal dimension, D_{fs}, that corresponds to the self-similarity between clusters of different mass. The radius of gyration, R_g, is defined as

$$R_g^2 = \frac{1}{s} \sum_i^s \left(\mathbf{r}_i - \langle \mathbf{r} \rangle\right)^2 \tag{2.5}$$

where $\mathbf{r}_i$ is the position of particle i in the aggregate and $\langle \mathbf{r} \rangle$ is the mean position of all particles. The latter is also equal to the centre of mass of the cluster. Figure 2.3 gives an example of how the fractal dimension of an assembly of clusters is obtained with the radius of gyration method.

The scaling between mass and radius of gyration results in a fractal dimension which should be treated with care, since particles far from the centre obtain a greater weight in equation (2.5). Consequently, the method is most suitable for symmetric clusters, which grow in the radial direction, *e.g.* classical clusters simulated by particle–cluster aggregation.[13,14] Hence, clusters that grow asymmetrically will be given a lower value of fractal dimension compared to clusters that grow symmetrically.

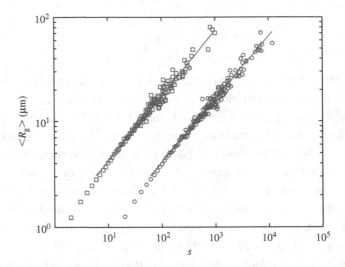

Figure 2.3 Double logarithmic plot of the radius of gyration, $\langle R_g \rangle$, as a function of the number of particles in the cluster, s, for octyl-coated ($\square$) and octadecyl-coated ($\bigcirc$) silica particles that have aggregated at the air–toluene surface. The octadecyl-coated system has been offset by a factor of 10 for clarity. The lines are least square fits and yield D_{fs} ($\square$) $= 1.57 \pm 0.04$ and D_{fs} ($\bigcirc$) $= 1.63 \pm 0.04$. Taken from Ref. [45]; with permission of Elsevier.

2.2.1.2 Box counting

The fractal dimension obtained from box counting, D_{fb}, is an estimate of the Hausdorff–Besicovitch dimension. The original method to identify D_{fb} (in 2-D) is to enclose the structure in a square of side length L_m. This square is further divided into smaller squares of side length l, such that $(L_m/l)^2$ squares are obtained. The number of small boxes, which contain sections of the cluster $N(L_m, l)$, scales as[15]

$$N(L_m, l) \propto (L_m/l)^{D_{fb}} \qquad (2.6)$$

For constant L_m, $a < l < L_m$ and $L_m/l \in \{1,2,3, \ldots\}$, the linear part of $-\log N(L_m, l)/\log l$ identifies D_{fb}. This approach is appropriate for structures filling up the available space, *e.g.* a gel, such that $L_m/a > 100$. However, if the structure is small compared to the size of the primary particle, a, the number of possible values of l becomes very limited, which may result in a large error in D_{fb}. This problem is minimized when the values of l are controlled and the length scale of the grid, L_m, is adjusted to contain a whole number of small boxes. Hence, a larger set of box sizes becomes available, resulting in more data points in the scaling of the cluster, and improvement of the D_{fb} estimate.

The box count dimension D_{fb} is fundamentally different from D_{fs}, in that the former identifies a scaling between mass and length scale within one object. The self-similarity dimension D_{fs} identifies a scaling relation between objects of different mass and length scale. However, the two measures are equal in many cases, especially for deterministic self-similar fractals.

2.2.2 Structural characterisation of ordered colloidal monolayers

The structural characterisation of ordered colloidal monolayers has traditionally focused on a determination of the degree of order. Various characterisation parameters, *e.g.* surface coverage, inter-particle distances, diffraction properties and Fourier space analysis have been used to evaluate the quality of colloidal films.[3,16–20] The characterisation method usually relies on an accurate determination of the position of each particle in the dense monolayer films, commonly obtained from processed images. Grier and Murray,[21] in a seminal work that was developed further by Hansen *et al.*,[22] showed how the degree of order of colloidal films can be characterised by analysing the dampening of the pair-distribution functions. The pair distribution, $g(r)$, can be determined by counting the number of particles covered by a ring with the radius, r, and thickness, δr, centred at particle i and normalised to the number density of particles within the ring, $\rho\ (=1/A_o)$. Averaging over the total number of particles, N, yields the pair-distribution function:

$$g(r) = \frac{1}{N}\sum_{i=1}^{N}\frac{A_o n_i(r)}{\pi\left(\delta r^2 + 2r\,\delta r\right)} \tag{2.7}$$

The radial distance r is represented in a discrete form, $r = m dr\ (m = 1, 2, \ldots)$ with dr smaller than the particle diameter. Comparison of the experimental pair-distribution function, $g(r)_{exp}$, obtained from the information in many images of a specific colloidal film, with an empirically modified pair-distribution function, based on a structure of an ideal triangular lattice, $g(r)_{triang}$, allows for a determination of the degree of order in the colloidal monolayers. By broadening the peaks in $g(r)_{triang}$ with a normal distribution, the statistical fluctuations of particle positions around the ideal lattice point can be accounted for. As the size of the ordered domains in an ideal lattice is infinite, an exponentially decaying function is introduced to account for the finite size of ordered domains in experimental particle films. The final expression is

$$g(r) = \left[\left|\int g_{triang}(r - x)\frac{\delta r}{(2\pi)^{1/2}\sigma(r)}\exp\left(-\frac{x^2}{2\sigma(r)^2}\right)dx - 1\right|\exp\left(-\frac{r}{\zeta}\right)\right] + 1 \tag{2.8}$$

where x is the distance from the position a particle would have in a lattice of perfect triangular order, $\sigma(r)$ is the standard deviation of particle positions and ζ is a measure of the correlation length, which is an estimate of the average size of the ordered domains.

Although this method gives a good estimate of the degree of order in a colloidal film, it is difficult to estimate the presence of irregularities in the structure and the method cannot distinguish between the different defect types in the ordered arrays. In a recent study, Hansen *et al.* proposed a complementary method for a quantitative analysis of the defect size distribution of 2-D crystal structures that is based upon a Delaunay triangulation procedure.[22] The method is based on tessellation of the space between the particles for characterising the pore sizes. The utility of a similar procedure has been demonstrated in investigations on the phase transitions of one-component liquids in 2-D.[21,23] Dual tessellations, *i.e.* a combination of Delaunay triangulation and Voronoi diagrams, have also been applied to obtain the volume, the area and the connectivity of pores in a medium.[24]

The combination of analysis of the pair-correlation function and the Delaunay triangulation procedure yields a complementary set of parameters that characterises an ordered structure, *e.g.* the size of ordered domains, the pore size distribution and the size distribution of a specific pore. The Delaunay triangulation procedure, in which all particle positions in an image are connected into a triangular lattice, is generated in such a way that no point in the lattice lies within any circle connecting the three corners of the triangle. Once the Delaunay triangulated lattice has been made, it is possible to identify different pores in the structure by the number of triangles enveloping one pore. Examples of different pores are shown in Figure 2.4. At hexagonal close packing, the triangles are equilateral with an inter-particle distance of two particle radii, $2a$, and one triangle will be sufficient to cover the pore between the particles. Defect triangles will have a centre-to-centre distance exceeding $2a\Delta$, where Δ is an error tolerance of the particle positions. Structural defects, such as stacking faults or larger pores, are identified by two or more linked defect triangles; two triangles corresponds to particles arranged in a square or rhombic geometry, and together with pores enveloped by three triangles, they constitute stacking faults in the film structure. A pore covered by four triangles corresponds most often to a missing particle in the film and is a point defect of the structure.

2.3 Colloidal Aggregation at Liquid Interfaces

Colloidal particles will aggregate when the effective interaction between the particles allows bonds to be formed when the particles collide. Aggregation is a non-equilibrium process where the aggregation rate and the structure of the aggregates are closely related.[2,14] Aggregation, also known by the terms coagulation, flocculation, clustering and agglomeration, is a process of large importance for numerous industrial applications in the chemical, environmental, electronics, mineral and biological sectors. Most

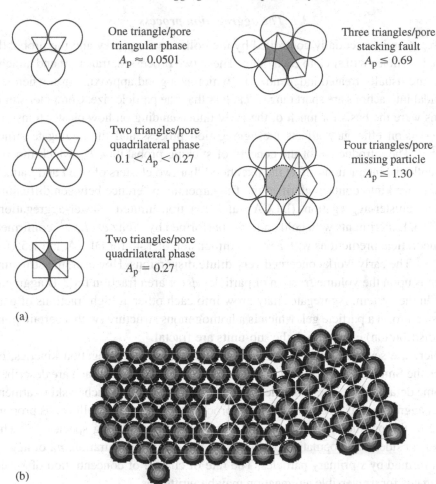

Figure 2.4 (a) Examples of the geometry of different particle arrangements together with the associated number of Delaunay triangles. The pores are shaded gray and the normalized pore area, A_p, is included. (b) A section of a triangulated particle network, with examples of different pores, characterised by the number of triangles that envelope them (highlighted). (a) taken from Ref. [22]; with permission of the American Chemical Society.

"real world" systems, involving aggregation of polydisperse, non-spherical particles with heterogeneous surface chemistry, are very complex and cannot be described in sufficient detail to enable a complete analysis. Model studies, providing systematic information on well-defined systems, *e.g.* experimental studies involving monodisperse latex or silica particles, can allow a more detailed analysis. The use of a 2-D system with the colloidal particles trapped at a liquid interface simplifies the observation of the temporal evolution of the clusters and also provides wider tests of theoretical predictions than are possible in 3-D alone.

2.3.1 The aggregation process

Aggregation is essentially controlled by the collision frequency and the probability that particles will stick to each other. These two processes, transport and attachment, are usually treated independently, which is a good approximation when the colloidal interactions are short ranged, *i.e.* less than the particle size. Computer simulations were the basis for much of the early understanding on how cluster transport and collision efficiency affects the aggregation process and the aggregate structure.[25,26] The introduction of the concept of sticking probability, P_{ij}, which models how colloidal interactions affect the likelihood that two clusters of size i and j attach when moved into contact, clarified the fundamental difference between diffusion-limited cluster-aggregation (DLCA) and reaction-limited cluster-aggregation, (RLCA). Experiments with gold colloids, performed by Weitz *et al.*,[27,28] confirmed the theoretical predictions with a fractal dimension of 1.44 for DLCA and 1.55 for RLCA.[*] The early work concerned very dilute dispersions. However, the structure depends upon the volume fraction of particles ϕ (or area fraction for 2-D aggregation) in the system. Aggregates may grow into each other at high fractions of particles and form a particle gel, which is a homogenous structure (with a certain pore size distribution) even though the sub-units are fractal.

There are several different approaches to describe the aggregation kinetics, of which the Smoluchowski approach and the dynamic scaling approach are described in some detail below. The fundamental assumption in the Smoluchowski treatment is that aggregation is a second-order rate process, *i.e.* the rate of collision is proportional to the product of the concentrations of the two colliding species.[29,30] The model considers the population balance of the number concentration, n_s, of aggregates formed by s primary particles. The rate of change of concentration of s-fold aggregates for irreversible aggregation may be written as

$$\frac{\mathrm{d}n_s}{\mathrm{d}t} = \frac{1}{2} \sum_{\substack{j=s-i \\ i=1}}^{i=s-1} k_{ij} n_i n_j - n_s \sum_{m=1}^{\infty} k_{ms} n_m \qquad (2.9)$$

where k_{ij} is the second-order rate constant. The time evolution of the total distribution of clusters with different sizes can be described as a set of coupled differential equations. The rate constants are often referred to as the reaction kernels, which include expressions relating to the transport mechanism. The structure of the cluster and the sticking probability was not incorporated in the classical treatment, but has received increasing attention in later years. The accuracy of the Smoluchowski approach to describe the aggregation process relies on the choice of an appropriate expression for the reaction kernels.

---

[*] Please note that RCLA has been changed to RLCA as per the abbreviation (reaction-limited cluster-aggregation).

The Smoluchowski approach characterises the aggregation process by the evolution of each cluster size. An alternative is to study the evolution of the distribution of cluster masses rather than the evolution of the individual sizes. The cluster–mass distribution, $n(s, t)$ is defined as the number of clusters of mass s at time t, normalized with the total volume or area, for 3-D and 2-D systems, respectively. The analysis of $n(s, t)$ builds on the statistics of the cluster-mass distribution[31] and allows a few parameters to describe the aggregation process without assumptions from underlying models.

There is a direct relation between $n(s, t)$ and the number fraction of particles, $\phi_n(t)$, given by

$$\phi_n(t) = \Sigma_s sn(s, t) \tag{2.10}$$

Furthermore, the number–average cluster mass, $N(t)$, defined as

$$N(t) = \frac{\sum_s sn(s, t)}{\sum_s n(s, t)} \tag{2.11}$$

is inversely proportional to the total number of clusters n_T formed by the total number of particles in the analysis. In simulations, the total number of particles is constant and $N(t)$ reflects the aggregation rate. The weight–average cluster mass, $S(t)$, is obtained from

$$S(t) = \frac{\sum_s s^2 n(s, t)}{\sum_s sn(s, t)} \tag{2.12}$$

and reflects the population change towards large clusters. A comparison between $S(t)$ and $N(t)$ gives a qualitative picture of the aggregation process.[32]

A more quantitative description of the aggregation kinetics is also available from the scaling properties of the cluster–mass distribution.[14] In the dynamic scaling theory, a few exponents describe the distribution of cluster masses around a characteristic cluster mass. In the limit of a low-area fraction of particles and large s and t, Vicsek and Family[33] showed by computer simulations that $n(s, t)$ is an algebraically decaying function of s for every t as well as an algebraically decaying function of t for every s. They proposed a dynamic scaling for the cluster–mass distribution of the form

$$n(s, t) \sim t^{-w} s^{-\tau} f(s/t^z) \tag{2.13}$$

where the cut–off function $f(x)$ limits the region where the scaling applies: $f(x) \approx 1$ for $x \ll 1$ and $f(x) \ll 1$ for $x \gg 1$. The relation (2.13) includes two exponents, w and z, describing the dynamic behaviour in addition to the static exponent τ. The term t^{-w} expresses the power law decay of $n(s, t)$ with time for every s. This is a specific feature of the cluster–cluster aggregation process, because at large t small clusters have been absorbed in the formation of larger clusters. The characteristic cluster size is determined by t^z, where the exponent z depends on the nature of the aggregation process.

2.3.2 2-D aggregation studies

The simple geometry of a 2-D system allows studies with imaging techniques, and has inspired thorough investigations of colloidal aggregation at air–liquid and liquid–liquid interfaces. The majority of the early work concerned the aggregation of latex and silica particles trapped at the air–water surface.[34–44] Hurd and Schaefer presented the first paper where small silica particles floating on the surface of an electrolyte solution of high ionic strength aggregated and formed strong, tenuous clusters.[35] Robinson and Earnshaw presented a sequence of papers[36–39,43] where they investigated the aggregation of colloidal particles trapped at the air–liquid surface and were the first to describe both the structural and kinetic aspects of the aggregation process in detail. Their procedure for studying charged particles at the air–water surface was based on spreading latex spheres from a dilute methanol dispersion onto the surface of a dilute electrolyte solution (see Figure 2.2) and aggregation was induced by increasing the ionic strength by addition of electrolyte into the sub-phase.[36–39] The addition of electrolyte reduces the electrostatic repulsive forces, allowing the particles to stick to each other due to the ubiquitous van der Waals attraction. Robinson and Earnshaw found that the fractal dimension of the clusters varied with the molarity of electrolyte in the sub-phase.[37] The clusters were more dense at low sub-phase molarities while the clusters became more tenuous when the molarity of the sub-phase was increased above $0.5\,M$ $CaCl_2$. The measured values of the fractal dimension suggested that the particles aggregate according to an RLCA process at low sub-phase molarities, thus indicating the presence of some type of repulsion even at these relatively high ionic strengths. Analysis of the kinetics showed that the aggregation process obeyed the scaling laws outlined above and can thus be interpreted within the dynamic scaling concept.[38]

The unexpected complexity of particle systems trapped at air–water surfaces motivated Bergström and co-workers to design interfacial systems where hydrophobic alkoxylated silica particles are spread and trapped on the surface of various organic liquids.[41,45] The main advantage of these systems is that the magnitude of the attractive interactions can be controlled. The total interaction energy is essentially controlled by the degree of particle immersion and the range of the steric repulsion. This results in a model system that is very flexible. It allows independent control of both the attractive and repulsive interactions through the choice of the organic liquid, which controls the degree of immersion, and the length of the grafted alkyl chains, which controls the range of the steric repulsion.

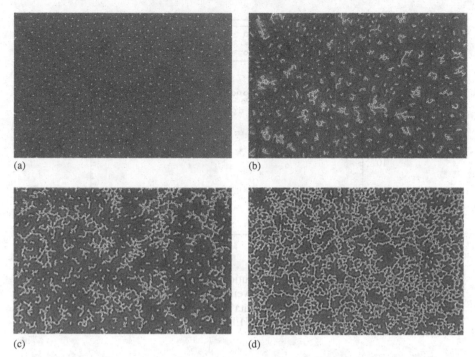

Figure 2.5 Images from the different stages of the aggregation process of octyl-coated silica particles floating on the air–toluene surface. (a) initial stage, $t = 4\,\text{min}$, where an ordered film characterized by a long-range repulsion has formed; (b) early clustering stage, $t = 90\,\text{min}$, where small clusters coexist with a substantial amount of discrete particles; (c) later clustering stage, $t = 270\,\text{min}$, where the cluster sizes have grown, but the size distribution is very wide; (d) final stage, $t = 400\,\text{min}$, where almost all clusters have joined into one large cluster or a particle network. Taken from Ref. [45]; with permission of Elsevier.

Hansen and Bergström used alkoxylated silica particles (diameter $2\,\mu\text{m}$) trapped at the air–toluene surface to study the cluster structures and the kinetics of the 2-D aggregation process.[45] Figure 2.5 gives one example of the evolution of the formation of 2-D clusters, with the clusters continuously growing in size and where eventually a large particle network is formed. The aggregation kinetics was analysed using dynamic scaling theory where the cluster-mass distribution was represented in a reduced form

$$n(s, t)s^2 \propto F(s/S(t)) \tag{2.14}$$

where the shape of the cut-off function characterises the aggregation behaviour. Figure 2.6 shows that the reduced cluster-mass distributions are somewhat scattered, but collapse into a master curve, $F(x)$. The bell shape with a clear peak in the

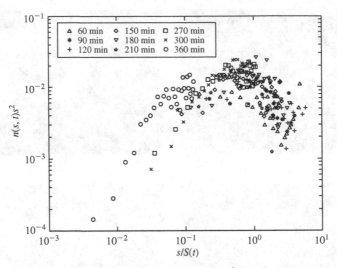

Figure 2.6 Reduced cluster–mass distribution, $n(s, t)s^2$, as a function of the reduced cluster mass, $s/S(t)$, at different times (given) for monolayers of octyl-coated silica particles at the air–toluene surface. The data shows clusters in the range $1 \leqslant s \leqslant 30$. Taken from Ref. [45]; with permission of Elsevier.

master curve indicates that small–small cluster attachments dominate the aggregation process. The cluster mass evolution, Figure 2.7, shows that the aggregation process is divided into two régimes; an initial régime where aggregation is slow, followed by a fast aggregating régime. The power law exponent, z, has a value below 0.5 in the slow aggregation region, which corresponds well with theoretical predictions, while the sudden increase in the aggregation and cluster growth rate by a factor of 10 in the fast aggregation régime does not agree with the classical theory of perikinetic aggregation. A simple analysis of the Péclet number, which identifies the importance of diffusion relative to convection for transport of particles and clusters, suggested that the second, fast aggregating régime, may be related to convection-limited aggregation of large clusters (CLCA).[45] Convection enhances the mobility of the large clusters and the overall collision frequency, thus the cluster growth rate increases because large clusters become more involved in the aggregation process. Recent work has shown that careful experimental design can avoid convective effects and yield kinetic exponents that vary between 0.3 and 0.6, which relates to RLCA and DLCA dominated aggregation processes, respectively.[46]

The need to add a significant amount of salt to the aqueous sub-phase to induce the aggregation cannot be explained by simply invoking a DLVO-type interaction force model. Earnshaw and Robinson suggested that the presence of dipoles at the particle surfaces can induce a dipole–dipole repulsion through the air phase that

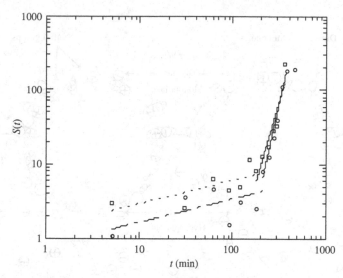

Figure 2.7 The weight–average cluster mass, $S(t)$, plotted against time for octyl-
($\square$) and octadecyl- ($\bigcirc$) coated silica particles at the air–toluene surface. Taken
from Ref. [45]; with permission of Elsevier.

can explain the enhanced stability for particles trapped at air–liquid surfaces compared to particles in a bulk liquid.[43] Recent work has tried to explain the aggregation behaviour of particles trapped at aqueous surfaces by relating experimental
observations to simulations or theoretical calculations. The issue that has gathered
most attention is the origin and the scaling of the long-range repulsion that controls
the colloidal stability and the aggregation behaviour at air–water surfaces.
Hidalgo-Alvarez and co-workers have presented a model where they combine
dipole–dipole and Coulombic (monopole– monopole) interactions between the
particles to explain the 2-D aggregation behaviour.[46,47] A simple sketch of the
complex nature of particle interactions is shown in Figure 2.8. Brownian dynamics
simulation results showed that the dipolar interaction controls aggregation at high
sub-phase salt concentrations while it is the Coulombic interaction that controls
the aggregation at low salt concentrations. Their results also suggest that it is
the fraction of charged groups at the particle surface in contact with air that essentially controls the 2-D aggregation kinetics, and that the surface critical coagulation concentration is defined by the salt concentration where this fraction becomes
negligible.[47]

2.4 Ordered Colloidal Systems

Colloidal crystals can be produced with different methods, ranging from the
evaporation-driven convective assembling technique[3] to various techniques to form

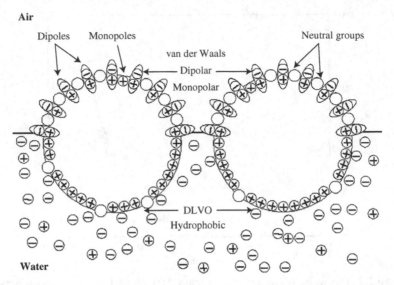

Figure 2.8 Sketch of two particles located at the air–water surface. The immersed parts of the particles interact through Derjaguin–Landau–Verwy–Overbeek (DLVO) and hydrophobic forces and the emergent parts interact through van der Waals, dipolar and monopolar interactions. Taken from Ref. [47]; with permission of the American Chemical Society.

colloidal films by self-assembly of particles attached at liquid interfaces. The general features of the different techniques will be described with some details on the importance of the inter-particle forces on the structural order.

2.4.1 Deposition on a substrate

The most common method to produce 2-D colloidal films with long-range order is by depositing colloidal particles on a substrate. Extensive experimental and theoretical studies (see Ref. [3] for a thorough description) have established a good understanding of the formation of the colloidal monolayer by the identification of two distinct stages: nucleation and crystal growth. The nucleus, which is a small, hexagonally ordered colloidal monolayer, forms when the thickness of the suspension film becomes sufficiently thin so that the particles start to protrude from the suspension film surface, see Figure 2.9. At this point, the particles are subjected to a strong and long-range attractive force, commonly called the immersion capillary force, F_{ca}, that originates from the deformation of the liquid surface. When drying proceeds, there is a convective flow of particles to the nucleus and the size of the 2-D colloidal crystal grows with time.

Denkov *et al.*[48] showed that the evaporation rate is an important factor controlling the structural quality of 2-D crystals. It has also been suggested that the 2-D

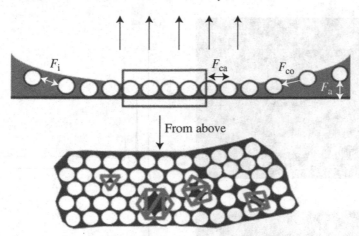

Figure 2.9 A schematic overview of the monolayer formation process when a dilute colloidal suspension is allowed to dry on a substrate. The dominating forces in the suspension are also denoted; immersion capillary force (F_{ca}), convection to the growing crystal (F_{co}), inter-particle forces (F_i) and the interaction with the substrate (F_a). The assembled 2-D colloidal crystal with typical defects is also shown.

crystal structure is affected by the particle–particle interactions, F_I, and the interactions between the substrate and the particles, F_a.[3,17,49] Changing the electrolyte concentration is a convenient way of controlling the inter-particle forces and the stability of an electrostatically stabilised colloidal dispersion. Rödner *et al.*[50] investigated the effect of added NaCl on the structural features when forming colloidal monolayers by drying dilute silica particle dispersions on a glass substrate. They observed that the colloidal structures are highly sensitive to the amount of added salt, ranging from network-like to well-ordered, see Figure 2.10. At the highest amount of added salt, Figure 2.10(a), the dried film does not display any long-range order, reflected by a pair-distribution function that decays within a couple of particle diameters. The system with no added salt, Figure 2.10(b), yields a 2-D colloidal crystal, with the $g(r)$ decaying slowly, corresponding to large ordered domains. Analysis of the pore size distribution using the Delaunay triangulation approach described previously showed that the concentration of larger defects, missing particles and large holes, increase significantly with increasing amount of added salt.[50] Hence, an increased adhesion between particles and substrate with increasing salt addition not only reduces the size of the ordered domains, but also results in a greater number of large defects.

2.4.2 Attractive colloidal systems

In a similar study on the assembly on a liquid interface, Hansen *et al.* varied the magnitude of the attractive inter-particle forces and studied the effect on the structural

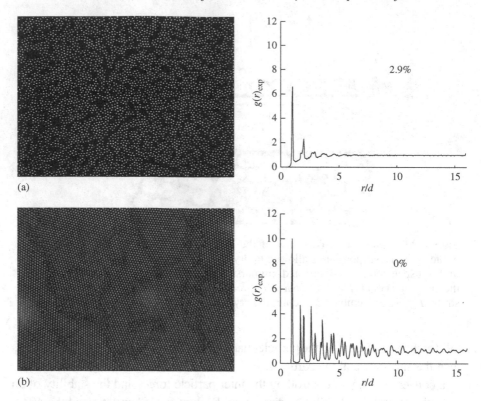

Figure 2.10 Representative images (139 × 105 μm) with the corresponding pair-correlation function, $g(r)_{\text{exp}}$, of colloidal monolayers formed when drying dilute silica dispersions. (a) Added salt (2.9% NaCl/silica ratio), (b) no added salt. The dimensionless distance in the pair-correlation functions is the radial distance, r, divided by the particle diameter, d. Taken from Ref. [50]; with permission of the American Chemical Society.

properties of monolayers of hydrophobic silica particles.[22] The inter-particle forces are mainly determined by the degree of immersion, *i.e.* the three-phase contact angle; a high contact angle results in a low degree of immersion, hence the van der Waals attraction is strong. The contact angles of silica particles, surface treated with octyltrichlorosilane (OcTS) and octadecyltrichlorosilane (OTS) were controlled by spreading them on different solvents, with increasing dielectric constants. The interaction energy, V_{min}, between the particles was calculated using the Williams and Berg model[42] for spheres at an air–liquid surface.

It was found, Figure 2.11, that the average size of the hexagonally ordered domains, estimated from the size of the ordered domains ζ, decreased exponentially with increasing inter-particle bond strength for the alkoxylated silica systems. Deviating values were seen for systems in which these particles were immersed in water, which

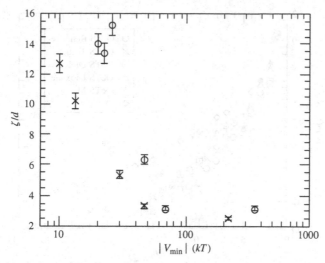

Figure 2.11 The normalised correlation length, ζ/d, plotted against the modulus of the particle interaction energy, V_{min}, for particle monolayers at the air–liquid surface formed from silica particles coated with alkyl chains of length 18 (x) and 8 (o) carbon atoms. Taken from Ref. [22]; with permission of the American Chemical Society.

was attributed to the very high contact angles. Hence, reducing the magnitude of the attractive interactions allows the particles to rearrange and attain a more favourable, densely packed structure. Onoda observed similar phenomena in his early work on 2-D clustering and ordering of polystyrene particles of variable size trapped at the air–water surface.[34] He found that large particles undergo a process of slow cluster growth where dense, ordered domains or clusters were formed by rearrangement of the particles in the cluster. Smaller particles did not order permanently, but formed relatively large, temporarily ordered domains by clustering and subsequent rearrangement and/or attachment/detachment. Onoda attributed the reversible dynamics to a delicate balance, with the attractive interactions being of a similar magnitude to the Brownian motion.

Hansen *et al.* also analysed the defect structure using the Delaunay approach.[22] Figure 2.12 shows that there is an exponential decrease in the number of pores with increasing n_t, which represents an increasing pore size as more triangles are needed to envelope a pore the larger the pore is. We find that there appears to be an increase in the number of large pores (represented by high n_t) with increasing bond strength, most pronounced for the aqueous system where the inter-particle bond strength is very high.

2.4.3 *Repulsive colloidal systems*

It was recognised early that a long-range repulsive interaction between particles trapped at a liquid interface can induce ordering and even lead to the formation of

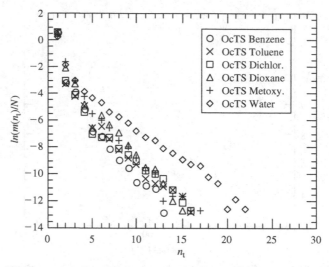

Figure 2.12 The normalized pore size distribution, $m(n_t)/N$, as a function of the number of triangles that envelopes a pore, n_t, for colloidal films of octyl-coated silica particles floating at the air–liquid surface for different liquids (given). Taken from Ref. [22]; with permission of the American Chemical Society.

2-D colloidal crystals.[51] Pieranski found that small latex particles spread on a low ionic strength aqueous medium form ordered structures (at the air–water surface) with a particle separation of up to 10 μm, *i.e.* more than 30 particle diameters.[51] As was already mentioned, much work has been devoted to understand the nature of the unusually long-range repulsion that occurs between colloidal particles trapped at liquid interfaces. The early realisation that the asymmetric nature of the colloidal system can result in dipole–dipole repulsion was complemented with the introduction of a Coulombic repulsion between particles through the non-aqueous phase (being either air or oil).[52] As discussed earlier, a combination of dipole–dipole and Coulombic repulsion (see Figure 2.8) was successfully used to explain the 2-D aggregation behaviour.[46,47]

Aveyard and co-workers have recently determined the distance dependence of the inter-particle repulsion between charged particles at an oil–water interface using a laser tweezer method.[53] They showed that a Coulombic repulsion, originating from dissociated groups on the particle surface that remain charged also in the oil phase, can explain the observed distance scaling and absence of any electrolyte dependence. Interestingly, it appears that the Coulombic repulsion may depend on the nature of the non-aqueous phase (being air or an oil) and also on the contact angle that the particles make with the interface.[52,54] Particles spread at an air–water surface form ordered structures at low ionic strength, but collapse into clusters when the ionic strength is increased. However, particles trapped at the octane–water interface

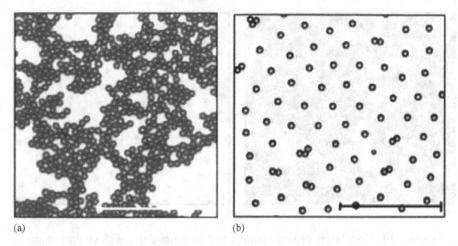

(a)   (b)

Figure 2.13 Comparison of the structure of monolayers of 2.6 μm diameter latex particles on 1 M aqueous NaCl in contact with (a) air and (b) octane. Scale bars represent 50 μm. Taken from Ref. [52]; with permission of the American Chemical Society.

remain highly ordered even at an ionic strength as high as 1 M, see Figure 2.13. Aveyard and co-workers attributed this striking difference in stability to a higher charge density at the particle–octane surface compared to the particle–air surface.[52] They also studied the compression and structure of latex particle monolayers assembled at the air–water and octane–water interfaces using a Langmuir trough.[52] The external pressure can result in structural transitions, *e.g.* from a hexagonal to a rhombohedral structure, for particles assembled at an octane–water interface, and the particle monolayers fold and corrugate when the pressure is increased beyond a critical pressure equal to the bare interfacial tension.[55]

The wettability of the particles, as quantified by the contact angle, can also have a profound effect on the formation and stability of ordered particle monolayers at an oil–water interface.[54] Figure 2.14 shows that very hydrophobic silica particles, with a contact angle measured into water of above 130°, resulted in well-ordered monolayers with large separation distances, while less hydrophobic particles form large aggregates. Horozov *et al.* explained this striking effect by assuming that the Coulombic repulsion of particles through the oil phase increases with the hydrophobicity of the particles.[54]

2.4.4 2-D colloidal foams and cellular systems

There have been a number of papers reporting the formation of complex colloidal structures, *e.g.* foams, cellular structures, ordered aggregates, striations, loops and voids, from particles spread at the air–water surface.[56–60] The spontaneous formation

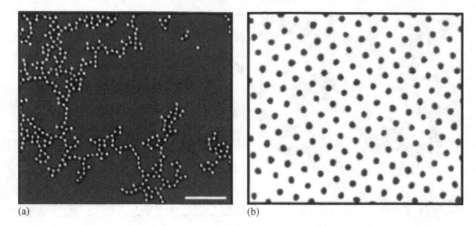

(a)   (b)

Figure 2.14 Monolayers of silica particles of diameter 3 μm with contact angles (through water) of (a) 118° and (b) 148° at the octane–water interface. The scale bar represents 25 μm. Taken from Ref. [54]; with permission of the American Chemical Society.

of these mesostructures at low surface coverage and the observations of the stability and dynamics of ordered colloidal monolayers with particles separated by distances up to several particle diameters suggest that the particles also may experience a long-range attraction in addition to the long-range repulsion discussed earlier. Hansen *et al.* reported that unconfined temporal ordered domains formed when hydrophobised silica particles had been spread from an ethanol dispersion onto the surface of a organic liquid.[22] Nikolaides *et al.* showed that unconfined small colloidal crystallites that are stable for an extended period of time can form when like-charged particles are spread at an oil–water interface.[61] They suggested that the fluid interface around small charged particles is deformed and that an electric-field-induced capillary force can induce a long-range attraction of sufficient strength to stabilise particles at large separation distances.[61,62] Stamou *et al.* however introduced an attractive capillary force that originates from a non-uniform wetting of the particles causing an irregular shape of the meniscus.[59] This is discussed in more detail in Chapter 3.

Recently, Hidalgo-Alvarez and co-workers questioned these explanations and showed in a series of experiments that the appearance of the mesostructures shown in Figure 2.15 could be explained by contamination of the air–liquid surface by silicone oil, which stems from the coating of needles and syringes used to deposit and spread the particle solutions at the air–water surface.[60] The early experiments, corresponding to a negligible release of silicone oil, resulted in the stable featureless particle monolayers, Figure 2.15(b), while repeated use of the syringe and needle resulted in the spontaneous formation of mesostructures, Figure 2.15(a). Control experiments with

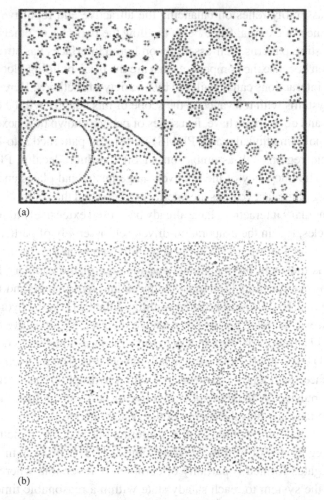

(a)

(b)

Figure 2.15 Different morphologies of colloidal monolayers made from 0.6 μm particles spread at an air–water surface from a dilute dispersion in a syringe: (a) mesostructured colloidal phases; (b) stable featureless colloidal monolayer. Taken from Ref. [60]; with permission of the American Chemical Society.

varying amounts of silicone oil at the air–water surface resulted in a rich plethora of structures, including voids, loops, striations and foam structures.

2.5 Directed Self-assembly at Liquid Interfaces

The need to find viable routes for the production of multi-functional nanomaterials with sensing, self-healing and "smart" properties has spurred an increasing interest in using the self-assembly of particles to direct the formation of different structures.[63]

Directed self-assembly relies on balancing the attractive and repulsive interactions with the influence of external force fields to allow the particles or other components to attain interesting patterns. Self-assembly on liquid interfaces has attracted a great deal of attention because the components that are the basic building blocks can move freely and the interactions can be tailored by a variety of methods. Here a few recent examples of systems with relevance to the scope of this chapter will be described.

Whitesides and co-workers have, in a series of papers, shown how hexagonal polymeric plates (polydimethylsiloxane, PDMS) with faces patterned into hydrophobic and hydrophilic regions self-assemble on an oil (perfluorodecalin, PFD) – water interface.[64-66] The buoyancy of the relatively large hexagonal plates and/or the non-uniform wetting of the faces causes the interface to bend and thus induces a lateral capillary force. Capillary interactions have already been used extensively to self-assemble colloidal particles, *e.g.* in the evaporation-driven self-assembly of particles from drying a dispersion on a substrate.[3] What is new in the work on the functionalised hexagonal plates is that both the magnitude and the sign of the capillary force can be manipulated by the functionalisation of the faces into hydrophilic and hydrophobic regions.[64] Figure 2.16 illustrates how the wetting of the PFD on the hydrophobic and hydrophilic faces can create positive or negative menisci. Two positive or two negative menisci will attract each other and try to force the faces together as decreasing the area of the PFD–H_2O interface is energetically favourable. The interaction between a hydrophobic face, with a positive meniscus, with a hydrophilic face with a negative meniscus, is repulsive. When these objects move toward one another, the area of the PFD–H_2O interface and thus the energy of the system increases.

It was shown that tuning of the capillary interactions provides a method of forming ordered aggregates of millimetre-sized objects.[64] The process in which these ordered aggregates formed is partially or completely reversible where agitation is used to force the system to reach steady state within a reasonable time. The magnitude of the interactions depended on the position of the hexagons relative to the mean plane of the interface. Figure 2.17 shows how the density difference between the PDMS hexagons and the liquids controls the magnitude of the capillary attraction and results in different patterns of the arrays. More complex patterns were obtained when the faces of the hexagons were functionalised in combinations that create a large number of possible assemblies with varying energy.[64-66]

It has been shown that calcium carbonate crystals precipitated from a calcium hydroxide solution may assemble and self-organise at the air–liquid surface.[67] The nucleation and growth of the calcite crystallites appears to result in the formation of hydrophobic facets that trap the particles at the air–water surface. Small crystallites aggregate into a fractal gel while larger crystals precipitated from aged solutions that organise into a lattice with an average separation distance exceeding 100 μm. Wickman and Korley attributed the self-organisation of the precipitated

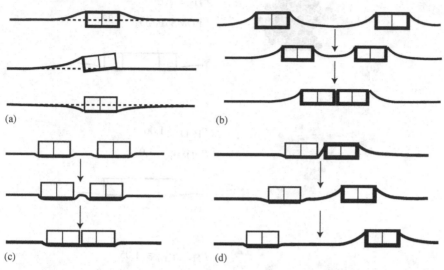

Figure 2.16 (a) Hexagons are pulled into the interface by the capillary forces on the hydrophobic faces. Dashed lines indicate the level of the oil (PFD, lower phase)-H_2O (upper phase) interface far from the objects. Hexagons where these forces are balanced float evenly at the interface; hexagons with an unbalanced distribution of forces are tilted at the interface. (b), (c) and (d) Three types of capillary interactions between hydrophobic and hydrophilic objects at the oil–H_2O interface. In (b), the oil wets the hydrophobic faces and forms positive menisci which pulls the hexagons together. In (c), the objects are hydrophilic and sink slightly into the oil–H_2O interface; this sinking creates a small negative meniscus which creates a weak attraction. In (d), a hydrophobic face with a positive meniscus is repelled by a hydrophilic face with a negative meniscus. Thick lines and thin lines indicate hydrophobic and hydrophilic faces, respectively. Taken from Ref. [64]; with permission of the American Chemical Society.

large crystals to a balance of attractive capillary forces, originating from the bending of the liquid interface by the weight of the large crystals, and repulsive electrostatic forces stemming from a charge at the surface of the particles exposed to air.[67]

Whitesides, Grzybowski and co-workers have used a system composed of millimetre-sized magnetised disks floating on a liquid interface to study the dynamics of the self-assembly process when the system is subjected to a rotating external magnetic field.[68–70] Figure 2.18 outlines the system where a permanent bar magnet rotates below a dish filled with a 3:1 by volume solution of ethylene glycol (EG) and water. The millimetre-sized polymeric disks (50–100 μm thick) are made of epoxy resin doped with 10–25% by weight of magnetite. Under the influence of the magnetic field produced by the rotating magnet, all of the disks floating on the EG/water–air surface experience a centro-symmetric force F_m directed towards the axis of rotation of the magnet. As the magnetic moments of the disks interact with

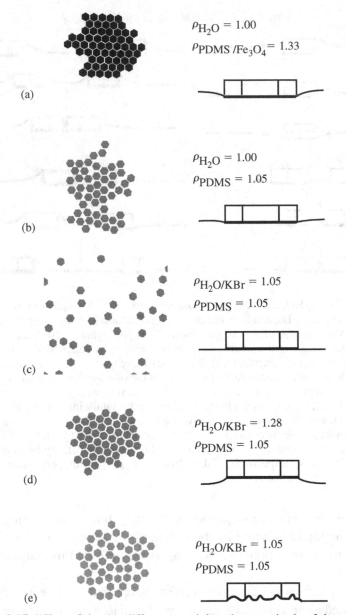

$\rho_{H_2O} = 1.00$

$\rho_{PDMS}/Fe_3O_4 = 1.33$

(a)

$\rho_{H_2O} = 1.00$

$\rho_{PDMS} = 1.05$

(b)

$\rho_{H_2O/KBr} = 1.05$

$\rho_{PDMS} = 1.05$

(c)

$\rho_{H_2O/KBr} = 1.28$

$\rho_{PDMS} = 1.05$

(d)

$\rho_{H_2O/KBr} = 1.05$

$\rho_{PDMS} = 1.05$

(e)

Figure 2.17 Effect of density difference and thus the magnitude of the attractive capillary forces on the patterns self-assembling hexagons form at PFD–H$_2$O interfaces. In (a), the hexagons sink into the PFD–H$_2$O interface and form a close-packed array. (b) Hexagons are only slightly denser than the water and barely perturb the interface, resulting in weak aggregates. (c) Densities of the PDMS and water are matched and the objects float on the interface without menisci. (d) PDMS is less dense than the water, and a positive meniscus forms at the interface. (e) Hexagons with a rough intersection between the bottom face and the side faces; micro-menisci form that cause the hexagons to assemble even when $\rho_{PDMS} = \rho_{H2O}$ (all ρ in g cm^{-3}). The density of PFD is 1.91 g cm^{-3}. Taken from Ref. [64]; with permission of the American Chemical Society.

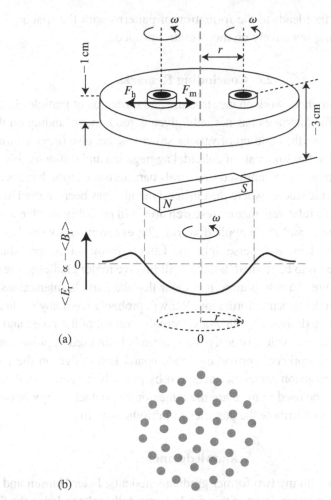

Figure 2.18 (a) Illustrates the experimental arrangement used in the dynamic self-assembly of rotating magnetic disks floating on a liquid interface. The magnetic force F_m attracts the disks towards the axis of rotation of the magnet, and the vortices these disks create in a surrounding liquid give rise to repulsive, pair-wise hydrodynamic forces F_h between them. The graph below the scheme has the profile of the average radial component of the magnetic induction, proportional to the energy of the magnetic field, in the plane of the interface. The photograph in (b) shows an aggregate formed by 37 rotating disks, 1.57 mm in diameter. Taken from Ref. [70]; with permission of Elsevier.

the magnetic moment of the external rotating magnet, the disks spin around their axes at an angular velocity equal to that of the magnet. The fluid motion associated with spinning results in repulsive hydrodynamic interactions F_h between the disks. The interplay of the attractive and repulsive interactions between the disks floating on the same interface leads to the formation of patterns with the spacing between the disks depending sensitively on their rotational speed.

2.6 Concluding Remarks

The substantial amount of work on the structure and formation of particle clusters and monolayers at liquid interfaces has not only given a good understanding on the controlling parameters for the build-up of various structures, but also provided important new insights into the fundamentals of colloidal aggregation and self-assembly. Important development of new and improved structural characterisation tools have been combined with systematic studies where the cluster dynamics has been studied in detail.

The future of this relatively mature research area will probably involve a combination of fundamental and more applied work. The experimental work has almost exclusively involved monodisperse spheres. Effects from anisotropic shape and multi-modality needs to be studied, but this will involve major challenges regarding quantitative structure characterisation. It is clear that the particle interactions have a profound effect on the structure formation. We will probably see many studies where colloidal systems are designed to allow a systematic control of the range and magnitude of the interactions. Optical tweezers have already been used in pioneering work by Aveyard and co-workers,[53] providing fundamental knowledge on the nature of long-range pair interaction sometimes acquired by partially immersed colloids. This method could also be used to measure the adhesion or contact energy between the particles and also to distribute the particles in periodic patterns.

Acknowledgements

The collaboration with my two former graduate students, Peter Hansen and Sandra Rödner, has been an integral part of this work. I gratefully acknowledge the Swedish Science Council (Vetenskapsrådet) and the Foundation for Strategic Research (SSF) for funding.

References

1. R.G. Larson, *The Structure and Rheology of Complex Fluids*, Oxford University Press, New York, 1999.
2. W.B. Russel, D.A. Saville and W.R. Schowalter, *Colloidal Dispersions*, Cambridge University Press, Cambridge, 1991.
3. P.A. Kralchevsky and K. Nagayama, *Particles at Fluid Interfaces and Membranes*, Elsevier, Amsterdam, 2001.
4. R.J. Pugh and L. Bergström (eds.), *Surface and Colloid Chemistry in Advanced Ceramics Processing*, Marcel Dekker, New York, 1994.
5. Special issue – From dynamics to devices: directed self-assembly of colloidal materials, *MRS Bulletin*, **23** (1998).

6. C.M. Soukoulis, *Photonic Crystals and Light Localisation in the 21st Century*, Kluwer Academic Publishers, Dordrecht, 2001.
7. P.A. Rundquist, P. Photinos, S. Jagannathan and S.A. Asher, *J. Chem. Phys.*, **91** (1989), 4932.
8. Special issue – Materials science aspects of photonic crystals, *MRS Bulletin*, **26** (2001).
9. Y. Xia, B. Gates and Z.Y. Li, *Adv. Mater.*, **13** (2001), 409.
10. B.B. Mandelbrot, *The Fractal Geometry of Nature*, W.H. Freema and Co., New York, 1983.
11. J. Feder, *Fractals*, Plenum Press, New York, 1988.
12. M.T.A. Bos and J.H.J. van Opheusden, *Phys. Rev. E*, **53** (1996), 5044.
13. T.A. Witten and L.-M. Sander, *Phys. Rev. Lett.*, **47** (1981), 1400.
14. F. Family and D.P. Landau (eds.), *Kinetics of Aggregation and Gelation*, Elsevier, Amsterdam, 1984.
15. A. Bunde and S. Havlin (eds.), *Fractals and Disordered Systems*, Springer, Berlin, 1991.
16. A.S. Dimitrov and K. Nagayama, *Langmuir*, **12** (1996), 1303.
17. C.D. Dushkin, G.S. Lazarov, S.N. Kotsev, H. Yoshimura and K. Nagayama, *Colloid Polym. Sci.*, **277** (1999), 914.
18. M. Semmler, E.K. Mann, J. Ricka and M. Borkovec, *Langmuir*, **14** (1998), 5127.
19. S.H. Park and Y. Xia, *Langmuir*, **15** (1999), 266.
20. K. Zahn and G. Maret, *Curr. Opin. Colloid Interf. Sci.*, **4** (1999), 60.
21. D.G. Grier and C.A. Murray, in *Ordering and Phase Transitions in Charged Colloids*, eds. A.K. Arora and B.V.R. Tata, VCH, New York, 1996, p. 69.
22. P.H.F. Hansen, S. Rödner and L. Bergström, *Langmuir*, **17** (2001), 4867.
23. A.H. Marcus and S.A. Rice, *Phys. Rev. E*, **55** (1997), 637.
24. S. Sastry, P.G. Debenedetti and F.H. Stillinger, *Phys. Rev. E*, **56** (1997), 5533.
25. P. Meakin, *Phys. Rev. Lett.*, **51** (1983), 1119.
26. M. Kolb, R. Botet and R. Jullien, *Phys. Rev. Lett.*, **51** (1983), 1123.
27. D.A. Weitz and J.S. Huang, in *Kinetics of Aggregation and Gelation*, eds. F. Family and D.P. Landau, Elsevier, Amsterdam, 1984, p. 19.
28. D.A. Weitz, J.S. Huang, M.Y. Lin and J. Sung, *Phys. Rev. Lett.*, **54** (1985), 1416.
29. M. von Smoluchowski, *Phys. Zeit.*, **17** (1916), 593.
30. M. von Smoluchowski, *Zeit. Phys. Chem.*, **92** (1917), 129.
31. A. Elaissari and E. Pefferkorn, *J. Colloid Interf. Sci.*, **143** (1991), 343.
32. R. Jullien and R. Botet, *Aggregation and Fractal Aggregates*, World Scientific, Singapore, 1987.
33. T. Vicsek and F. Family, *Phys. Rev. Lett.*, **52** (1984), 1669.
34. G.Y. Onoda, *Phys. Rev. Lett.*, **55** (1985), 226.
35. A.J. Hurd, and D.W. Schaefer, *Phys. Rev. Lett.*, **54** (1985), 1043.
36. J.C. Earnshaw and D.J. Robinson, *Prog. Colloid Polym. Sci.*, **79** (1989), 162.
37. D.J. Robinson and J.C. Earnshaw, *Phys. Rev. A*, **46** (1992), 2045.
38. D.J. Robinson and J.C. Earnshaw, *Phys. Rev. A*, **46** (1992), 2055.
39. D.J. Robinson and J.C. Earnshaw, *Phys. Rev. A*, **46** (1992), 2065.
40. J. Stankiewicz, M.A. Cabrerizo-Vílchez and R. Hidalgo-Alvarez, *Phys. Rev. E*, **47** (1993), 2663.
41. M. Kondo, K. Shinozaki, L. Bergström and N. Mizutani, *Langmuir*, **11** (1995), 394.
42. D.F. Williams and J.C. Berg, *J. Colloid Interf. Sci.*, **152** (1992), 218.
43. J.C. Earnshaw and D.J. Robinson, *Langmuir*, **9** (1993), 1436.
44. A.T. Skjeltorp, *Phys. Rev. Lett.*, **58** (1987), 1444.
45. P.H.F. Hansen and L. Bergström, *J. Colloid Interf. Sci.*, **218** (1999), 77.

46. A. Moncho-Jorda, F. Martinez-Lopez and R. Hidalgo-Alvarez, *J. Colloid Interf. Sci.*, **249** (2002), 405.
47. A. Moncho-Jorda, F. Martinez-Lopez, A.E. Gonzalez and R. Hidalgo-Alvarez, *Langmuir*, **18** (2002), 9183.
48. N.D. Denkov, O.D. Velev, P.A. Kralchevsky, I.B. Ivanov, H. Yoshimura and K. Nagayama, *Langmuir*, **8** (1992), 3183.
49. J.J. Guo and J.A. Lewis, *J. Am. Ceram. Soc.*, **82** (1999), 2345.
50. S. Rödner, P. Wedin and L. Bergström, *Langmuir*, **18** (2002), 9327.
51. P. Pieranski, *Phys. Rev. Lett.*, **45** (1980), 569.
52. R. Aveyard, J.H. Clint, D. Nees and V.N. Paunov, *Langmuir*, **16** (2000), 1969.
53. R. Aveyard, B.P. Binks, J.H. Clint, P.D.I. Fletcher, T.S. Horozov, B. Neumann, V.N. Paunov, J. Annesley, S.W. Botchway, D. Nees, A.W. Parker, A.D. Ward and A.N. Burgess, *Phys. Rev. Lett.*, **88** (2002), 246102.
54. T.S. Horozov, R. Aveyard, J.H. Clint and B.P. Binks, *Langmuir*, **19** (2003), 2822.
55. R. Aveyard, J.H. Clint, D. Nees and N. Quirke, *Langmuir*, **16** (2000), 8820.
56. J. Ruiz-Garcia and B.I. Ivlev, *Mol. Phys.*, **5** (1998), 371.
57. J. Ruiz-Garcia, R. Gamez-Corrales and B.I. Ivlev, *Phys. Rev. E*, **58** (1998), 660.
58. F. Ghezzi, J.C. Earnshaw, M. Finnis and M. McCluney, *J. Colloid Interf. Sci.*, **238** (2001), 433.
59. D. Stamou, C. Duschl and D. Johannsmann, *Phys. Rev. E*, **62** (2000), 5263.
60. J.C. Fernandez-Toledano, A. Moncho-Jorda, F. Martinez-Lopez and R. Hidalgo-Alvarez, *Langmuir*, **20** (2004), 6977.
61. M.G. Nikolaides, A.R. Bausch, M.F. Hsu, A.D. Dinsmore, M.P. Brenner, C. Gay and D.A. Weitz, *Nature*, **420** (2002), 299.
62. K.D. Danov, P.A. Kralchevsky and M.P. Boneva, *Langmuir*, **20** (2004), 6139.
63. G.M. Whitesides and B. Grzybowski, *Science*, **295** (2002), 2418.
64. N. Bowden, I.S. Choi, B.A. Grzybowski and G.M. Whitesides, *J. Am. Chem. Soc.*, **121** (1999), 5373.
65. N. Bowden, S.R.J. Oliver and G.M. Whitesides, *J. Phys. Chem. B*, **104** (2000), 2714.
66. N. Bowden, F. Arias, T. Deng and G.M. Whitesides, *Langmuir*, **17** (2001), 1757.
67. H.H. Wickman and J.N. Korley, *Nature*, **393** (1998), 445.
68. B.A. Grzybowski, H.A. Stone and G.M. Whitesides, *Nature*, **405** (2000), 1033.
69. B.A. Grzybowski, H.A. Stone and G. M. Whitesides, *Proc. Nat. Acad. Sci.*, **99** (2002), 4147.
70. B.A. Grzybowski and C.J. Campbell, *Chem. Eng. Sci.*, **59** (2004), 1667.

Lennart Bergström is Professor in Materials Chemistry in the Department of Physical, Inorganic and Structural Chemistry at Stockholm University. He was previously manager of the Materials and Coatings section at the Institute for Surface Chemistry, YKI, and has also held a position as director of the Brinell Centre: Inorganic Interfacial Engineering at the Royal Institute of Technology in Stockholm. His current research interests include the synthesis, functionalisation and assembly of particulate materials and the development of novel processing methods for ceramic and functional materials. He received an M.Sc. in Chemical Engineering (1984) and a Ph.D. in Physical Chemistry (1992) at the Royal Institute of Technology, KTH, in Stockholm. He has been a visiting scientist at the University of Washington, USA, and at the Tokyo Institute of Technology, Japan. He is the author of more than 90 papers and holds 4 patents.

Tel. +46-8-162368, *Fax*: +46-8-152187, *E-mail*: lennartb@inorg.su.se

3

Theory for Interactions between Particles in Monolayers

Juan C. Fernández-Toledano, Arturo Moncho-Jordá,
Francisco Martínez-López and Roque Hidalgo-Álvarez

*Grupo de Fisica de Fluidos y Biocoloides,
Departamento de Fisica Aplicada, Universidad de Granada,
Campus de Fuente Nueva 18071, Granada, Spain*

3.1 Introduction

Particle monolayers are formed when small colloidal solid particles adsorb at liquid–vapour or liquid–liquid interfaces. Typical examples are latex monolayers at the air–aqueous salt solution[1] and oil–water interfaces.[2,3] The interaction between particles within the monolayer is dependent on both the properties of the fluids that make up the interface and on the nature of the adsorbed particles. Therefore, a detailed analysis of the interactions in colloidal monolayers is quite complex and distinctions must be made to take into account the different components of the monolayer.

The total interaction between particles in the monolayer determines their stability behaviour. Thus, examples of stable monolayer systems with particles that remain independent for a long time have been reported,[4] in spite of the fact that in a thermodynamic sense, colloidal particles are not stable because of their great surface to volume ratio. Some monolayer systems showed a triangular structure suggesting the existence of long-ranged particle interactions. In other reported systems, however, it was found that particles are unstable and aggregate to form fractal structures[5] or even became organized to form the so-called mesostructures.[6,7] When fractal structures appear, the particle interaction potential is short ranged and has a minimum at very short distances. In the other cases, the formation of mesostructures can be explained if the interaction energy between the particles has a minimum at a typical average distance between particles of the order of a few times the particle diameter.[8]

The different colloidal stability behaviour shown experimentally by particle monolayers has fundamental importance in a wide set of industrial applications. Typical examples are the manufacture of emulsion polymers in stirred-tank reactors[9] and separation processes such as froth and solvent extraction. In these cases, it is found that 2-D colloidal aggregation can occur with different consequences. When manufacturing emulsion polymers, it is important to prevent aggregation because it has a negative effect on the characteristics of the synthesized particles, increasing the particle polydispersity. However, in separation processes, the aggregation of the colloidal particles is accelerated by the re-dispersion of aggregates formed at the bubble surfaces. This foaming action[10] of colloidal particles at the bubble surfaces is dependent on the degree of surface aggregation[11] and finally on the interfacial properties of the particles.

Thus, a correct description of interactions between colloidal particles in monolayers is of great interest and enables us to understand these kinds of processes and to improve their use for industrial applications. In this chapter, a detailed discussion of the theory of interactions in colloidal monolayers is presented. The theory considers Derjaguin–Landau–Verwey–Overbeek (DLVO), capillary, hydrophobic, monopolar and dipolar interactions between the particles. The contribution of these terms to the total interaction energy has been computed numerically using typical values for most of the parameters included in the theory.

3.2 Theoretical Model

As stated in the introduction, the interaction between particles forming monolayers is dependent on the particle characteristics and on the properties of the fluids defining the interface where colloidal particles are trapped. But also, the whole procedure of particle accommodation at the interface must be taken into account because this can affect the properties of the interfacial particles, as could their charge and type of surface groups. This fact might be important in explaining the different stability behaviour that has been observed experimentally for latex particles.[12,13] However, until now, it was never treated in scientific works. Here, we use a model that only considers particle and fluid properties to compute the total interaction between particles. It neglects history effects as previously mentioned.

3.2.1 Model of the colloidal particles

The shape, size, chemical composition and internal and surface structural properties of the particles must be given in order to have a complete model of them. Although it is not infrequent to find works reporting experimental results using ellipsoidal particles, here we only consider spherical particles with radius a. This does not represent an important limitation as most applications make use of spherical particles since they are easily synthesized.[14]

Due to their surface properties, colloidal particles are strongly trapped at an interface due to capillary and electrostatic potentials.[15] Thus, the contact angle θ is the parameter that determines the position of the particles at the interface, *i.e.* the immersed fraction of the particles. Its value is given by Young's law equation

$$\cos \theta = \frac{\gamma_2 - \gamma_1}{\gamma} \tag{3.1}$$

Here, γ is the interfacial tension between the fluid phases, and γ_1 and γ_2 are the interfacial energies between the colloidal particle and phases 1 and 2, respectively. Usually, ionic surface groups of colloidal particles can dissociate when they are in polar media and then produce dipoles or even monopoles that give rise to long-ranged interactions. The formed dipoles originate, at least in part, from the counterions in the polar phase and they can become exposed to the non-polar phase because of thermal fluctuations that are able to rotate the colloidal particles. For example, this is the case of colloidal latex particles at the air–water surface. As the dipole number depends on the number of ionic surface groups on the particles, it is expected that their maximum amount is not directly dependent on the ionic strength of the polar phase. The monopoles, like the dipoles, originate from the dissociated surface groups of the particles and they can be exposed to the non-polar phase across the interface; however, it is not so clear how they can remain stable without attaching counter charges as some experimental works report.[16]

Internal properties of the colloidal particles, *i.e.* chemical composition and structure, affect their interaction both in dispersions and at the interface when they form monolayers. The chemical composition and structure of the particles determine the significance of the dispersion forces and even the existence of a dipolar type interaction, as might be the case for magnetic particles. There is also the possibility that particles do not have a smooth surface but a surface that can be structured or rough. This fact can affect the particle arrangement at the interface and the inter-particle interactions and consequently the colloidal stability. Despite the number of possible choices, the particle model considers spherical particles with radius a, contact angle θ, with a uniform chemical composition and carrying some dipoles and monopoles on their surfaces.

3.2.2 The model of the fluid phases

The simplest model for the fluid phases assumes a uniform continuous polar or non-polar medium characterized by some physical average properties like the mass density, ρ and the dielectric constant, ε.

3.3 The Interaction Energy

The calculation of the total pair energy is quite difficult and it is mainly determined by the particle accommodation at the interface. Thermodynamics limits the form that the interface makes with the particle surface, in such a way that the interface near the particle surface forms a curved meniscus contacting the surface of the particle at a well-defined angle, the contact angle θ; θ depends on the properties of the particle surface and on the nature of the fluids that form the interface as indicated by Young's equation. Nevertheless, for colloidal particles whose size ranges from a few nanometres to some micrometres, the meniscus can be considered approximately flat. This is true because the Bond number ($=\Delta\rho g a^2/\gamma$) for these systems is very small,[17] usually of the order of 10^{-8} for colloidal particles of many hundred nanometres radius at the air–water surface. Here $\Delta\rho$ is the difference of densities between the two fluid phases and g is the gravity constant. The flat meniscus approximation helps us to simplify the computation of the immersed part of the particle. It also implies that capillary interaction is practically negligible for particles of colloidal size.

3.3.1 Terms of the interaction energy

There are different terms in the interaction energy that must be taken into account for interfacial particles in order to explain their stability properties. Usually, the DLVO theory[18,19] allows studying the stability behaviour of colloidal particles in bulk (3-D stability). However, this is not the case for colloidal particles at an interface (2-D stability). Furthermore, in two dimensions there are new elements like capillary attraction, intrinsic to interfacial phenomena, which have no analogy in bulk aggregation. Besides, the DLVO expressions for the double layer and dispersion interactions have to be corrected in order to account for the presence of the interface that reduces the degrees of freedom for the movement of the colloidal particles.

The different terms of the inter-particle interaction energy that we will consider here are: double layer and dispersion interactions (2-D DLVO), capillary interaction, hydrophobic interaction and monopolar and dipolar interactions. These terms have a different order of contribution to the total interaction energy, which depends on the specific values of the parameters included in the theoretical model. Here, typical values from the literature have been assigned to the parameters in order to account for most of the experimentally reported situations and to have as correct as possible weighting of the different energetic contributions.

3.3.2 Considerations and approximations in the calculation of the interaction energy

The numerical results presented here have been obtained using different approximations and considerations. In order to allow the reader to assess clearly the limitations and applicability of the results, they are indicated below. First, the addition of the different terms of the interaction energy has been assumed so that the total energy of interaction is the sum of all the contribution terms previously mentioned. Some interaction terms have been computed using the approximation of a flat meniscus as for colloidal particles where the Bond number is very small. This is the case of the dispersion and double-layer interactions. The flat meniscus approximation does not imply that capillary interaction is neglected in the global analysis of the interactions but that the meniscus deformation does not change considerably other interaction terms so there is no need to modify their computation procedure.

For the sake of simplicity, some of the interaction terms between particles at the interface have been computed with the assumption that interaction occurs only between their emergent parts and between their immersed parts, respectively, and that the interaction between the emergent part of one particle and the immersed part of another one can be neglected. This is also the case of the dispersion and double layer interaction (DLVO interaction) and of the dipolar and monopolar interactions.

As the model considers spherical colloidal particles, the computation of their potential of interaction, $V_{sph-sph}$, can be done in some cases using the Derjaguin approximation[20]

$$V_{sph-sph}(h_o) = \int V_{flat}(h)\, dS(h) \qquad (3.2)$$

which makes use of the corresponding interaction potential per unit area between infinite half spaces, V_{flat}. The integration must be done over the particle surface. Here h_o is the minimum distance between the particles and h is the local distance between the different surface elements. The Derjaguin approximation is useful to estimate the total interaction when the range of interaction is small compared to the radius of curvature of the particle.

To compute the pair interaction potential by using the Derjaguin approximation presents two main problems. The first one is related to the determination of the particle surface fraction above and below the interface which must be taken into account to calculate the interaction. The second one relates to the specific expression of the interaction potential per unit area between half spaces, V_{flat}, that depends on the kind of the interaction. The first problem is solved with the help of the flat meniscus approximation. With this assumption, the calculation of the particle surface fraction becomes easier. Figure 3.1 sketches the variables used to calculate both emergent and immersed parts of the particle. The relative amount of both parts depends on the value of the contact angle θ. Indeed, the immersed part increases as the contact angle decreases. For $\theta > \pi/2$ (90°) the emergent part of the particle is larger than the immersed one (a) while for $\theta < \pi/2$ the inverse occurs (b). This dependence with the contact angle will have a great importance on the value of the different terms of the interaction.

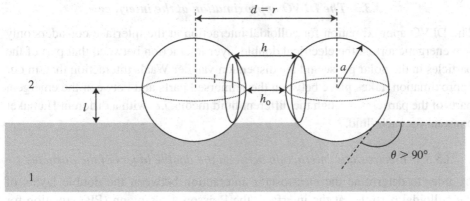

(a)

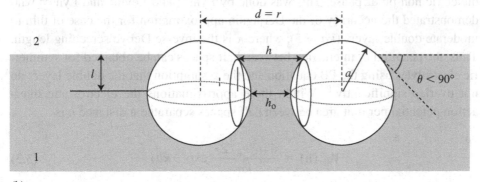

(b)

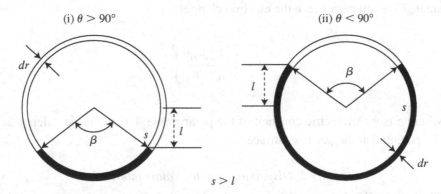

(c)

Figure 3.1 Sketch of the colloidal particle arrangement at the interface between phases 1 and 2 for (a) $\theta > 90°$ and (b) $\theta < 90°$. The particle surface fraction included into each phase depends on the particle hydrophobicity, *i.e.* the contact angle θ. The immersed part is indicated in (c) by a thick black line for $s > l$.

3.3.3 *The DLVO approximation at the interface*

The DLVO approximation for colloidal interaction at the interface considers only two energetic terms: the electrical double layer interaction between that part of the particles in the polar phase and the dispersion van der Waals interaction that, in our approximation, takes place between the immersed parts and between the emergent parts of the particles through the different fluid media, *i.e.* with a different Hamaker constant for each fluid.

3.3.3.1 *Electrostatic interaction between the double layers of the particles*

In order to determine the electrostatic interaction between the double layers of the colloidal particles at the interface, the Poisson–Boltzmann (PB) equation for the polar phase should be solved simultaneously with the Laplace equation in the dielectric non-polar phase. This was done by Lyne[21] and Levine and Lyne[22] who demonstrated the accuracy of the Derjaguin approximation for the case of thin to moderate double layers ($\kappa a > 5$), where κ is the inverse Debye screening length. Thus, the potential of interaction between half spaces can be obtained for symmetric electrolytes using the PB equation and the assumption that the double layers do not overlap significantly.[19] Within these approximations the electrostatic interaction potential per unit area between half spaces separated a distance h is

$$V_{\text{flat}}(h) = \frac{64 n^0 k_B T Z^2}{\kappa a} \exp(-\kappa h) \tag{3.3}$$

where ($Z = \tanh(v e \varphi_0 / 4 k_B T)$), n^0 is the salt concentration, k_B the Boltzmann constant, T the temperature, e the electron charge.

$$\kappa = \left(\frac{e^2 \, 2 v^2 n^0}{\varepsilon k_B T} \right)^{1/2},$$

where ε is the dielectric constant of the polar phase, v is the ionic valence and φ_0 the potential at the particle surface.

3.3.3.2 *Dispersion van der Waals interaction*

The van der Waals interaction is completely understood under quantum mechanics formalism. It originates from the fluctuations of the electron clouds around the atomic nucleus. These fluctuations produce temporary dipoles that are able to induce the formation of new dipoles in the neighbouring atoms, giving rise to an attraction between them. To calculate the van der Waals interaction energy we have used the Derjaguin approximation. The expression for the interaction potential per unit area between half spaces separated a distance h was obtained from Gregory[23] and Overbeek[24]

$$V_{\text{flat}}(h) = -\frac{A}{12\pi h^2} \frac{1}{1 + bh/\lambda}$$

(3.4)

It takes into account the retardation effect due to the limit on the speed of light. In this expression, the constants b and λ have values 5.32 and 100 nm, respectively. Furthermore, the Hamaker constant A was assigned different values for both fluid phases. For the case of polystyrene particles at the air–water surface the values of the Hamaker constants were taken to be $A_{\text{air}} = 6.6 \times 10^{-20}$ J and $A_{\text{water}} = 0.95 \times 10^{-20}$ J.

3.3.4 The hydrophobic interaction

The interaction of the colloidal particles with the surrounding fluid molecules is the origin of the so-called structural forces. This interaction affects the structure of the fluid near the particle surface, giving rise to an increase or decrease of its order as compared with the fluid structure far away from the particles surface. Therefore, structural forces can be regarded as entropic interactions. The character of this interaction allows solid particles to be classified as lyophilic or lyophobic depending on whether they have or do not have affinity for the liquid in which they are dispersed, respectively. When the fluid is water, the colloidal particles are designated as hydrophilic or hydrophobic.

Hydrophobic particles repel water from their surface so when dispersed in water they experience an attraction. However, the interaction of hydrophilic particles with the fluid produces a layered arrangement of the nearest fluid molecules that prevent the approach of the colloidal hydrophilic particles, *i.e.* they feel a net repulsion. Besides the dependence of the structural forces on the chemical properties of the particles surface and the fluid, they are also affected by the surface roughness[25] and the wetting degree of the particles. For example, colloidal particles at the air–water surface that are partially wetted will experience less interaction than fully wetted ones. As our model for the particles considers that they have smooth surfaces, the effect of roughness will not be explicitly taken into account, although the effect of the wetting degree is considered through the contact angle.

Experimentally, the hydrophobic attraction is detectable when the contact angle is greater than 64°,[26] while the hydrophilic repulsion is important when the contact angle is lower than 15°.[27] For contact angle values ranging from 15° to 64° the hydrophobic interaction is negligible and the DLVO theory is sufficient to explain the colloidal stability in solution. Christenson and Claesson[28] have shown that the

Table 3.1 *Typical parameter values for the hydrophobic interaction potential between half spaces. Taken from Ref. [29]; with permission of Academic Press.*

Parameter	Hydrophilic particles	Hydrophobic particles
λ_o (nm)	0.6–1.1	1–2
W_o (mJ m^{-2})	3–30	−20 to −100

hydrophobic interaction decays exponentially for plane surfaces. Thus, because of the short range of this interaction, the Derjaguin approximation is useful to estimate the hydrophobic interaction between two partially immersed particles.

The expression used for the hydrophobic interaction potential per unit area between half spaces separated by a distance h is

$$V(h) = W_o \exp(-h/\lambda_o) \qquad (3.5)$$

where W_o and λ_o are constants related to the strength and range of the interaction. Typical values for these parameters obtained from Ref. [29] are given in Table 3.1.

3.3.5 The monopolar and dipolar interactions

The proposed model for the particle at the interface includes the possibility of some kind of charges, monopoles and dipoles, in the non-polar fluid phase. The most important question now is to understand how these dipoles and residual charges can arise in this part of the particle surface. The origin of this charge and dipoles can be understood if one takes into account the rough character of the particles at a molecular level and the process of monolayer formation, *i.e.* the so-called spreading process.

Usually, particle monolayers are formed from a particle suspension using a spreading agent like methanol, which helps the formation of a uniform surface distribution of colloidal particles at the interface. The spreading process is rather turbulent because of the spreading agent effect that allows the colloidal particles to rotate and even penetrate into the sub-phase, *i.e.* the polar fluid phase that is most frequently water. This process is so vigorous that particles can trap traces of polar fluid at their surfaces around the hydrophilic surface charged groups.[2,13,30] Therefore, there is not a complete dewetting of the part of the particle lying above the water level. This water could be in the form of either small droplets or as a thin layer of water around the particle surface in the non-polar medium (air, oil, *etc.*). The presence of the polar liquid at this part of the particle surface can affect the

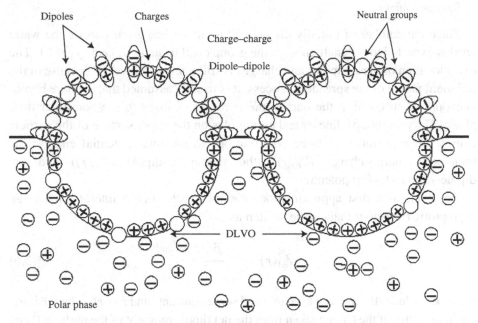

Figure 3.2 Illustration of the particle–particle interactions through the non-polar phase (charge–charge and dipole–dipole) and the polar phase (DLVO forces) for particles at a liquid interface.

configuration of their surface chemical groups, which can be found in the following three possible configurations in the non-polar phase (see Figure 3.2):

1. *Non-dissociated*: Most of the surface groups remain neutral and do not contribute to the interaction potential.
2. *Forming dipoles*: Some wetted surface groups, even though dissociated, can form dipoles together with the counterions from the polar phase. These dipoles are effectively screened in the polar phase as a cloud of ions is formed around the particle. However, this does not occur in the non-polar phase, as usually it has no dissolved salt. Hence, the dipoles at the surface of particles can give rise to a total dipole moment and to quite strong dipole–dipole repulsion between particles that is mediated through the non-polar phase. To account for the number of dissociated groups on the particle surface we define the fraction of dipoles, f_d, as the percentage of them forming dipoles in relation to the total surface groups above the polar liquid (water) level.
3. *Totally dissociated*: Only a few of the total surface groups acquire a net charge and form monopoles. Analogously to f_d, the fraction of monopoles, f_c, is defined as the percentage of dissociated groups forming monopoles. Since the non-polar phase has generally a low dielectric constant, these charges generate a monopole Coulombic

repulsive potential which is very intense and long ranged, even for very small monopole fraction values f_c.

Since the number of initially dissociated ionic surface groups above the water level is expected to be small, most of the groups will be neutral, *i.e.* $f_c, f_d \ll 1$. The exact location of these dipoles and charges is still unclear. However, because of the turbulent nature of the spreading process, it is usually assumed that they are homogeneously distributed at the part of the particles exposed to the non-polar fluid phase. The presence of dipoles and monopoles on the upper surface of the particle causes the appearance of three contributions to the total potential energy: the monopole charge–charge ($V_{cc}(r)$), the monopole–dipole ($V_{cd}(r)$) and the dipole–dipole ($V_{dd}(r)$) potentials.

However, in a first approximation, the monopole–dipole interaction has an asymptotic behaviour that can be written as

$$V_{cd}(r) = k' \frac{\vec{P} \cdot \hat{r}}{r^2} \tag{3.6}$$

where k' includes the monopole charge and some constants and $\hat{r} = \vec{r}/r$ where $\vec{r}$ is the position vector of the charge taken from the net dipole moment of the particle $\vec{P}$ and r is its module. If the particle has a homogeneous surface dipole distribution, $\vec{P}$ is always normal to the interface and also to the vector $\vec{r}$. Therefore, under this approximation, the monopole–dipole interaction potential term is zero and does not contribute to the total energy. Thus we will only consider the contribution of the dipole–dipole interaction, $V_{dd}(r)$, and monopole–monopole interaction, $V_{cc}(r)$.

It should be emphasized that the total surface area exposed to the non-aqueous phase depends on the contact angle (see equation (3.4)). Therefore, the dipole–dipole and charge–charge repulsions change if the particle hydrophobicity is varied. This effect has been experimentally observed in colloidal monolayers formed by polystyrene spheres trapped at the air–water surface, where adding small amounts of surfactant to the aqueous phase (around 10^{-5} M) leads to a reduction of the stability of the monolayer, inducing aggregation.[13] At the oil–water interface, the particle hydrophobicity has also a significant role in the spatial distribution of the colloidal particles.[31] Very hydrophobic particles with large contact angles (above 129°) lead to well-ordered hexagonal monolayers. On the contrary, the monolayers of less hydrophobic particles with contact angles below 115° are completely disordered or aggregated. Between 115° and 129°, a disorder–order coexistence region occurs. Of course, these intervals can be different if we change the non-polar medium or the surface density of charged groups in the colloidal particles.

The repulsion produced by the charge–charge and dipole–dipole pair interactions is in general very strong, and dominates over the other energetic contributions even for very small values of f_c and f_d. Consequently, the stability of colloidal monolayers for a given contact angle will be essentially ruled by the surface charge and surface dipole percentages f_c and f_d, respectively.

3.3.5.1 Dipolar electrostatic interaction between the emergent parts of the particles

To estimate numerically this dipolar interaction we have supposed that the part of the particle exposed to the non-polar phase has a uniform distribution of dipoles. As the dipoles are originated from the surface groups of the particles and the counterions from the solution, we can consider them as two unit charges with opposite sign separated by a distance, d_{dip}, equal to the sum of the radius of typical anions and cations, *i.e.* of the order of 0.3 nm. This parameter can be changed in the calculation but there is no apparent reason to do it.

To calculate the dipolar interaction energy the emergent surface has been divided into different parts with a surface dS and charge $f_d \sigma \, dS$, where σ is the surface charge density and f_d is the fraction of the surface groups forming dipoles, as was indicated above. At a distance d, equal to the dipole length from these surface elements, we have considered the existence of a charge of equal magnitude and opposite sign that forms part of the dipole. Therefore, the total charge on the emergent dipole is zero. The interaction between these pairs of charges for two particles allows the dipolar interaction energy to be determined. This calculation implies a 4-D integration given by

$$V(h_o) = \int d\theta_1 \int d\theta_2 \int d\phi_1 \int d\phi_2 V(\theta_1, \theta_2, \phi_1, \phi_2) \qquad (3.7)$$

where $V(\theta_1, \theta_2, \phi_1, \phi_2)$ depends on the particle radius a, the distance between particles h_o, the spherical coordinates θ_1, θ_2, ϕ_1 and ϕ_2, taken with origin at each particle, the surface charge density σ, the dipole fraction f_d and the dipole length d_{dip}. The expression for $V(\theta_1, \theta_2, \phi_1, \phi_2)$, obtained using Coulomb's law, has the form

$$V(\theta_1, \theta_2, \phi_1, \phi_2) = \frac{f_d^2 \sigma^2 a^4}{4\pi\varepsilon\varepsilon_o} \sin\theta_1 \sin\theta_2 \, G(a, d_{dip}, \theta_1, \theta_2, \phi_1, \phi_2) \qquad (3.8)$$

where

$$G(a, d_{dip}, \theta_1, \theta_2, \phi_1, \phi_2) = \frac{1}{F(a, a)} - \frac{1}{F(a + d_{dip}, a)}$$
$$- \frac{1}{F(a, a + d_{dip})} + \frac{1}{F(a + d_{dip}, a + d_{dip})} \qquad (3.9)$$

and the function F reads

$$F(A, B) = \sqrt{X^2 + Y^2 + Z^2} \tag{3.10}$$

where

$$
\begin{aligned}
X &= A \sin \theta_1 \cos \phi_1 - B \sin \theta_2 \cos \phi_2 - 2a - h_o \\
Y &= A \sin \theta_1 \sin \phi_1 - B \sin \theta_2 \sin \phi_2 \\
Z &= A \cos \theta_1 - B \cos \theta_2
\end{aligned} \tag{3.11}
$$

Here ε is the dielectric constant of air and ε_o is the permittivity of a vacuum.

3.3.5.2 Monopole–monopole interaction

The dipole–dipole interactions arising between colloidal particles trapped at interfaces account for the higher electrolyte concentrations needed to induce aggregation in such a kind of 2-D system, as compared to colloidal systems in bulk dispersion. For large surface packing fractions of particles, the dipole–dipole interactions seems to give a fair description of the spatial ordering and the experimental surface pressure–area curves obtained by compressing the colloidal monolayers at the air–water and oil–water interfaces. However, for dilute systems, *i.e.* large inter-particle distances, the particle monolayers exhibit a long-range repulsion, which cannot be explained assuming dipole repulsive interactions between the emergent parts of the particles.[2,30]

In this respect, Sun and Stirner[32] performed molecular dynamic simulations including charge–charge interactions besides the dipole–dipole contribution in order to account for the experimental results obtained by Aveyard *et al.*[2,33] in colloidal monolayers of polystyrene particles trapped at the octane–water interface. The net dipole moment and total charge densities were taken as fitting parameters. They observed that each mechanism gives rise to a specific surface pressure–area behaviour. The dipole–dipole interactions were only able to reproduce the experimental data for high coverage (above 0.45) but not for more dilute regimes. Analogously, they showed that the charge–charge interactions could describe the data for low particle densities. Clearly, all these results point out that the short-range repulsion between colloidal particles is mainly mediated by dipole–dipole interactions, depending asymptotically on the inter-particle distance r as $V_{dd}(r) \propto 1/r^3$. Analogously, they found that the long-range repulsion is governed by charge–charge Coulombic interactions through the oil phase, $V_{cc}(r) \propto 1/r$. According to their calculations, about 0.4% of the maximum charge on the upper part of the particle is sufficient to explain the long-range repulsion through the oil phase.

The existence of Coulombic long-range forces between colloids at interfaces has also been supported by other experiments in stable polystyrene particle monolayers at the air–water surface, even at quite low particle surface packing fractions (around

0.01). The effective colloid–colloid interaction potential was calculated by Quesada-Pérez et al.[16] by means of the inversion of the experimentally measured radial distribution function $g(r)$ through the Ornstein–Zernike equations[34]

$$h(r) = c(r) + \rho_p \int h(r') \, c(\vec{r} - \vec{r}') \, d\vec{r}' \qquad (3.12)$$

where ρ_p is the particle number density, $h(r) = g(r) - 1$ and $c(r)$ is the so-called direct correlation function. $c(r)$ is linked to $g(r)$ and the interaction potential $V(r)$ through a closure relationship. The inversion process was performed assuming the hypernetted chain (HNC) approximation as the closure, which has been proved to be very accurate for long-range repulsive interactions[34]

$$\beta V(r) = h(r) - c(r) - \ln{[h(r) + 1]} \qquad (3.13)$$

They reported a long-range repulsive effective potential that accounted for the great stability observed for the colloidal particles.[16] Although the dipolar interaction could contribute to the total repulsion, it did not explain these results alone. Indeed, even with the assumption that all the surface charged groups at the upper part of the colloidal particle are forming dipoles, the resulting pair potential is not long ranged enough in order to explain the experimental radial distribution function. The electrostatic forces between the immersed parts of the particles could not even justify this strong inter-particle interaction. In fact, some of these experiments were done at moderate sub-phase ionic strength (1 mM), where the electric double layers are sufficiently screened by the counterions to avoid long-range interactions.[35] Moreover, these interactions propagating through the aqueous phase decay exponentially. In conclusion, charge–charge interactions through the non-aqueous medium seem to be again the only reasonable explanation for such high stability observed in this kind of 2-D system, irrespective of the ionic strength of the sub-phase.

The dipole–dipole interaction energy dominates at short distances, and it has to be calculated integrating numerically over the whole distribution of dipoles. In principle, $V_{cc}(r)$ should also be determined in the same way. However, since this potential has a longer range than the dipole–dipole contribution, it is worth simplifying the calculations assuming that the total charge of the upper part of the spherical particle is concentrated in its center. This approximation fails at short inter-particle distances, where the specific details of the surface charge and dipole distributions have to be considered in order to obtain accurate values for the interaction potentials. Nevertheless, for such small distances, $V_{dd}(r)$ dominates over the other terms and the correction introduced by the exact calculation of $V_{cc}(r)$ becomes negligible. According to this, the charge–charge interaction between two colloidal particles reads[2]

$$V_{cc}(r) = \frac{q^2}{4\pi\varepsilon_r\varepsilon_0} \left[\frac{1}{r} - \frac{1}{\sqrt{4\zeta^2 + r^2}} \right] \qquad (3.14)$$

Here, r is the distance between the centers of the particles, ε_r is the relative dielectric constant of the non-polar phase, and $\zeta = a(1 - \cos\theta)/2$. The net charge of the particle above the water level, q, is given by

$$q = A\sigma f_c = 2\pi a^2[1 - \cos\theta]\sigma f_c \qquad (3.15)$$

Here, A is the area of the particle exposed to the non-polar medium and σ is the maximal surface charge density. The first term in equation (3.14) is the usual Coulombic interaction energy through the non-aqueous phase between two identical point charges of value q, separated by a distance of r. The second term corresponds to the interaction between the second particle and the image charge of the first one, located at a distance $\sqrt{4\zeta^2 + r^2}$ from the second one, symmetrically with respect to the interface that divides both media. For large inter-particle distances, equation (3.14) takes the form

$$V_{cc}(r) = \frac{q^2}{4\pi\varepsilon_r\varepsilon_0}\frac{2\zeta^2}{r^3} \qquad (3.16)$$

One of the most important experimental results that support the presence of unscreened Coulombic pair inter-particle interactions acting through the non-polar phase is that obtained using optical tweezers.[3] In such experiments, a couple of particles at the interface are originally entrapped inside two laser beams well removed from each other. Then, the inter-particle distance is decreased until one of the particles is released from the beam. At this point, the inter-particle force is equal to the trapping force of the laser. By repeating the same procedure for several laser powers, the force can be determined as a function of the particle separation. The experimental data showed that colloidal particles at the oil–water interface interact with a long-range repulsive force that scales with the inter-particle distance as $F(r) \propto r^{-4}$, but does not depend on the ionic strength in the water phase. These results agree with the interaction potential given in equation (3.16) and points out the fact that charge–charge interactions are involved in the stability of the colloidal monolayers at the oil–water interface.

3.3.5.3 Influence of V_{cc} and V_{dd} on the aggregation process

In contrast to the above-mentioned observations at the oil–water interface (where the inter-particle Coulombic interactions are insensitive to the ionic strength of the aqueous phase), it is still possible to provoke coagulation processes for high-enough concentrations of salt at the air–water surface.[2,13] This difference between both interfaces was explained in terms of the smaller contact angle of the particles at the air–water interface, causing a decrease of the net charge at the particle–air surface.[2] However, this justification does not answer the question of why the electrolyte concentration reduces the effective repulsion between colloidal particles. Obviously, the total number of dissociated surface charges and dipoles placed in the air phase has to be influenced somehow by the amount of salt in the water. In order to give an explanation for this effect, the aggregation has been studied and interpreted using the above-proposed model for the interaction energy.

The dipole and charge contributions influence the critical coagulation concentration, CCC (defined as the salt concentration where the initial aggregation rate reaches the maximum value) so that it is usually two orders of magnitude above the one in bulk dispersions. It is not just the CCC which is affected by these contributions, but also the kinetic and structural properties of the clusters formed during the aggregation process. The kinetic behaviour is featured by means of the time evolution of the so-called cluster-size distribution, $\{n_i(t)\}_{i=1}^{\infty}$, defined as the number of clusters of size i, *i.e.* containing i particles, at time t. For dilute systems, this time evolution is well described by the Smoluchowski equation[36]

$$\frac{dn_i}{dt} = \sum_{j=1}^{i-1} k_{j,(i-j)} n_j(t) n_{i-j}(t) - \sum_{j=1}^{\infty} k_{ij} n_i(t) n_j(t) \tag{3.17}$$

The first term on the right-hand side of equation (3.17) refers to the formation of i-size clusters due to aggregation of smaller entities. The second term accounts for the disappearance of i-size clusters after coagulation with another one. The aggregation kernel k_{ij} quantifies the mean rate at which two smaller clusters of size i and j combine and form a cluster of size $i + j$. It contains all the physical information about the coagulation mechanism. Of course, k_{ij} has to be understood as an orientational and configurational average of the exact aggregation rate for two clusters colliding under a specific orientation.

For large clusters and long aggregation times, the solution of the Smoluchowski equation can be expressed in terms of a time-independent scaling function, $\phi(x)$, as[37,38]

$$n_i(t) \sim S_w(t)^2 \phi(i/S_w(t)) \tag{3.18}$$

where $S_w(t)$ is the weight average cluster size at time t, given by

$$S_w(t) = \frac{\sum_{i=1}^{\infty} i^2 n_i(t)}{\sum_{i=1}^{\infty} i n_i(t)} \tag{3.19}$$

This phenomenon is called dynamic scaling and it has been observed in both experiments and computer simulations for 2- and 3-D systems. When the colloidal system undergoes aggregation, it finally reaches the long-time scaling behaviour, where the average cluster size adopts a power law dependence on the aggregation time given by

$$S_w(t) \propto t^z \qquad (3.20)$$

where z is the kinetic exponent and gives an estimate of the global aggregation rate. The aggregation between two clusters occurs as soon as they come closer and overcome the repulsive barrier of the interaction potential. Consequently, z is strongly sensitive to the main aspects of the inter-particle interaction potential, *i.e.* the height and range of the repulsive barrier. It is a well-known result that for reaction limited cluster aggregation (RLCA) the kinetic exponent z increases as the repulsive barrier is raised. The range of the interaction also plays an important role in the aggregation kinetics. Indeed, for long-range particle–particle interactions, one particle in a cluster "feels" the repulsion between all the particles in the other approaching cluster. Therefore, long-range interactions make large clusters less reactive than small ones, and therefore

$$z_{\text{long-range}} < z_{\text{short-range}} \qquad (3.21)$$

In conclusion, the global aspects of the aggregation process can be described by means of the *CCC* and the kinetic exponent z. The exact value of these parameters is a function of the particle characteristics (size, surface charge density, hydrophobicity, contact angle) and the nature of the interface (oil–water or air–water). However, the two key parameters that play the main role in the control of the aggregation kinetics for a fixed contact angle are the fraction of surface charges and surface dipoles in the non-aqueous phase, f_c and f_d.

Experimental observations of polystyrene particles trapped at the air–water surface show that the electrolyte concentration in the sub-phase significantly affects the aggregation rate.[39] Figure 3.3 shows the value of z as a function of the salt concentration [KBr]. For electrolyte concentrations above the *CCC* ([KBr] > 1 M), z does not depend on the ionic strength and is given by $z \approx 0.6$. However, for [KBr] below 1 M, the exponent z decreases continuously from 0.6 to 0.3 as the salt concentration is reduced. These changes in the kinetic and structural properties of the coagulation process suggest that f_c and f_d are in fact dependent on the electrolyte concentration in the aqueous phase. Indeed, the decrease in z at low ionic strength indicates that the total interaction potential is now long ranged (see equation (3.21)), which could be due to the increase of the surface charge fraction f_c on the air-exposed part of the colloidal particles perhaps at the expense of a reduction in f_d. In the limit of very low electrolyte concentrations, these surface charge repulsive

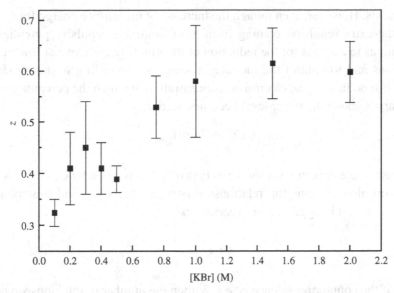

Figure 3.3 Kinetic exponent, z, as a function of the salt concentration in the water phase for the aggregation of polystyrene particles (diameter = 0.6 μm) at the air–water surface.

interactions prevent the coagulation, and are the main reason responsible for the high-ordered spatial ordering observed in such systems.

The physical explanation of the increase in f_c with decreasing salt concentration lies again in the initial preparation of the monolayer. A high salt concentration in the aqueous phase emphasizes the trapping of counterions by the dissociated charged groups and the subsequent formation of surface dipoles on the upper part of the particle during the initial spreading and rotation of the colloidal particles. Above the *CCC*, the counterion concentration is so high that all dissociated groups in the air phase have a corresponding counterion inside the hydration water layer. Hence, in this configuration, the percentage of charges is zero ($f_c = 0$) and the inter-particle repulsion is mainly due to the dipole–dipole interaction V_{dd}. However, for counterion concentrations below the *CCC*, this mechanism becomes less effective and the dissociated surface groups above the water level have more chance of losing their counterions completely. In turn, this gives rise to some degree of ionization of the surface groups, f_c. The long-range charge–charge interaction V_{cc} then becomes the one that controls the reactivity between the colloidal clusters as the aggregation proceeds.

Since all experimental data depicted in Figure 3.3 involve relatively large salt concentrations, the number of dipoles that turn into single charges is expected to be small. Therefore, f_d may be considered nearly constant for the whole set of experiments. However, even though the increase of the surface charge fraction f_c is small, the extra repulsion coming from the Coulombic repulsive potential is as important as to account for the reduction of the kinetic exponent z at low salt concentrations and to control the monolayer stability. According to this model, the *CCC* is just defined as the electrolyte concentration at which the percentage of surface charges above the water level becomes zero

$$CCC = [\text{Salt}]_{f_c=0} \tag{3.22}$$

In the case of the experiments shown in Figure 3.3, it is given by $CCC = 1$ M. This conclusion also explains the relationship between the *CCC* and the counterion valence, v, found in aggregation experiments[1]

$$CCC \approx \frac{1}{v} \tag{3.23}$$

Indeed, if the counterion valence is, *e.g.* 3, then the number of total ions we need to counterbalance the entire dissociated surface groups in the non-polar phase would be three times less. Consequently, the *CCC* is reduced to one third.

In conclusion, the dipole–dipole repulsive interaction, V_{dd}, controls the aggregation process at high salt concentrations (above the *CCC*). For low ionic strength, the charge–charge term, V_{cc}, is the one that dominates and rules the stability and the kinetics of the coagulation. f_c and f_d were obtained by fitting the above-mentioned experiments using Brownian dynamic simulations. The best agreement between experiment and simulation was found for a surface dipole concentration of $f_d \approx$ 3.1%, which leads to a repulsive barrier with a height close to the thermal energy $k_B T$. The percentage of surface charges changes from 0.0011% to zero as the salt concentration increases from 0.1 to 1 M. It should be noted that the dipolar and monopolar interaction energies used in these simulations were obtained assuming that the particles were located in one dielectric medium. That is, any effect on the electrostatic interactions coming from the existence of an interface was neglected, which represents a simplification of the real problem. Therefore, the real values for f_c and f_d can differ from the ones determined through the previous comparison, although the differences are expected to be small. In any case, the percentage of charges at the air–water surface is small compared to the values obtained for the oil–water interface, which explains why it is easier to induce coagulation at the air–water surface.

3.3.5.4 Contact angle dependence of the interaction potential

As stated previously, the contact angle value is the main parameter controlling the behaviour of colloidal particles at the interface. So, as a practical application, we study

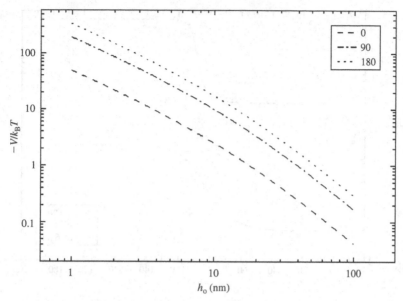

Figure 3.4 Retarded van der Waals interaction energy between particles at the air–water surface as a function of the inter-particle distance for three values of the contact angle (given in degrees) and particle radius $a = 300\,\text{nm}$.

the different terms of the interaction and the total pair energy for the case of colloidal particles with different hydrophobic properties and the same radius (300 nm) at the air–water surface. For the sake of simplicity, we have not included the monopolar interaction, analysed in detail in the previous sub-section.

The theoretical conditions of the computed potential values correspond with experimental systems for which it is hoped that no monopoles appear; *i.e.* for high salt concentration. In the computation we have used standard constant values as those already indicated. In this section we will not consider the capillary interaction because it is negligible for colloidal particles, as will be shown in Section 3.3.6.

The first analysed interaction is the van der Waals interaction. In Figure 3.4 we show the dependence of the van der Waals interaction energy on the separation distance between particles for different contact angles taking into account the retardation effect. The effect of the contact angle on the van der Waals interaction for different fixed values of the separation distance is shown in Figure 3.5. This interaction potential changes abruptly at short distances when the contact angle goes from values lower than 90° to values greater than 90°, contrary to the predicted soft dependence reported by Williams.[12]

The dependence of the double layer electrostatic interaction between the immersed parts of the particles on the contact angle at fixed separation is shown in Figure 3.6.

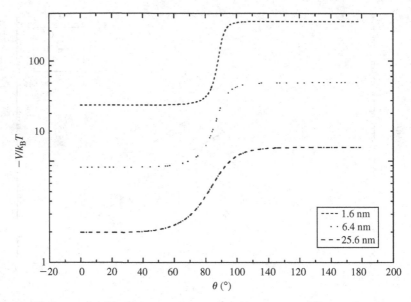

Figure 3.5 van der Waals interaction energy between particles at the air–water surface as a function of contact angle for different inter-particle separation values h_o (given). It should be noted that this interaction is a very sensitive function of the contact angle.

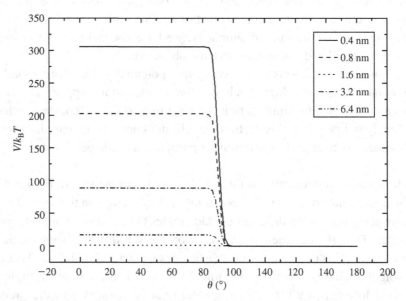

Figure 3.6 Double layer interaction energy between immersed parts of the colloidal particles as a function of contact angle for different inter-particle distances (given). The dependence on the contact angle has a similar critical behaviour to that of the van der Waals interaction at $\theta = 90°$.

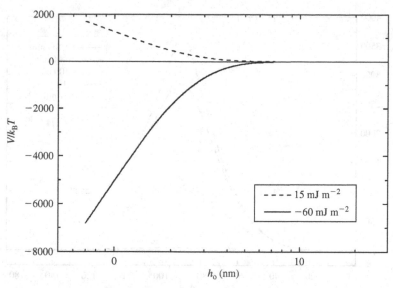

Figure 3.7 Hydrophobic interaction *versus* inter-particle distance for two values of W_o (given) with $\lambda_o = 1$ nm. The interaction range extends to a few nanometres.

In this case, there is also a critical dependence with the contact angle value. The interaction strength is considerable for contact angles below 90°, while for values above 90° it is negligible. This energetic term decays strongly with the separation between the particles (compare 0.4 and 6.4 nm).

The decaying behaviour of the hydrophobic interaction with the separation between particles is shown in Figure 3.7. This behaviour is similar to that of the electrostatic interaction in the sub-phase as could be expected, since both interactions have the same dependence on the inter-particle distance. The dipolar repulsion is shown in Figures 3.8 and 3.9 as a function of the contact angle at different distances and as a function of the separation between particles for different contact angles, respectively. As a general behaviour, the curves have a maximum near $\theta = 90°$.

For partially immersed particles at the air–liquid surface, the van der Waals interaction between colloidal particles is enhanced due to the partial particle exposure to the air phase. The increase of the van der Waals attraction coupled with the decrease of the electrostatic repulsion due to the electrical double layer overlapping does not account for the great stability experimentally observed in colloidal monolayers. The sum of these terms, *i.e.* electrostatic interaction between the sub-phase double layers and van der Waals interaction that constitute the DLVO approximation, is shown in Figure 3.10. As can be seen, the change of the contact angle can drastically reduce the repulsive energy barrier of the interaction potential between colloidal particles at the interface. This reduction is so important that increasing the surface

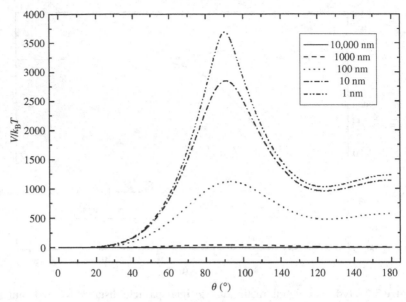

Figure 3.8 Dependence of the dipolar interaction energy on the contact angle θ for different inter-particle distances (given). A typical maximum near $\theta = 90°$ is found in all cases. The values used in the computation were $\sigma = 0.6 \, \mathrm{e^- \, nm^{-2}}$, $f_\mathrm{d} = 1$ and $d_\mathrm{dip} = 0.3 \, \mathrm{nm}$.

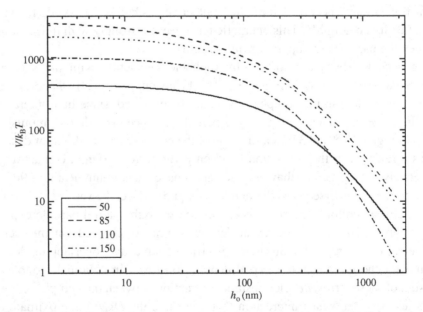

Figure 3.9 Dipolar interaction energy as a function of the inter-particle distance for different values of the contact angle (given in degrees). The asymptotic dipolar behaviour as h_o^{-3} is observed at long distances. The values used in the computation were $\sigma = 0.6 \, \mathrm{e^- \, nm^{-2}}$, $f_\mathrm{d} = 1$ and $d_\mathrm{dip} = 0.3 \, \mathrm{nm}$.

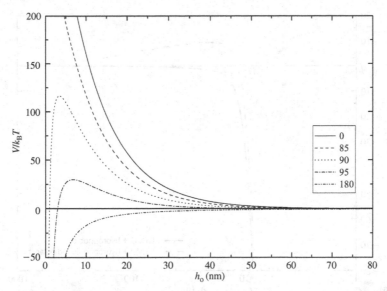

Figure 3.10 Dependence of the DLVO interaction energy between interfacial particles on the inter-particle distance for different values of the contact angle (given in degrees). When changing θ from 0 to 180°, the repulsive barrier disappears. The plotted data were obtained for a 1 mM salt concentration and $\varphi_0 = -40$ mV. The Hamaker constants were those indicated in Section 3.3.3.2.

potential and decreasing the Hamaker constant values in the DLVO interaction potential are not enough to justify the great stability of colloidal particles at the interface even for 1 M salt concentration.

This difficulty is overcome by adding the hydrophobic and dipolar interaction terms to the DLVO interaction. Figures 3.11(a) and (b) show this total contribution for both hydrophobic and hydrophilic particles, respectively. As expected, the hydrophobic interaction gives a potential with a minimum and repulsion at short distances for hydrophilic particles while there is a stronger attraction at short distances for hydrophobic ones. When the dipolar repulsion term is included, the interaction pair energy shows a repulsive barrier that prevents the aggregation of the colloidal particles. The dipolar interaction strength can be changed by varying the dipole fraction that gives the interaction. This effect is shown in Figure 3.12 where the dependence of the total pair energy on the dipole fraction is indicated for hydrophobic particles. The number of dipoles is indicated as a percentage of the maximum quantity that corresponds with all the surface groups forming dipoles. For 100% of dipoles, it is practically impossible that the potential barrier could be overcome by the particles and so they remain stable without aggregating. Thus, it can be concluded that the dipolar interaction is the most important contribution to the stability of colloidal particles at the air–liquid surface for high salt concentration in the sub-phase.

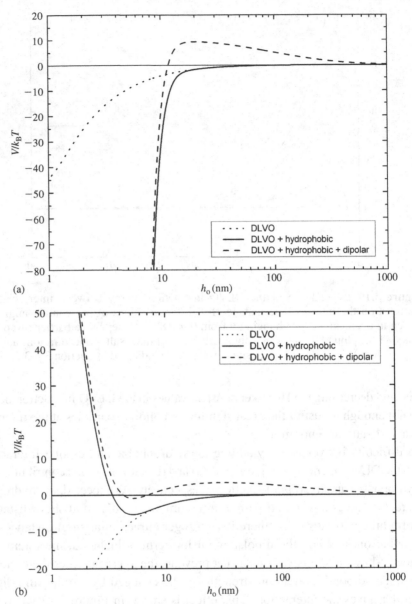

(a)

(b)

Figure 3.11 Addition of the hydrophobic and dipolar interactions to the DLVO interaction energy for (a) hydrophobic and (b) hydrophilic particles respectively. The parameters used were: (a) $\theta = 82°$, salt concentration = 1 M of 1:1 electrolyte, $\varphi_o = -40\,\text{mV}$, surface density of dipoles = 0.18 dipoles nm^{-2}. Hydrophobic character is characterized by $W_o = -60\,\text{mJ}\,\text{m}^{-2}$ and $\lambda_o = 1\,\text{nm}$. (b) $\theta = 50°$, salt concentration = 1 M of 1:1 electrolyte, $\varphi_o = -56\,\text{mV}$, surface density of dipoles = 0.18 dipoles nm^{-2}. Hydrophobic character is characterized by $W_o = 3\,\text{mJ}\,\text{m}^{-2}$ and $\lambda_o = 1\,\text{nm}$.

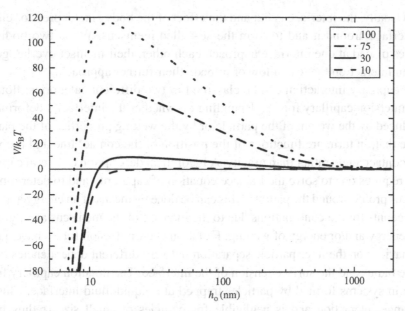

Figure 3.12 Effect of the dipole fraction on the total pair potential for hydrophobic particles at the air–water surface. The values given of 100, 75, 30 and 10 correspond to the percentages of the groups forming dipoles at the surface. The contact angle used and the surface potential were 90° and −40 mV, respectively. The salt concentration corresponds to 1 M of a 1:1 electrolyte. Full coverage corresponds to $\sigma = 0.6\,\mathrm{e^-\,nm^{-2}}$ forming dipoles in the air phase.

The dipolar interaction allows us to explain quantitatively why the *CCC*, *i.e.* the lower salt concentration for which fast colloidal aggregation occurs, is much lower for colloidal particles in dispersion than for aggregation at the interface. Thus, this term justifies the great stability found for colloidal particles at the air–liquid interface.

The form of the pair interaction potential curves is different for hydrophobic and hydrophilic particles (Figure 3.11). Hydrophobic particles can aggregate at a primary minimum while for hydrophilic ones aggregation occurs at a secondary minimum. This difference could explain the re-structuring phenomena observed by Hórvölgyi *et al.* for glass beads of lower hydrophobicity,[40] which aggregate at larger distances. The aggregates formed by these particles are re-structured and become more compact.

3.3.6 The capillary interaction

The capillary interaction has no equivalent in bulk aggregation. When a particle or other body contacts the boundary between two fluid phases their different interaction with both phases causes the perturbation of the interface from its original

form. Usually, the interface is flat and the effect of the body is to depress or elevate
the interface around it and to form the so-called meniscus. When two bodies or
particles placed at one interface approach each other, their menisci overlap generating an interaction that can allow or impede their further approach.

The capillary interaction can be classified in two different categories: flotation
and immersion capillary forces, depending on whether the interfacial deformation
is produced by the weight of the particle or by the wetting properties of the particle
surface that, in turn, are functions of the position of the contact line and the value
of the contact angle rather than gravity. A theoretical description of lateral capillary
forces requires one to solve the Laplace equation of capillarity and to determine the
meniscus profile around the particle. This can be done by means of an energy approach
that accounts for the contributions due to the increase of the meniscus area, gravitational energy and/or energy of wetting. Flotation and immersion forces have a similar behaviour on the inter-particle separation but very different dependencies on the
particle radius and the surface tension of the interface. The flotation capillary forces
appear in systems formed by particles trapped at a liquid–fluid interface. This is a
long-range interaction and is negligible for particles of small size (radius below
$10\,\mu m$). Immersion forces arise when particles are trapped at a liquid interface over
a substrate or in a thin liquid film, and they are long-range interactions that can be
very important even for small particles (radius between few nanometres to
micrometres). Recent studies have shown that immersion forces can also arise in
floating particles when these particles display an irregular meniscus over their surface[41,42] or when there is an external electric field that pushes the particle into the
liquid sub-phase, so-called electrodipping.[43–46] These immersion interactions
between particles trapped at interfaces were used to explain the unexpected long
range attractive interaction reported in recent experiments of colloidal particles
spread on interfaces.[6–8,41,43,47,48]

3.3.6.1 Calculation of the capillary interaction potential

The thermodynamic approach to calculate the capillary interaction potential, V_{men},
between particles at an interface uses the change of free energy that occurs when
the colloidal particles approach each other. Its value is given by

$$V_{men}(d) = \hat{F} - F_1 - F_2 = \hat{F} - 2F_1 \qquad (3.24)$$

where $\hat{F}$ is the free energy of the equilibrium configuration for the two colloidal particle system and F_1 ($=F_2$) is the single particle equilibrium free energy. In order to
calculate the pairwise capillary potential, the terms F_1 (due to the presence of one
particle alone at the interface) and $\hat{F}$ have to be determined. These three contributions
depend, in turn, on the interface profile near the particles, characterized through the

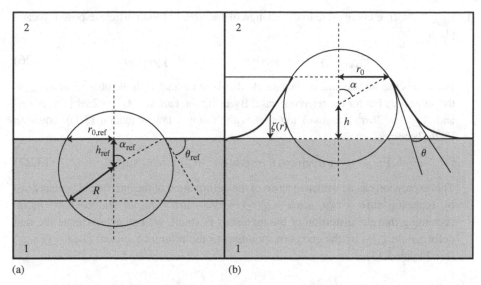

Figure 3.13 (a) Description of the reference state for a colloidal particle confined at an interface where θ_{ref} is the equilibrium contact angle. (b) Description for the deviations from the reference state where ζ is the meniscus profile and h is the height from the particle center to the interface.

distortion of the meniscus, $\zeta(r)$. $\zeta(r)$ is defined as the difference between the real profile $z(r)$ and the profile of a flat meniscus, z_{ref}: $\zeta(r) = z(r) - z_{ref}$, where r is the distance between any point of the interface and the center of the particle. We detail the calculations of these three contributions for the free energy below.

3.3.6.1.1 One particle at the interface To compute the capillary free energy of a colloidal particle at the interface it is necessary to consider a reference state, with respect to which changes in free energy will be measured. Usually, the reference state has a planar meniscus with the colloidal particle at such a height h_{ref} that Young's law is satisfied (Figure 3.13), *i.e.* the contact angle θ formed by the planar interface and the single particle surface is given by Young's law in the equilibrium state: $\cos \theta = (\gamma_2 - \gamma_1)/\gamma$.

The capillary free energy, F_1, of a single particle at a liquid interface will depend on the meniscus profile $\zeta(r)$, *i.e.* on the distortion of this meniscus from the reference state. According to Ref. [46], the free energy may be separated into the following terms

$$F_1 = F_{cont} + F_{men} + F_{grav} + F_{\Pi} + F_{coll} \qquad (3.25)$$

Next, we analyse each of these five contributions separately.

(i) F_{cont} is the free energy due to the change of the areas of the particle exposed to each phase

$$F_{cont} = \gamma_1 A_1 + \gamma_2 A_2 - (\gamma_1 A_{1,ref} + \gamma_2 A_{2,ref}) \qquad (3.26)$$

Here, A_i is the surface area of the particle that is in contact with the phase i, and $A_{i,ref}$ is the same area but for the reference case. By using the fact that $A_{1,2} = 2\pi a^2(1 \pm \cos\alpha)$ and $A_{1,2,ref} = 2\pi a^2(1 \pm \cos\theta)$ together with Young's law (equation (3.1)), equation (3.26) becomes

$$F_{cont} = -\pi a^2 \gamma[(\cos\theta - \cos\alpha)^2 + \sin^2\alpha - \sin^2\theta] \qquad (3.27)$$

This expression can be written in terms of the deformation of the interface at contact $\zeta(r_0)$ by replacing $\sin\alpha = r_0/a$, $\cos\alpha = (\zeta(r_0) - h)/a$, $\sin\theta = r_{0,ref}/a$ and $\cos\theta = -h_{ref}/a$. Assuming that the distortion of the meniscus is small, we can approximate the real deformation $\zeta(r_0)$ by the one corresponding to the reference system $\zeta(r_0) \approx \zeta(r_{0,ref})$ (see Figure 3.13)

$$F_{cont} \approx \pi\gamma\left[\zeta(r_{0,ref}) - \Delta h\right]^2 + \pi\gamma(r_0^2 - r_{0,ref}^2) \qquad (3.28)$$

where $\Delta h = h - h_{ref}$.

(ii) F_{men} is the free energy contribution due to the change of the meniscus area, given by

$$F_{men} \approx \gamma \int_{S_{men}} dA\sqrt{1 + \left|\nabla\zeta(r)\right|^2} - \gamma \int_{S_{men,ref}} dA \qquad (3.29)$$

where S_{men} and $S_{men,ref}$ are the meniscus areas (excluding the area occupied by the particle) in the perturbed and the reference system, respectively. For small slopes of the meniscus deformation $\zeta(r)$, *i.e.* for $|\nabla\zeta(r)|^2 \ll 1$, and with the approximation $S_{men} \approx S_{men,ref}$, we have

$$F_{men} \approx \gamma\pi(r_{0,ref}^2 - r_0^2) + \frac{1}{2}\gamma \int_{S_{men,ref}} \left|\nabla\zeta(r)\right|^2 dA \qquad (3.30)$$

(iii) F_{grav} represents the changes in the gravitation potential energy with respect to the reference state due to displacements of the volumes of the fluid phases

$$F_{grav} \approx \frac{1}{2}\gamma \int_{S_{men,ref}} q^2\zeta(r)^2 dA, \qquad q = \sqrt{\Delta\rho g/\gamma} \qquad (3.31)$$

Here, q^{-1} symbolizes the characteristic capillary length which determines the range of action of the capillary forces. The capillary length has the typical value $q^{-1} = 2.7$ mm for the air–water surface (and it is higher than any other characteristic longitude of the system) and $\Delta\rho = \rho_1 - \rho_2$ is the difference between the densities of the phases 1 and 2.

(iv) F_Π accounts for the presence of external fields acting perpendicularly to the fluid interface. Its origin can be diverse. It can be produced by the disjoining pressure between two adjacent phases across the liquid film, by the electrostatic pressure caused by an electric external field, *etc.* By defining $\Pi(r)$ as the force per unit area, this energetic contribution reads

$$F_\Pi \approx - \int\limits_{S_{\text{men,ref}}} \Pi(r)\zeta(r)\,dA \simeq -2\pi \int\limits_{r_{0,\text{ref}}}^{\infty} r\Pi(r)\zeta(r)\,dr \qquad (3.32)$$

(v) F_{coll} is the free energy change due to the vertical displacement of the particles

$$F_{\text{coll}} = -F\Delta h \qquad (3.33)$$

where F is the total force acting on the particle perpendicularly to the interface.

By adding these five terms, we finally obtain the total free energy F_1, calculated with the approximation of small slope of the meniscus ($|\nabla\zeta|^2 \ll 1$) and assuming $S_{\text{men}} \approx S_{\text{men,ref}}$

$$F_1 \approx 2\pi\gamma \int\limits_{r_{0,\text{ref}}}^{\infty} r \left(\frac{1}{2}\left(\frac{d\zeta}{dr}\right)^2 + \frac{q^2\zeta^2}{2} - \frac{1}{\gamma}\Pi\zeta \right) dr + \pi\gamma \left(\zeta(r_{0,\text{ref}}) - \Delta h \right)^2 - F\Delta h$$

$$(3.34)$$

We can find the shape of the meniscus that minimizes the free energy by solving the so-called linearized Young–Laplace equation[49]

$$\left.\frac{dF_1}{d\zeta}\right|_{\zeta_{\min}} = 0 \Rightarrow \frac{d^2\zeta}{dr^2} + \frac{1}{r}\frac{d\zeta}{dr} = q^2\zeta - \frac{\Pi(r)}{\gamma} \qquad (3.35)$$

The last expression is the key equation giving a theoretical description of the capillary interactions. This equation governs the meniscus shape for a given configuration of particles at an interface. In order to solve this differential equation, two boundary conditions are necessary. First, the slope of the meniscus height just at the contact line $r = r_0 \approx r_{0,\text{ref}}$ has to be[46]

$$\left.\frac{d\zeta}{dr}\right|_{r=r_{0,\text{ref}}} = \frac{\zeta(r_{0,\text{ref}}) - \Delta h}{r_{0,\text{ref}}} \qquad (3.36)$$

The second requirement comes from the fact that the meniscus has to be asymptotically flat at sufficiently large distances from the particle surface

$$\lim_{r\to\infty} \zeta(r) = 0 \qquad (3.37)$$

The solution of the differential equation (3.35) with the boundary conditions (equations (3.36) and (3.37)) may be expressed in the following general form[46]

$$\zeta(r) = \frac{1}{\gamma} I_0(qr) \int\limits_r^\infty s\Pi(s)K_0(qs)ds + \frac{1}{\gamma} K_0(qr) \left[C + \int\limits_{r_{0,\text{ref}}}^r s\Pi(s)I_0(qs)ds \right] \quad (3.38)$$

where $K_0(x)$ and $I_0(x)$ are the modified Bessel functions of zero order of first and second kind, respectively,[50] and C is an integration constant which can be obtained with the boundary condition given by equation (3.36). By replacing the explicit form for $\zeta(r)$ into equation (3.25) we finally obtain the free energy of one particle, F_1.

3.3.6.1.2 Two particles at the interface In order to calculate the effective pairwise interaction potential, $V_{\text{men}}(d)$, between two identical particles at a liquid interface at a fixed lateral distance d in equilibrium, it is necessary to find out the free energy of this configuration. According to Oettel *et al.*[46] this free energy may be approximated as

$$\hat{F} \approx \gamma \int\limits_{S_{\text{men,ref}}} dA \left(\frac{1}{2}\left|\nabla\hat{\zeta}\right|^2 + \frac{q^2\hat{\zeta}^2}{2} - \frac{1}{\gamma}\hat{\Pi}\hat{\zeta} \right)$$

$$+ \sum_{\alpha=1,2} \left[\frac{\gamma}{2r_{0,\text{ref}}} \oint\limits_{\partial S_\alpha} dl(\Delta\hat{h}_\alpha - \hat{\zeta})^2 - \hat{F}\Delta\hat{h}_\alpha \right] \quad (3.39)$$

Here, $\hat{\zeta}(r)$ is the meniscus shape in the presence of two particles, $\Delta\hat{h}_\alpha$ are the corresponding heights of each particle, $\hat{\Pi}$ is the vertical force per unit area acting on the interface due to external fields and $\hat{F}_\alpha$ is the force acting perpendicularly to the interface over each particle. S_α is the area of the circular disk defined by the contact line around each particle while ∂S_α are the contact lines (counter-clockwise). In this context, the area of the meniscus in the reference system (flat interface) is defined as $S_{\text{men,ref}} = R^2 - (S_1 \cup S_2)$.

By minimizing $\hat{F}$ with respect to ζ as in equation (3.35) the following second order partial differential equation is found

$$\nabla^2\hat{\zeta} = q^2\hat{\zeta} - \frac{\hat{\Pi}}{\gamma} \quad (3.40)$$

This equation has to be solved again imposing two boundary conditions. The first one is transversality conditions at the boundary defined by the contact lines:[46]

$$\partial S_1 \cup \partial S_2$$

$$\frac{\partial \hat{\zeta}}{\partial n_\alpha} = \frac{\hat{\zeta}(\vec{r}) - \Delta \hat{h}_\alpha}{r_{0,\text{ref}}} = \frac{\hat{\zeta}(\vec{r}) - \bar{\zeta}_\alpha}{r_{0,\text{ref}}} - \frac{F}{2\pi\gamma r_{0,\text{ref}}} \tag{3.41}$$

where $\partial/\partial n_\alpha$ is the derivative in the outward normal direction of ∂S_α, and

$$\bar{\zeta}_\alpha \equiv \frac{1}{2\pi r_{0,\text{ref}}} \oint_{\partial S_\alpha} dl \hat{\zeta} \tag{3.42}$$

is the mean height of the distortion of the meniscus at the contact line. The second boundary condition tells us that the meniscus has to be asymptotically flat at sufficiently large distances from the particle surface

$$\lim_{r \to \infty} \hat{\zeta}(\vec{r}) = 0 \tag{3.43}$$

The final solution of equation (3.40) with the boundary conditions given by equations (3.41) and (3.43) is a very complex analytical problem that can only be solved in the homogeneous case, *i.e.* for $\hat{\Pi} = 0$.[51] The general problem is solved introducing the approximation of superposition, valid in the limit of large inter-particle separation $(d \gg a)$[52]

$$\begin{aligned} \hat{\zeta}(r) &\approx \zeta_1(r) + \zeta_2(r) \\ \hat{\Pi}(r) &\approx \Pi_1(r) + \Pi_2(r) \\ F(r) &\approx F_1(r) = F_2(r) \end{aligned} \tag{3.44}$$

where $\zeta_\alpha, \Pi_\alpha, F_\alpha$, with $\alpha = 1, 2$, are the meniscus profile, the stress of the interface and the force acting on the interface due to each particle respectively, analogous to the case of one particle at the interface. The effective potential between two particles due to the interface deformation by the presence of the two particles can be calculated now using equation (3.24). Oettel *et al.*[46] calculated an approximate general form for the effective potential due to the meniscus deformation using the assumption of superposition

$$V_{\text{men}}(d) \approx - \int_{S_{\text{men,ref}}} \Pi_1 \zeta_2 dA + \int_{S_1} \Pi_2 \zeta_2 dA - \gamma \oint_{\partial S_1} \frac{\partial(\zeta_1\zeta_2)}{\partial n_1} dl - \frac{1}{2}\gamma \oint_{\partial S_1} \frac{\partial(\zeta_2^2)}{\partial n_1} dl$$

$$+ \frac{\gamma}{r_{0,\text{ref}}} \oint_{\partial S_1} \left[\bar{\zeta} - \zeta_2\right]^2 dl - 2F\bar{\zeta} \tag{3.45}$$

where $\bar{\zeta} \equiv \frac{1}{2\pi r_{0,\text{ref}}} \oint_{\partial S_1} \zeta_2 dl$, is the mean height of the single particle meniscus with respect to the contact line of the other colloidal particle. There are two different asymptotic behaviours of the effective potential ($a \ll d \ll q^{-1}$), depending on the relation between the dimensionless quantities ε_Π and ε_F, defined as

$$\varepsilon_F \equiv \frac{-F}{2\pi\gamma r_{0,\text{ref}}}, \qquad \varepsilon_\Pi \equiv \frac{1}{2\pi\gamma r_{0,\text{ref}}} \int_{S_{\text{men,ref}}} \Pi dA. \qquad (3.46)$$

All the forces acting over particles are included in ε_F, while the stress over the interface is included in ε_Π. Thus, the relation between ε_F and ε_Π is the relation between F and Π. If $\varepsilon_\Pi \neq \varepsilon_F$, V_{men} reads

$$V_{\text{men}}(d) \approx -2\pi\gamma r_{0,\text{ref}}^2 (\varepsilon_F - \varepsilon_\Pi)^2 K_0(qd)$$

$$\approx -2\pi\gamma r_{0,\text{ref}}^2 (\varepsilon_F - \varepsilon_\Pi)^2 \ln \frac{2e^{-\gamma_E}}{qd} \qquad (3.47)$$

where $\gamma_E = 0.577216$ is Euler's constant. This is a long-range (logarithmic) attractive effective potential that does not depend on the sign of the forces and stress acting on the system. If $\varepsilon_\Pi = \varepsilon_F$, then

$$V_{\text{men}}(d) \approx -2\Pi(d) \int_{S_{\text{men,ref}}} \zeta dA \quad \text{with } S_1 = S_2 = S \qquad (3.48)$$

In this case, $V_{\text{men}}(d)$ has the same dependence on the distance as the stress. It is a short-ranged interaction that can be attractive or repulsive depending on the form of $\Pi(d)$. Typically $\Pi(d) \sim d^{-n}$ so $V_{\text{men}}(d) \sim d^{-n}$.

3.3.6.2 Capillary flotation forces

In this particular case, there are no stresses acting at the meniscus ($\Pi = 0$) and the force F on the particles is due to their weight and the buoyancy force. Thus, we are in the case $\varepsilon_\Pi \neq \varepsilon_F$ with $\varepsilon_\Pi = 0$. Therefore, using equation (3.47) we obtain

$$V_{\text{men}}(r) \approx -2\pi\gamma r_{0,\text{ref}}^2 \varepsilon_F^2 K_0(qr) = -2\pi\gamma Q^2 K_0(qr) \qquad (3.49)$$

where $Q = r_{0,\text{ref}} \varepsilon_F$ is the so-called "capillary charge"[42,49,52] by analogy with the 2-D electrostatic problem. The dependence of the capillary charge on the particle radius a is easily obtained from equation (3.46)

$$Q = r_{0,\text{ref}} \varepsilon_F = \frac{-F_g}{2\pi\gamma} \propto m \propto a^3 \Rightarrow Q^2 \propto a^6 \qquad (3.50)$$

Thus, the capillary flotation interaction strongly depends on the particle radius a. The "capillarity charges" for floating particles can be expressed in terms of a and the three-phase contact angle θ through the expression[53]

$$Q = \frac{1}{6} q^2 a^3 (2 - 4D + 3 \cos \theta - \cos^3 \theta) \qquad (3.51)$$

where $D = (\rho_S - \rho_2)/(\rho_1 - \rho_2)$ and ρ_S is the density of the colloidal particle. The values for the flotation capillary interaction for two spherical particles at contact ($d = 2a$) trapped at an air–water surface are plotted in Figure 3.14 versus the particle radius. As can be seen, the flotation capillary interaction is negligible, *i.e.* $V(2a) \ll k_B T$, for particles with $a < 10 \, \mu m$. In other words, the weight of floating micrometre (or sub-micrometre) sized particles is too small to create any surface deformation.

3.3.6.3 Capillary immersion forces

Long range attractive forces between micrometre-sized particles at a fluid interface were experimentally detected[6–8,41,43,47,48] and attributed to immersion capillary interactions. Usually, these kind of capillary interactions come out when colloidal

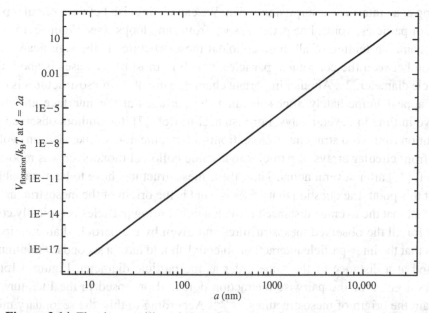

Figure 3.14 Floating capillary interaction energy between two spherical colloidal particles in contact ($d = 2a$) trapped at the air–water surface *versus* the particle radius a for a contact angle $\theta = 82°$. As can be seen, the interaction is negligible for particles with radius under $10 \, \mu m$.

particles are trapped in a thin liquid film[49,54–56] or at a liquid interface formed over a substrate.[54] Nevertheless, it has been reported recently that the immersion capillary interactions between floating particles may also appear when the three-phase contact line (particle–liquid–fluid) is irregular rather than flat due to the non-homogeneous wetting of the particle surfaces[41,42] or when there is an electrical stress that causes an immersion of the particles into the liquid phase.[43–45] However, some of these models are still nowadays incomplete and none of them could explain all the characteristics of the mesostructures found experimentally. Here, we will comment on these immersion capillary interactions between floating particles. Then, we will focus on the long-range behaviour of those interactions used here to explain the attractions reported experimentally for large inter-particle distances.

3.3.6.4 Colloidal mesostructures

Experimental results reported since 1995 have shown the existence of new colloidal structures (loosely bound, internally ordered, *etc.*) when particles are spread at an air–water surface[6–8,47,48] or an oil–water interface.[43] In order to distinguish such ordered structures from the random, fractal objects formed in colloidal aggregation processes (commonly called clusters), Ghezzi and Earnshaw[47] proposed to call them colloidal mesostructures. The diverse morphology of the mesostructures arising at an interface is surprisingly rich. We can distinguish between circular patterns of particles, voids, line patterns, soap froths and loops[8] (see Figure 3.15(a)).

A common feature of all these colloidal mesostructures is the significant separation between the constituent particles, which in most of the cases is about the particle diameter, $2a$. Another important characteristic of the mesostructures is that they appear immediately after spreading the particles at the interface and then evolve in time in several ways. For instance, in Ref. [7], the authors observed an evolution from void structures to soap froths after some hours. Likewise, an evolution from circular arrays of particles to a stable colloidal monolayer was reported in Ref. [47] after several hours. Thus, these mesostructures have to be metastable.

At this point, the question that arises is what is the origin of the mesostructures? The fact that the average distance between neighbouring particles is typically constant for all the observed mesostructures and given by the particle diameter, indicates that the inter-particle interaction potential should have a secondary minimum located at a distance of the same order as the particle diameter. Figure 3.15(b) shows a guess of the pairwise interaction potential proposed in the literature to explain the origin of mesostructures.[7,8,43,47] According to this, the secondary minimum explains why the inter-particle distance is roughly constant, while the small potential barrier situated at larger distance prevents the collapse of the individual structures. The process of spreading the monolayer is quite turbulent, and the

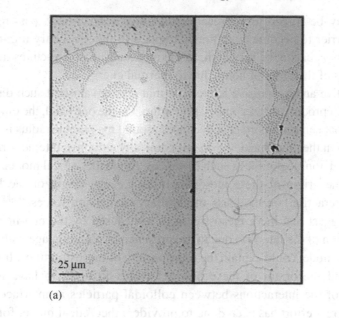

(a)

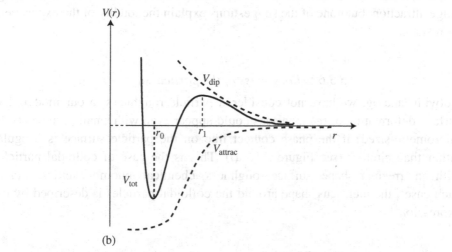

(b)

Figure 3.15 (a) Colloidal mesostructures formed by charged spheres of polystyrene of radius $a = 600\,$nm at the air–water surface. These mesostructures arrange in circular domains (left images), "foam" and voids (top right image), loops (bottom right image), *etc.* (b) Shape of pairwise interaction potential proposed in the literature. It has a secondary minimum at distances r_0 around the particle diameter, which could explain the characteristic distance between particles in the mesostructures.

particles may be forced to approach each other very closely, passing over the potential barrier to become temporarily trapped in the secondary minimum at r_0. The barrier at r_1 cannot be very high, as the metastable mesostructures disaggregate on timescales of the order of hours due to thermal energy.

An attractive and a repulsive interaction that do not vanish at such distances are necessary to reproduce this secondary minimum. On the one hand, the only repulsive interaction that can be important at distances around the particle radius is the dipolar interaction. On the other hand, we do not know any attractive interaction that could be considered non-negligible at such distances. Some theoretical models have tried to explain the origin of these mesostructures, *i.e.* the origin of the long range attractive interaction, on the basis of capillary immersion forces.[41–43,45,46] Other authors[57] suggest that these mesostructures are caused by the non-homogeneous surface tension of the interface due to the presence of polluting agents like silicone oil. This last model could explain the great variety of mesostructures. It is essential to understand the origin of these new structures in order to have a complete description of the interactions between colloidal particles at interfaces. For this purpose, a great effort has been done to provide a theoretical model for this long-range attraction, but none of the suggestions explain the totality of the experimental results.

3.3.6.5 Roughness of the particle surface

Notwithstanding, we have not considered particle roughness in our model. The surface deformation of the meniscus could appear even with small particles (sub-micrometre-sized) if the phase contact line on the particle surface is irregular rather than circular (see Figure 3.16(a)). This is the case of colloidal particles with an irregular shape, surface roughness, chemical inhomogeneities, *etc*. In such cases, the meniscus shape around the colloidal particles is described by the expression[41]

$$\zeta(r,\varphi) = \sum_{m=1}^{\infty} K_m(qr)(A_m \cos(m\varphi) + B_m \sin(m\varphi)) \tag{3.52}$$

which is a solution of the linearized Young–Laplace equation for small meniscus slopes in cylindrical coordinates (r, φ), where A_m, B_m are integration constants and $K_m(x)$ is the modified Bessel function of order m. For $qr \ll 1$, $K_m(qr) \propto (qr)^{-m}$, equation (3.52) reduces to a multipolar expansion (analogue to an electrostatic multipolar expansion in 2-D).

The capillary forces between particles with an irregular contact line could be considered as a kind of immersion force insofar as they are related to the particle

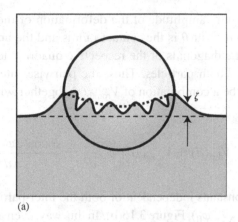

(a)

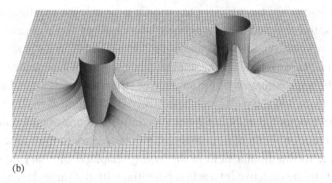

(b)

Figure 3.16 (a) Representation of the contact line for a colloidal particle at a fluid interface. The dashed line is the ideal flat contact line, the continuous line is the representation of the undulated contact line and the dotted line is the quadrupoles contribution of the irregular contact line. (b) Representation of the quadrupolar term for two colloidal particles trapped at a fluid interface. The total interface deformation, which is due to the presence of the particles, depends on the mutual orientation between the particles.

wettability, rather than to the particle weight. A theoretical description of the lateral capillary force between colloidal spheres with an undulated contact line was recently given by Stamou et al.[41] They realized that the leading multipole order in the capillary force between such particles is the quadrupole–quadrupole interaction ($m = 2$ in equation (3.52)) (see Figure 3.16(b)). The interaction energy of this energetic contribution is given by

$$V_{\text{rough}}(d) \approx -12\pi\gamma H^2 \cos(2(\varphi_A + \varphi_B))\left(\frac{r_c}{d}\right)^4 \quad \text{for m} = 2; d \gg 2r_c \quad (3.53)$$

where H is the maximum amplitude of the deformation of the contact line at the particle surface, $r_c = a = \sin\theta$ is the contact radius and the angles φ_A and φ_B are subtended between the diagonals of the respective quadrupoles and the line connecting the centers of both particles. Thus, the pairwise interaction potential at long distances could be a combination of $V_{rough}(d)$ together with the dipolar interaction, $V_{dip}(d)$

$$V_{d \gg a}(d) = V_{dip}(d) + V_{rough}(d) \approx \frac{A}{d^3} - \frac{B\cos(2\varphi_A + 2\varphi_B)}{d^4} \qquad (3.54)$$

where A and B are constants independent of both the inter-particle distance (d) and particle orientation ($\varphi_A + \varphi_B$), Figure 3.16(b). In this way, even a minimal roughness of the contact line could be sufficient to produce a significant capillary attraction.[41] For multipoles, the sign and the magnitude of the capillary force depend on the particular mutual orientation. So, quadrupoles tend to assemble in a square lattice.[42]

This model is, however, rather incomplete since the meniscus deformation is calculated with the approximation of the absence of stress and perpendicular forces acting on the interface and the particles, respectively ($\Pi = 0$ and $F = 0$). According to this, the effective potential associated with this deformation is obtained and added to the dipolar interaction. For a correct analysis, it is necessary to introduce the electrostatic stress due to the dipolar field in the Young–Laplace capillary equation. This adds new terms to the effective interaction potential with the same dependence on the distance as the term found by Stamou *et al.*[41] Nevertheless, even introducing this new term in the total interaction potential, the dependence of the dipolar interaction with the distance is like $1/r^3$ and the attractive interaction has an asymptotic behaviour as $1/r^4$. Hence, the combination of both interactions cannot explain the shape of the total interaction potential shown in Figure 3.15(b). Therefore, this model does not account for the origin of the colloidal mesostructures.

3.3.6.6 Meniscus deformation due to an electric field

The magnitude of the interfacial deformation, characterized by the contact angle (Figure 3.13(a)), is determined from the condition $2\pi r_c \gamma \sin\theta = F$, so a normal external force counterbalances the surface tension force F. In the case of floating capillary forces, the normal force is due to gravity and the magnitude of the surface distortion decreases rapidly when decreasing the particle size, becoming negligible for particles with a radius $a < 10\,\mu m$.

Nevertheless, for such small particles, Nikolaides *et al.*[43] suggested that the normal force F could have an electric origin, F_{el}, rather than gravitational, F_g. They also proposed that the asymmetric charging of the particles adsorbed at interfaces could produce a dipolar field. The electrostatic force associated with this field

could be responsible for the interfacial deformation and the lateral capillary interactions. With these considerations, and neglecting the electrostatic stress acting on
the interface ($\Pi_{el}(r) = 0$), they finally obtained a long-range (logarithmic) capillary attraction. However, if we take this stress into account the situation changes
drastically. Then, the new dependence[44,58] on the inter-particle distance is $1/d^6$ rather
than logarithmic, *i.e.* it is a short-range interaction that does not explain completely
the origin of the secondary minimum.

In turn, Danov *et al.*[45] reported experimental results showing that the interfacial
deformation around glass particles (with radius between 200 and 300 μm) at the
oil–water interface is dominated by an electrical force, called electrodipping by
them.[45] In that work, they suggested that this force is due to charges at the particle–oil
boundary (or at particle-non-polar phase boundary in general). They solved numerically the electrostatic boundary problem which gives rise to a long-range (logarithmic) electric field, inducing the capillary attraction between two floating particles.
However, they calculated the pairwise effective potential solving the Young–Laplace
capillary equation only in the presence of gravity ($\Pi = 0$, $F_{el} = 0$, $F_g \neq 0$) with
the superposition approximation, while the solution is used even in the case of
$\Pi \neq 0$. Hence, the solution given in Ref. [45] appears inaccurate because the
approximation of superposition is not correctly applied and new terms arise when
the capillary equation with $\Pi \neq 0$ is formally solved. Moreover, Danov *et al.* solve
the electrostatic problem using the asymptotic behaviour ($r \gg r_{0,\text{ref}}$) of $\Pi(r)$, but the
dominant contribution to the interaction potential due to the meniscus deformation
stems from points $r \approx r_{0,\text{ref}}$. As a consequence, they obtain a non-vanishing pre-
factor of the logarithmic term in the interaction potential. Therefore, the presence
of charges at the particle-non-polar phase boundary cannot explain an attractive
logarithmic interaction.

Oettel *et al.*[46] proposed a logarithmic long-range attractive attraction tuned by a
small external electrostatic field. However, this attraction is only possible if the net
force acting on the system does not vanish. In order to show this, they considered
the total stress tensor $\tilde{T}$ which consists of the Maxwell stress tensor (due to the
electrostatic field) and a diagonal osmotic pressure tensor (due to the electrolyte).
The total force acting on the whole system (interface and particles) perpendicularly to the interface can be calculated through

$$F_z = \hat{e}_z \cdot \oint_{S_{\text{tot}}} d\vec{A} \cdot \tilde{T} = 2\pi\gamma r_{0,\text{ref}}(\varepsilon_\Pi - \varepsilon_F) \tag{3.55}$$

where $S_{\text{tot}} = S_{\text{men}} + S_1 + S_2$.

In the absence of external forces, $F_z = 0$, we are in the case of $\varepsilon_\Pi = \varepsilon_F$ and
we have to apply equation (3.48) which gives a short-range interaction. Thus, the

existence of a net force acting on the interface is necessary to have an attractive logarithmic (long-range) interaction. If $F_z \neq 0$, then $\varepsilon_\Pi \neq \varepsilon_F$ and so we have to use equation (3.47) which gives a long-range interaction. Therefore, in the presence of an external electric field, the values of ε_Π and ε_F can be calculated numerically solving the electrostatic boundary problem. Subsequently, the interaction potential due to the interfacial deformation, $V_{men}(d)$, can be theoretically obtained by using equation (3.45).

Oettel *et al.* calculated the external electric field necessary to obtain the long-range attractive interaction shown in Ref. [43] and they obtained a relatively small value $-E \approx 1.8 \times 10^{-4}\,\mathrm{V\,m^{-1}}$. Indeed, an electric field $E \sim 10^{-3}\,\mathrm{V\,m^{-1}}$ is enough to provoke a secondary minimum of about $1\,kT$. These calculations indicate how sensitive the system can be to spurious external electric fields that give the required long-range attraction. This model, however, is not able to explain such an attraction for isolated systems $(F_z = 0)$.[8] In view of these conclusions, further effort (theoretical and experimental) is needed to achieve a complete understanding of these amazing mesostructures and to have a complete description of the interactions that appear when colloidal particles are spread at a liquid interface.

3.4 Concluding Remarks

The interactions between particles in monolayers have been theoretically discussed. These interactions are dependent on both the properties of the fluids that make up the interface and on the nature of the adsorbed particles. We can distinguish two different stability behaviours: stable monolayers with particles that remain independent for a long time, and unstable monolayers with aggregates of fractal structure or the so-called mesostructures. In the first case some very regular geometrical structures have been observed suggesting the existence of long-ranged particle interactions. When the structures are fractal in character the particle interaction potential is short ranged and has a minimum at very short distances. The third case, the formation of mesostructures, is still a controversial subject.

The different terms of the inter-particle interaction energy are double layer and dispersion interactions (2-D DLVO), capillary interactions which are intrinsic to interfacial phenomena and have no analogy in the 3-D case, structural forces and monopolar and dipolar interactions. These terms have a different weight in the total interaction energy. The capillary interaction involved the flotation and immersion capillary forces, depending on whether the interfacial deformation is produced by the weight of the particle or by the wetting properties of the particle surface that, in turn, are functions of the position of the contact line and the value of the contact angle rather than gravity. Immersion forces are long-range interactions and play an important role in the behaviour of colloidal monolayers even for small particles.

Acknowledgements

The authors acknowledge financial support from the "Ministerio de Educación y Ciencia, Plan Nacional de Investigación (I+D+i), MAT2003-08356-C04-01" and from the European Regional Development Fund (ERDF). Helpful discussions with Dr. Alvaro Dominguez are also acknowledged.

References

1. A. Moncho-Jordá, F. Martínez-López and R. Hidalgo-Álvarez, *J. Colloid Interf. Sci.*, **249** (2002), 405.
2. R. Aveyard, J.H. Clint, D. Nees and V.N. Paunov, *Langmuir*, **16** (2000), 1969.
3. R. Aveyard, B.P. Binks, J.H. Clint, P.D.I. Fletcher, T.S. Horozov, B. Neumann, V.N. Paunov, J. Annesley, S.W. Botchway, D. Nees, A.W. Parker, A.D. Ward and A.N. Burgess, *Phys. Rev. Lett.*, **88** (2002), 246102.
4. P. Pieranski, *Phys. Rev. Lett.*, **45** (1980), 569.
5. D.J. Robinson and J.C. Earnshaw, *Phys. Rev. A*, **46** (1992), 2045.
6. J. Ruiz-Garcia, R. Gámez-Corrales and B. I. Ivlev, *Phys. Rev. E*, **58** (1998), 660.
7. J. Ruiz-Garcia, R. Gámez-Corrales and B. I. Ivlev, *Physica A*, **236** (1997), 97.
8. F. Ghezzi, J.C. Earnshaw, M. Finnis and M. McCluney, *J. Colloid Interf. Sci.*, **238** (2001), 43.
9. V. Lowry, M.S. El-Aasser, J.W. Vanderhoff, A. Klein and C.A. Silebi, *J. Colloid Interf. Sci.*, **112** (1986), 521.
10. J.T. Long, *Engineering for Nuclear Fuel Reprocessing*, Gordon and Breach Science Publishing Inc., New York, 1967.
11. R.J. Pugh and L. Nishkov, Paper at *28th Conference of Metallurgists*, ed. G.S. Dobby and S.R. Rao, Montreal, 1990.
12. D.F. Williams, *Ph.D. Thesis*, University of Washington, 1991.
13. D.J. Robinson and J.C. Earnshaw, *Langmuir*, **9** (1993), 1436.
14. A. Martín, *Ph.D. Thesis*, Universidad de Granada, 1993.
15. J.C. Earnshaw, *J. Phys. D*, **19** (1986), 1863.
16. M. Quesada-Pérez, A. Moncho-Jordá, F. Martínez-López and R. Hidalgo-Álvarez, *J. Chem. Phys.*, **115** (2001), 10897.
17. D.Y.C. Chan, J.D. Henry Jr. and L.R. White, *J. Colloid Interf. Sci.*, **79** (1981), 410.
18. R.J. Hunter, *Foundations of Colloid Science*, Vol. 1, Clarendon Press, New York, 1987.
19. E.J.W. Verwey and J.Th.G. Overbeek, *Theory of Stability of Lyophobic Colloids*, Elsevier, Amsterdam, 1948.
20. B.V. Derjaguin, *Kolloid Zeit.*, **69** (1934), 155.
21. M.P. Lyne, *M.S. Thesis*, University of British Columbia, 1989.
22. S. Levine and M.P. Lyne, Paper at *63rd ACS Colloid and Surface Science Symposium*, Seattle, 1989.
23. J. Gregory, *J. Colloid Interf. Sci.*, **83** (1981), 138.
24. J.Th.G. Overbeek, *Proc. Königs Akad. Wetensch. B*, **69** (1966), 501.
25. M.L. Gee and J. Israelachvili, *J. Chem. Soc. Faraday Trans.*, **86** (1990), 4049.
26. J. Israelachvili and R.M. Pashley, *J. Colloid Interf. Sci.*, **98** (1984), 500.
27. N.V. Churaev, Uspekhi Kollidnoj Khimii, *FAN Tashkent*, **70** (1987), 250.
28. H.K. Christenson and P.M. Claesson, *Science*, **239** (1988), 390.

29. J. Israelachvili, *Intermolecular and Surface Forces*, 2nd edn., Academic Press, New York, 1992.
30. J. Sun and T. Stirner, *Langmuir*, **17** (2001), 3103.
31. T. S. Horozov, R. Aveyard, J.H. Clint and B.P. Binks, *Langmuir*, **19** (2003), 2822.
32. J. Sun and T. Stirner, *Phys. Rev. E*, **67** (2003), 051107.
33. R. Aveyard, J.H. Clint, D. Nees and N. Quirke, *Langmuir*, **16** (2000), 8820.
34. J.-P. Hansen and I.R. McDonald, *Theory of Simple Liquids*, 2nd edn., Academic Press, London, 1986.
35. A. Moncho-Jordá, M. Quesada-Pérez, F. Martínez-López and R. Hidalgo-Álvarez, *Progr. Colloid Polym. Sci.*, **123** (2004), 119.
36. A. Schmitt, G. Odriozola, A. Moncho-Jordá, J. Callejas-Fernández, R. Martínez-García and R. Hidalgo-Álvarez, *Phys. Rev. E*, **62** (2000), 8335.
37. P. Meakin, T. Vicsek and F. Family, *Phys. Rev. B*, **31** (1985), 564.
38. T. Vicsek and F. Family, *Phys. Rev. Lett.*, **52** (1984), 1669.
39. A. Moncho-Jordá, F. Martínez-López, A.E. González and R. Hidalgo-Álvarez, *Langmuir*, **18** (2002), 9183.
40. Z. Hórvölgyi, M. Máté and M. Zrínyi, *Colloids Surf. A*, **84** (1994), 207.
41. D. Stamou, C. Duschl and D. Johannsmann, *Phys. Rev. E*, **62** (2000), 5263.
42. P.A. Kralchevsky, N.D. Denkov and K.D. Danov, *Langmuir*, **17** (2001), 7694.
43. M.G. Nikolaides, A.R. Bausch, M.F. Hsu, A.D. Dinsmore, M.P. Brenner, C. Gay and D.A. Weitz, *Nature*, **420** (2002), 299.
44. M. Megens and J. Aizenberg, *Nature*, **424** (2003), 1014.
45. K.D. Danov, P.A. Kralchevsky and M.P. Boneva, *Langmuir*, **20** (2004), 6139.
46. M. Oettel, A. Domínguez and S. Dietrich, *Phys. Rev. E*, **71** (2005), 051401.
47. F. Ghezzi and J.C. Earnshaw, *J. Phys. Condens. Matter*, **9** (1997), L517.
48. J. Ruiz-García and B.I. Ivlev, *Mol. Phys.*, **95** (1998), 371.
49. P.A. Kralchevsky and K. Nagayama, *Adv. Colloid Interf. Sci.*, **85** (2000), 145.
50. M. Abramowitz and J.A. Stegun, *Handbook of Mathematical Functions*, Dover Press, New York, 1974.
51. P.A. Kralchevsky, V.N. Paunov, I.B. Ivanov and K. Nagayama, *J. Colloid Interf. Sci.*, **151** (1991), 79.
52. V.N. Paunov, P.A. Kralchevsky and N.D. Denkov, *J. Colloid Interf. Sci.*, **157** (1993), 100.
53. F. Martínez-López, M.A. Cabrerizo-Vílchez and R. Hidalgo-Álvarez, *J. Colloid Interf. Sci.*, **232** (2000), 303.
54. P.A. Kralchevsky, N.D. Denkov, V.N. Paunov, O.D. Velev, I.B. Ivanov, H. Yoshimura and K. Nagayama, *J. Phys. Condens. Matter*, **6** (1994), A395.
55. P.A. Kralchevsky, C.D. Dushkin, V.N. Paunov, N.D. Denkov and K. Nagayama, *Prog. Colloid Polym. Sci.*, **98** (1995), 12.
56. P.A. Kralchevsky and N.D. Denkov, *Curr. Opin. Colloid Interf. Sci.*, **6** (2001), 383.
57. J.C. Fernández-Toledano, A. Moncho-Jordá, F. Martínez-López and R. Hidalgo-Álvarez, *Langmuir*, **20** (2004), 6977.
58. M.G. Nikolaides, A.R. Bausch, M.F. Hsu, A.D. Dinsmore, M.P. Brenner, C. Gay and D.A. Weitz, *Nature*, **424** (2003), 1017.

Juan Carlos Fernández-Toledano (far left) is a graduate in Physics and currently a Ph.D. student at the University of Granada. He is studying colloidal systems confined at interfaces, considering the different kinds of structures that appear when colloidal particles are spread on interfaces. He is also interested in the study of the distribution of fractal aggregates over a plane.

Arturo Moncho-Jordá (second left) is Assistant Professor in the Department of Applied Physics and member of the Biocolloid and Fluid Physics Group at the University of Granada. His current topics are the study of 2-D aggregation processes by means of experiments, simulations and kinetic rate equations, and inhomogeneous properties of polymer–colloid mixtures using density functional methods.

Francisco Martínez-López (second right) is a Lecturer in the Department of Applied Physics at the University of Granada. He is interested in the experimental and theoretical study of colloidal stability of disperse systems in both 2- and 3-D and also in the computer simulation techniques used to characterize the processes that these systems undergo.

Roque Hidalgo-Álvarez (far right) is Professor of Physics and currently Head of the Biocolloid and Fluid Physics Group in the Department of Applied Physics at the University of Granada. He has expertise in the experimental characterization of the colloidal stability of disperse systems in both 2- and 3-D using optical techniques, but his research interests also include electrokinetic processes, liquid matter topics and the thermodynamics of nonequilibrium systems.

Tel.: +34-958-243213; *Fax*: +34-958-243214; *E-mail*: rhidalgo@ugr.es

4

Particle-Assisted Wetting

Werner A. Goedel

Department of Physical Chemistry, Chemnitz University of
Technology, Chemnitz, 6209111 Germany

4.1 Introduction

Wetting and de-wetting of surfaces by a liquid are fascinating phenomena of great importance for scientific and technological problems including the self-protection of living organisms,[1] the production of microstructures,[2] as well as the integrity and uniformity of decorative, lubricating and protective coatings and the prevention of fogging.[3] Wetting is influenced by short-range forces, such as hydrogen bonding and donor/acceptor interactions, and by long-range dispersion forces. Depending on the relative strength of these forces, one can observe de-wetting, complete wetting or partial wetting. In the first case, the liquid forms lenses co-existing with the bare surface. In the second case, any amount of liquid applied to the surface will spread out as an even layer with a thickness given by the volume of the applied liquid per area of the surface. In the last case, the competition between favourable short-range forces and unfavourable long-range forces gives rise to the formation of a wetting layer of limited thickness, which often is a monolayer but may in principle have any thickness, that co-exists with lenses formed by excess of the liquid. While most of us are familiar with the technological importance of wetting of solid surfaces, it is worth noting that the wetting of liquid surfaces is technologically important as well, *e.g.* for the production of thin uniform sheets of material in float cast processes or in the context of slowing down the evaporation of water from open reservoirs in arid regions. Wetting layers of unlimited thickness may form if the dispersion forces are favourable. These dispersion forces can be estimated from Lifschitz theory.[4,5] As a rule of thumb, a liquid with a low refractive index is likely to wet a medium with a high refractive index, while a liquid with a high refractive index is unlikely to wet a medium with a low refractive index. Given the relatively low refractive index of water, most organic liquids do not wet a water surface.[6,7] Only some low molar mass materials of low refractive index, such as pentane, wet a water surface and some liquids have been shown to undergo a wetting transition

152

when the physical properties of the liquid sub-phase are adjusted by heating[8] or by addition of soluble substances that alter the refractive index.[9,10]

One obvious difference between a liquid and a solid interface is the fact that a solid interface is not deformed by an applied liquid and, if lenses are formed, they have a flat bottom. The contact angles can be described by Young's equation. Lenses at a liquid interface, on the other hand, can easily deform that interface and thus lenses with two curved surfaces are formed.

Another striking difference between liquid and solid interfaces is the fact that the latter can be rough. This roughness further influences wetting. The most prominent examples are the hydrophobic surfaces of plant leaves or artificial "super"-hydrophobic surfaces. In these cases, a solid surface is roughened[11] or decorated with hydrophobic solid particles.[1] In the case of complete contact between an applied water droplet and the solid surface, increasing the roughness increases the interaction between the liquid and the surface. Thus, the apparent contact angle, observed at length scales exceeding the length scales of the particles or protrusions, is considerably higher (if the surface is hydrophobic) or lower (if the surface is hydrophilic) than the local contact angle at the air–water–solid contact line.[12] In the case of high local contact angles, the liquid retracts further from the surface and only incompletely covers it, in effect "lifting" the liquid droplets off the surface, creating an air–liquid surface and placing only the protrusions of the solid surface or the apex of the particles that rest on the surface into the liquid.[13,14] The position of the protrusions or particles within the newly created air–liquid surface is given by the shape of the protrusion and the contact angle between the solid particle and the liquid–air surface.

If particles are not applied to a solid interface as mentioned above, but to a liquid interface, we see again a decisive difference. In contrast to solid surfaces, particles can *penetrate* liquid interfaces and by doing so they reduce the area of the bare liquid interface. This gives rise to a significant reduction in total interfacial energy. Thus, particles have a strong tendency to adsorb to liquid interfaces, can be used to stabilize emulsions, *e.g.* oil-in-water or water-in-oil[15–18] and form well-defined monolayers at planar liquid interfaces.[19–21] This affinity of particles for a liquid interface can also be utilized to mediate the wetting of the surface of one liquid (for simplicity from now on called "water") by a second one (from now on called "oil"). The following two sections briefly summarize the theoretical basis of such particle-assisted wetting, and give some examples of experiments that confirm these expectations.

4.2 Theory

If a liquid surface (of say, water) is covered by a layer of a second liquid (of say, oil), the change in energy per unit area, $\Delta E/A$, is given by the sum of the interfacial tensions of the newly generated interfaces, γ_{ao} and γ_{wo}, minus the interfacial tension of the liquid without the layer, γ_{aw}; (see Figure 4.1)

$$\Delta E/A = \gamma_{ao} + \gamma_{wo} - \gamma_{aw} \qquad (4.1)$$

On a water surface, however, this scenario is more the exception than the rule. Most oils applied to a water surface form lenses instead of a thin layer[6] and the value of $\Delta E/A$ calculated for a hypothetical layer of oil on water is positive. In the case of partial wetting, one has a layer of limited thickness on the water surface. All phases and interfaces are considered to be in equilibrium and saturated with each other. Thus, in the case of partial wetting, the interfacial tension γ_{aw} is not the surface tension of pure water, but the effective tension of water covered by the thin equilibrium layer.

On the other hand, placing a particle from one of the adjacent phases into a fluid–fluid interface is associated with a negative change in interfacial energy. This gain in energy is due to the fact that part of the fluid–fluid interface is replaced by the particle. This gain in energy per particle is equal to $\Delta E = -\pi r^2 \gamma (1 - \cos \theta)^2$. In this equation, r is the radius of the particle, γ is the fluid–fluid interfacial tension and θ is the angle made by the particle and the fluid–fluid interface (measured into the fluid in which the particle was incorporated before contacting the interface). If the particle approaches the interface from the other side, the term $-\cos \theta$ changes

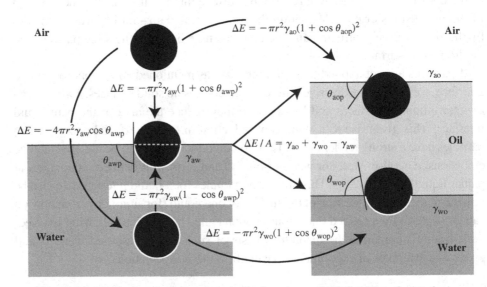

Figure 4.1 Placing particles from any bulk phase into a fluid–fluid interface reduces the interfacial energy. The particles may be placed into any of the interfaces, *e.g.* air–water (left), oil–water (bottom right) and oil–air (top right), involved in the formation of a wetting layer. This can give rise to a reduction of the total interfacial energy large enough to induce wetting. The symbols are defined in the text.

sign as indicated in Figure 4.1 on the upper left hand side. This equation can be applied to all interfaces involved in the formation of a wetting layer, *i.e.* oil–air, oil–water and air–water. In the following, it is assumed that a mixture of oil and particles applied to a water surface can adopt the five scenarios schematically depicted in Figure 4.2: formation of lenses with (i) complete incorporation of the particles, (ii) complete separation between particles and oil, or the formation of a laterally uniform mixed layer with the particles adsorbed to (iii) the top, (iv) the bottom or (v) both interfaces of the layer. One can use the above equations to calculate the total interfacial energies of all these scenarios at given interfacial tensions, contact angles and areas per particle, and decide which of these scenarios is most favourable.[22] The contact angles and interfacial tensions are correlated with each other through Young's equation. In order to describe the system completely, only five independent parameters are needed: the area per particle, the contact angles at two out of the three 3-phase contact lines involved and the ratio of two interfacial tensions to the third one.

The results of these calculations for a hypothetical liquid with identical interfacial tensions at the upper and lower interfaces of the wetting layer and with dense

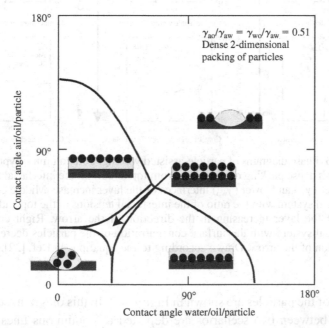

Figure 4.2 Theoretical phase diagram of particle-assisted wetting of an oil on a water surface. Five scenarios are depicted depending on the magnitude of the contact angles of the particles at the oil–water and the oil–air interfaces (both measured through the oil phase). One of these scenarios is energetically more favourable than the other four.

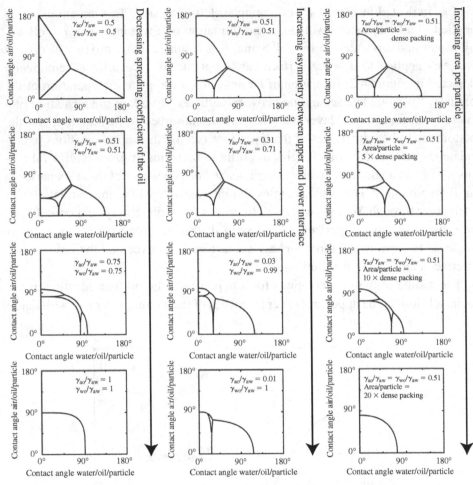

Figure 4.3 Phase diagrams of particle-assisted wetting. Left column – hypothetical system with dense packing of particles. From top to bottom the interfacial tensions at the upper (γ_{ao}) and lower (γ_{wo}) interfaces of the layer increase. Middle column – hypothetical system with the ratio of the interfacial tensions at the top and bottom surface of the layer increasing in the direction of the arrow. Right column – hypothetical system with the surface concentration of the particles decreasing in the direction of the arrow. Drawn according to the equations in Ref. [22].

2-D packing of the particles are shown in Figure 4.2. In this diagram, conditions of co-existence between two scenarios are depicted as continuous lines. The areas confined by the lines indicate regions in which the scenario depicted by the corresponding graphics is most favourable. At very high or very low contact angles at the interfaces with oil (measured through oil), no particle-assisted wetting is expected. Instead, the formation of lenses is predicted. These lenses may either be separated from the particles (high contact angles at both interfaces) or completely engulf them (low contact angles at both interfaces). Around contact angles of 90° ($\pi/2$), the diagram predicts that particle-assisted wetting will occur, with the particle being preferentially adsorbed to the interface at which the contact angle is closest to 90°, or adsorbed to both interfaces if the contact angles are comparable. The conditions leading to particle-assisted wetting depend not only on the contact angles, but also on the interfacial tensions with the oil and on the area per particle. These dependencies are illustrated in Figure 4.3. The relative position of regions of stability of each scenario in these diagrams is the same as in Figure 4.2. To avoid overloading the graphs with schematic drawings, Figure 4.3 was drawn without the graphical illustrations depicted in Figure 4.2. The first column of Figure 4.3 depicts "symmetric" cases in which the surface tension of the upper surface of the oil layer γ_{ao} is identical to the lower one γ_{wo} and the particles are closely packed at the interface. From top to bottom, the ratio of these interfacial tensions to the interfacial tension of the water surface γ_{aw} is gradually increased. As can be seen from the graphs, the regions of particle-assisted wetting gradually decrease with increasing interfacial tensions of the oil and finally are completely gone in the most unfavourable case. The second column of Figure 4.3 depicts what happens if the interfacial tension at the oil–air surface is made smaller than the interfacial tension at the oil–water interface. The particles preferentially adsorb to the interface with the higher interfacial tension until finally the region representing a layer with the particles at the top surface vanishes. The last column of Figure 4.3 shows the effect of decreasing the particle coverage at constant interfacial tensions. With decreasing particle coverage, the predicted regions of particle-assisted wetting decrease and finally vanish again. It is worth noting in this example that even at a coverage of the surface by particles of only 10% of full coverage, particle-assisted wetting is still expected to occur.

4.3 Experiments

Particle-assisted wetting has been investigated by applying mixtures of particles and polymerizable oils to water surfaces.[23] The use of a polymerizable oil has the advantage that one can solidify the oil, transfer it to solid substrates and image it with scanning electron microscopy (SEM). The oils alone do not form mesoscopic wetting layers on a water surface but retract into liquid lenses. The presence of the polar oils reduces the surface tension of water. This indicates that the lenses co-exist with a layer of equilibrium thickness, presumably with an Angstrom-scale monomolecular layer. Examples of such lenses are shown for the oil trimethylol-propane trimethacrylate (TMPTMA) in Figure 4.4(a). If colloidal silica particles of 140 nm diameter coated with polyisobutene chains (SiO$_2$–PIB) are mixed with

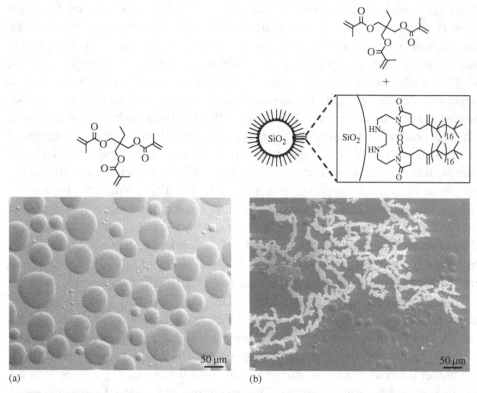

Figure 4.4 SEM images of oils and their mixtures with silica particles after application to a water surface, UV-irradiation and transfer to mica substrates. The chemical structures are depicted above each image. (a) Lenses formed by the oil TMPTMA alone. (b) Particle aggregates and oil lenses formed by a mixture of TMPTMA and silica particles coated with PIB chains. Taken from Ref. [23]; with permission of the American Chemical Society.

the oil TMPTMA and applied to a water surface, one again observes the formation of small lenses already by the unaided eyes. The particles form irregular 3-D aggregates on the surface that co-exist with oil lenses (see Figure 4.4(b)). On the other hand, mixtures of the oil with silica particles that were coated with methacrylate-terminated silanes (SiO_2–TPM) spread evenly on the water surface without forming lenses. A close-up of a cross-section of such a layer is shown in Figure 4.4(c). For the sake of good sample preparation, the particles were removed via etching with hydrofluoric acid before cutting. If a slightly different oil, pentaerythrol tetraacrylate (PETA), is used, layers can be obtained in which the particles adhere to both interfaces of the layer (see Figure 4.4(d)). As discussed before, the formation of the mixed layers is expected to be a function of the contact angles at the fluid–oil-particle interfaces. It was possible to estimate these contact angles by embedding the particles in the relevant interfaces, photo cross-linking the oil and analysing the

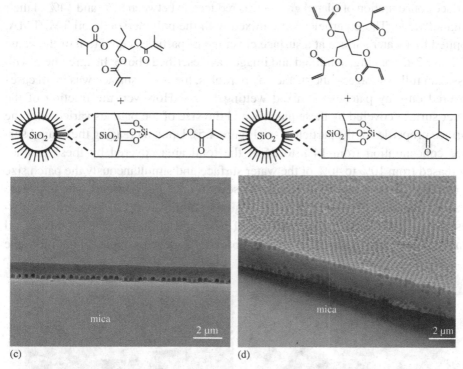

(c)   (d)

Figure 4.4 (*continued*) (c) Side view of a wetting layer formed from silica particles coated with methacrylate-terminated silane and TMPTMA. Particles only adsorb at the lower oil–water interface. (d) Wetting layer formed from a mixture of silica particles coated with methacrylate-terminated silane and PETA oil. Particles adsorb at the oil–air surface and the oil–water interface. For the sake of sample preparation, the particles were removed by etching with hydrofluoric acid before cutting. Taken from Ref. [23]; with permission of the American Chemical Society.

height traces of atomic force microscope (AFM) pictures of the protruding parts of the embedded particles. It is instructive to compare the contact angles at the upper and lower interface for the system composed of TMPTMA and particles coated with methacrylate-terminated silanes (25° and 50° respectively) with the corresponding contact angles in the system PETA/methacrylate-terminated silane-coated particles (23° and 20° respectively). The dissymmetry of both contact angles in the former case is in agreement with the location of the particles at the lower interface only, while the symmetric conditions in the latter case is in agreement with the observed adsorption of the particles at both interfaces.

The influence of the wettability of the particles has been tested further using silica particles with systematically varied surface properties.[24] Fumed silica particles (irregular clusters composed of spherical primary particles of ≈20 nm diameter) were reacted to various extents with dichlorodimethylsilane in such a way that the surface concentration of silanol groups was reduced to between 87% and 14% of their original value. These particles were mixed with the polymerizable oil TMPTMA, applied to a water surface at a surface coverage of particles equal to half the value of dense 2-D packing, solidified and imaged as described above. In agreement with less than full coverage of the surface by particles, the water surface was in all cases covered only by patches of mixed wetting layers. However, the fraction of the water surface covered by these patches and the size of the patches varied with the surface properties of the particles (see Figure 4.5). Upon increasing the silanol surface concentration from 14% to 36%, the total area covered by these patches increased from 10% to 40% of the water surface and simultaneously the patch size is increased from a few hundred μm^2 to several cm^2. For particles with a surface silanol content of 51%, the patch size begins to diminish again. For particles possessing a silanol surface concentration of 67% and 80%, the co-existence of oil lenses and patches of mixed layers was observed for each sample. In addition, the

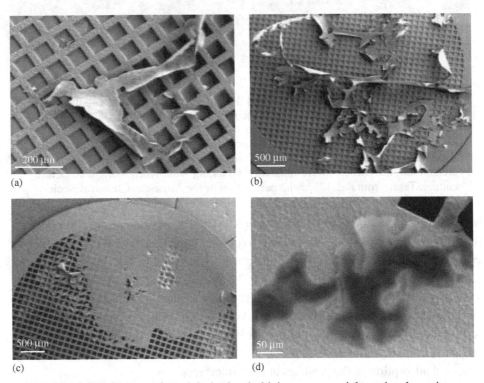

Figure 4.5 Influence of particle hydrophobicity upon particle-assisted wetting. Patches of wetting layers formed from mixtures of the oil TMPTMA and fumed silica particles treated to various extents with dichlorodimethylsilane and bearing (a)14%, (b) 24%, (c) 36% and (d) 51% surface silanol groups. Taken from Ref. [24]; with permission of the American Chemical Society.

area covered by the patches (approximately 20%) is much smaller than those formed from more hydrophobic particles. Thus, particles with a silanol content of 36% have an optimum compatibility with the oil and thus facilitate wetting more efficiently than the other particles. Unfortunately, the non-spherical shape and the inability to determine exact contact angles make a quantitative comparison to the theory presented in Figure 4.2 impossible. Nevertheless, this result is qualitatively in agreement with the expectations.

The above experiments were conducted at a surface coverage very near to a close packed monolayer of the particles. The mathematics depicted above in principle allows any surface concentration of particles lower than full coverage. As can be seen from the last column in Figure 4.3, even at surface concentrations considerably smaller than close packing one can expect particle-assisted wetting. In reality, however, particles embedded in a thin layer of oil are subject to comparatively strong- and long-range attractive capillary forces.[25,26] Thus, if the above experiments are repeated at lower surface concentrations of the particles, one obtains patches of close packed particles that are embedded in oil lenses (see Figure 4.6(a)). Thus, in order to overcome this attraction, one needs repulsive forces between the particles. These forces could in principle be electrostatic repulsion between highly charged particles.[27,28] However, addition of high charge to the particle surface will change the contact angles as well and thus the influence of inter-particle forces on particle-assisted wetting cannot be clearly separated from other effects. Therefore, we chose to use paramagnetic particles[29] and induce repulsive forces[30] by applying an external magnetic field perpendicular to the water surface. This allows us to investigate the differences between attractive and repulsive particles without altering any other properties of the system. Figure 4.6 shows a mixture of polystyrene particles of 2 μm diameter containing 22% of magnetic iron oxide and of the polymerizable oil TMPTMA, applied at sub-monolayer coverage to a water surface. To facilitate the observation, the volume of oil was chosen in such a way that the particles penetrate both interfaces of an assumed oil layer and thus are visible in top view. Without an external field, Figure 4.6(a), the particles form 2-D, comparatively densely packed aggregates. The oil is associated with the particles and forms lenses that do not evenly cover the water surface. An external magnetic field, applied perpendicular to the water surface, induces magnetic dipoles within the particles and thus renders them repulsive. This repulsion is large enough to break up the islands of particles and to spread them over the water surface, taking the oil with them in a thin layer (see Figure 4.6(b)). The oil layer is almost featureless and therefore barely visible in the electron microscopy image on the right. It is more visible if it is solidified and transferred to an electron microscopy grid (inset on the right). Although the layer was shattered in the process, it is clearly seen that the fragments cover the openings of the electron microscopy grid.

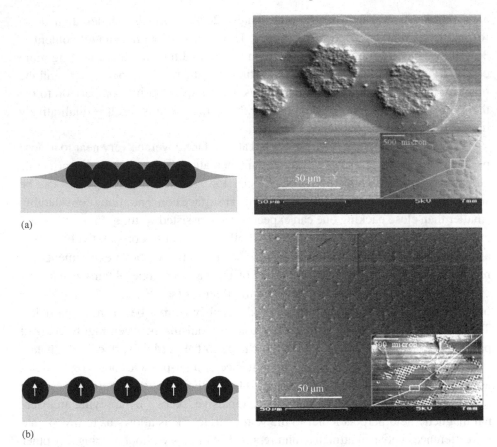

Figure 4.6 Mixtures of the polymerizable oil TMPTMA and paramagnetic polystyrene particles applied to a water surface and subsequently photo cross-linked
and transferred to a mica substrate. A schematic is shown on the left and an SEM
image on the right: (a) In the absence of an external field, (b) in the presence of a
magnetic field perpendicular to the water surface. Taken from Ref. [29]; with permission of the American Institute of Physics.

As shown in Figure 4.4(c), the mixed wetting layers can be prepared with an
amount of oil large enough that the layer thickness is considerably larger than the
dimensions of the particles. If the amount of oil is reduced, one can obtain mixed
layers in which the particles protrude the upper and lower interfaces. Figures 4.7(a)
and (b) show such a solidified layer from the bottom and top view respectively
after cross-linking. If in these layers the particles are selectively removed after
cross-linking, one obtains thin solid membranes composed of the solidified oil
bearing dense arrays of uniform holes.[31] The resulting porous membranes can be
easily transferred to any desired substrate such as mica, silicon wafers and even to

(a)

(b)

(c)

(d)

Figure 4.7 (a) Bottom view and (b) top view of a composite membrane obtained from mixed layers, with a mixing ratio between the photo cross-linkable oil TMPTMA and the silica particles low enough to allow the particles to penetrate through both interfaces of the oil layer. (c) A porous layer obtained from the photo cross-linked oil after removal of the particles. (d) A freely suspended porous membrane transferred to a microscopy grid. The dark dot with an arrow at the bottom of the image is a defect, showing the contrast between a covered and a non-covered area. Taken from Ref. [31]; with permission of Wiley-VCH.

macroporous supports like electron microscopy grids with 100 μm wide openings. Figures 4.7(c) and (d) depict a side view of a resulting porous membrane transferred to a mica plate (part of the membrane was peeled off by scotch tape) and a top view of a freely suspended porous membrane transferred to an electron microscopy grid, respectively. The membrane covers the openings of the grid completely. The pores are visible from the top and the bottom of the membrane and one can see the underlying grid through the pores.

The size of the pores is given by the size of the particles used as templates. It is therefore possible to tune the pore size by using particles of appropriate sizes.[32] Figure 4.8 shows examples of porous membranes obtained by the method described above with various pore sizes. It has to be noted that not only the pore size but also

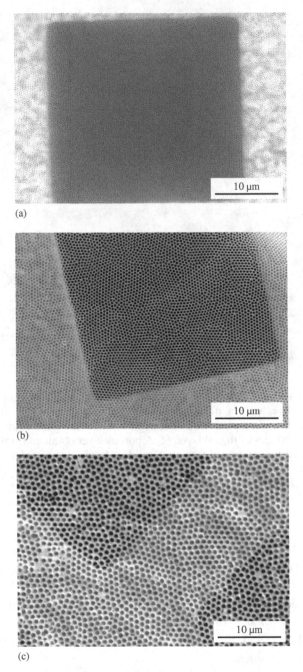

(a)

(b)

(c)

Figure 4.8 Freely suspended porous membranes of various pore size, obtained *via* particle-assisted wetting of a water surface using particles of (a) 0.33, (b) 0.56 and (c) 1.2 μm diameter, photo cross-linking of the oil, removal of the particles and transfer to a support with openings. Taken from Ref. [32]; with permission of the American Chemical Society.

the membrane thickness decreases with the size of the particles. Thus, the membranes with micrometer pore sizes can be handled without any support, whereas membranes with sub-micrometer pores need to be supported by a grid or other (porous) substrate.

4.4 Conclusions

Particle-assisted wetting is a phenomenon which can be utilized to induce wetting of a liquid by a second liquid. It enables us to perform the "mission impossible" of forming wetting layers on a water surface. The concept might be extended further to other liquid surfaces in technologically important areas, like compatibilization in 3-phase blends or flotation in the presence of oil. It furthermore opens up new and technologically interesting applications like the switching of wettability by external fields, and the preparation of regular surface structures and membranes of controlled and uniform pore sizes.

References

1. W. Barthlott and C. Neinhuis, *Planta*, **202** (1997), 1.
2. H. Gau, S. Herminghaus, P. Lenz and R. Lipowsky, *Science*, **283** (1999), 46.
3. R. Wang, K. Hashimoto, A. Fujishima, M. Chikuni, E. Kojima, A. Kitamura, M. Shimohigoshi and T. Watanabe, *Nature*, **388** (1997), 431.
4. J. Israelachvili, *Intermolecular and Surface Forces*, Academic Press, London, 1994, p. 179.
5. W.B. Russel, D.A. Saville and W.R. Schowalter, *Colloidal Dispersions*, Cambridge University Press, New York, 1989, pp. 153–156.
6. A.W. Adamson, *Physical Chemistry of Surfaces*, 5th edn., Wiley, New York, 1990, pp. 110–112.
7. J.T. Davies and E.K. Rideal, *Interfacial Phenomena*, Academic Press, London, 1961, pp. 21–25.
8. E. Bertrand, H. Dobbs, D. Broseta, J. Indekeu, D. Bonn and J. Meunier, *Phys. Rev. Lett.*, **85** (2000), 1282.
9. T. Pfohl, H. Möhwald and H. Riegler, *Langmuir*, **14** (1998), 5285.
10. T. Pfohl and H. Riegler, *Phys. Rev. Lett.*, **82** (1999), 783.
11. T. Onda, S. Shibuichi, N. Satoh and K. Tsujii, *Langmuir*, **12** (1996), 2125.
12. R.N. Wenzel, *Ind. Eng. Chem.*, **28** (1936), 988.
13. P.G. de Gennes, *Rev. Mod. Phys.*, **57** (1985), 827.
14. J. Bico, C. Marzolin and D. Quere, *Europhys. Lett.*, **47** (1999), 220.
15. O.D. Velev, K. Furusawa and K. Nagayama, *Langmuir*, **12** (1996), 2374.
16. B.P. Binks and S.O. Lumsdon, *Phys. Chem. Chem. Phys.*, **1** (1999), 3007.
17. B.P. Binks and S.O. Lumsdon, *Langmuir*, **16** (2000), 8622.
18. S. Arditty, C.P. Whitby, B.P. Binks, V. Schmitt and F. Leal-Calderon, *Eur. Phys. J. E*, **11** (2003), 273.
19. F.C. Meldrum, N.A. Kotov and J.H. Fendler, *Langmuir*, **10** (1994), 2035.
20. M. Szekeres, O. Kamalin, P.G. Grobet, R.A. Schoonheydt, K. Wostyn, K. Clays, A. Persoons and I. Dékány, *Colloids Surf. A*, **227** (2003), 77.
21. J.H. Clint and S.E. Taylor, *Colloids Surf.*, **65** (1992), 61.

22. W.A. Goedel, *Europhys. Lett.*, **62** (2003), 607.
23. H. Xu and W.A. Goedel, *Langmuir*, **19** (2003), 4950.
24. A. Ding, B.P. Binks and W.A. Goedel, *Langmuir*, **21** (2005) 1371.
25. P.A. Kralchevsky, V.N. Paunov, I.B. Ivanov and K. Nagayama, *J. Colloid Interf. Sci.*, **151** (1992), 79.
26. P.A. Kralchevsky, V.N. Paunov, N.D. Denkov, I.B. Ivanov and K. Nagayama, *J. Colloid Interf. Sci.*, **155** (1993), 420.
27. R. Aveyard, J.H. Clint, D. Nees and V.N. Paunov, *Langmuir*, **16** (2000), 1969.
28. R. Aveyard, B.P. Binks, J.H. Clint, P.D.I. Fletcher, T.S. Horozov, B. Neumann, V.N. Paunov, J. Annesley, S.W. Botchway, D. Nees, A.W. Parker, A.D. Ward and A.N. Burgess, *Phys. Rev. Lett.*, **88** (2002), 246102.
29. P. Tierno and W.A. Goedel, *J. Chem. Phys.*, **122** (2005), 094712.
30. K. Zahn, J.M. Mendez-Alcaraz and G. Maret, *Phys. Rev. Lett.*, **79** (1997), 175.
31. H. Xu and W.A. Goedel, *Angew. Chem. Int. Ed.*, **42** (2003), 4694.
32. F. Yan and W.A. Goedel, *Chem. Mat.*, **16** (2004), 1622.

Werner Goedel studied Chemistry at the University of Cologne from 1983–1988. He obtained his Ph.D. in 1992 at the Max-Planck-Institute for Polymer Research and the University of Mainz and then did postdoctoral work at Stanford University. From 1993–1999 he worked at the Max-Planck-Institute for Colloids and Interfaces (Berlin). After his habilitation at the University of Potsdam, he moved his group in 2000 to the University of Ulm. In 2003, he joined the Polymer Physics Department of BASF in Ludwigshafen as research associate. Since 2004, he has been a Full Professor and holds the Chair of Physical Chemistry at Chemnitz University of Technology. His current research interests include self-assembled monolayers, wetting phenomena, thin and porous membranes and chemical vapour deposition.

Tel: + 49-371-5311713;   *Fax*: + 49-371-5311371;   *E-mail*:   werner.goedel@chemie.tu-chemnitz.de

Section 2

Particles at Curved Liquid Interfaces

5

Particle-Laden Interfaces: Rheology, Coalescence, Adhesion and Buckling

Gerald G. Fuller,[a] Edward J. Stancik[a] and Sonia Melle[b]

[a] Department of Chemical Engineering, Stanford University,
Stanford, CA 94305-5025, USA

[b] Departamento de Óptica, Facultad de Ciencias Físicas,
Universidad Complutense de Madrid, Ciudad Universitaria,
Madrid 28040, Spain

5.1 Introduction

Particulate additives are found in the formulations of a great many high-surface area products in the form of emulsions and foams. Their presence is normally desirable for the purposes of stability. In the case of ice cream foams, tiny fat globules can attach themselves to the surfaces of the air pockets and hinder the process of coarsening by Ostwald ripening.[1] In this case, the particles are a natural ingredient. In other instances, colloidal particles are deliberately added and Pickering emulsions are an important example.[2,3] The occurrence of particles leading to stabilization can also be unwelcome, as in the case of emulsions formed when seawater and crude oils vigorously mix. This environmental problem can lead to very stable emulsions as a result of particles formed by asphaltenes or clay collecting at the oil–water interface.[4]

The presence of particles at a fluid–fluid interface leads to numerous, profound consequences. Since a very large amount of energy is normally required to remove a particle from an interface (see Chapter 1 for a detailed explanation of this point), particles in these monolayers are normally irreversibly attached. Furthermore, as described in Chapter 2, these systems can be modified by exquisite tuning of inter-particle forces, particle chemistry and particle size to create a wide range of morphologies of these "2-D suspensions". These range from flocculated networks,[5] 2-D (two dimensional) crystals[6] and froth-like arrangements.[7] As a result, particle mono-layers offer a wide range of mechanical responses and the interfacial rheology of these systems will generally be highly viscoelastic.[8]

The mechanical behavior of an interface will largely control the draining process that occurs when two droplets approach one another during the process of coalescence. This squeezing flow of the continuous phase between the droplets will necessarily force a surface flow of the respective surfaces and the interfaces will experience surface flow gradients that will be resisted by its rheological character. This response is captured using a rheological constitutive equation of the general form

$$\tau_{ij}^{s} = f_{ij}(\nabla \mathbf{u}^{s}) \tag{5.1}$$

where τ_{ij}^{s} is the surface stress tensor with dimensions of force per unit length and $\nabla \mathbf{u}^{s}$ is the velocity gradient tensor of the surface flow field, $\mathbf{u}^{s}$. The linear form of this relationship is the Boussinesq–Scriven equation for Newtonian interfaces[9]

$$\tau_{ij}^{s} = 2\mu^{s}D_{ij}^{s} + (\nu^{s} - \mu^{s})D_{kk}^{s}\delta_{ij} - \Pi\delta_{ij} \tag{5.2}$$

where $D_{ij}^{s} = (1/2)[(\partial u_{i}^{s}/\partial x_{j}) + (\partial u_{i}^{s}/\partial x_{i})]$ is the rate of strain tensor, μ^{s} is the surface shear viscosity, ν^{s} is the surface dilatational viscosity and Π is the surface pressure of the interface. The second term on the right-hand side recognizes that most interfaces are able to expand and contract in area and cannot be assumed to be incompressible, which is an assumption that is readily adopted in the case of most bulk liquids ($D_{kk} = 0$). Although this linear relationship can describe the behavior of a limited number of interfaces and serves as a convenient introduction to the concepts of interfacial viscosities, most particle-laden interfaces are non-Newtonian and require a non-linear connection between surface stresses and surface velocity gradients.

Equation (5.2) was written using simple Cartesian coordinates, and this brings up another important distinction between the rheology of bulk materials and the rheology of interfaces. Since interfaces have a shape, it is normally necessary to utilize generalized coordinates in the description of vectors and tensors except for the limiting case of planar interfaces. However, this level of complexity is not required for the discussion in this chapter. One must also realize that since interfaces can bend, mechanical properties, such as the bending modulus, κ, must be considered, whereas this is normally only necessary for special bulk materials, such as liquid crystals. The bending modulus has long been recognized as an important consideration for vesicles[10] and has more recently been taken up for particle monolayers by Aveyard et al.[11] and Kralchevsky et al.[12]

In addition to droplet coalescence, the interfacial rheology of particle-laden interfaces will have an important influence on the process of desiccation of sessile droplets and Ostwald ripening. These important problems are ubiquitous to both natural phenomena and industrial applications. Consider, e.g., the drying of a droplet consisting of a suspension of particles, which has been studied in detail by the group of Allain.[13] These authors observe a number of curious and distinct shape transitions of such droplets as they lose water through evaporation.

This chapter will first present recent findings on the fluid mechanical response and interfacial rheology of particle monolayers. This will include simple observations of how such layers react to well-controlled surface flows, as well as quantitative measurements of interfacial viscoelasticity. Both repulsive particles that form well-ordered lattices and attractive, network-forming particles will be discussed. The coalescence and adhesion of particle-laden droplets will then be discussed. The critical role of the contact angle adopted by the particles at an oil–water interface will be evident. Indeed, the drop-to-drop adhesion event will be used as a means of determining this important property. Finally, the problem of shape transitions in isolated, sessile and pendant droplets covered by particles and subject to volume changes will be discussed. This latter problem serves as a model of desiccation and Ostwald ripening processes.

5.2 Interfacial Shear Rheology and Fluid Dynamics of Particle Monolayers

In this section, the flow behavior and rheology of particle monolayers will be examined. We will restrict ourselves to situations where the applied surface flow is accomplished at constant surface area. Polystyrene spheres offer a convenient system for study since they are commercially available with narrow size distributions and many different particle diameters. In the majority of results discussed here, the particles have a diameter of 3 μm and reside between decane and water. Figure 5.1 shows a regular array of such particles. The result is a hexagonal pattern representing a balance of electrostatic repulsion forces and attractive forces. The nature

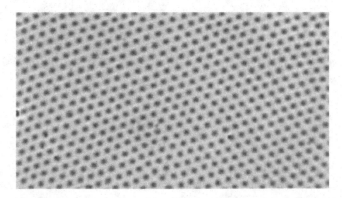

Figure 5.1 A hexagonal array of 3μm diameter polystyrene spheres residing at the planar interface between decane and water.

of these inter-particle forces is not well understood, but there are convincing arguments that the electrostatic forces are most likely carried through the oil phase.[11] The explanation for the attractive forces normally invokes capillary forces. However, this in turn requires a deflection of the interface in the vicinity of the particles, and it is the source of this deflection that has been the subject of debate. Normally the particle sizes are too small to allow gravitational forces to cause the effect. This apparent dilemma has been addressed by a number of recent papers and attributes the distortion of the interface to a pressure on the particle that originates from the differences in the dielectric constants of the two fluids comprising the interface.[14] This has been discussed in detail in Chapter 3.

These layers are mobile and able to deform. Nonetheless, it is expected that the crystalline morphology will resist flow-induced deformation as a result of the inter-particle forces that place the particles in potential wells. The depth of these wells will increase with the particle area fraction so that hydrodynamic forces will find it increasingly difficult to shear these 2-D suspensions as their concentration increases. This phenomenon was explored in Refs. [15] and [16]. In these papers, a parallel band device was used to produce a simple shearing flow of crystalline arrays at the interface between water and decane. The size of the polystyrene spheres (3 μm diameter) made it possible to use simple optical microscopy to track the particle motions as functions of concentration and velocity gradient. Two types of particle motion were observed and the phase diagram pictured in Figure 5.2 was constructed.

At low concentrations and high shear rates, hydrodynamic forces are sufficiently large to induce columns of particles to shift in registry relative to adjacent columns as shown in the inset cartoon. Fourier transformation of the particle positions at any given time (not shown) indicate that the hexagonal structure shown in Figure 5.1 becomes distorted and aligned relative to the flow direction. This anisotropic structure adopts a time-independent appearance once steady state is achieved. Upon cessation

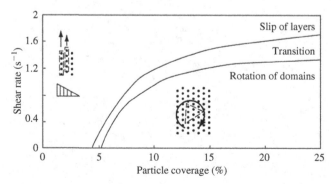

Figure 5.2 Phase diagram of particle motion as a function of shear rate and particle surface concentration for the system in Figure 5.1.

of flow, this anisotropy quickly relaxes and the array of particles assumes its original isotropic, hexagonal form. As the shear rates are reduced or as the concentration is increased, a transition was observed away from a slip flow and towards a domain-driven flow. The domains consist of 2-D clusters of particles in hexagonal arrangements. Fourier transformation of images of the particle positions in this flow regime reveal that the hexagonal arrangement of the particles is not distorted, but melting of the crystalline lattice occurs from the hexagonal rest state towards a hexatic structure. However, the microstructure oscillates indefinitely in time and this is manifested in two ways. First, although the particles within the domains retain a local hexagonal arrangement, they rotate in time in the same sense as the vorticity of the simple shear flow. However, the angular velocity of this rotation is not constant and speeds up and slows down every 60° on the lattice. Second, the melting is also oscillatory and the structure fluctuates at regular intervals between a hexagonal crystal and a hexatic.

Using an interfacial stress rheometer (ISR),[17] the interfacial rheology of these layers can also be measured. This instrument is built around a Langmuir trough so that either insoluble or soluble monolayers can be studied. In the case of particle monolayers, which are essentially insoluble, this platform provides a convenient way to systematically vary the area fraction. Residing within the trough is a channel that is formed from a rectangular tube that has been cut in half. This U-shaped channel is placed in the trough so that its parallel sides point upward. Furthermore, the channel is cut to a height so that its sides are coincident with the depth of the trough. In this case, the interface formed between decane (or air, depending on the problem being studied) and water simultaneously meets the top edge of both the channel and the trough. Within the channel, a slender magnetic needle is placed at the interface and is held there by the force of interfacial tension. Because a meniscus normally forms within the channel, gravitational forces normally self-center the needle so that it is in the center of the channel and parallel to its sides. Surrounding the Langmuir trough, two large electromagnetic coils are placed in the Helmholtz condition so that a constant magnetic field gradient can be applied to the needle. This gradient will apply a force to the needle that will cause it to push against the resistance offered by the monolayer. As it glides parallel to the channel, its position within the channel is continuously monitored using an inverted microscope. Two types of experiments are normally performed: dynamic tests where oscillatory forces are applied and creep compliance measurements where a constant force is established. In the first case, the simultaneous measurement of the stress on the needle and the resulting strain in the monolayer yields the dynamic modulus, $G^*(\omega)$, where ω is the applied frequency. The dynamic modulus is normally divided into two contributions – the elastic modulus, $G'(\omega)$, and the viscous modulus, $G''(\omega)$. These two quantities capture the viscoelastic nature of the monolayer. One important criterion that is often considered is the relative magnitude of these two quantities. Roughly, if $G'(\omega) > G''(\omega)$, the interface is considered to be more elastic (and solid) than viscous. In creep compliance measurements, one normally reports the compliance (the ratio of strain to the applied stress) as a function of time. This is often fit to simple mechanical models to extract the surface elastic modulus, relaxation time and surface viscosity. Note that this instrument measures the surface viscoelasticity by controlling the applied stress. This is in contrast to alternative rheological methods where one controls the applied strain and measures the resulting stress. For this reason, the present instrument is referred to as a stress rheometer.

A comprehensive study of the viscoelasticity of particle monolayers using dynamic testing can be found reported in Ref. [8]. This paper presents measurements of the elastic and viscous moduli as functions of frequency. Measurements were taken over a wide range of concentration of polystyrene particles and a time-concentration scaling was found to successfully collapse the data for all concentrations onto single master curves. In analogy to time-temperature superposition that is used to collapse data for polymeric liquids, this normalization of the data indicates that the intrinsic relaxation time scale of the layers is intimately connected to the concentration. At high frequencies, the scaled values of $G''(\omega)$ exceeded those of $G'(\omega)$ and this is explained by the fact that the primary source of dissipation at high frequency arises from the purely viscous liquid–liquid interface between the particles.

The results of creep compliance measurements of the surface viscosity are reported in Figure 5.3 for 3 μm polystyrene spheres at the decane–water interface. In the vicinity of 75–80% coverage, the viscosity increases dramatically from levels that are not much larger than the value for a clean decane–water interface. This concentration is much higher than the concentration ranges reported in Figure 5.2 for the transition between slip and domain flow. The transition at the higher area fraction appears to be a "jamming" effect where the mobility of the interface is rapidly reduced towards a solid-like response.

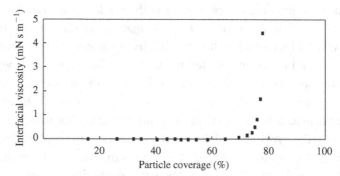

Figure 5.3 Interfacial viscosity of a monolayer of 3μm diameter polystyrene spheres at the decane–water planar interface as a function of particle surface concentration. Taken from Ref. [8]; with permission of the American Physical Society.

5.3 Coalescence and Adhesion of Emulsion Droplets

One important aim of placing particles at liquid interfaces is to stabilize droplets against coalescence. However, the success of this application depends on a number of factors, such as particle concentration, inter-particle forces and the particle contact angle. The rule of thumb for the contact angle is that the majority of the particle surface area resides within the continuous phase. For polystyrene spheres at the interface between decane and water, the contact angle measured into the aqueous phase is 130°. This measurement was accomplished by spin coating a layer of the particles onto glass and fusing them together by raising the temperature slightly above the glass transition of polystyrene. A contact angle goniometer can then be used to measure the contact angle by placing a droplet of water under decane on the slide. Because the contact angle is greater than 90°, these particles will more naturally stabilize water-in-oil emulsions.

The process of coalescence of two drops can be mimicked with the experimental setup shown in Figure 5.4.[6,18] The case of the oil droplet in part (a) of this figure is expected to be less stable and more apt to coalesce as the droplet is brought against the planar interface. Indeed, this is what is observed in experiment, as shown in Figure 5.5 where a droplet of decane of approximately 1 mm diameter is brought up

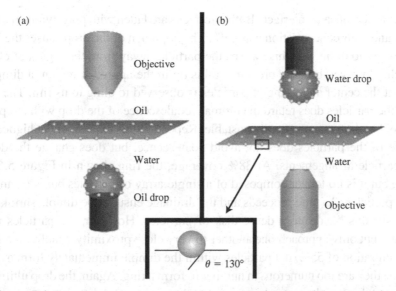

Figure 5.4 Experimental arrangement for the examination of the approach of a droplet against a planar liquid–liquid interface. Both the droplet and planar interface are laden with particles. (a) An oil droplet is brought from below against a water–decane interface and viewed using an upright microscope. (b) A water droplet is brought from above against a decane–water interface and viewed using an inverted microscope.

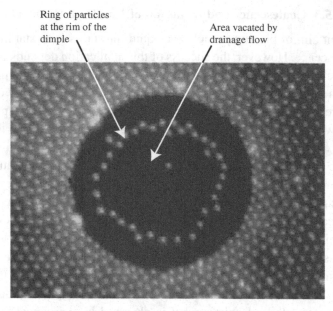

Ring of particles
at the rim of the
dimple

Area vacated by
drainage flow

Figure 5.5 Rearrangement of polystyrene spheres as a result of a drainage flow that occurs when a drop of decane is brought against a water–decane interface. The area fraction of the spheres on the interfaces is 13%.

against a water–decane interface. Both interfaces are laden with polystyrene particles of 3 μm and their concentration was 13%. The approach of the drop causes the intervening water to drain, and this sweeps the particles away from the region of closest approach. However, because pressure builds up in the approach region, a dimple is formed at the center and a ring of particles is observed to cling to its rim. The presence of the particles does retard the ultimate coalescence of the drop with the planar interface, but the system is finally unstable. Repeating this experiment at higher area coverage of the particles does not avoid coalescence, but does change the details of the particle arrangements. At 38% coverage, the ring shown in Figure 5.5 still appears, but it is no longer composed of a single array of particles but is an annular band of particles. As time proceeds and the drainage ensues, the dimple shrinks and compresses this band into a dense disk of particles. However, the particles never aggregate, but only approach one another to very close proximity. Finally, at a particle concentration of 55%, the particles within the dimple immediately form a dense disk since they are too numerous in number to form a ring. Again, the drop ultimately coalesces at these higher concentrations, but the process is delayed.

If the decane drop rising from the bottom against the planar interface is replaced by a water drop being lowered from above, as shown in Figure 5.4(b), the result is qualitatively different. In this case, there is the possibility of bridging of the particles across the two decane–water interfaces, which causes the water drop to become fastened

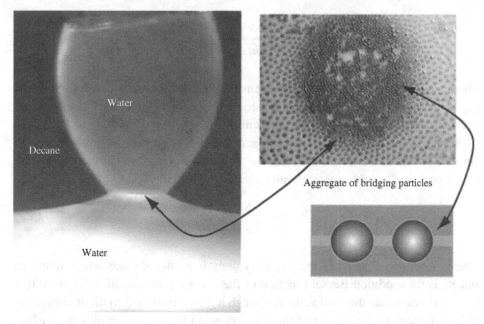

Figure 5.6 Adhesion of a water droplet to a decane–water interface. All surfaces are laden with polystyrene particles. The picture on the left shows the water droplet stretched as it is pulled away from the interface. On the right we see the aggregate of bridging particles. The cartoon depicts the bridging of the particles (side view).

against the planar interface. This is due to the large contact angle of the particles, which allows this stable geometry to occur for a sandwich of water–decane–water interfaces, but does not permit particles to straddle decane–water–decane interfaces. This is shown in Figure 5.6, which shows a droplet from that side that is adhered to the decane–water interface and stretched as it is pulled away. The adhesion is caused by particles that are able to bridge across a thin layer of decane from one body of water to the other and this is depicted in the cartoon in Figure 5.6.

The coalescence and bridging processes described above have also been recently revealed using precision-controlled film draining methods.[19] In a very comprehensive set of experiments, Horozov *et al.* studied the draining of vertical films of both water and oil surrounded by the complementary liquid. In addition to confirming that hydrophobic particles can bridge across thinning oil films, this work also provides convincing evidence that electrostatic repulsion forces operate through the oil phase.

The deformed pendant drop and bottom interface can be analyzed to predict the adhesive force that binds the drop to the interface. The easier of the two interfaces to analyze is the bottom interface since its deformation can be analyzed in the limit of small curvatures. If h is the height of this interface above its position far from the adhered droplet, the differential equation describing the shape of the interface is

$$\frac{1}{r}\frac{d}{dr}\left(r\frac{dh/dr}{\sqrt{1+(dh/dr)^2}}\right) = \frac{\Delta\rho g}{\gamma_{ow}}h \tag{5.3}$$

where r is the radial position, $\Delta\rho$ is the density difference between decane and water, γ_{ow} is the bare oil–water interfacial tension and g is the gravitational acceleration. The boundary conditions for this problem are $h(r \to 0) = 0$ and $h(r = R_{agg}) = 0$, where R_{agg} is the radius of the particle aggregate that is formed. Because the curvatures are small for the lower interface, the term $(dh/dr)^2$ can be neglected within the square root sign and this equation can be readily solved to yield

$$h = h_0 \frac{K_0(r/L)}{K_0(R_{agg}/L)} \tag{5.4}$$

where $L = \sqrt{\gamma_{ow}/\Delta\rho g}$ is the capillary length of the decane–water interface and K_0 is the modified Bessel function of the second kind and of order zero. It is now left to calculate the total vertical force that must be applied to lift the interface. This is the sum of the weight of the water beneath the aggregate plus the surface tension force projected onto the vertical axis. In the limit of small deformations, this force is

$$\begin{aligned}
F &= \Delta\rho g \pi R_{agg}^2 h_0 - 2\pi\gamma_{ow}R_{agg}\left(\frac{dh}{dr}\right)_{r=R} \\
&= \Delta\rho g \pi R_{agg}^2 h_0\left(1 + \frac{2K_1(R_{agg}/L)}{(R_{agg}/L)K_0(R_{agg}/L)}\right)
\end{aligned} \tag{5.5}$$

where K_1 is the modified Bessel function of the second kind and of order one. The force of adhesion, F, can be measured directly by placing a force transducer above the droplet. This can be compared directly against the prediction of equation (5.5) as the height h_0 is varied. This was accomplished in Ref. [18].

Another calculation that can be performed is the "pull-off" force necessary to detach the droplet. As the droplet is raised, the adhesive disk such as the one shown in the upper right corner of Figure 5.6 will be compressed to form a dense, circular monolayer disk of bridging particles. To overcome the adhesive force and detach the droplet, it is only necessary to pull the particles on the outer perimeter of the aggregate from the interface. The force to remove one sphere from a liquid–liquid interface is

$$F_p = 2\pi\gamma_{ow}a\cos^2\left(\frac{\theta}{2}\right) \tag{5.6}$$

where a is the sphere radius and θ is the contact angle. This must be multiplied by the number of spheres on the aggregate perimeter, $N_p = \pi R_{agg}/r$ to calculate the "pull-off" force, F_T. This leads to the following equation for the contact angle

$$\theta = 2 \cos^{-1} \left(\frac{F_T}{2\pi^2 \gamma_{ow} R_{agg}} \right)^{\frac{1}{2}} \tag{5.7}$$

This result offers another route to determine the contact angle of colloidal particles residing at liquid–liquid interfaces, which can be a difficult procedure using other means, such as direct observation using optical microscopy. This possibility was first recognized by Ashby *et al.*[20] Figure 5.7 shows the results of this exercise for the polystyrene sphere-water–decane system. The dashed line is the result of 130° measured using a contact angle goniometer and is in good agreement with the data obtained using equation (5.7). This procedure produces results that are largely independent of the surface coverage of the particles since the aggregate is compressed to form a dense, closest packing of the particles as the droplet is pulled upward.

The results of this section indicate that choosing the proper contact angle for particles attached to the surface of droplets can lead to a stable adhesion of the droplets if they come into contact with one another. However, it is important to note that these studies were performed using repulsive particles at concentrations below closest packing. The design of Pickering emulsions with optimal stability normally employs attractive particles that form weakly flocculated 2-D networks or higher concentrations (or both). In these circumstances, the particle layers may not be able to readjust their structure as two droplets approach. As a result, the particles will not have an opportunity to bridge across interfaces and induce adhesion. However, it has been observed that the adhesion effect can be very important in enhancing stability in the

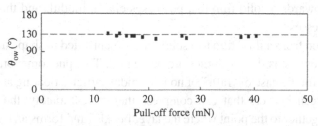

Figure 5.7 The contact angle (measured through water) of polystyrene particles at the decane–water interface as a function of the pull-off force. The dashed line is the result obtained using a contact angle goniometer for a sessile water drop residing on a thin film of the polystyrene particles spuncoat onto glass and surrounded by oil. The square symbols are data obtained using equation (5.7).

event that the particle concentration is not very large. This was reported by Vignati *et al.*[21] who observed that droplets covered with modest amounts of particles avoided coalescence through a collection of the particles at the point of closest approach that followed bridging events.

5.4 Buckling of Collapsing Droplets

Two important processes undertaken by emulsions are Ostwald ripening and desiccation. The former phenomenon can also occur in foams and describes the slow evacuation of droplets and bubbles by diffusion of the interior fluid into the bulk. Since the Laplace pressure of smaller drops and bubbles exceeds that of their larger counterparts, they are depleted more rapidly. For this reason, Ostwald ripening causes a coarsening of emulsions and foams and is normally an undesirable effect. Additives, such as particles, can help to stabilize the interfaces so that the droplets and bubbles can resist collapsing beneath a minimum size, see discussion in Chapter 1.

A second process that occurs with emulsions is desiccation or drying. This can occur with either drops attached to surfaces in the form of pendant or sessile drops, or those suspended in air, as in the case of spray drying. The majority of previous studies on the collapse of droplets has considered the problem of dessication,[13,22] and report on remarkable shape transitions by evaporating droplets composed of complex liquids consisting of polymer solutions or particulate suspensions. A droplet of liquid encapsulated by a bare interface that is only characterized by a surface tension will be reduced in size in a manner that is completely described by the Young–Laplace equation. In the absence of gravity, a suspended droplet will remain spherical and the internal pressure will increase in inverse proportion to the decreasing drop radius. If the droplet is sessile, it will shrink as a spherical cap and its contact angle will remain constant. The situation is completely changed if the droplet contains a polymer solute[23] or suspended particles,[24] which can be driven to the interface. This can induce a transition from an interface that is fluid and only characterized by a surface tension towards a solid film that possesses elastic moduli and that can sustain anisotropic stresses.

The transition from a fluid film to a solid film is manifested in shape transitions and buckling of droplets as their volumes are decreased. This phenomenon has recently been analyzed for the case of "rafts" of non-colloidal particles floating at the air–water interface between barriers that can compress them.[25] Ultimately, the particles are compressed together to the point where the layer buckles and forms an undulating surface with a characteristic wavelength. This is quite a general pattern, and had been observed previously for colloidal particles at oil–water interfaces by Aveyard *et al.*[26] Mahadevan *et al.*[25] have developed a simple mechanical model for such rafts that predicts the Young's modulus, $E = 4.54\ \gamma_{aw}/d$, where γ_{aw} is the air–water surface tension and d is the particle diameter. This assumes perfectly spherical, hard spheres exhibiting a contact angle of 90° that are close packed in a hexagonal arrangement. Furthermore, by modeling this problem as analogous to the buckling of an elastic beam, the wavelength of the undulations is determined to be $\lambda = \pi(4Ed^3/(3\Delta\rho g(1 - \nu^2)))^{1/4}$, where ν is the Poisson ratio. Using these results, they developed a "buckling" assay where measurement of the wavelength of wrinkling instabilities can be used to estimate the Young's modulus and thereby test microstructural theories for this property. Indeed, they were able to obtain quantitative agreement between the predictions of the model and a series of experiments involving a wide range of particle types and sizes.

Figure 5.8 shows a simple experiment where a sessile droplet of water adhered to a plastic, Delrin surface is drained from below. The droplet is surrounded by decane and covered with a monolayer of polystyrene particles with a diameter of 3 μm. The initial coverage of particles is sufficiently low to create an interface that is very fluid. However, at some point during the drainage of the droplet, it is expected that the particles will approach one another and cause the droplet to become a solid shell. The successive images in Figure 5.8 show the droplet in various states as the volume is decreased. Ultimately, the shell is observed to collapse, but prior to that event a number of important processes is observed. First, the adhesive area of the droplet remains constant as the height of the droplet is diminished. This effect is similar to the case of desiccation of droplets composed of polymer solutions and particle suspensions, and the pinning has been suggested as arising from adsorption of the solutes onto the substrate. Second, because of the pinning of the droplet, the contact angle (as measured from within the droplet) decreases as the droplet is evacuated.

The buckling of the shell defining the droplet surface first appears at the top of the droplet. Because of the large size of the droplet, it is deformed by gravity and the curvature is a function of position and achieves a minimum value at the top. For this reason, the droplet is most likely to undergo buckling at this position. The shape instability produces a flattened, circular disk that proceeds downward. Although the internal pressure of the droplet has not been measured, the flatness of the top portion indicates that the pressure drop across the interface is zero. The morphology of the flattened region is revealed in Figures 5.8(e)–(h). This indicates that a disk is formed with a crumpled topology with undulations that are roughly in the form of concentric, annular ridges. Measurement of the wavelength of these instabilities was used to estimate the Young's modulus of these particle monolayers using the "buckling" assay described above. The theory described in Ref. [25] produces predictions of the correct order of magnitude, but were consistently too large. However, the theory assumes hard sphere interactions and the experiments reported here employed repulsive spheres and effectively soft interactions.

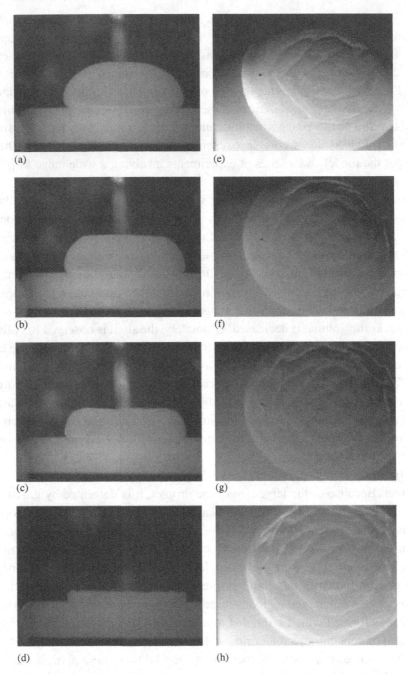

(a)   (e)

(b)   (f)

(c)   (g)

(d)   (h)

Figure 5.8 Sequence of images of a collapsing droplet of water (few millimeters in diameter) surrounded by decane. The interface is covered with polystyrene spheres of 3 μm in diameter. Images (a)–(d) show the side view and images (e)–(h) are the corresponding top views.

5.5 Concluding Remarks

In this chapter, we have reviewed the mechanical properties of liquid–liquid interfaces that are laden with colloidal particles. The presence of these particles leads to a number of important changes in these properties, and these affect the process of drop coalescence and the processing of emulsions. Since these operations invariably involve flow processes and interfaces are mobile, their rheology becomes important. However, interfacial rheology is far from a mature field, and experimental techniques and methodologies need to be developed. A limited number of rheological material functions can be acquired with existing equipment, such as dynamic shear moduli and surface shear viscosities, but other properties remain difficult to obtain. These include dilatational properties, the measurement of which is often confounded by Marangoni stresses, and bending moduli.

These interfaces offer fascinating challenges because of liquid-to-solid transitions that can occur even during rather mundane processes such as desiccation. These transitions have fundamental consequences on many experimental techniques that can confound their use. For example, the measurement of surface tension using the pendant drop method can be compromised if the presence of particles at the interface endows it with mechanical properties such as a Young's modulus. Standard methods rely on fitting a measured shape profile to shape equations that only incorporate surface tension.

The consequences of these enhanced mechanical properties reveal themselves in numerous ways. The non-Newtonian flow properties of particle-laden interfaces mean that predicting coalescence and particle deformation processes require constitutive equations embedded with the boundary equations associated with momentum balance equations. As these interfaces approach solid-like behavior, droplets also become prone to buckling and cracking of their surfaces.

Acknowledgments

The authors acknowledge support from the National Science Foundation and from Unilever.

References

1. A.W. Adamson and A.P. Gast, *Physical Chemistry of Surfaces*, 6th edn., Wiley, New York, 1997.
2. S.U. Pickering, *J. Chem. Soc.*, **91** (1907), 2001.
3. B.P. Binks, *Curr. Opin. Colloid Interf. Sci.*, **7** (2002), 21.
4. P.M. Spiecker and P.K. Kilpatrick, *Langmuir*, **20** (2004), 4022.
5. H. Hoekstra, J. Vermant, J. Mewis and G.G. Fuller, *Langmuir*, **19** (2003), 9134.
6. E. Stancik, M. Kouhkan and G.G. Fuller, *Langmuir*, **20** (2004), 90.

7. J.C. Fernandez-Toledano, A. Moncho-Jorda, F. Martinez-Lopez, R. Hidalgo-Alvarez, *Langmuir*, **20** (2004), 6977.
8. P. Cicuta, E.J. Stancik and G.G. Fuller, *Phys. Rev. Lett.*, **90** (2003), 236101.
9. D.A. Edwards, H. Brenner and D.T. Wasan, *Interfacial Transport Processes and Rheology*, Butterworth-Heinemann, Boston, 1991.
10. U. Seifert, *Adv. Phys.*, **46** (1997), 12.
11. R. Aveyard, J.H. Clint and T.S. Horozov, *Prog. Colloid Polym. Sci.*, **121** (2002), 11.
12. P.A. Kralchevsky, I.B. Ivanov, K.P. Ananthapadmanabhan and A. Lips, *Langmuir*, **21** (2005), 50.
13. L. Pauchard, F. Parisse and C. Allain, *Phys. Rev. E*, **59** (1999), 3737.
14. K.D. Danov, P.A. Kralchevsky and M.P. Boneva, *Langmuir*, **20** (2004), 6139.
15. E.J. Stancik, G.T. Gavranovic, M.J.O. Widenbrant, A.T. Laschitsch, J. Vermant and G.G. Fuller, *Faraday Disc.*, **123** (2003), 145.
16. E.J. Stancik, A.L. Hawkinson, J. Vermant and G.G. Fuller, *J. Rheol.*, **48** (2003), 159.
17. C.F. Brooks, G.G. Fuller, C.W. Frank and C.R. Robertson, *Langmuir*, **15** (1999), 2450.
18. E.J. Stancik and G.G. Fuller, *Langmuir*, **20** (2004), 4805.
19. T.S. Horozov, R. Aveyard and J.H. Clint, *Langmuir*, **21** (2005), 2330.
20. N.P. Ashby, B.P. Binks and V.N. Paunov, *Chem. Commun.*, (2004), 436.
21. E. Vignati, R. Piazza and T.P. Lockhart, *Langmuir*, **19** (2003), 6650.
22. Y. Gorand, L. Pauchard, G. Calligari, J.P. Hullin and C. Allain, *Langmuir*, **20** (2004), 5138.
23. L. Pauchard and C. Allain, *Europhys. Lett.*, **62** (2003), 897.
24. N. Tsapis, E.R. Dufresne, S.S. Sinha, C.S. Riera, J.W. Hutchinson, L. Mahadevan and D.A. Weitz, *Phys. Rev. Lett.*, **94** (2005), 018302.
25. D. Vella, P. Aussillous and L. Mahadevan, *Europhys. Lett.*, **68** (2004), 212.
26. R. Aveyard, J.H. Clint, D. Nees and V.N. Paunov, *Langmuir*, **16** (2000), 1969.

Gerald Fuller (left) is Professor of Chemical Engineering at Stanford University. He received his Ph.D. from Caltech in 1980 and his B.Sc. from the University of Calgary. His research interests concern the rheology of complex interfaces and the use of applied optical methods to reveal flow-induced structure. He has served as the Chairman of the Department of Chemical Engineering at Stanford and the President of the Society of Rheology. He is a Fellow of the American Physical Society and was awarded the Bingham Medal from the Society of Rheology. He has authored and co-authored over 180 papers and one book.

Edward Stancik (centre) is a Research Engineer in the Materials Science and Engineering Division of Central Research and Development at E.I. DuPont de Nemours and Co. He earned a B.S. degree from the University of Illinois at Urbana-Champaign and M.S. and Ph.D. degrees from Stanford University, all in chemical engineering. His research interests centre around complex fluids and particularly on garnering novel material properties from multi-component liquids. His current efforts seek to develop applications for these fluids through non-contact transfer methods such as inkjet printing.

Sonia Melle (right) received the B.Sc. degree in physics from Complutense University of Madrid in 1995, and the Ph.D. degree in physics from Universidad Nacional de Educación a Distancia (Spain) in 2002 for her work on the dynamics of magneto-rheological suspensions subject to the action of rotating magnetic fields. Her research was co-advised by Professor Fuller at Stanford University. Since then, she is collaborating with Fuller's group researching on interfacial physics. She worked for 2 years at the National Institute of Microelectronics in Madrid working on magneto-optical properties of nanostructured systems. She is currently Assistant Professor in the Optics Department, Complutense University of Madrid.

Tel.: + 1-650-7239243; *Fax*: + 1-650-7257294; *E-mail*: ggf@stanford.edu

6

Solids-Stabilized Emulsions: A Review

Robert J.G. Lopetinsky, Jacob H. Masliyah and Zhenghe Xu

Department of Chemical and Materials Engineering,
University of Alberta, Edmonton, Alberta TG6 2C7, Canada

6.1 Introduction

An emulsion is a system of dispersed droplets of one immiscible liquid in another. Simple emulsions are either oil-in-water (o/w) or water-in-oil (w/o). Emulsions can be defined as colloidal systems, although emulsion droplets are usually larger than the range specified for a colloidal system, *i.e.* diameter $> 1\,\mu m$.[1] Emulsions are encountered in many industries and scientific disciplines. Multi-disciplinary study is required for a better understanding of emulsion behaviour and better control over industrial emulsions. In this review, solids-stabilized emulsions are reviewed and they are defined as an emulsion that is stabilized by fine solid particles. Some finely divided solids assist in the emulsion formation, and/or improve its stability. These types of emulsions have widespread applications in industrial settings and have a history of being studied, dating back to 1903.[2]

6.1.1 Objective of review

Over the last decade, solids-stabilized emulsion experimentations are becoming increasingly more sophisticated and focused on microscopic level understanding. Some recent work has been performed to study the structure of particles at the droplet interfaces. In this review we will summarize important experimental and theoretical studies related to solids-stabilized emulsions. Considering the vast literature on solids-stabilized emulsions, this review aims at a selective, not comprehensive, overview of the progress in the field, with emphasis on key factors affecting the stability of solids-stabilized emulsions and the structure of emulsion drop interfaces. First an introduction into the applications and history of solids-stabilized emulsions is warranted, followed by a brief description of the basic characteristics of emulsions.

186

6.1.2 Industrial relevance

The role of fine particles in the stabilization of emulsions has been recognized in a number of important industrial processes. The underlying principle of how solids stabilize emulsions has been under study for a long period of time. Important industrial applications have been a driving force for increased interest of studying emulsion stabilization by fine solids. Tambe and Sharma[3] listed many industries such as the food, cosmetic, pharmaceutical, petroleum and agrochemical industries where emulsions are important. Industrial applications of emulsions can be divided into groups in terms of whether the occurrence of an emulsion is desirable or undesirable. For both cases, understanding the principles of emulsion stabilization is an important step to achieving desired industrial scale techniques for dealing with emulsions.

6.1.2.1 Desirable emulsions

In applications where emulsion stability is desirable, effort is made to enhance the stability and prevent emulsion breakdown. A few examples of products where stable emulsions are required include ice cream, skin moisturizers, drilling fluids and herbicides. Although their applications may be less widespread, solids-stabilized emulsions are just as important as emulsions without solids as stabilizers. An example of an industry where an increased emulsion stability is usually desirable is the food processing industry. Rousseau[4] published a review article summarizing the importance of emulsions stabilized by solid fat crystals in the food industry. It was found that other solid particles such as ice crystals and egg yolk are important stabilizers for a number of food emulsions.[4] Food emulsions are discussed to a large extent by McClements[5] and also in detail in Chapter 8.

6.1.2.2 Undesirable emulsions

In some applications, emulsions are undesirable and means to destabilize emulsions are pursued in order to separate the dispersed and continuous phases. A major challenge of emulsions in the petroleum industry is that undesirable w/o emulsions formed during various stages of oil recovery disrupt downstream oil processing.[6] Fine particles can play a major role in stabilizing these emulsions. Therefore, removing the particles would help demulsify these mixtures to improve recovery of oil without water and help avoid problems caused by fine particles themselves.[6] In bitumen processing, clay particles are found to be capable of stabilizing bitumen-in-water emulsions during extraction processes[7] and to contribute to the stability of water-in-diluted bitumen emulsions in bitumen froth treatment.[8] Another area of the petroleum industry where solids-stabilized emulsions are important is in the area of spill technology. Lee[9] provided a review of w/o emulsions formed with seawater and crude oil as in spills. In his review, Lee listed a number of agents that help stabilize emulsions and acknowledged the importance of sea particulates, clay particles and wax particles in this respect.

6.1.3 History of solids-stabilized emulsions studies

Ramsden was the first to report emulsion droplets stabilized by particles in 1903.[2] However, Pickering[10] conducted the first systematic study on solids-stabilized emulsions and recognized the role of finely divided insoluble emulsifiers. Pickering's study has been widely recognized as the first study on solids-stabilized emulsions. The contribution of his work earned the name of "Pickering emulsions" for solids-stabilized emulsions. These two classic studies demonstrated that fine solid particles would remain at an oil–water interface and promote droplet stability. Finkle *et al.*[11] noted that an interface covered by bi-wettable particles should curve towards the phase that more poorly wets the particles. This observation suggests that the dispersed phase in an emulsion, governing emulsion type, is determined by the contact angle which the particle makes with the oil–water interface. The role of solids wettability in emulsion stabilization was further studied by Schulman and Leja.[12] In their study, the contact angle of barium sulphate particles was varied systematically. They concluded that particles with an oil–water contact angle slightly less than 90° would stabilize o/w emulsions and those with angles slightly greater than 90° would stabilize w/o emulsions. Particles with contact angles that deviate greatly from 90° (very hydrophobic or very hydrophilic) are not observed to stabilize emulsions. The study pertaining to solids-stabilized emulsions has been progressed greatly since these early accounts. Several review articles including those by Menon and Wasan,[6] Tambe and Sharma[3] and Aveyard *et al.*[13] were published in more recent years.

6.1.4 Criteria of emulsion stabilization

In emulsion studies, an emulsion is considered stable if it is resistant to physical changes over a practical length of time. Several physical processes can indicate instability in an emulsion. The different methods by which an emulsion can become unstable or breakdown are outlined in Figure 6.1, which is a combination of suggested mechanisms from Robins and Hibberd[14] and Auflem.[15]

During destabilization by the flocculation of an initially well-dispersed emulsion as illustrated in Figure 6.1(a), creaming and sedimentation processes shown by Figure 6.1(b) and (c) can take place where the size and size distribution of emulsion droplets do not change. During the flocculation process, droplets come together and form aggregates without losing their original size. Creaming and sedimentation are caused by gravity, creating a concentration gradient due to density differences of the two immiscible liquids. For example, oil droplets from an o/w emulsion may

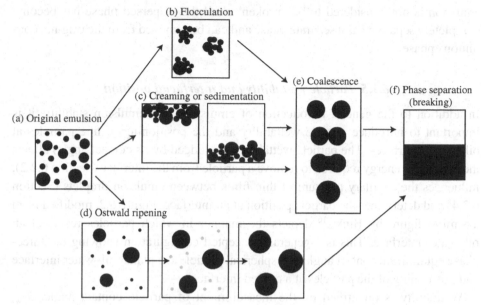

Figure 6.1 The different processes involved in the breakdown of an unstable emulsion.

be subject to creaming when the oil has a lower density than the aqueous phase. Ostwald ripening shown in Figure 6.1(d) occurs in emulsions where the dispersed phase has a limited solubility in the continuous phase so that large drops grow as smaller drops decrease in size due to transport of the soluble liquid from the small droplet to the large droplet through the continuous phase. Coalescence shown in Figure 6.1(e) is a phenomenon where many droplets merge to create fewer larger droplets, thereby reducing the total interfacial area of the system. Another process by which an emulsion is transformed is phase inversion. During this process the dispersed phase becomes the continuous phase and *vice versa*. In order for phase separation in Figure 6.1(f) to occur, a progression of the above processes as shown in Figure 6.1 must occur.

The kinetic stability of an emulsion can be considered in a number of ways. Emulsions can be stable in terms of flocculation, creaming and sedimentation where the entire dispersion remains homogeneous and no separation is observed. Alternatively, an emulsion can be stable in terms of coalescence only, so that each emulsion droplet is maintained at a certain size but the emulsion as a whole changes in physical appearance. Depending on the application of emulsions, the level of emulsion breakdown has varying importance. For example, for many food emulsions, creaming is unacceptable and an emulsion that does so would be considered "broken". For some w/o emulsions in the petroleum industry, on the other hand, an emulsion is not considered to be "broken" until the dispersed phase has become completely separated as a separate phase and can be removed from the original continuous phase.

6.1.5 Particle wettability and interfacial position

In addition to the general introduction of emulsions and emulsion stability, it is important to introduce particle wettability and the positioning of fine particles at oil–water interfaces. The particle wettability (as judged by its contact angle) affects the amount of energy required to remove a particle from the interface (Section 6.2.2), influences the capillary pressure of thin films between emulsion droplets (Section 6.2.4) and determines the particle position at an interface. Figure 6.2, modified from a similar figure by Binks,[16] depicts the manner in which particles reside at an oil–water interface. This is a generally accepted configuration showing the three-phase equilibrium contact angle of a spherical particle at a planar oil–water interface and positioning of the particles at a curved interface.

Wettability is quantified by the measurement of particle contact angle, θ_{ow}, which is, by convention, measured through the water phase as shown in Figure 6.2. The terms "hydrophilic" and "hydrophobic" usually refer to particles with contact angles less than and greater than 90°, respectively. Due to contact angle hysteresis,

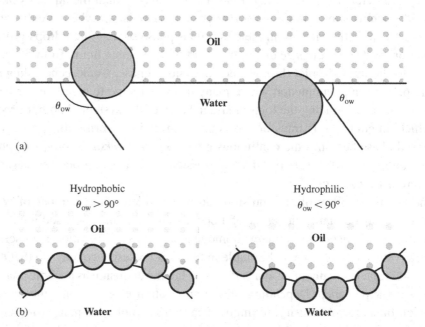

Figure 6.2 Preferential position of small particles at a (a) planar and (b) curved oil–water interface. Modified from Ref. [16].

particles in emulsion systems may not establish the same position at an interface as that dictated by the equilibrium contact angle. Contact angle hysteresis is the difference between the contact angle formed by a liquid advancing, θ_a, and receding, θ_r, over a solid surface (*i.e.* $\theta_h = \theta_a - \theta_r$). This hysteresis is caused by several phenomena including surface roughness, surface chemical heterogeneity and adsorption of impurities from the liquid phases.[17]

The contact angle of very small particles (<0.1 μm) is difficult to measure but several methods have been used including the capillary rising method based on Washburn's equation and pressing the powder into tablet form for goniometric measurement. For a fuller description, refer to Xu and Masliyah.[18] One method that has been used for silica nanospheres involves measuring the heat of immersion of sample particles with a calorimeter and then calculating the equilibrium contact angle from the measured heat of immersion.[19] In experiments that use particles with tailored wettability, it is common to measure the contact angle of a solid plate that has the same surface characteristics as the fine particles. Unless otherwise noted, the equilibrium contact angle will be discussed in this review, and the particles will be described in the terms of hydrophilic and hydrophobic as outlined above.

6.1.6 Thermodynamics of emulsification

Emulsions consist of a liquid phase that is in droplet form, dispersed in a continuous liquid phase with which it is immiscible. In general, in the absence of a surface active agent, emulsions are thermodynamically unstable because the interfacial area between the two phases is larger when dispersed droplets are present as opposed to a single interface between the two bulk phases. McClements[5] gave a simplified explanation (previously presented by Hunter[20]) of the thermodynamics involved in emulsion formation by comparing the free energy of a system before and after emulsification. The initial and final states of the system and the energy changes are shown in Figure 6.3. The creation of dispersed droplets increases the interfacial area by an amount ΔA (and therefore the interfacial free energy) and the configurational entropy of the system.[5] In the absence of solids or surfactants at the interface, the increase in free energy, which opposes emulsification, is much greater than the increase in configurational entropy, so the term entropy can be neglected. Thus the overall free energy change is given by[5]

$$\Delta G_{emul} = \gamma_{ow} \Delta A \qquad (6.1)$$

in which γ_{ow} is the oil–water interfacial tension. The free energy of emulsification, ΔG_{emul}, is positive and the emulsion is therefore thermodynamically unstable.

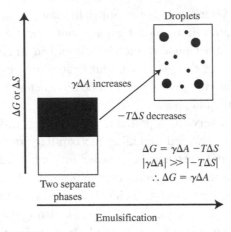

Figure 6.3 Diagram of the free energy and entropy change of a system during emulsification. The total change in free energy is positive so the system with droplets (*i.e.* emulsion) is thermodynamically unstable.

For systems that contain an emulsifier (surfactants or solids), the thermodynamics becomes more complicated as the surfactant molecules or solid particles will now be present at the interface. The interfacial tension is generally lowered by the presence of surfactants and so the free energy of emulsification is smaller.[20] Unlike surfactants, solid particles of micron size do not affect the interfacial tension.[21] For emulsions stabilized by solids alone, the driving force for formation of emulsion droplets is the preference of particles to reside at liquid–liquid interfaces. Thus small droplets form in order for more particles to adsorb from the bulk phases to obtain a position at an interface. However, systems with mixtures of surfactants and fine solids are more complex. Surfactant adsorption on interfaces is enhanced by the presence of oppositely charged particles leading to a reduction in interfacial tension.[21] The charged particles result in charge neutralization of the oil–water interface, which reduces repulsion between surfactant molecules thereby allowing for an increase in surfactant adsorption.[21] The reduction in interfacial tension caused by surfactants reduces the energy required for the formation of new liquid–liquid interface (*i.e.* emulsion droplets). In some cases, surfactants reduce the interfacial tension sufficiently that spontaneous emulsification occurs.[22] The emulsions formed under such conditions, often known as microemulsions, are thermodynamically stable and are not discussed here. Although macroemulsions are thermodynamically unstable, they may be kinetically stable (or metastable) to physical changes like droplet coalescence. Thus droplets will stay dispersed for an extended period of time, relative to some practical limit, *i.e.* processing time of oil. The source of the energy barrier responsible for emulsion stability is largely unexplained for solids-stabilized emulsions and is therefore often investigated.

6.2 Emulsion Stabilization by Particles

Many theoretical studies have been conducted in order to explain how solid particles promote emulsion stabilization. As discussed previously, emulsion stability can be classified in a number of ways. Ultimately, an emulsion is unstable when droplets coalesce readily. Thus, an explanation of emulsion stability should start with an explanation of the factors that cause droplets to be resistant to coalescence. Most often droplet stability is attributed to the steric hindrance provided by a particulate layer surrounding the droplet. However, some investigations have considered the influence of other phenomena on emulsion stability, such as the stability of the thin film between emulsion droplets and the influence of particles on the rheological properties of the interface.

6.2.1 Macroscopic configurations and stabilization mechanisms

In general for solids-stabilized emulsions, several configurations may exist that would prevent droplet coalescence and promote emulsion stability. These configurations and four underlying mechanisms are outlined in Figure 6.4. Most often in solids-stabilized emulsions, complete coverage of droplets by particles is observed so that when two droplets contact each other, the two particle layers prevent coalescence. This is named bilayer stabilization and is shown in Figure 6.4(a). Another possible configuration is the formation of a single layer of particles between two emulsion droplets. Stability from "bridging" by a single layer of particles is shown in Figure 6.4(b). Particle bridging has been observed between planar interfaces and the surface of pendant drops covered with particles.[23,24] The contact angle of the particles needs to be greater than 90° before bridging can occur between water droplets in oil.[23] The wettability of the particle would dictate its thermodynamically favourable position. Particles need to have an interfacial position with a majority portion of the particle in the continuous phase, otherwise the interfaces of the emulsion droplets will collide. Strong particle adhesion to the interface prevents the particle from being pushed out of the interfacial position. In this case, droplets with a single particle layer can be stabilized.

In both bilayer and single layer configurations, steric hindrance can prevent particle displacement from the interface and lateral displacement at the interface, as depicted in Figures 6.4(a)i and ii, respectively. Another mechanism contributing to the prevention of coalescence is the stability of the thin film of continuous phase formed between the emulsion droplets. The thin film stability is influenced by the maximum capillary pressure preventing film thinning (Figure 6.4(a)iii), and the rheological properties affecting film drainage (Figure 6.4(a)iv). All of these mechanisms can be related to the properties of the particles and emulsion systems.

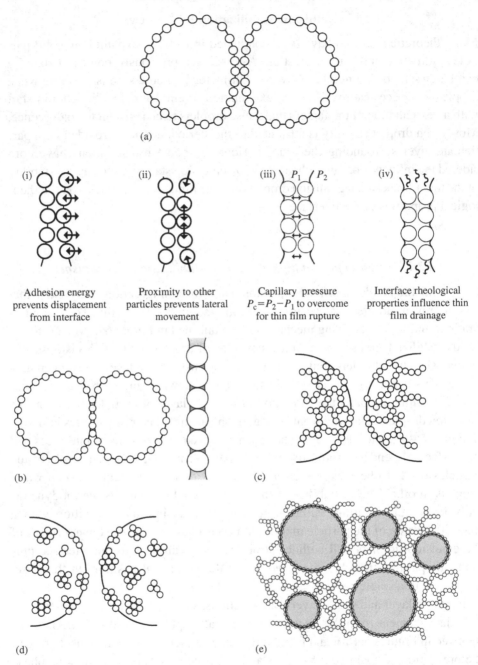

Figure 6.4 Possible configurations of particles in solids-stabilized emulsions, (a)–(e), and the underlying mechanisms responsible for stability (i)–(iv).

Other configurations observed in experiments include the formation of a 2-D network of aggregated particles on the droplet surface as shown in Figure 6.4(c) and stabilization of droplets sparsely covered by domains of particles as shown in Figure 6.4(d). Steric hindrance should be the major mechanism responsible for emulsion stabilization when a 2-D network is formed at the interface because the energy of adsorption prevents the particles from being removed from the interface and the strength of the particle aggregation prevents displacement at the interface. The stability exhibited by sparsely covered droplets is not well understood, but the formation of a single or double layer bridge at the droplet contact area as observed by Vignati *et al.*[25] may be responsible. The same mechanisms (i–iv) would then be the underlying factors preventing coalescence.

The final configuration exhibited by stable solids-stabilized emulsions is the formation of a 3-D network of particles between emulsion droplets when an abundance of particles is present as shown in Figure 6.4(e). However, it has been reported that when a 3-D network forms, a dense particle layer on droplet surfaces still exists[26] and so the mechanisms responsible for configuration Figure 6.4(a), *i.e.* i–iv, are ultimately responsible for stability but the particle network extending through the continuous phase improves stability by preventing droplet–droplet contact.

The following sections focus on a review of studies that were carried out to theoretically investigate the configurations and mechanisms discussed above. Experimental evidence for the existence of these configurations is discussed later. Many studies consider idealized systems, focusing on theoretical treatment of one of the stabilization mechanisms or experiments with well-characterized systems. It is very likely that in industrial emulsions a combination of many configurations exists and many underlying mechanisms are responsible for emulsion stability.

6.2.2 Particle removal from interfaces

The so-called steric hindrance preventing droplet coalescence is caused by the formation of a dense particle layer around droplets and the high energy required to expel particles from the interface. Coalescence of droplets completely covered by particles and the manner in which the particle layer provides an energy barrier to coalescence are described in Figure 6.5. When droplets coalesce, the total interfacial area is reduced and the particles at the interface are removed or desorbed. If the particles spontaneously adsorb at the interface, removing them requires energy.[27] If the free energy increase associated with desorption of the particles is larger than the free energy decrease caused by a reduction in interfacial area, then coalescence would not be spontaneous.[27] The inset in Figure 6.5 is a schematic of how the energy required to remove material from the interface provides a barrier to coalescence.

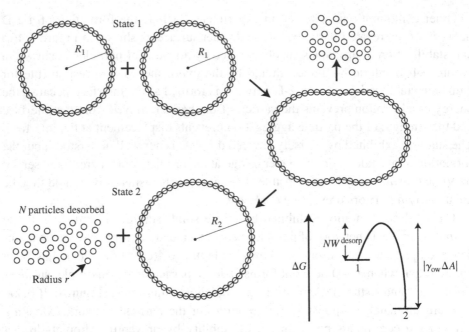

Figure 6.5 Coalescence of two emulsion droplets that are covered with particles. Removal of adsorbed material from the interface (upper right) can be an energy barrier to coalescence, as shown on the lower right.

To further understand the nature of the steric hindrance we can pursue a thermodynamic model of the coalescence of two solids-stabilized emulsion droplets. We consider the case of two water drops of radius R_1 completely covered with N_{initial} particles of radius r. For further simplification we ignore the effects of curvature, *i.e.* the particles are much smaller than the droplet, and particle–particle interactions. Let the two water droplets coalesce resulting in one water drop of radius R_2 as shown in Figure 6.5. Upon coalescence, Δn particles will be desorbed from the interface. A positive free energy change occurs when the oil–water interfacial area is reduced but a negative free energy change occurs due to particle desorption from the interface. The total Gibbs free energy change due to coalescence is given by

$$\Delta G_{\text{coal}} = \gamma_{\text{ow}} \cdot \Delta A - W^{\text{desorption}} \cdot \Delta n \tag{6.2}$$

where

$$\Delta n = N_{\text{final}} - N_{\text{initial}} \tag{6.3}$$

The term $W^{\text{desorption}}$ is the energy needed to remove one particle from the interface. The volume of water is conserved leading to

$$2\left(\frac{4\pi}{3}R_1^3\right) = \frac{4\pi}{3}R_2^3 \tag{6.4}$$

$$R_2 = 2^{\frac{1}{3}}R_1 \tag{6.5}$$

The change in interfacial area is given by

$$\Delta A = A_{\text{final}} - A_{\text{initial}} = 4\pi R_2^2 - 8\pi R_1^2 \tag{6.6}$$

Combining equations (6.5) and (6.6) gives

$$\Delta A = 4\pi R_1^2 (2^{\frac{2}{3}} - 2) \tag{6.7}$$

To a first approximation, a change in interfacial area leads to the following expression for the number of particles desorbed from the interface

$$\Delta n \approx \frac{\Delta A}{\pi r^2} \tag{6.8}$$

To use this expression we do not include a packing factor for the particles so that all of the interfacial area is covered by particles. Thus, when a particle is removed from the interface, the interfacial area can reduce by the full cross-sectional area of the particle. Substituting for Δn, equation (6.2) can now be expressed as

$$\Delta G_{\text{coal}} = \gamma_{\text{ow}} \cdot \Delta A - \frac{W^{\text{desorption}} \Delta A}{\pi r^2} \tag{6.9}$$

and upon simplification, equation (6.9) becomes

$$\Delta G_{\text{coal}} = \Delta A \left(\gamma_{\text{ow}} - \frac{W^{\text{desorption}}}{\pi r^2}\right) \tag{6.10}$$

Substituting for ΔA using equation (6.7) leads to

$$\Delta G_{\text{coal}} = 4\pi R_1^2 (2^{\frac{2}{3}} - 2)\left(\gamma_{\text{ow}} - \frac{W^{\text{desorption}}}{\pi r^2}\right) \tag{6.11}$$

The work needed to remove a spherical particle from a planar oil–water interface is given by[16,28–30]

$$W^{\text{desorption}} = \pi r^2 \gamma_{\text{ow}} (1 + \cos\theta)^2 \tag{6.12}$$

for particle removal into the oil phase, and

$$W^{\text{desorption}} = \pi r^2 \gamma_{\text{ow}} (1 - \cos \theta)^2 \qquad (6.13)$$

for particle removal into the water phase.

The particle radius r is assumed to be small enough such that gravity can be neglected and θ is the equilibrium contact angle measured through the water phase (see Figure 6.2). Examination of equations (6.12) and (6.13) reveals that bi-wettable particles are thermodynamically favoured to sit at the interface and the energy required to remove them is high, of the order of thousands of kT.[29] In addition, particles with contact angles close to 90° have the highest energy of adsorption and will be the least likely to be removed from the interface.

Since we are using water droplets, it is reasonable to assume that upon coalescence excess particles are ejected from the water drop interface into the bulk oil phase. Substituting equation (6.12) into equation (6.11), the change in free energy is given by

$$\Delta G_{\text{coal}} = 4\pi R_1^2 (2^{\frac{2}{3}} - 2) \left(\gamma_{\text{ow}} - \frac{\pi r^2 \gamma_{\text{ow}} (1 + \cos \theta)^2}{\pi r^2} \right) \qquad (6.14)$$

which, after simplification, leads to

$$\Delta G_{\text{coal}} = 4\pi R_1^2 \gamma_{\text{ow}} (2 - 2^{\frac{2}{3}})(\cos \theta (\cos \theta + 2)) \qquad (6.15)$$

Similarly, for the case of oil droplets in water, the change in free energy for ejection of particles from the interface into the bulk water phase is given by

$$\Delta G_{\text{coal}} = 4\pi R_1^2 \gamma_{\text{ow}} (2 - 2^{\frac{2}{3}})(\cos \theta (\cos \theta - 2)) \qquad (6.16)$$

A plot of scaled free energy for $0° < \theta < 180°$ is shown in Figure 6.6. The free energy change is positive when $\theta < 90°$ and negative when $\theta > 90°$. This means that coalescence is not thermodynamically favourable if the surface is covered by hydrophilic particles and they are brought across the interface into the bulk oil phase, as shown in Figure 6.6. However, it is far more likely that a water droplet in oil would be surrounded by hydrophobic particles. In this case the particles straddle the interface so that a larger area resides in the oil phase. When these hydrophobic particles are ejected from the interface, a smaller volume of particle needs to cross the interface, as shown in Figure 6.6. Coalescence is therefore thermodynamically favourable for $\theta > 90°$. However there is an energy barrier which prevents the droplets from coalescing. As described here, the energy barrier is equal to the work of desorption required to remove N particles from the interface. This work of desorption is far greater than that supplied by thermal energy. This energy

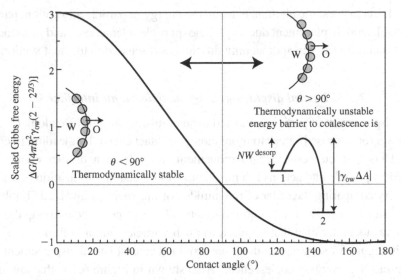

Figure 6.6 Gibbs free energy change for the coalescence of two water droplets in oil covered with particles as a function of particle contact angle. Also given are pictures showing the method of particle removal into the oil phase and a sketch of the energy barrier to coalescence.

barrier leads to steric hindrance between droplets. The particles are strongly held at the interface and cannot be removed. Thus, when droplets covered by particles collide, the interfacial layer strongly resists displacement of the particles and coalescence is prevented.

More complicated theoretical investigations have been conducted in order to explain the nature of the steric hindrance more fully. The thermodynamics of an emulsion system has been investigated to determine the free energy change associated with particles partitioning from a bulk phase to an interface.[7,29,31] Levine *et al.*[29] calculated other energies involved, such as the dipole–dipole interactions between particles, and showed that particles at interfaces are thermodynamically favoured to stay there. They demonstrated that the energy of particle adsorption is not counterbalanced by particle interaction energies provided by double layer repulsion, solvation forces, solid elastic forces, capillary and van der Waals forces. This means that the thermodynamic stability (metastability) of the system could not be theoretically explained. A similar approach is taken by Aveyard *et al.*[13,31] but they consider smaller particles (nanosized) and a curved interface. Their analysis includes the effects of line tension and curvature energies which are applicable to nanometer size particles. Their results showed that these energies could provide a barrier to particle adsorption and thus may explain the thermodynamic stability. A new thermodynamic description of emulsification using solid particles was given in outline in Chapter 1.

Solids-Stabilized Emulsions: A Review

In addition to the steric hindrance due to the energy of particle adsorption, particles also resist lateral displacement due to particle–particle interactions, and influence the thin film stability between approaching droplets as discussed in the next section.

6.2.3 *Lateral displacement of particles at an interface*

Lateral displacement of particles attached to an emulsion drop surface along the interface allows for "exposed" areas to more closely contact each other, leading to coalescence. Thus, particle resistance to movement away from a droplet contact area provided by particle interactions is a mechanism of emulsion stabilization (see Figure 6.4(b)). By comparing the order of magnitudes of the energies involved, Tambe and Sharma[3] showed that the force required to laterally displace particles along the interface is much smaller than that required to push particles into a bulk phase. Results shown in Figure 6.7 suggest that the force provided against lateral displacement is the limiting factor preventing coalescence.[3] Also shown in Figure 6.7 is that significant hindrance to movement of particles at the interface is not provided until there is a high concentration of particles at the interface. Particle interactions play an important role. Particles at the interface that are loosely packed (deflocculated) will provide little resistance to lateral displacement from droplet contact areas.[3]

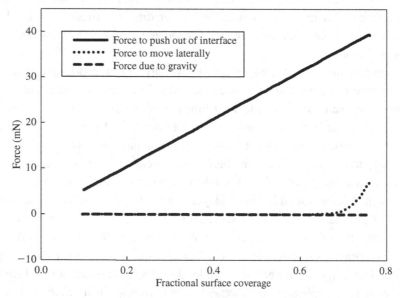

Figure 6.7 Forces involved in moving 1 μm particles adsorbed at a liquid–liquid interface as a function of the interface coverage by particles. Other model parameters are: drop size 0.1 mm, water density 1.0 g cm^{-3}, oil density 0.74 g cm^{-3}, interfacial tension 50 mN m^{-1} and particle contact angle 150°. Modified from Ref. [3].

This mechanism would be significant for emulsion stabilization only in droplets that are densely covered by particles so that a bilayer of particles is the structure resisting droplet coalescence as shown in Figure 6.4(a). In addition, if a 2-D network of particles was formed on the droplet surface as shown in Figure 6.4(c), the strength of particle interactions would also provide a hindrance to particle movement at the interface. If the network was fairly rigid, then the hindrance to particle movement at the interface could be greater than loose particles even when densely packed. As presented by Tambe and Sharma,[3] the energy barrier to lateral displacement would not be responsible for emulsion stability unless the surface concentration of particles is sufficiently high. This mechanism cannot explain emulsion stability when the surface of droplets is only partially covered by close-packed domains with "exposed" interfaces or very low particle coverage (*i.e.* incomplete coverage without a 2-D network). Experimental evidence of stable emulsions (droplets) with low particle coverage of droplet surfaces will be discussed later.

6.2.4 Capillary pressure in the thin film between droplets

6.2.4.1 Two layers of particles

The stability of the thin film between two emulsion droplets stabilized by solid particles governs emulsion stability. Particle layers on the surface of emulsion droplets approach and contact each other instead of being removed from the interface because of the high energy required for displacement. The particle layers trap a film of continuous phase between the layers. The film of finite thickness formed in the pore space of the particle layers thins until a critical thickness is reached and ruptures when the pressure inside the film exceeds the maximum capillary pressure.[32] The maximum capillary pressure $P_{c,max}$ can be calculated as[33]

$$P_{c,max} = \frac{2\gamma_{ow}\cos\theta}{br} \qquad (6.17)$$

where b is a constant dependent on the packing exhibited by the particles. As $P_{c,max}$ increases, so does the emulsion stability. Thus predictions of $P_{c,max}$ will show the effect of contact angle and particle radius on emulsion stability.

The validity of calculated capillary pressures was tested with model emulsion films[32] and the response of emulsion droplets to gravitational and centrifugal fields.[33] In the model films formed between macroscopic spheres, calculations were in agreement with the measured values. When solids-stabilized emulsions were subjected to an applied pressure difference, agreement of experimental and theoretical maximum capillary pressures was limited. Emulsions became unstable at experimental capillary pressures lower than the calculated values. The lower threshold pressures were attributed to uneven packing in the solids layer and reduced film elasticity when the film is close to the critical thickness.[33] Experimental results show that emulsions formed with multilayers of particles are more stable with regard to the threshold capillary pressure they can withstand.[33]

The difference in the advancing and receding contact angle (known as contact angle hysteresis) can result in different values of the critical capillary pressure depending on whether the particle is moving to the interface from the continuous or dispersed phase.[32] If moving from the continuous phase the receding contact angle is used to calculate the capillary pressure, whereas if moving from the dispersed phase the advancing contact angle is used. As a result, the maximum capillary pressure and thus emulsion stability changes with contact angle hysteresis.[32] From equation (6.17) it can be noted that the capillary pressure in the thin film between particle layers is 0 when the contact angle is equal to 90°. This leads to the condition that contact angles not equal to 90° are required for emulsion stabilization. These capillary pressure calculations are limited to densely packed layers, and the pressure limit model is less important when loosely packed layers approach each other. Also, these calculations do not consider the effect of a 2-D network at the interfaces. Furthermore, the two-layer theory does not account for the existence of stable emulsions where droplet surfaces are partially covered with close-packed domains of particles.

6.2.4.2 Single particle layer (bridging)

As was shown in Figure 6.4(b), a single particle layer between emulsion droplets also provides stability by preventing thinning of the film of continuous phase between them. A pressure difference between the outside and inside of the droplets (*i.e.* capillary pressure), exists when the oil–water interface curves around the particles[34] (see Figure 6.8). A maximum capillary pressure must be reached in order for the intervening liquid film to rupture and coalescence to occur. Denkov *et al.*[34] modelled the capillary pressure of the thin film around a single layer of monodispersed, spherical particles evenly distributed between two emulsion droplets. They showed the influence of particles on the resistance to thinning and rupture of the thin film formed between two droplets. The model considers the influence of particle size, contact angle, contact angle hysteresis and particle spacing on the maximum capillary pressure that the film can resist. It was shown that particles of smaller sizes result in a higher maximum pressure and thus greater film stability. In addition, contact angles near 0° (and also near 180° when the opposite emulsion type is considered) and/or greater contact angle hysteresis result in a higher maximum pressure. Interestingly, for particles of contact angle equal to 90°, the model reports the maximum capillary pressure that the film can withstand as 0. This suggests that emulsion stability would be greatest for particles with contact angles not too close to 0° (or 180°) or 90° so that the combination of thin film stability and the energy of particle adsorption is the most synergistic.[34]

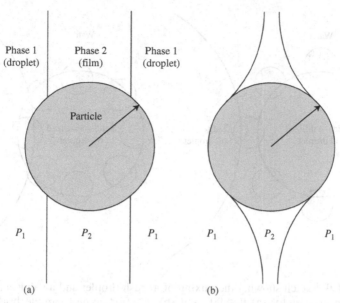

Figure 6.8 A spherical particle bridging two dispersed phases. Curvature of the interface around the particle produces a pressure difference, $P_c = P_1 - P_2$. (a) $P_c = 0$, (b) $P_c > 0$. Modified from Ref. [34].

Comparison of different inter-particle distances in the model shows that densely packed particles create films with a very large maximum capillary pressure that are more stable than films between loosely packed particles. The model presented by Denkov *et al.*[34] does not consider particle aggregation. However, when uneven separation occurs between particles (formation of "holes"), the film becomes unstable at lower capillary pressure (greater film thickness) which may result in decreased emulsion stability.

In order for particle bridging to occur, the particle wettability must be such that particles protrude from the emulsion droplet. This was considered in the model by Denkov *et al.*[34] In the manner they analysed, particles with contact angles greater than 90° are not able to form films. Since wettability dictates the particle position on the dispersed phase side of the interface, single layer particle bridging cannot occur when an emulsion is stabilized by particles that are protruding into the droplet. The inability of particles to form a bridge between emulsion droplets can be exploited to provide a means of demulsification. Addition of fresh amounts of the dispersed phase, *e.g.* fresh oil added to an o/w emulsion, is an effective demulsification process when particles are protruding into the droplets.[35] "Scavenging" of emulsion droplets would occur in the manner shown by Figure 6.9. When a fresh globule collides with a stabilized droplet and the particles are unable to form a bridge between the droplets, the fresh globule engulfs the emulsion droplet. Demulsification of an o/w emulsion

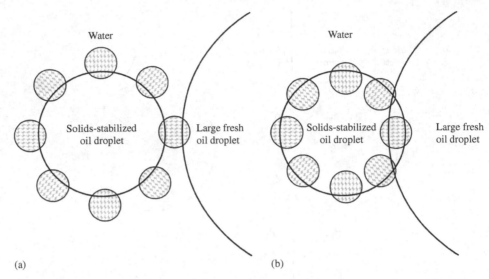

(a)   (b)

Figure 6.9 Sketch showing the mixing of a fresh droplet and a solids-stabilized emulsion droplet for (a) $\theta < 90°$ and (b) $\theta > 90°$. When particle bridging is absent, (b), the large globule engulfs the droplet leading to demulsification. Modified from Ref. [35].

using this scavenging method was investigated in several articles including Refs. [35–38].

6.2.5 Rheological properties of droplet interfaces

In addition to the maximum capillary pressure in intervening thin films, the rate of film drainage may contribute to emulsion droplets' resistance to coalescence. Tambe and Sharma[3,39] have discussed the rheological properties of interfaces in the presence of small particles. They predicted the elasticity and viscosity (both shear and dilational) of a model interface containing spherical particles in cubic[39] and hexagonal close-packed arrangements.[3] The models take into consideration the contribution of particle–particle interactions. They showed that colloid-laden interfaces exhibit viscoelastic behaviour at high particle concentrations and the elastic contribution increases substantially with particle concentration. The increase in viscoelastic properties in relation to particle concentration happens only on droplet surfaces with densely packed particle layers.

The dilational elasticity and viscosity have been shown to affect the rate of thin film drainage, as illustrated in the model developed by Tambe and Sharma in 1991.[40] Figure 6.10 shows how an increase in particle concentration increases the duration of film drainage. The contribution of particles to the interfacial rheology of an emulsion droplet surface appears to cause a substantial decrease in drainage rate of

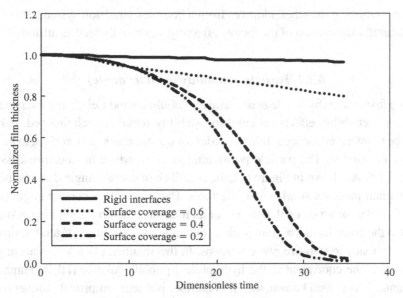

Figure 6.10 Rate of drainage of the thin film between two colloid-laden interfaces for different interface coverage by particles. Modified from Ref. [39].

thin films between approaching droplets. This should result in more stable emulsions as it slows down droplet coalescence. It is likely that a reduced rate of film drainage would be present in all of the configurations (a)–(e) outlined in Figure 6.4. However, the models presented by Tambe and Sharma only apply to a bilayer of densely packed particles (Figure 6.4(a)). The effect of particles on interfacial rheology and thin film drainage has not been modelled for emulsion droplets that exhibit single particle bridging, 2-D networks and sparsely covered interfaces.

6.3 Experimental Factors in Emulsion Stability

Several key factors involved in stabilization of solids-stabilized emulsions are the wettability, concentration, size and location of particles. Other factors that influence the stability of a solids-stabilized emulsion include particle shape, oil type (polarity), rheology of the liquid phases and, of course, the presence of additives in emulsion systems. In most laboratory investigations a model emulsion system is created to contain only three phases – oil, water and particles. Often, other additives such as electrolytes,[3,13] surface active agents[12] and flocculating agents[41] were introduced in the system to alter molecular forces. In industrial applications, many chemical species including emulsifying and thickening agents can accompany the fine solids to promote emulsion stability. For example, surface active asphaltene molecules and fine solids were found to contribute to stabilization of water-in-diluted bitumen emulsions.[8]

However, removing the other additives from a complex emulsion system allows for a more focused examination of the factors affecting solids-stabilized emulsions.

6.3.1 Particle wettability (contact angle)

Since the first studies by Finkle *et al.*[11] and by Schulman and Leja,[12] the interpretation of particle wettability effects on emulsion stability remains unchallenged. Particles should be of intermediate wettability in order for positioning at an interface to be energetically favourable. The particle positioning at an interface has been discussed in Section 6.1.5. As shown in Figure 6.2, the equilibrium contact angle dictates the position of small particles at oil–water interfaces. Particles with a contact angle greater than 90° tend to protrude on the oil side of an interface, thereby stabilizing w/o emulsions. On the other hand, particles with a contact angle less than 90° tend to protrude on the water side to stabilize o/w emulsions. In this manner, particle contact angle or wettability is the equivalent to the hydrophile–lipophile balance (HLB) number for surfactants.[16] As is well known, the HLB number is a semi-empirical number used to assess the emulsifying characteristics of surfactants based on knowledge of their molecular structure. Determining the emulsion type that would form in the presence of a certain surfactant is not as straightforward as reading from an HLB table.[42] Similarly, the contact angle of particles alone does not determine the type of emulsions that will be stabilized by the particles. In general, intermediate contact angles about, 90°, are optimal for stabilizing emulsions, as these particles tend to straddle an interface. However a particle's wettability in some cases does not solely depend on the initial contact angle or the contact angle of a flat surface of the same solid. Surface roughness or particle shape will have an effect on the particle wettability, leading to contact angle hysteresis.[13] Schulman and Leja[12] were the first to experimentally demonstrate that contact angles near 90° produced the most stable emulsions. As was pointed out earlier, see equation (6.13), the energy required to remove a particle from an interface depends on the particle contact angle.[29] The energy is highest when contact angle equals 90° and drops sharply as the angle moves away from 90°. Yan *et al.*[43] reported further experimental evidence that intermediate contact angles produced the most stable emulsions. They used fumed silica nanoparticles with various contact angles, treated clay particles and latex spheres to form various toluene–water emulsions. Only particles with intermediate hydrophobicity produced stable emulsions. Highlighted in Figure 6.11 are the results obtained by Yan *et al.*[43] for emulsions formed with silica nanoparticles of toluene–water contact angles ranging from 0–96°.

The results obtained by Yan *et al.*[43] and presented in Figure 6.11 show that particles with contact angles significantly less than 90° (*i.e.* 0° and 60°), cannot stabilize emulsions. Also shown in Figure 6.11 is that hydrophilic particles ($\theta < 90°$) can stabilize w/o emulsions, which is contrary to observations from other studies. The reason for stabilizing w/o emulsions by hydrophilic particles is discussed later (Section 6.3.3).

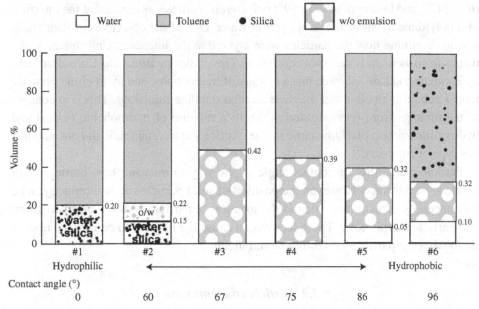

Figure 6.11 Appearance of vessels in systems containing initially 20 vol.% water, toluene and fumed silica particles (30 nm diameter) for particles of different wettability, given as the contact angle (°) measured into water at the oil–water interface. Taken from Ref. [43]; with permission of Elsevier.

By changing the concentration of stearic acid in the oil phase, Tambe and Sharma[3] were able to vary the wettability of several types of solids. Phase inversion occurred when the stearic acid concentration caused the normally water-wet particles to become oil-wet. The presence of the surface active stearic acid molecules was believed to contribute to the action of the solid particles at the interface. Binks and Lumsdon[44] varied the silanol content of silica particles (wettability) to cause emulsion inversion. Unlike the introduction of stearic acid, changing surface silanol content of the particles does not introduce any other species at the interface so that the role of particle surface wettability in emulsion inversion can be unambiguously demonstrated. Binks and Lumsdon[45] studied solids-stabilized emulsions of toluene–water systems where a mixture of hydrophobic and hydrophilic silica was used to stabilize the emulsion. They showed that transitional phase inversion occurs when the average wettability of particles changes from hydrophobic to hydrophilic and *vice versa*. The average position of the particles at the interface determines both the type of emulsion and emulsion stability. As the average contact angle changes in such a way that, on average, particles move from protruding on the water side to protruding on the oil side, phase inversion from o/w to w/o emulsions occurs. This inversion would occur even though the interfacial position of individual particles contrasts that of the bulk film. In a separate study, Tarimala and Dai[46] used a confocal microscope to observe a mixture of hydrophobic

($\theta \sim 117°$) and hydrophilic ($\theta \sim 59°$) polystyrene particles assembled at the interface of a poly(dimethylsiloxane) oil droplet in water. One of the objectives of their study was to determine how the particles were layered at the interface. Unfortunately, the results from their study were not conclusive. The studies by Binks and Lumsdon[45] and by Tarimala and Dai[46] showed that a mixture of hydrophobic and hydrophilic particles would remain at an oil–water interface and can stabilize emulsions. This is in contrast to the findings from previous studies where a mixture of hydrophobic (silica) and hydrophilic (carbon black/mercuric iodide) particles at a certain ratio did not stabilize any type of emulsion.[47]

Particle wettability or contact angle is one way to measure how particles will reside at an interface. However, the position of solid particles at an interface can be influenced by other factors such as the location of particles prior to emulsification and particle morphology. These factors can cause the particle to behave the opposite to what the particle wettability would dictate.

6.3.2 Particle concentration

Apart from a few recent studies showing that emulsion droplets will be stable against coalescence even with low surface coverage of the droplets by particles, for most solids-stabilized emulsions particle concentration is an important factor in emulsion stability. In some cases increasing the particle concentration will increase the emulsion volume and/or cause the emulsion droplets to decrease in size. This phenomenon has a limit however, dictated by the size of the droplets. In all cases, sufficiently high particle concentration is needed in order to achieve high particle coverage of droplets. However, high concentrations do not necessarily mean a densely packed monolayer, as in some emulsions stable droplets are observed without dense coverage.

In many solids-stabilized emulsion systems a complete coverage of droplets by solids is considered necessary for emulsion stabilization. Excess particles may increase stability by providing a 3-D network of particles surrounding droplets as shown in Figure 6.12. It has been shown that as the concentration of particles increases, the size of emulsion droplets decreases to accommodate more particles at the interface.[3,6,13] Tambe and Sharma[3] showed the existence of a limiting concentration such that any increase in concentration of particles above this limit would not result in any smaller droplets or increased emulsion stability. The limit of particle concentration is assumed to correspond to a full coverage of entire droplets with a dense particle film. Using transmission X-ray microscopy, Thieme et al.[26] obtained the images shown in Figure 6.12 of a solids-stabilized emulsion that exhibits a 3-D network of particles surrounding emulsion droplets. Emulsions with a network of particles exhibited an enhanced stability over emulsions without the network. However, their study demonstrated that an encapsulating film of particles

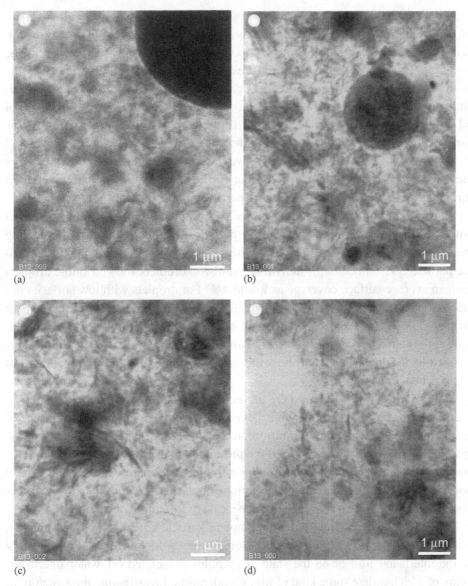

Figure 6.12 Transmission X-ray microscope images of o/w emulsion droplets stabilized by particles (montmorillonite and magnesium aluminum hydroxide ~200 nm diameter) showing the network of particles in the continuous phase surrounding the droplet. Taken from Ref. [26]; with permission of Elsevier.

was necessary for the emulsion stability. Abend and Lagaly[48] showed that an increase in the concentration of particles above the amount necessary for droplet surface coverage provided an increase in emulsion stability by the formation of a particle network surrounding the droplets.

Recent studies showed that droplets could be stabilized without a full coverage of the droplet by particles. Midmore[41] was able to obtain a stable o/w emulsion using colloidal silica particles and hydroxypropyl cellulose as a co-stabilizer. The co-stabilizer alone was unable to produce a stable emulsion and contributed to the solids-stabilized emulsion by flocculating the particles. To produce a stable emulsion with particles and co-stabilizer, a minimum surface coverage of only 29% of oil droplets by particles was necessary. To effectively stabilize the droplets by low surface coverage it was proposed that the flocculated particles were considered to form a "2-D gel structure" that effectively kept droplets from coalescing.[41] For the low surface coverage by particles, the surface area of the emulsion increased (droplet size decreased) with increasing particle concentration.

Vignati *et al.*[25] provided other evidence that emulsion stability can be achieved with low surface coverage of droplets by particles without a dense film protecting the droplet. In their study, stable o/w emulsions were produced using fluorescent silica particles. Visualization of individual emulsion droplets showed stable droplets with an average surface coverage as low as 5%. For droplets with low surface coverage of particles, the particles re-distributed on the interface and concentrated near droplet–droplet contact areas. Although it is not clear why particles tend to concentrate at the contact areas, the preferential distribution of particles on the droplet surface within the contact area may be responsible for stabilizing these droplets against coalescence.

Yan and Masliyah[35] observed the presence of particles in bulk emulsion phases even when the emulsion droplet size dictates that complete monolayer coverage of the emulsion droplets would require more particles than what is available in the system. Levine *et al.*[29] also reported the presence of solid particles in the continuous phase when insufficient amounts for complete coverage were present. These observations show that higher concentrations do not induce complete droplet coverage and that complete coverage is not necessary for emulsion stability.

The reasons why both partial and complete particle coverage lead to emulsion stability are the subject of further study. Some recent work has focused on the study of the interfacial film or on the state of particles at isolated oil–water interfaces, many of which utilize some kind of film visualization. Investigating the structure of emulsion drop interfaces should provide information as to how the interfacial region of emulsion droplets prevents coalescence and promotes emulsion stability.

6.3.3 Particle location prior to emulsification

Although particle wettability has generally been shown to determine the type of solids-stabilized emulsions, another dominant factor is the original location of the particles. From studies reviewed in the particle wettability section, it is generally accepted that hydrophobic particles will stabilize w/o emulsions. However, Yan *et al.*[43] showed that to stabilize an emulsion, solids should reside in the continuous phase prior to emulsification. Hydrophobic particles originally submerged in a water phase have little chance to cross an oil–water interface and enter the oil phase. As a result, these particles were unable to stabilize w/o emulsions. Experimentally, particles of intermediate hydrophobicity when introduced through the water phase have been shown to move across an oil–water interface to attain the expected thermo-dynamically favourable position straddling the interface. However, very hydrophobic particles initially in water would only cross the interface when a significant amount of energy was added.[49] Another challenge is that hydrophobic particles in water are unlikely to be well dispersed. As a result, their transfer across interfaces is impeded. It is clear that the way a particle sits at an interface is equally if not more important than measured particle wettability. Binks *et al.*[13,50,51] have also shown that particles should initially be in the liquid forming the continuous phase of the preferred emulsions. Binks and Rodrigues[51] linked the change in the preferred emulsion type to contact angle hysteresis that occurs when particles are placed in a fluid and migrate across an interface. Interactions between the solid and fluid, perhaps enhanced by particle roughness, change the effective contact angle and therefore the preferred emulsion type.

Recently the importance of the initial particle location in determining the preferred emulsion type was demonstrated experimentally. The findings provided a legitimate justification for earlier observations. For example, Yan and Masliyah[35] were able to stabilize o/w emulsions with hydrophobic particles. In their study, kaolinite clay particles were coated with asphaltenes to obtain hydrophobic particles with oil–water contact angles between 102° and 143°. Since these hydrophobic particles were dispersed into the aqueous phase prior to emulsification, they were able to stabilize o/w emulsions, as anticipated.

Although the effect of particle location provides an extra complication when predicting what type of stable emulsion can be expected from a particular system, it does provide an additional factor, which can be controlled when producing emulsions. The maximum stability of an emulsion can be obtained by placing particles with appropriate wettability in the appropriate continuous phase. For example, to obtain the most stable w/o emulsion, hydrophobic particles should be dispersed in the oil phase prior to emulsification.

6.3.4 Oil type

For industrial applications where solids-stabilized emulsions are undesirable, choosing the oil phase is not an option. However, some information describing the effect of oil type on emulsion stability has been gained. Oil type affects interfacial

tension, the contact angle of solids, chemical interactions between the particle surface and the liquids, and the energy of particle attachment to an interface.[13,50] For
silica particles with the same surface treatment (silanol content), the continuous
phase of a produced emulsion was dependent on the oil characteristics. Systems
with non-polar oils tend to produce o/w emulsions whereas w/o emulsions are preferred for systems with polar oils.[50] A complete explanation of these findings in
terms of the surface energies of all the components involved was given by Binks
and Clint.[52]

Considering contact angle hysteresis, both particle location and oil type can be
related back to the effective contact angle of the particle in an emulsion system. The
ultimate effect that particle wettability, particle location and oil type have is to determine how the particle resides at an emulsion droplet interface and the energy holding
the particle at the interface. In addition to these factors, which basically alter the surface characteristics of the solid particles, the particle shape and size also play a role
in determining emulsion stability.

6.3.5 *Particle size*

For a given material, the size of particles determines the ability of a particle to
remain suspended and hence at an oil–water interface. Experimentally it has been
shown that decreasing particle size would lead to increased emulsion stability and
decreased droplet size until a critical particle size is reached.[3,53] More recently,
Binks and Lumsdon[54] studied the effect of particle size on emulsion stability, and
demonstrated that larger particles produce less stable emulsions. Tambe and Sharma[3]
used alumina particles with average sizes of 4 and 37 μm in preparing emulsions
from the same decane–water mixture. As shown by the results in Figure 6.13, the
larger particles could not stabilize emulsions while small ones could. This was true
at various concentrations of stearic acid, used to change the wettability of the particles. A particle mixture of two different sizes at the same solid content produced
a less stable emulsion.

In a recent article by Tarimala and Dai,[46] images of poly(dimethylsiloxane)-
in-water emulsion droplets stabilized by polystyrene particles were provided. They
found that particle polydispersity disrupts the surface coverage of an oil droplet by
the particles in water. The image reproduced in Figure 6.14 shows that when the
large (4 μm diameter) particles are present there is no ordered arrangement of particles, in contrast to the case when only small (1 μm diameter) particles are present.

Although the emulsions studied by Tarimala and Dai were considered stable,
linking the visualization with the previous work by Tambe and Sharma[3] leads us to
conclude that polydispersity of fine particles reduces the stability of emulsions by
disrupting their ability to form a protective barrier resistant to coalescence. In general,

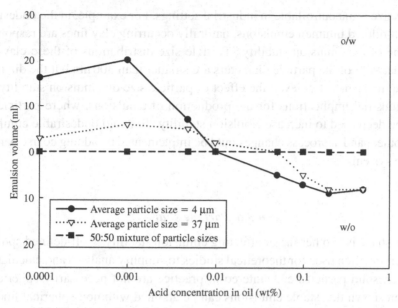

Figure 6.13 Variation of emulsion volume with stearic acid concentration in systems of water, decane and alumina particles for different particle sizes. Modified from Ref. [3].

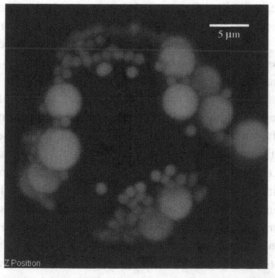

Figure 6.14 Confocal microscopy image of the surface of an o/w emulsion droplet stabilized by a mixture of 1 and 4 μm polystyrene particles of different wettability. Scale bar is 5 μm. Taken from Ref. [46]; with permission of the American Chemical Society.

particle size is uncontrollable in industrial settings. For example, in the undesirable water-in-diluted bitumen emulsions, naturally occurring clay fines are responsible for some of the emulsion stability.[8] Particle size distributions of these clay fines show that 90% of the particle diameters are smaller than 300 nm but the diameters range up to 1 μm.[55] However, the effect of particle size on emulsion stability may have industrial implications for the production of emulsions, where particle size could be decreased to increase emulsion stability. To avoid undesirable emulsions on the other hand, a process change could be implemented by adding coarse particles into the system.

6.3.6 *Particle shape*

Particle shape is another factor affecting emulsion stability. Well-defined spherical particles are often used for theoretical studies to simplify analysis and calculations. However, solid particles encountered in practice are not necessarily spherical. It has been shown that stable emulsions can be attained with non-spherical fine clay particles[43] and fumed silica.[43,56] To date no systematic study to elucidate the role of particle shape in emulsion stabilization has been performed. However, studies have alluded to the role of particle roughness. Considering particle roughness as a major cause of contact angle hysteresis, the particle roughness can affect the apparent contact angle of particles and hence influence emulsion stability in the same manner as particle wettability. By comparing smooth spherical particles to smaller particles with noticeable surface roughness, Vignati *et al.*[25] determined that rough particles produce less stable emulsion droplets and were unable to stabilize the same emulsions as was the case when larger, smoother particles were used. This could be due to the reduced particle-interface contact resulting in a smaller energy of adsorption of the rough particles. Reduced surface contact of rough particles appears to considerably reduce the interfacial potential well holding the particles at the interface.[25] The surface coverage of emulsion droplets by the rough particles was very different from the coverage by smooth particles. For rough particles, the surface coverage of particles was much lower and particles were either grouped together or randomly distributed across the surface. In contrast, a high surface coverage and an even particle distribution were observed for the smooth particles. Vignati *et al.*[25] attributed the low surface coverage of droplets by rough particles to decreased adsorption kinetics.

It should be noted that among all the parameters described above, particle shape seems to be the least important factor, or least understood, in determining emulsion stability. However, the influence of particle shape on emulsion stability has not been fully studied and the possibility of industrial implications exists.

6.4 Structure of Particles at Emulsion Drop Interfaces

Earlier studies investigating solids-stabilized emulsions have mostly been macroscopic in nature, where bulk emulsion stability is the focus. More recently, attention has been shifted to directly monitoring the interfacial structure. The techniques used in these microscopic studies include freeze fracture transmission electron microscopy (TEM)[57] and scanning electron microscopy (SEM),[58] laser confocal microscopy[46] and video-microscopy of fluorescent particles.[25] The visualization of oil–water interfaces and particle behaviour at these interfaces has resulted in many new insights into solids-stabilized emulsions. One notable aspect of these visualization experiments is to reveal the orientation of particles at the interface. Freeze fracture SEM micrographs of both w/o and o/w emulsion droplets showed a complete monolayer of particles at the interface.[58] The SEM image in Figure 6.15 shows a close-packed arrangement of silica particles on the interface of an o/w (Miglyol 810N) emulsion droplet. Similar particles were found to be flocculated and formed aggregates at the interface in a cyclohexane-in-water emulsion, as shown in Figure 6.16. The interfacial structure here is similar to that proposed by Midmore[41] to explain the stabilization of an emulsion by an estimated particle surface coverage of 29%.

For other solids-stabilized o/w or w/o emulsion systems, a more densely packed particle layer was observed at the interface. A similar observation was made with larger particles and relatively low magnification imaging techniques. Tarimala and Dai[46] obtained some clear images of fluorescent microspheres at the oil–water interface of an emulsion droplet. As shown in Figure 6.17, 1 µm polystyrene spheres form a hexagonal close-packed structure but do not completely cover the oil droplet interface.

Figure 6.15 Freeze fracture SEM micrograph of fumed silica particles at the oil–water interface in a Miglyol 810N-in-water emulsion. The polydisperse particles exhibit a close-packed arrangement. Taken from Ref. [58]; with permission of the Royal Society of Chemistry.

With freeze fracture TEM, Midmore[57] showed areas of close-packed particles at the interface of oil emulsion droplets stabilized by silica particles and surfactants. In an emulsion system containing castor oil, water, silica and a nonionic surfactant with polyoxyethylene chains (Synperonic NP30) the particle coverage of an emulsion drop was found to be about 72%, mostly in a single interfacial layer. In addition, multilayer particle domains were observed on the oil droplets from a system containing a surfactant with fewer polyoxyethylene chains (Synperonic NP2). The interfacial structure observed by Midmore in the presence of surfactants was similar to the structures reported by Binks and Kirkland.[58]

Recently we studied interfacial films of asphaltenes and fumed silica particles deposited from a planar oil–water interface.[59] Interfacial Langmuir films of silica alone and a mixture of asphaltenes and silica particles were formed at a toluene–water interface. The Langmuir film was compressed to a desired interfacial pressure and was then deposited onto a silicon substrate by lifting the vertically oriented thin

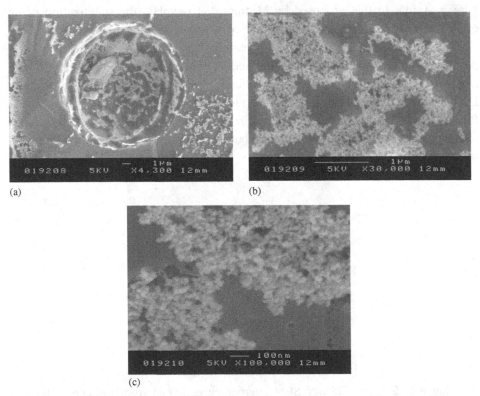

(a)   (b)

(c)

Figure 6.16 Freeze fracture SEM micrographs at different magnifications of the interfacial structure of a fumed silica-stabilized cyclohexane-in-water emulsion. Note the presence of silica clusters and particle-free regions. Taken from Ref. [58]; with permission of the Royal Society of Chemistry.

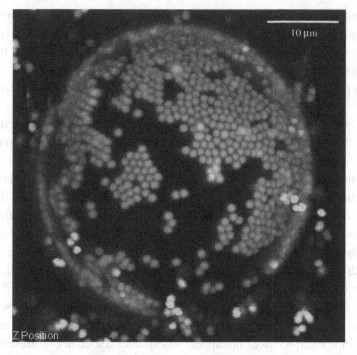

Figure 6.17 Partial coverage of an o/w emulsion droplet by 1 μm polystyrene spheres. Note the hexagonal close-packed arrangement of particles in covered areas. Scale bar is 10 μm. Taken from Ref. [46]; with permission of the American Chemical Society.

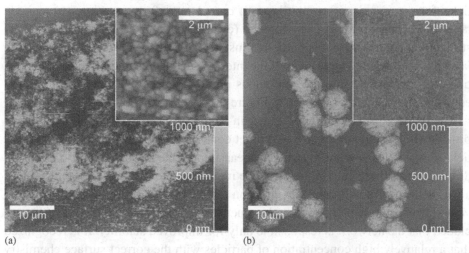

Figure 6.18 AFM height images of single layer particle films deposited onto silicon wafers from a planar toluene–water interface. (a) Silica only, (b) 50 wt.% asphaltene − 50 wt.% silica mixture. Taken from Ref. [59]; with permission of the University of Alberta.

wafer through the interface (*i.e.* the Langmuir–Blodgett deposition technique). The deposited films were then imaged with an atomic force microscope (AFM) and with a SEM. Shown in Figure 6.18 are the AFM images of transferred films formed by silica alone and from a mixture of silica and asphaltene.

The film formed with silica alone (a) appears to be less closely packed on the planar interface than the films observed with freeze fracture SEM of an emulsion droplet surface reported by Binks and Kirkland.[58] When the film is constructed of particles and asphaltenes (b), particles are flocculated into groups or domains at the interface. Higher magnification imaging of the films formed with the asphaltene-silica mixture reveals that areas outside of the silica clusters are occupied by asphaltene aggregates. The interfacially active asphaltene molecules behaved differently from the two surfactants studied by Midmore.[57] The asphaltene molecules created their own interfacial structure, in contrast to the polyoxyethylene surfactants which only contributed to the structure of the aggregated silica particles. The silica clusters formed with asphaltenes were large and circular in shape, which is significantly different from the smaller, uneven clusters shown by Binks and Kirkland.[58] Whether the planar interface caused this distinct difference remains to be established. Considering the size of emulsion droplets and the size of the particle clusters, the curvature of the interface does not appear to cause such a distinct difference in particle morphology at interfaces.

These studies indicate the absence of a universal interfacial structure in solids-stabilized emulsions. Experimental evidence shows that particles can be present in the five major configurations outlined previously in Figure 6.4: (a) individual particles form a monolayer or multilayer of complete surface coverage, (b) a single layer of particles bridges two droplets, (c) particles cluster together into a 2-D structure at the interface, (d) particles cluster together into closely packed, single layer domains leaving large areas of the interface uncovered and (e) multilayer particle domains extend into the continuous phase forming a 3-D network between emulsion droplets, as was shown in Figure 6.12.

The densely packed particles, either in single or multilayer configuration, can provide strong steric hindrance to droplet coalescence. The partially covered particle domains leave some interfacial areas unprotected. Conceptually, emulsion droplets under these conditions would be more likely to coalesce. The presence of a 3-D structure should enhance emulsion stability as emulsion droplets are prevented from contacting each other by a network of particles. Thieme *et al.*[26] concluded that the network is responsible only for improved stability. It is recognized that a relatively high concentration of particles with the correct surface chemistry is required in order for a network to form. Therefore, the 3-D network is considered as a secondary structure and the droplet coverage by an interfacial structure of particles is the primary emulsion stabilizing feature.

6.4.1 Dynamic interfaces

Other studies have been performed to capture the movement of particles at interfaces, or the structure at a more dynamic interface. In these studies the response of perturbations to contact with another interface was investigated. Asekomhe et al.[49,60] studied the behaviour of a macroscopic water drop (~3 mm diameter) covered with silica particles (12 μm diameter) suspended in an oil phase. The drop was expanded and retracted by increasing and decreasing the volume of a water droplet hanging on a capillary tube. The behaviour of the interfacial particulate layer was monitored as the interfacial area was altered. Still images from video recordings of droplet retraction for cases with the hydrophobic and hydrophilic particles present are shown in Figures 6.19 and 6.20, respectively. Upon significant reduction of the interfacial area, the interface containing particles of intermediate

(a)   (b)

(c)   (d)

Figure 6.19 Images showing the collapse of a layer of hydrophobic particles (diameter = 13μm, $\theta = 125°$) covering the surface of a pendant water drop (diameter ≈ 3 mm) in oil. (a) Particles injected in water, (b) water drop crumpling on retraction but particles do not leave the interface, (c) water drop continues to crumple on retraction and (d) crumpled interface only. Taken from Ref. [49]; with permission of the American Chemical Society.

Figure 6.20 Images showing the collapse of a layer of hydrophilic particles (diameter = 13 μm, θ = 45°) covering the surface of a pendant water drop (diameter ≈ 3 mm) in oil. (a) Particles injected in water, (b) water drop remains spherical on retraction, (c) particles get sucked in with water and (d) particles completely sucked in with very little crumpling. Taken from Refs. [49 and 60]; with permission of the American Chemical Society.

hydrophobicity crumpled as shown in Figure 6.19. Particles remained attached to the oil–water interface due to their bi-wettable nature and strong association between the particles. The crumpling is reversible, as the interfacial layer became smooth upon drop expansion. This behaviour demonstrates the strength of densely packed, particulate interfacial films. In the case of hydrophilic particles, the droplet appeared to deform with very little crumpling (Figure 6.20). For an emulsion droplet with a similar interfacial structure, one can anticipate that the particles would remain at the oil–water interface, even under significant drop deformation. This could lead to enhanced emulsion stabilization since the interfacial layer is largely resistant to disruption, preventing the water droplets from coalescing.

The manner that interfaces only partially covered with solids can provide emulsion stability is not well established. Some recent information about the interfacial structure of colliding interfaces may shed some light on the subject. Particles on sparsely covered droplet interfaces are able to re-distribute on the interface. They tend to collect near droplet contact areas (from interface or bulk) and may transfer between droplets.[25] The accumulation of particles near the interface of contact could prevent droplet coalescence. Particles were observed to form a single layer between the droplet interfaces and "bridge" the droplets.[25] This is a possible mechanism for prevention of droplet coalescence. A similar bridging effect was visually observed when examining a water droplet contacting a planar interface, both covered with polystyrene particles.[23,24] Particles were observed to bridge the two interfaces and form a disc of densely packed particles over the area of contact. The bridging prevented coalescence and was sufficiently strong to cause deformation of the droplet upon approach to and retraction from the planar interface.

Investigation into the interfacial structure of emulsion drop and model oil–water interfaces has led to improved understanding of stabilization mechanisms of emulsions by fine solids. Interfaces of emulsion droplets can be densely covered by particles. Spherical or near spherical particles will form a layer of a hexagonal close-packed structure. In addition to densely covered interfaces, particles can become flocculated and form aggregates at the interface. Incomplete surface coverage results in the formation of aggregated particle domains at the interface. Surface active molecules, such as asphaltenes, will influence the behaviour of solids at the interface, contributing to the formation of round shaped clusters of particles at a planar oil–water interface.[59] Emulsion stabilization by solids will certainly be affected by the interfacial structure of solids. Re-distribution and bridging of particles at partially covered interfaces help explain the stabilization mechanism of particles without a dense interfacial structure. Theoretical studies of emulsion stability generally involve the assumption that the interfaces are covered with a close-packed monolayer of particles. Recent study of interfacial structures has shown that this is not always the case for many systems.

6.5 Conclusions

The study of solids-stabilized emulsions has progressed substantially since the first work of Ramsden[2] and Pickering.[10] The single most important parameter influencing emulsion formation is the wettability of the particles involved. The wettability will determine the preferred position of the particle at an oil–water interface and influences the type of emulsion that is preferentially formed in a system. Particles that are hydrophilic and have an oil–water contact angle of $\theta < 90°$ have been shown in many studies to stabilize o/w emulsions. Conversely hydrophobic particles ($\theta > 90°$) tend to stabilize w/o emulsions. Particle roughness, their initial location and the properties of the oil phase can all influence the effective wettability of particles and are factors in the emulsion stabilization. Emulsions will tend to form with the continuous phase being that in which particles are initially dispersed. In part, this is due to the change in contact angle (hysteresis) experienced by the particle surface when it is dispersed in a liquid phase but is also due to their inability to migrate across the interface from one phase to another. Hysteresis can be increased by particle roughness affecting the final emulsion stability. Rough particles have been shown to form less stable emulsions than similar particles with smoother surfaces.

The theoretical explanations for stability of solids-stabilized emulsions have historically focused on providing reasons for the steric barrier which solid particles furnish. Advances in the thermodynamic treatment of particle adsorption/desorption free energy have been made to include the contributions of line tension. In addition to steric hindrance, interfacial particles will also influence the hydrodynamic properties of the interface and emulsion droplet contact area. Theoretical investigations have shown that particles provide significant resistance to liquid film drainage and hence provide emulsion stability.

Many investigations have been performed to record the nature of emulsion droplet interfaces and particle behaviour at interfaces. Several macroscopic configurations of solids-stabilized emulsions have been observed in experiments. These include emulsion droplets with closely packed layers of particles, single particle layers bridging between droplets, 2-D aggregation on droplet surfaces, 3-D aggregation between droplets and the observation of stable droplets poorly covered by domains of close-packed particles. The stabilization of emulsions with droplets poorly covered by particles is remarkable and represents a significant challenge for the development of an adequate theory to explain solids-stabilized emulsions. To date, the substantial increase in solids-stabilized emulsion research has provided many interesting experimental results and a firm basis for theory to help explain such results.

References

1. A.D. McNaught and A. Wilkinson, *Compendium of Chemical Terminology: IUPAC Recommendations (The IUPAC "Gold Book")*, 2nd edn., Blackwell Science, Oxford, 1997.
2. W. Ramsden, *Proc. Roy. Soc.*, **72** (1903), 156.
3. D.E. Tambe and M.M. Sharma, *Adv. Colloid Interf. Sci.*, **52** (1994), 1.
4. D. Rousseau, *Food Res. Int.*, **33** (2000), 3.
5. D.J. McClements, *Food Emulsions: Principles, Practice and Techniques*, CRC Press, Boca Raton, 1999.
6. V.B. Menon and D.T. Wasan, *Sep. Sci. Technol.*, **23** (1988), 2131.
7. S. Levine and E. Sanford, *Can. J. Chem. Eng.*, **63** (1985), 258.
8. Z.L. Yan, J.A.W. Elliott and J.H. Masliyah, *J. Colloid Interf. Sci.*, **220** (1999), 329.
9. R.F. Lee, *Spill Sci. Technol. Bull.*, **5** (1999), 117.
10. S.U. Pickering, *J. Chem. Soc.*, **91** (1907), 2001.
11. P. Finkle, H.D. Draper and J.H. Hildebrand, *J. Am. Chem. Soc.*, **45** (1923), 2780.
12. J.H. Schulman and J. Leja, *Trans. Far. Soc.*, **50** (1954), 598.
13. R. Aveyard, B.P. Binks and J.H. Clint, *Adv. Colloid Interf. Sci.*, **100–102** (2003), 503.
14. M.M. Robins and D.J. Hibberd, in *Modern Aspects of Emulsion Science*, ed. B.P. Binks, Royal Society of Chemistry, Cambridge, 1998, pp. 115–144.
15. I.H. Auflem, *Ph.D. Thesis*, Norwegian University of Science and Technology, Trondheim, 2002.

16. B.P. Binks, *Curr. Opin. Colloid Interf. Sci.*, **7** (2002), 21.
17. A.W. Adamson and A.P. Gast, *Physical Chemistry of Surfaces,* 6th edn., Wiley-Interscience, New York, 1997.
18. Z. Xu and J.H. Masliyah, in *Encyclopedia of Surface and Colloid Science*, ed. A.T. Hubbard, Marcel Dekker, New York, 2002, pp. 1228–1241.
19. N.X. Yan, Y. Maham, J.H. Masliyah, M.R. Gray and A.E. Mather, *J. Colloid Interf. Sci.*, **228** (2000), 1.
20. R.J. Hunter, *Foundations of Colloid Science*, 2nd edn., Oxford University Press, New York, 2001.
21. W. Wang, Z. Zhou, K. Nandakumar, Z.Xu and J.H. Masliyah, *J. Colloid Interf. Sci.*, **274** (2004), 625.
22. R.J. Hunter, *Foundations of Colloid Science*, Vol. 2, Oxford University Press, Oxford, 1989.
23. N.P. Ashby, B.P. Binks and V.N. Paunov, *Chem. Commun.*, **4** (2004), 436.
24. E.J. Stancik, M. Koukhan and G.G. Fuller, *Langmuir*, **20** (2004), 90.
25. E. Vignati, R. Piazza and T.P. Lockhart, *Langmuir*, **19** (2003), 6650.
26. J. Thieme, S. Abend and G. Lagaly, *Colloid Polym. Sci.*, **277** (1999), 257.
27. I.D. Morrison and S. Ross, *Colloidal Dispersions: Suspensions, Emulsions, and Foams*, Wiley-Interscience, New York, 2002.
28. J.H. Clint and S.E. Taylor, *Colloids Surf.*, **65** (1992), 61.
29. S. Levine, B.D. Bowen and S.J. Partridge, *Colloids Surf.*, **38** (1989), 325.
30. T.F. Tadros and B. Vincent, in *Encyclopedia of Emulsion Technology*, Vol. 1, eds. P. Becher, Marcel Dekker, New York, 1983, pp. 129–285.
31. R. Aveyard, J.H. Clint and T.S. Horozov, *Phys. Chem. Chem. Phys.*, **5** (2003), 2398.
32. A.V. Nushtayeva and P.M. Kruglyakov, *Colloid J.*, **65** (2003), 341.
33. P.M. Kruglyakov, A.V. Nushtayeva and N.G. Vilkova, *J. Colloid Interf. Sci.*, **276** (2004), 465.
34. N.D. Denkov, I.B. Ivanov, P.A. Kralchevsky and D.T. Wasan, *J. Colloid Interf. Sci.*, **150** (1992), 589.
35. Y. Yan and J.H. Masliyah, *Colloids Surf. A*, **75** (1993), 123.
36. N.X. Yan and J.H. Masliyah, *Colloids Surf. A*, **96** (1995), 243.
37. N.X. Yan and J.H. Masliyah, *Colloids Surf. A*, **117** (1996), 15.
38. N.X. Yan, C. Kurbis and J.H. Masliyah, *Ind. Eng. Chem. Res.*, **36** (1997), 2634.
39. D.E. Tambe and M.M. Sharma, *J. Colloid Interf. Sci.*, **162** (1994), 1.
40. D.E. Tambe and M.M. Sharma, *J. Colloid Interf. Sci.*, **147** (1991), 137.
41. B.R. Midmore, *Colloids Surf. A*, **132** (1998), 257.
42. H.T. Davis, *Colloids Surf. A*, **91** (1994), 9.
43. N.X. Yan, M.R. Gray and J.H. Masliyah, *Colloids Surf. A*, **193** (2001), 97.
44. B.P. Binks and S.O. Lumsdon, *Langmuir*, **16** (2000), 8622.
45. B.P. Binks and S.O. Lumsdon, *Langmuir*, **16** (2000), 3748.
46. S. Tarimala and L.L. Dai, *Langmuir*, **20** (2004), 3492.
47. T.R. Briggs, *Ind. Eng. Chem.*, **13** (1921), 1008.
48. S. Abend and G. Lagaly, *Clay Min.*, **36** (2001), 557.
49. S.O. Asekomhe, R. Chiang, J.H. Masliyah and J.A.W. Elliott, *Ind. Eng. Chem. Res.*, **44** (2005), 1241.
50. B.P. Binks and S.O. Lumsdon, *Phys. Chem. Chem. Phys.*, **2** (2000), 2959.
51. B.P. Binks and J.A. Rodrigues, *Langmuir*, **19** (2003), 4905.
52. B.P. Binks and J.H. Clint, *Langmuir*, **18** (2002), 1270.
53. H. Bechhold, L. Dede and L. Reiner, *Koll. Zeit.*, **28** (1921), 6.
54. B.P. Binks and S.O. Lumsdon, *Langmuir*, **17** (2001), 4540.
55. D.M. Sztukowski and H.W. Yarranton, *J. Disp. Sci. Technol.*, **25** (2004), 299.

56. B.P. Binks and S.O. Lumsdon, *Phys. Chem. Chem. Phys.*, **1** (1999), 3007.
57. B.R. Midmore, *Colloids Surf. A*, **145** (1998), 133.
58. B.P. Binks and M. Kirkland, *Phys. Chem. Chem. Phys.*, **4** (2002), 3727.
59. R.J.G. Lopetinsky, *M.Sc. Thesis*, University of Alberta, Canada, 2005.
60. S.O. Asekomhe, *M.Sc. Thesis*, University of Alberta, Canada, 2004.

Robert Lopetinsky (left) graduated from the University of Alberta with a B.Sc. in Materials Engineering in 2002. Undergraduate research in oil sands extraction caused him to join the Oil Sands Research Group in the Department of Chemical and Materials Engineering. He is currently pursuing an M.Sc. in Chemical Engineering, studying interfacial films related to emulsion stability. He will join Dynatec Metallurgical Technologies Division in Fort Saskatchewan, Alberta.

Jacob Masliyah (centre) received his B.Sc. in chemical engineering from University College London and his Ph.D. from the University of British Columbia. He joined the University of Alberta in 1977. He is the recipient of many awards including The Gordin Kaplan Award for research excellence. More recently, he received the Alberta Centennial Medal for his contribution to Alberta's economy. Dr. Masliyah is a Fellow of the Canadian Academy of Engineering and the Royal Society of Canada. His work has ranged from fundamental research in transport phenomena to highly applied research related to bitumen extraction from oil sands. Presently, Prof. Masliyah holds the NSERC Industrial Research Chair in Oil Sands Engineering and Canada Research Chair.

Zhenghe Xu (right) obtained his B.Sc. and M.Sc. degrees in Minerals Engineering from Central-South University of Technology, China in 1982 and 1985. His Ph.D. was in Materials Engineering from Virginia Polytechnic Institute and State University, USA in 1990. After postdoctoral work in the USA, he became an Assistant Professor in Metallurgical Engineering at McGill University in Montreal in 1992. He then moved to the Department of Chemical and Materials Engineering at the University of Alberta as an Associate Professor in 1997 and became Full Professor in 2000. His research areas include interfacial phenomena, materials technology, mineral processing and coal cleaning. Dr. Xu has an NSERC-EPCOR-AERI Industry Research Chair in Advanced Coal Cleaning and Combustion Technology.

Tel.: +1-780-492-4673; *Fax*: +1-780-492-2881; *E-mail*: jacob.masliyah@ualberta.ca

7

Novel Materials Derived from Particles Assembled on Liquid Surfaces

Krassimir P. Velikov[a] and Orlin D. Velev[b]

[a]FSD, UFHRI, Unilever R&D Vlaardingen, Olivier van Noortlaan, 3133 AT Vlaardingen, The Netherlands
[b]Department of Chemical and Biomolecular Engineering, North Carolina State University, Raleigh, NC 27695, USA

7.1 Introduction

7.1.1 Functional materials

The research in novel nano- and microstructured materials with custom designed properties, composition and structure (symmetry) has been driven by the potential of using such materials in forthcoming advanced technologies. There is specific strong interest in complex one dimensional (1-D) and periodic two dimensional (2-D) and three dimensional (3-D) structures from colloidal particles. Complex 1-D assemblies have been considered as novel building blocks for materials with advanced hierarchical structure.[1] Periodic 2-D and 3-D structures from colloidal particles display several unique and potentially usable properties resulting from the existence of a long-range order.[2-9] The formation of such materials is dependent on the interactions of particles with liquid–solid or liquid–liquid interfaces and/or their confinement in thin liquid films. The need to understand the forces involved in the assembly of structures from colloidal particles has led to important fundamental insights in the field of colloidal forces and self-organization.

The materials formed by organization and assembly of colloidal particles have a number of unique and potentially utilizable features. Coatings and surface patterning with colloidal particles allow for modification of properties like wetting and reflectivity. The long-range organization of particles in coatings brings a number of significant advantages. Particle arrays often display strong interaction with light and other electromagnetic radiation. The remarkable optical properties of the 3-D colloidal crystals, such as diffraction, interference, scattering and adsorption, have for a long time been a subject of interest and fascination. Such 2-D and 3-D colloidal crystals can serve as self-assembled diffraction gratings, filters, reflectors,

waveguides and photonic crystals. The 2-D particle and protein arrays immobil-
ized onto solid surfaces have the property of being "pre-formatted" substrates and
have been considered as media for information storage, where data can be written
if a practical way of flipping the particles in different states is found. Such nano-
magnetic materials could make possible data storage in ever smaller domains.

The structural stability and flexibility (elasticity) of colloid-based materials
depends on the number of adhesive contacts between the particles in the material,
symmetry of the structure and the type of material building the particles. Naturally,
the number of contacts per particle is maximized in the case of close packed arrays;
that can be further increased or decreased by using binary systems of particles with
different size and interaction. Some binary crystals that have been obtained *via* the
layer-by-layer assembly process display even higher packing than the ones obtained
in bulk crystallization. Catalytic materials formed from aggregated or fused particles should have high surface/volume ratio, small equally sized pores and good
mechanical properties, which will maximize their catalytic/reactivity throughput
and stability. The assembly of particles onto or near non-planar surfaces can also
impart a number of interesting materials. Making of "supra-particles" with complex
shape by droplet templating makes possible the creation of a whole new range of
material building blocks. Colloidal supra-particles with complex shape could allow
model studies on interaction and self-assembly of "colloidal molecules". An additional advantage for materials science is opened by the use of functional and responsive colloidal particles, which can assemble into novel responsive (smart) materials.
Such materials can change their properties upon changing environmental conditions
such as pH, temperature and composition of the medium.

The goal of this chapter is to summarize the principles used in the assembly of
colloidal materials (proteins, nanoparticles and microparticles) at liquid interfaces, to
discuss the underlying mechanisms and to present examples of materials obtained
and their potential applications. We will focus our attention mainly on close packed
structures formed on or at the vicinity of a liquid interface(s). The reader interested
in long-ranged 3-D crystals formed in the bulk of suspensions may refer to recent
reviews.[5–13]

7.1.2 *Forces acting during colloidal assembly*

The materials formed by colloidal assembly could be of value only when the processes
for their fabrication are well understood and controlled. This requires characterization
and control of the colloidal interactions between the particles, and between the particles
and the interface(s). The interactions between colloidal particles are governed by a
wide variety of forces, which may promote or prohibit the formation of ordered structures. In addition, the formation of ordered arrays of particles is often accompanied by

processes such as drying (dewetting) and coating where the particles are pressed against or confined between (moving) interfacial boundaries. The assembly of structures from proteins is often based on highly specific interactions such as molecular complementarity, and recognition and immuno-binding. A brief survey of the various interactions encountered in colloidal assembly is given below. For more comprehensive information on particle interactions the reader is referred to Chapter 3 and to some other recent reviews.[11,12,14]

Electrostatic repulsion and van der Waals attraction are the two omnipresent components of the intermolecular interactions that always need to be considered in the analysis of the factors in colloidal assembly.[15,16] However, these two components are rarely the single or the most important driving forces of particle assembly. Long-ranged electrostatic repulsion in highly deionized environments increases the effective volume per particle, which may result in a phase transition to a crystalline state. Short-ranged electrostatic repulsion is often responsible for the adsorption of particles on interfaces, which can then lead to crystallization or particle sticking and aggregation. In almost all assembly processes, the electrostatic repulsion is opposing the disruptive role of the van der Waals forces. The van der Waals attraction typically works against the formation of high-quality-ordered assemblies, as it tends to cause particle aggregation or permanent sticking to the substrate. Strong electrostatic repulsion is present in many systems to counteract coagulation or adhesion. A powerful alternative to the electrostatic repulsion, as a means of protecting the particles against coagulation, is the coating of their surfaces with a protective layer of (non-)ionic surfactant or polyelectrolyte that give rise to steric repulsion.[15,16] Once the particles are assembled, however, the presence of attractive interactions may be desirable to bond them in the structure formed.

Another attractive component, the depletion (or structural) force, can appear for entropic reasons in concentrated systems containing mixtures of colloidal particles of different size[14,17] or close to micropatterned surfaces.[12] The presence of a large fraction of small particles can be used as a means of segregating the larger particles in a separate phase and compressing them into an ordered array. The depletion interaction is unique as an assembly tool, as it is the only attractive interaction that could induce phase separation reversibly and without coagulation.

The processes of colloidal crystallization often include adsorption of particles onto fluid interfaces or protrusion of particles through liquid interfaces. The deformation of the interfaces leads to the emergence of strong capillary forces that can move or rearrange the particles.[18,19] The first type of capillary force is induced by deformation of a single surface by particles subjected to external fields. Such forces occur when particles adsorbed on the surface of a semi-infinite liquid phase are subjected to a normal force due to an external field such as gravity, electrostatics or magnetic. This force is usually insignificant for sub-micrometre particles.

The second type of capillary force is induced by particles incorporated in a thin liquid film and protruding from the surface. At least one of the interfaces of the film should be fluid, while the other may either be fluid or solid. The same types of forces can also occur when vertically immobile particles or objects protrude from a single fluid surface (*e.g.* on top of a drying colloidal crystal). As the normal position of the particles with respect to the interface is sustained, these "immersion" forces can be large in magnitude even for small particles and could play a role in the 2-D crystallization of colloidal particles in thin films.[18,19]

The last type of capillary force is induced by liquid bridges between the particles and between particles and the surface. This case is realized when the liquid forms a meniscus between the pairs of particles instead of a continuous film around them. These are by far the strongest forces that may emerge during the dewetting or drying of colloidal particles adsorbed on surfaces. The capillary bridges contribute strongly to the re-arrangement and adhesion of particles in drying crystal arrays.

In most cases, colloidal assembly on liquid interfaces is either driven or strongly affected by the hydrodynamic forces[20] that result from particle and fluid motion. Other fundamental hydrodynamic problems related to colloidal assembly processes include the flow of liquid through close packed arrays of particles, the rheological properties of the colloidal arrays under flow, and shear and the thinning of thin liquid films containing colloidal particles. Many of these problems have been studied theoretically but explicit solutions for all of the real hydrodynamic interactions encountered are often not available.

A variety of structures from biological molecules such as protein crystals, nucleic acids and protein–lipid complexes are formed when the molecules specifically orient each other based on short-ranged recognition and binding; these can be referred to as bio-specific interactions.[15,21,22] The forces underlying the biomolecular assembly are essentially the ones outlined above – electrostatic, hydrophobic and steric, plus hydrogen bonding. What makes these interactions unique in biological molecules is the intrinsic coupling of the intermolecular forces and shape complementarity of the molecules, their complex, short-ranged nature and the high energies of association. Examples of such interactions include protein complementarity during 2-D and 3-D crystallization and immuno-specific and bio-specific recognition. Another highly selective and widely used biocolloidal interaction is DNA and RNA complementarity recognition. Matching strands of DNA and RNA can be used as a lock-and-key tool in the assembly of colloidal structures. The main advantage of this approach is that the number of matching polynucleotides is virtually unlimited and they can be synthesized relatively easily. The advances in protein and nucleic acid biophysics, and genetic engineering and expression, warrant the expanding application of bio-specific molecules in the assembly of novel colloidal structures.

7.1.3 Classification of assembly processes on liquid surfaces

The classification of the mechanisms of colloidal assembly at liquid interfaces can be done in different ways. In this chapter, we will classify the material assembly processes on the basis of the type and number of liquid interfaces involved. Further, we will consider two classes of materials. The first class is formed by particle assembly onto or near flat interfaces, which could be liquid–solid, liquid–liquid or liquid–gas, or flat films between any of these types of flat interfaces. The second class of materials is assembled inside or near spherical surfaces (typically droplets). A schematic summary of the classification based on the number, types and curvature of the interfaces is presented in Figure 7.1.

The simplest case, which will be considered in Section 7.2, involves only one gas–liquid interface (Figure 7.1(a)). Colloidal particles under the action of various forces can organize into close packed or loose structures floating on the surface. The second assembly scheme is realized when instead of a gas phase a second liquid phase, immiscible with the first one, is present (Figure 7.1(b)). Probably the most technologically relevant assembly techniques are the ones on solid substrates, where the particles are assembled in the liquid film between a gas–liquid and a liquid–solid interface (Figure 7.1(c)). This approach typically produces coatings from crystals of single size and binary crystals from particles of two sizes. A third possible scheme is to assemble particles in a thin film between two free liquid surfaces. This can be done either using a wetting film from a colloidal dispersion on a heavier liquid (Figure 7.1(d)) or using a free-standing foam (Figure 7.1(e)) or emulsion film (Figure 7.1(f)).

Assembly on or near spherical interfaces such as emulsion droplets and spherical lipid layers is gaining more attention due to its potential for making materials of novel structure and symmetry. In this type of assembly both the interface and the confined volume by the droplet play an important role. Two major approaches will be considered here: (i) particles initially present in the dispersed phase and (ii) particles initially present in the continuous phase. In the first case the assembly process is driven by removing liquid, *e.g.* by evaporation, from a droplet that may be dispersed in a gas phase (Figure 7.1(g)), another liquid phase (Figure 7.1(h)) or at the liquid–gas interface (Figure 7.1(i)). In the second case, when the particles collect around the droplet surface from the surrounding continuous phase, adsorption and formation of shells may take place at the gas–liquid interface of a bubble (Figure 7.1(j)) or at the liquid–liquid interface of an emulsion droplet (Figure 7.1(k)). All of these assembly schemes will be discussed in detail in the following sections.

7.2 Materials from Colloidal Structures Assembled on a Single Liquid Interface

Colloidal particles can be confined at interfaces by two mechanisms. First, the particles can be physically adsorbed by attraction to the interface and/or adopting a

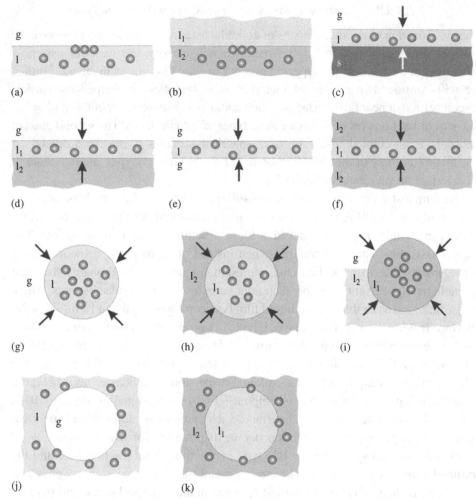

Figure 7.1 Schematic classification of colloidal assembly schemes involving liquid interfaces. Colloidal assembly schemes involving flat liquid interfaces include: (a) assembly at a single gas–liquid interface; (b) assembly at a single liquid–liquid interface between two immiscible fluids; (c) assembly in a wetting film between a solid and a liquid interface; (d) assembly between two liquid interfaces – wetting films on heavier liquid; (e) assembly in a free-standing foam and (f) an emulsion film. Colloidal assembly schemes involving spherical liquid interfaces include: (g) assembly in a droplet in a gas phase; (h) in a droplet dispersed in another liquid phase or (i) in a droplet at a liquid–gas interface and (j) at the gas–liquid interface of a gas bubble or (k) at the liquid–liquid interface of an emulsion droplet. The arrows show the direction of movement of the liquid interface.

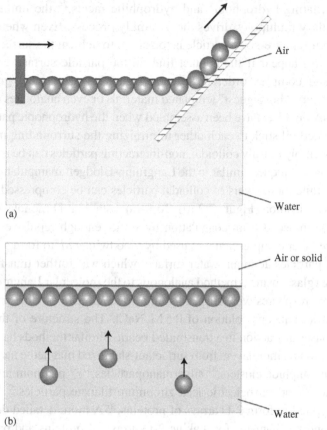

Figure 7.2 Schematic of the two most common approaches to the assembly of 2-D crystals onto a single gas–liquid surface. (a) Langmuir–Blodgett type of deposition. (b) Adsorption or attraction to an interface.

configuration where they protrude through the interface and are partially wetted by both phases. The mechanisms of particle adsorption at interfaces are described in detail in Chapter 1 of this book. The particles can be confined at an interface even if they do not adsorb when they float up towards the top surface of a heavier liquid (Figure 7.2) or when gravity pulls them towards the bottom liquid interface. The formation of a material from particles in either case requires developing controllable techniques for their organization and assembly into thin layers at the interface.

Techniques for the 2-D assembly at gas–liquid interfaces have been studied extensively by a number of investigators. On the macroscopic scale, Whitesides *et al.* investigated the self-assembly of topologically complex, millimetre-sized objects floating on the gas–liquid interface.[23–27] The particles attract each other laterally due to the action of flotation capillary forces. In the system of floating hexagonal plates of alternating hydrophobic and hydrophilic facets,[24] the immersion force between capillary multipoles drives the assembly process. Even when the weight of a floating micrometre-sized particle is too small to deform the surface, attractive forces could still appear if the contact line on the particle surface is irregular.[19] Fascinating and complex ordered structures, such as arrays and gratings with different symmetry, "host–guest" templated materials or even analogues of biorecognition and DNA doublets have been assembled when the hydrophobic patches wetted by the fluorinated oil stick to each other minimizing the surrounding menisci.[25–28]

The 2-D assembly of truly colloidal, non-interacting particles can be accomplished by compression techniques similar to the Langmuir–Blodgett manipulation of molecular films.[29] Dilute monolayers of colloidal particles can be compressed to the point of a 2-D phase transition (Figure 7.2(a)). To form a stable 2-D monolayer, the particles have to be impeded from coagulation by strong enough repulsive interactions. The method of layer compression was used by Goodwin *et al.* to form a 2-D ordered array of latex particles at an air–water surface, which was further transferred onto a solid substrate (glass) using a method analogous to the molecular Langmuir–Blodgett films.[29] The microspheres were suspended in a water/ethanol mixture and spread on the surface of an aqueous solution of 0.5 M NaCl. The structure of the layer was examined by laser diffraction in a transmitted beam. Similar methods have been used to prepare mono- and multilayers from surfactant-stabilized magnetite nanoparticles,[30] from cadmium sulphide clusters,[31] silver nanoparticles,[32,33] platinum and palladium nanocrystallites[34] and ferroelectric lead zirconium titanate particles.[35] The method can also be applied to form 2-D arrays of proteins.[36] A more detailed description of the methods and mechanisms for making 2-D arrays of proteins and particulates in liquid films is given in the next section of this chapter.

Dunsmuir *et al.*[37] patented a method whereby adsorbed latex microspheres are compressed using a Langmuir trough to form a close packed ordered structure and the particulate layer is then deposited onto a substrate withdrawn from the water surface (Figure 7.2(a)). To prepare the original monolayer, the latex particles from suspension are adsorbed on the surface of alumina-modified glass (in a random adsorbed disordered coating), dried in air and then transferred to the air–water surface by slowly dipping the substrate. The dried particles remain floating on the surface and after being ordered by compression in the trough are deposited onto another glass substrate. The method is simple and rapid enough to be of practical value. The coating consists of randomly oriented micrometre-sized domains of hexagonal ordering. The specific structure of similar compressed particulate layers has been investigated by Fulda and Tieke,[38] who measured the surface pressure isotherm of latex monolayers and found that the effective area per particle is larger than the one in an ordered close packed array. A possible reason for this might have been lateral coagulation of the particles upon compression in a close packed monolayer (expressed by the lack of reversibility in the isotherm).

Organized particulate monolayers may also be formed from particles exhibiting long-range electrostatic repulsion on the surface. A 2-D array of interacting particles out of contact, similar to the 3-D crystals in deionized suspensions, is formed. A similar system was first studied by Pieranski.[39] The long-ranged repulsion between the particles in the arrays does not come from trivial charge–charge repulsion, as the microspheres protrude out of the water surface and the distribution of charges across the interface is highly asymmetrical. Pieranski suggested a model of dipole–dipole electrostatic repulsion between the particles.

Aveyard *et al.*[40] have characterized in detail the surface pressure isotherms and the structure of sulphate latex monolayers on air–water and octane–water interfaces. Due to the low contact angle, θ_{aw}, of the particles at the air–water surface ($\approx 30°$), the microspheres are almost completely submerged in the water and the repulsive forces are relatively weak. The particles form a hexagonal long-ranged ordered structure, but it is relatively easy to induce collapse and aggregation by adding electrolyte or by compressing them laterally. The situation, however, changes dramatically when the particles are adsorbed at the octane–water interface, where the contact angle θ_{ow} is higher (≈ 70–$80°$), and large portions of the microspheres are located in the oil phase, where they strongly repel each other mostly *via* a Coulombic force. The microspheres form a very long-ranged hexagonal structure with centre-to-centre separations that are many times larger than the particle diameter (Figure 7.3(a)). Interestingly, when this 2-D array is compressed in one direction, it can change into

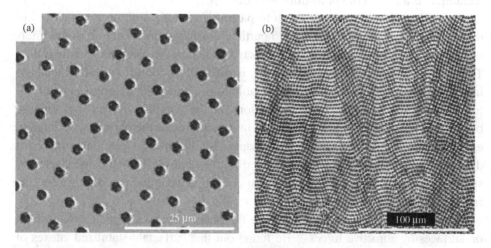

Figure 7.3 Examples of the structure of monolayers of 2.6 μm diameter latex particles adsorbed at an octane–water interface. (a) Non-close packed hexagonal array formed at low surface pressures due to the long-ranged electrostatic repulsion. (b) Anisotropic rhombohedral array of particles compressed close to the collapse pressure of the monolayer where folding occurs. Taken from Ref. [40]; with permission of the American Chemical Society.

an anisotropic rhombohedral 2-D array (Figure 7.3(b)). The authors derived formulae for the long-ranged interaction potential of the partially submerged particles electrostatically repelling each other through the oil phase. A theoretical surface pressure–area isotherm was derived that describes the experimental data well. The extraction and immobilization of such a loose structure formed at an oil–water interface is more complicated. However, the observed highly ordered and eventually anisotropic 2-D arrays may well be of materials science interest.

An advanced method for the formation of compressed and ordered particulate layers that are directly transferred to solid substrates has been developed by Picard *et al.*[41,42] The particles in the "dynamic thin laminar flow device" are adsorbed on the surface of the thin liquid film surrounding a rotating glass cylinder. The shear force of the liquid in the film compresses the particles into a densely packed monolayer against a Teflon barrier. At first, the monolayer had been transferred onto the surface of a wetting film on a hydrophilic glass substrate, and gently deposited on the surface of the substrate by drying.[41] Subsequently, the apparatus was modified to the direct deposition of monolayers onto solid substrates.[42] This modification allowed the use of both hydrophilic and hydrophobic plates as substrates, and the formation of multilayers by subsequent deposition of monolayers on top of each other. The method has also been applied with some success to the deposition of latex microspheres as small as 50 nm and of protein monolayers.[41] More recently, Reculusa and Ravaine have demonstrated the fabrication of 3-D colloidal crystals of various thicknesses from silica particles by means of multiple use of the Langmuir–Blodgett technique as a means of controlling the thickness.[43]

A qualitatively different approach to particle crystallization on interfaces is to use attractive interactions that lead to particle clustering and ordering at the interface. The ubiquitous van der Waals attraction could make particles stick together. The van der Waals interaction, however, is difficult to control and often leads to irreversible aggregation; thus it is preferable to use controllable long-range attractive forces. One class of such interactions is the capillary forces that emerge because of the deformation of the liquid interface around the particles. A detailed explanation of these forces is given in Chapter 3 of this book. Colloidal particles of diameter bigger than 5–10 μm can be directly deposited on the interfaces by spreading a drop of suspension in an evaporating solvent and can deform the surfaces enough to bring them together.

Onoda was one of the first to describe the phenomenon of particle organization on surfaces by attractive forces.[44] He found out that surfactant-stabilized latexes of 1–15 μm in size spontaneously form ordered clusters that become stable above a certain number of microspheres in the "nucleus" ($\sim$14 for 2 μm in diameter). This process bears resemblance to molecular 3-D crystallization from solution where a certain size of the nuclei is required before the crystal growth can proceed. The author points out that van der Waals and capillary interactions may exist in his system. The van der Waals attraction is possibly the predominant cause of particle clustering while the capillary forces may be of significance only in cluster–cluster interactions.

A similar process of spontaneous 2-D crystallization of floating silica microspheres was reported by Kondo *et al.*[45] This study has been performed with silica particles coated with alkoxyl chains floating on the surface of benzene (see Chapter 2). The quality of the 2-D arrays strongly depends on the length of the alkoxyl chain in the particle coatings. Disordered, fractal aggregates are formed with untreated silica, changing to short-range-ordered incomplete layers with butyl and decyl chain-coated spheres and to a dense ordered multi-crystalline monolayer when the silica microspheres are covered by dodecyl chains. Again the attractive van der Waals forces are expected to be predominant and the attractive capillary interactions are deemed important during the approach and merging of the initially formed clusters. This behaviour is widely analogous to the 3-D crystallization of proteins and colloids that takes place only within a certain window of slightly attractive interactions.[46–49]

The particle collection and assembly at surfaces can be speeded up and controlled by magnetic forces. Dimitrov *et al.*[50] studied the behaviour of paramagnetic polystyrene microspheres attracted by external magnetic fields to air–water and glass–water interfaces. The major forces behind the formation of the patterns appear to be the lateral component of the magnetic field opposed by the induced dipole–dipole repulsion. Lateral magnetic forces have also been used recently for assembling arrays from floating millimetre-size particles.[51] Thus, while a large variety of techniques for particle collection on free liquid interfaces has been reported, many alternative methods for faster and controlled assembly and deposition could still be developed. This may also make possible the deposition of layers from more complex particles with magnetic, electronic or photonic functionality.

7.3 Materials from Colloidal Structures Assembled between a Solid and a Liquid Interface

The most simple and facile method of assembling particulate materials from liquid films is the deposition of coatings on solid surfaces from wetting films of particle suspensions (Figure 7.1(c)). Ordered structures formed upon drying of latex suspensions[52] or protein solutions[53,54] have been observed and reported more than half a century ago. Assembly of super-lattices from nanocrystals has been observed more recently.[55,56] In this section we will first discuss the mechanism of convective assembly, driven by the evaporation of the solvent in the films, and then present the wide variety of structures formed by modifications of the method.

7.3.1 Basics of convective assembly

The mechanism of convective assembly in wetting films on solid substrates (Figure 7.1(c)) has been established following detailed investigations on particle dynamics during the controlled drying or dewetting process in model cells (Figure 7.4).[57,58] The schematics of the original experimental cell, where the trajectories of the suspended particles were observed directly, is shown in Figure 7.4(a). A drop containing latex microspheres is spread over a hydrophilic glass surface. The wetting film is formed inside a Teflon ring with an inclined edge on the inner side. The inclined edge guides the formation of a concave meniscus which is very important for the assembly process. The particles begin to assemble in the centre of the cell when the thickness of the drying film becomes approximately equal to their diameter. The drying of the film in the centre brings about water influx from the surrounding meniscus towards the 2-D particle crystal (Figure 7.4(a)). Particles from the bulk suspension are dragged by this convective flow to the front of the arrays and become incorporated into the growing crystals. The front of the ordered array moves on as new particles arrive and the thin dried film expands outwards. The water evaporation is facilitated by the large surface of the growing crystal. A well-known snapshot of the growing 2-D array with the trajectories of the particles moving towards is shown in Figure 7.4(b).[58] A more detailed view of this process is presented in Figure 7.5. Slower evaporation and thicker menisci lead to the formation of ordered multilayer deposits, while thinner films and quick evaporation lead to the rapid growth of hexagonal monolayers.[59]

The solvent flowing from the bulk suspension transports the particles to the crystal, which acts as a filter for the incoming particles. Due to its concave shape, the meniscus cross-section decreases close to the crystal front. As a result of decreased volume, the particles go through a strong concentration gradient, which most likely induces a pre-ordering (or pre-crystallization) close to the crystal front. The evaporating solvent in the drying crystal forms capillary necks between the particles in the arrays. In addition to shrinkage of the particles, these capillary necks could cause strong attraction between the particles and may be responsible for crack formation after drying.[60]

In contrast to the 3-D assembly in the bulk of strongly deionized suspensions, the long-ranged electrostatic repulsion and the other classic intermolecular forces are less important in the convective assembly. These colloidal forces, however, play an important role in the case of nanoparticles and protein assembly. The electrostatic repulsion between the charged microspheres, and between the microspheres and the substrate, also assures that the particles do not coagulate or adhere to the surface. The lateral capillary forces are important mostly at the nuclei formation stage when the first particles protrude the surface of the thinning film.[58,61] By using paramagnetic microparticles, Helseth and Fischer have demonstrated that the capillary forces are of importance for understanding the attraction between the

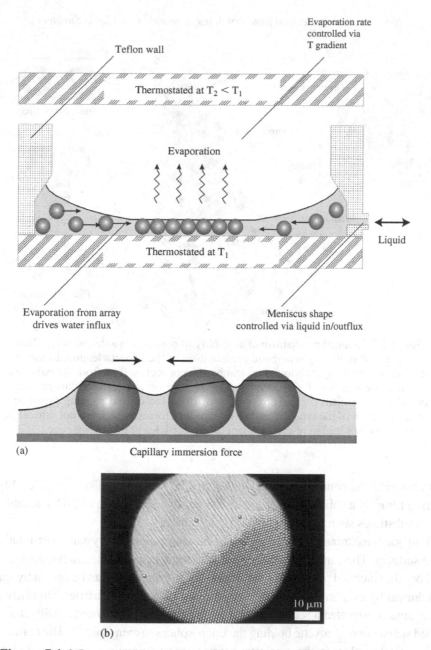

(a) Teflon wall

Evaporation rate controlled via T gradient

Thermostated at $T_2 < T_1$

Evaporation

Thermostated at T_1

Liquid

Evaporation from array drives water influx

Meniscus shape controlled via liquid in/outflux

Capillary immersion force

(b)

10 μm

Figure 7.4 2-D assembly of colloidal particles by liquid convection in a drying/dewetting liquid film. (a) Schematic of the original cell for 2-D assembly of latex particles and the possible routes to control the process by the evaporation rate and liquid film thickness.[58,65] The side view below shows the action of the capillary forces that leads to the final arrangement of the crystal. (b) Optical micrograph (top view) of the "live" process of 2-D convective assembly. The latex microspheres are carried by the flux caused by the evaporation of the solvent and are incorporated in the growing crystal (bottom right). The worm-like traces on the top left are the actual trajectories of the particles moving during the exposure period. Taken from Ref. [57]; with permission of the Nature Publishing Group.

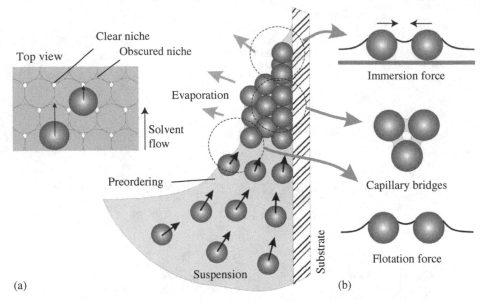

Figure 7.5 Detailed mechanism of assembly of particles in a thin wetting film on a vertical substrate. (a) Schematic presentation of the process leading to the formation of a multilayer colloidal crystal by the convective flow from the bulk suspension towards the drying crystal. The picture on the far left illustrates the possible mechanism for the preferred growth of fcc crystal packing (see text). (b) Illustration of the types of attractive capillary forces, which are not critical in the process, but lead to re-arrangement of the particles in the drying zone.

particles near the contact line of liquid drops.[62] The mechanisms of assembly in wetting films on a solid substrate are also operative in the case of 2-D assembly on liquid substrates such as fluorinated oil[63] or mercury.[61]

A major breakthrough has been the linear deposition of crystals over relatively large surfaces. The convective assembly process can be performed in horizontal, vertical or other inclined positions of the substrate. The meniscus can be driven by liquid withdrawal by evaporation only or with the aid of a moving barrier. Dimitrov and Nagayama constructed a set-up where a glass plate is withdrawn vertically at a controlled speed out of a cuvette holding the microsphere suspension.[64,65] The convective assembly takes place in the meniscus between the liquid surface in the cell and the vertical plate. The authors have performed a simplified balance of the aqueous and particle fluxes leading to the growth of the crystalline area and have suggested the following formula for the rate of substrate withdrawal, v_w (experimentally adjusted equal to the rate of crystal growth)[65]

$$v_w = \frac{\beta l J_e \phi}{h_k (1 - \varepsilon)(1 - \phi)} \tag{7.1}$$

where β is a coefficient of proportionality between the velocity of the particles and the velocity of the aqueous flux around them, l is the "evaporation length" or the equivalent distance where the evaporation takes place from within the particulate array, ϕ and ε are the volume fraction of the particles in the suspension and in the ordered array respectively, h_k is the thickness of the deposited crystalline film (for monolayers h_k is equal to the particle diameter, D) and J_e is the evaporation flux per unit area in the drying film. Prevo and Velev[59] have simplified this formula further and have proved that it is valid for a wide range of withdrawal velocities and arbitrary thicknesses on the coatings deposited. They investigated the thickness and structure of crystal multilayers deposited at systematically varied meniscus withdrawal rate and particle volume fraction and constructed an "operational phase diagram" that relates these parameters to the number of layers and symmetry of the deposited crystal layers. Detailed theoretical calculations of the speed of 2-D crystal growth have been attempted by Dushkin *et al.*[66,67] Independently, a theoretical study of the general dynamics of the film thickness profile, particle accumulation and movement of drying fronts in latex films was published by Routh and Russel.[68]

Recently, Noris *et al.*[69] suggested a possible mechanism for the preferred face-centred cubic (fcc) stacking observed in colloidal crystals formed by this mechanism. They hypothesized that the solvent flow through the pore space of close packed spheres plays a major role in the evaporation-driven crystal formation. The placement of particles on the growing crystal may be directed by the difference in the flow strength between "clear" and "obscured" niches (Figure 7.5(a), left inset). An fcc packed crystal may be formed if each subsequent layer of spheres is deposited exclusively at the clear niches where the flow rate is higher.

7.3.2 *Applications of convective assembly*

The convective assembly technique has been applied to the fabrication of 2-D and 3-D structured materials on all length scales, from microparticles to nanoparticles and protein molecules. The simplest example is a drying wetting film from latex suspensions on solid substrates; this process is interesting due to its practical application in painting and coating. Latex microspheres can be assembled into monolayers of 2-D arrays of particles, complex 2-D patterns, thin layers and thin colloidal crystals. Convective assembly has been performed both on homogeneous substrates and on substrates that have been chemically or physically patterned. Applied sequentially, it allows the creation of supra-crystals and binary crystals of particles of two different sizes. Several strategies for the fabrication of 2-D and 3-D assemblies are discussed below.

7.3.2.1 *2-D and 3-D colloidal crystals*

Even during simple evaporation of a drop of a suspension on a solid surface, one observes that the particles contained in the droplet are deposited in the vicinity of the receding contact line of the drying drop. Adachi *et al.*[70] have experimentally studied the circular patterns deposited by drying latex droplets on glass surfaces. A non-linear model of the jump-wise contact line movement that leads to the formation of "striped" deposits is developed. The formation of such "coffee ring" deposits by drying drops has been examined by Deegan *et al.*[71] In a similar way, Maenosono *et al.*[72] obtained and investigated rings of semi-conductor nanoparticles. Ring-shaped structures from particles can also be created during expansion of the contact line; when a thinning suspension film breaks, the receding contact line around the growing hole may collect the particles into an annular ring-like array.[73] The simplest way to obtain uniform monolayers from droplets, instead of a multilayered "coffee ring", is to spread out the droplet before drying.[74] This can be achieved by spin-coating.[75–78] By varying the particle concentration and spin speed, monolayers and bilayers of hexagonal packing could be fabricated.[79]

The cell for radial growth of 2-D crystals has been applied by Dushkin *et al.*[80,81] to the formation of crystalline multilayers from sub-micrometre (55- and 144 nm-sized) particles on glass and mica substrates, and study of their optical properties. A simple combination between the drying of spread droplets and the convective assembly method was suggested by Micheletto *et al.*[82] Droplets of latex suspension were spread onto the surface of a slightly tilted ($\sim$9°) glass plate. The evaporation starts from the top side of the droplet and the crystal grows downwards until a transition to disordered multilayers occurs about halfway through the original droplet length. The technique was used to estimate the role of particle polydispersity in the formation of latex arrays on silicon substrates.[83]

Dimitrov and Nagayama have shown how, by withdrawing a glass plate vertically from a cuvette with a suspension, one can construct a quasi-stationary process that allows coating of large areas.[64,65] A convenient and simple modification of this technique was later developed by Jiang *et al.*[84] The substrates were immersed into vials with a dispersion of silica microspheres in ethanol (1–4 vol.%). After allowing slow evaporation of the ethanol from the suspension at a constant evaporation rate, homogeneous coatings of multilayer colloidal crystals up to a few centimetres in size were fabricated. Several examples of crystals from different types of colloidal particles fabricated by this method are shown in Figure 7.6. The number of crystal layers depends linearly on the particle volume fraction and on the inverse particle diameter, *i.e.* the bigger the particles, the smaller the number of layers in the crystal. The method has been applied to the assembly of different types of colloidal particles including core-shell ones (Figure 7.6(d)).[85] The vertical convective assembly can be used for the successive deposition of multiple layers on top of the crystal formed from a previous deposition. This allows the fabrication of crystals many tens of layers thick.[86,87] The method was further extended to a layer-by-layer deposition of particles of two different sizes.[88] The growth of binary crystals will be discussed later in this chapter.

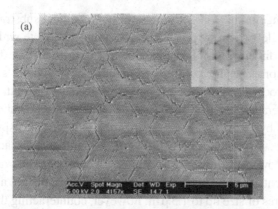

Figure 7.6 SEM micrographs of colloidal crystals assembled from different close packed colloidal particles. (a) Silica particles (diameter 220 nm) building a 12 layer thick crystal assembled on a glass substrate. Top view showing the (111) crystal plane (cracks formed after drying are clearly visible). The inset shows a Fourier transform of a $10.76 \times 10.76 \, \mu m^2$ region. Courtesy of K.P. Velikov. (b) Side view of a thin colloidal crystal as seen from the top in (a). (c) Large silica spheres (diameter 855 nm) assembled directly on a silicon wafer. Taken from Ref. [90]; with permission of the Nature Publishing Group. (d) ZnS (core)-SiO$_2$ (shell) colloidal particles of total radius 128 nm with a 44 nm shell. The scale bar is 2 μm. Taken from Ref. [85]; with permission of the American Institute of Physics.

A major challenge in the convective assembly on vertical substrates is the deposition of large particles where solvent evaporation must compete with sedimentation. Large particles sediment rapidly and cannot be confined in the drying meniscus. Three principal approaches for keeping sedimenting large spheres suspended during the deposition process have been proposed: mechanical agitation,[89] convection induced by a temperature gradient[90] and isothermal heating evaporation.[91] Mechanical agitation was used by Ozin *et al.* for crystallization of large silica spheres within surface relief patterns on optical microchips.[89,92] In another approach developed by the same group, low pressure was used to accelerate ethanol evaporation and to counteract sedimentation.[93] This method appears successful for assembling polystyrene latex spheres of size up to 1.5 μm; however, counteracting the sedimentation rate of faster sedimenting silica colloids of *ca.* ⩾ 1 μm in size seems to be quite challenging.

Vlasov *et al.*[90] addressed this problem in coating silicon wafers placed vertically in a vial containing a suspension of silica spheres (855 nm in size, 1% by volume) in ethanol. Vertical flow was generated by a temperature gradient across the vial (from 80°C at the bottom to 65°C near the top). This approach yields up to 20-layer-thick crystals that coat centimetre-scale areas of the Si wafer (Figure 7.6(c)). The authors speculated that the improved quality of these samples is due to a meniscus-induced shear that aligns the close packed layers into an fcc crystal during deposition.[94] Several other groups have reported similar results obtained by heating sphere dispersions.[95–97] Wong *et al.* proposed an isothermal heating method,[91] whereby heating the solvent (ethanol) very close to the boiling point (79.8°C) is able to keep spheres suspended in the dispersion by convective flows, and the time of growth is shortened to about 1 h for centimetre-size films. In order to obtain thick colloidal crystal films, high concentrations of silica sphere dispersions (up to 25 wt. %) had to be used. The thickness of the crystals produced by isothermal heating is about a factor of 2 smaller than that reported by Jiang *et al.* for natural ethanol evaporation.[84]

The quality of the crystals deposited by various convective assembly techniques has always been a major research problem. Abundant defects such as vacancies, crystal boundaries and stacking faults form during the self-assembly process. It is commonly accepted that the polydispersity of colloidal spheres should be smaller than 2% to minimize crystal imperfections, as required for demanding optical applications. Wong *et al.* demonstrated that the polydispersity may not be a very indicative predictive parameter for the order of the colloidal crystal films containing a minor amount of impurity particles with appreciably different sizes.[91] They conclude that exhaustive purification of silica spheres is necessary for producing high-quality colloidal crystals.

The second major problem on the road to the elusive photonic materials is the formation of cracks in the colloidal crystals due to shrinkage during drying. Chabanov *et al.*[60] reported that this can be avoided by sintering of the silica spheres as dry powders at temperatures as high as 600°C; cracks do not appear in a subsequent infiltration step. In addition, they found that the heat treatment of the silica spheres can significantly change their refractive index, depending on the temperature and their solvent history.

Most of the approaches described work well only for low volume, laboratory-scale crystal deposition. Their scaling-up to an industrial-scale mass fabrication is difficult because the process is very slow, and it is difficult to control the uniformity or crystalline thickness over large areas. To address this problem, Velev and Prevo developed an operational basis for rapid and controlled deposition of crystal coatings from particles of a wide size range.[59] They fabricated colloidal crystal coatings by dragging with constant velocity a small volume of highly concentrated suspension confined in a meniscus between two plates. The two major process parameters that allow control over the coating thickness and structure were the deposition speed and particle volume fraction. The evaporation rate was not found to affect the process to a large extent. The method has been proven to work for both latex microspheres and gold nanoparticles.[98] Jiang and McFarland[99] reported the use of the spin-coating technique for rapid fabrication of high-quality wafer scale size colloidal crystals. They used a dispersion of monodisperse silica colloids in triacrylate monomer that was spin coated onto a variety of substrates. The method allows the fabrication of colloidal crystal–polymer nanocomposites with highly uniform thickness that can be controlled simply by changing the spin speed and time.

7.3.2.2 *Convective assembly of nanoparticles and nanocrystals*

The field of assembly of nanoparticles and nanocrystals is developing rapidly, driven in part by the unique physical properties of nanoscale domain materials. The important quantum size effects in these materials are discussed in many recent reviews.[100–109] The first 2-D and 3-D assemblies were observed with Ag_2S and CdSe nanocrystals.[55,56,110–112] Since then, several groups have succeeded in preparing various self-organized lattices of silver, gold, cobalt and cobalt oxide nanoparticles (*e.g.* see Refs. [100–109] and references therein). An example of a highly ordered 2-D crystal from gold nanoparticles[113] is presented in Figure 7.7(a). Most structures have been assembled from nanocrystals stabilized with alkanethiols deposited by evaporation of the suspension on a solid substrate or by using the Langmuir–Blodget technique.[32–34,114,115]

A variety of nanoparticle super-lattice structures has been observed at different deposition conditions. For example, circular monolayer domains surrounded by regions of bare substrate have formed from silver,[116,117] gold,[112] CdS[72] and barium ferrite[118] nanocrystals. The formation of these structures has been correlated to either substrate and nanoparticle wetting characteristics[72,112,116,117] or magnetic[118]

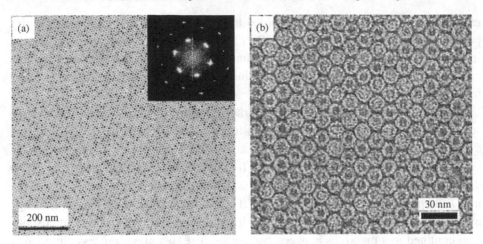

Figure 7.7 TEM micrographs of 2-D arrays from nanoparticles and proteins. (a) Highly ordered gold nanocrystal monolayer. The inset shows a fast Fourier transform of the 2-D crystal. Taken from Ref. [113]; with permission of the American Institute of Physics. (b) Apoferritin grown on the surface of a glucose solution.[36,153,155] Courtesy of H. Yoshimura, Nagayama Protein Array Project.

properties. Shafi *et al.*[119] have observed that barium ferrite nanoparticles can combine into "Olympic rings" (intersecting rings of diameter 0.6–5 μm). The intersection of two rings is not immediately understandable, as it contradicts the proposed mechanism for ring formation in drying holes. The mechanism of intersected ring building has been attributed to the interplay of magnetic forces with regular particle–substrate interactions. Maillard *et al.*[120,121] found that the formation of rings of nanoparticles (silver, copper, cobalt and silver sulphide) in the micrometre range is related to Benard–Marangoni instabilities in the liquid films. Tripp *et al.*[122] showed that Co nanoparticles can self-assemble in similar rings as a consequence of the following processes: dipole-directed self-assembly (typically 5–12 particles, ring diameter between 50 and 100 nm) and evaporation-driven hole formation in viscous wetting layers (ring diameter ranging from 0.5 to 10 μm). Govor *et al.*[123] presented experimental evidence for phase separation in a thin film of a binary mixture that includes polymer and $CoPt_3$ nanoparticles with a stabilizer. This phase separation was found to give rise to self-organized rings of nanoparticles of diameter ranging from 0.6 to 1.5 μm.

Wyrwa *et al.*[124] have described 1-D arrangements of metal nanoparticles at the phase boundary between water and dichloromethane. Under other assembly conditions, large "wires" composed of silver nanoparticles were observed.[125] The degree of self-organization has been found to vary with the length of the alkyl chains stabilizing the nanoparticles.[126] Stowell and Korgel[127] reported that sterically stabilized gold nanocrystals self-assemble into hexagonal networks or 2-D honeycombs during drop casting of a dispersion onto a substrate. By changing the particle concentration, the microscopic morphology of self-assembled structures can be tuned from rings at high concentrations to honeycombs at low concentrations.

It has been demonstrated that 3-D super-lattices of nanocrystals are often organized in an fcc structure.[55,56,110,111,128,129] In other cases, the particles pack in a hexagonal lattice.[130,131] The physical properties of 2-D and/or 3-D assemblies of silver nanocrystals differ from those of isolated nanoparticles.[112,129,132–136] These changes in the physical properties are due to the close vicinity of nanocrystals in the super-lattice.[135,137,138] Furthermore, the electron transport properties change drastically with the nanocrystal organization.[118,139]

It is accepted that these self-assembly processes occur at the liquid–substrate interface,[109] where the competing effects from the diffusion of nanocrystals along the substrate and solvent dewetting could lead to pattern formation at conditions far away from equilibrium.[140] Narayanan *et al.*[113] found that highly ordered 2-D assemblies are formed at the liquid–air surface of droplets during solvent evaporation. Changing the solvent evaporation rate can influence whether 2-D or 3-D assemblies will be formed using the same colloidal nanocrystals. A slower initial evaporation rate could allow the nanocrystals to diffuse away from the gas–liquid surface before the concentration reaches its critical value for 2-D crystallization. The formation of 3-D assemblies will then occur in the bulk of the droplet. At a faster initial evaporation rate the nanocrystals will accumulate at the 2-D liquid–air surface and eventually induce 2-D crystallization. Obviously, convective processes are present in such systems. The control of the convective assembly process can lead to the deposition of large areas of uniform films from as-synthesized "naked" Au nanoparticles.[98] These films have controllable electrical and optical properties.

Shah *et al.*[141] have studied the time-dependent structural reorganization of monolayers of gold nanocrystals deposited from liquid carbon dioxide. By comparing the translational and orientational correlation functions and a translational order parameter with a computer-simulated equilibrium state, they found that the nanocrystal organization kinetics is slower than single particle diffusion limited assembly and is most likely dominated by crystal organization. Compressed carbon dioxide is an interesting assembly medium as it has variable density and does not exhibit dewetting instabilities; this may allow deposition of spatially continuous monolayers at fast evaporation rates.

7.3.2.3 Assembly of protein layers in wetting films

The convective assembly and monolayer adsorption techniques can also be used for the assembly of 2-D layers of proteins. This vast field will only be exemplified here. The proteins behave to some extent as colloidal particles, but also have a rich functionality and range of bio-specific interactions. The strategies for 2-D protein crystallization are diverse and include chemical and physical modification of the surfaces, by *e.g.* using charged lipid layers, orienting of the proteins by specific binding and modifying both the non-specific and specific-colloidal interactions.

Price *et al.*[53] found by electron microscopy that an aqueous suspension of a purified plant virus evaporated on microscopy grids had formed 2-D (and sometimes 3-D) crystals. Crystals from other plant viruses were obtained by this simple method.[142,143] Horne and Ronchett[144] later formulated the negative staining carbon film protocol for preparing 2-D arrays of viruses and proteins. As a necessary step this method involves spreading and drying of a suspension on a solid substrate. Many 2-D crystals of various viruses[145] and proteins[146–148] have been observed.

Ferritin has been one of the most studied model proteins for 2-D crystallization. This high molecular weight protein has a spherical shape and core from biomineralized iron (Figure 7.7(b)).[149] The first report on 2-D ferritin crystallization was based on protein attraction to lipid monolayers of opposite charge at an air–water surface.[150] More recently ferritin has been crystallized on clean water surfaces by the "spreading" method of Yoshimura *et al.*[36,151–153] The protein solution is injected below the surface of a denser glucose-containing sub-phase, upon which it spontaneously spreads by buoyancy. The existence of a thin adsorption layer of denatured and unfolded protein segments on the surface has been observed and believed of importance for the crystallization process.[36,151]

The large spherical shape of ferritin makes it possible to treat this molecule as a small colloidal particle and apply the convective crystallization technique to form 2-D crystals. Ferritin has been crystallized in 2-D by using convective assembly in thin wetting films on solid and liquid supports.[152,154] However, ferritin also has some specific and directional interactions that can be triggered in appropriate environments. Typically, the crystallization is enhanced by the presence of *ca.* 10 mM of Cd^{2+} ions that specifically contribute to the short-range electrostatic binding of the neighbouring molecules in the arrays.[153,155] Site-directed mutagenesis of apoferritin, the iron-less precursor of ferritin, can lead to the change of the original hexagonal lattice to ones of oblique and square symmetry.[151,153]

Streptavidin is another protein that has been widely crystallized on liquid interfaces. Streptavidin and its homologous glycoprotein, avidin, are well known for their high affinity to biotin, a low molecular weight compound.[22,156] Biotin can easily be conjugated to various lipids. Thus lipid monolayers and bilayers with biotin on their surface can be formed, that can attach avidin and streptavidin.[22,157,158] The tetragonal streptavidin crystals grow in a variety of dendritic and "butterfly" shapes. These streptavidin arrays can serve as precursors for the formation of multilayered and functionalized arrays *via* the two binding sites on the opposite side of the monolayer. Various assemblies and devices based on 2-D streptavidin crystals have been reported.[159]

A big group of "S-layer" proteins, used by bacteria to build cell envelopes, has been designed by nature as self-assembling 2-D arrays. These proteins possess "side-on" specific-binding sites (operating mostly by H-bonds) that guide their spontaneous assembly into stable, highly crystalline lattices.[160–164] The symmetry of the lattices varies within the different species from oblique to square and hexagonal.[160–164] These proteins can be easily re-crystallized in suspension or onto various supporting surfaces including Langmuir–Blodgett films, liposomes, silicon wafers, *etc.*[160–165]

The assembly of 2-D protein crystals is key to the study of the fundamentals of protein–lipid and protein–protein interactions.[22,166] In structural biology, 2-D protein arrays are used for determining the 3-D structure of the proteins by electron crystallography.[167] The processing of the diffraction image by computers allows reconstruction of the 3-D conformation of the protein molecules from the X-ray diffraction pattern of the 2-D crystal. The method has specific advantages for characterizing membrane proteins, protein–lipid complexes and complex membrane structures that are not prone to 3-D crystallization. Protein arrays have found application in biosensors,[168] ultrafiltration membranes,[161,163,165] surface patterning and functionalization,[161] and have potential in advanced catalysis, memory storage and energy conversion devices.[152] Details on the numerous 2-D crystallization strategies for many proteins are available in the literature.[22,147,163,166,169]

7.3.2.4 Binary particle crystals

A mixture of particles of two sizes can self-organize into 2-D[170] and 3-D[171–174] binary crystals, which can have different stoichiometries and crystal symmetries depending on the size ratio and concentration. Binary colloidal crystals of large (L) and small (S) particles were first observed in nature. Two types of structures, with stoichiometry LS_2 (atomic analogue AlB_2) and LS_{13} (atomic analogue $NaZn_{13}$), were found in Brazilian opals.[171] Later, binary crystals with particles of different size ratio (R_S/R_L) were observed in suspensions of charge-stabilized polystyrene[175] and of hard-sphere-like polymethylmethacrylate (PMMA) particles.[174] Geometrical packing arguments can be used to predict which crystal phase will form if volume fractions are significantly higher than the equilibrium volume fractions.[171,172] Computer simulations have also shown that entropy alone is sufficient to cause bulk crystallization of LS_2 and LS_{13} crystals from a mixture of hard spheres of two sizes.[176] Interestingly, binary crystals from proteins have not been observed yet.

Several observations of assembly of 2-D and 3-D binary crystals in dewetting/ drying films on solid substrates have been reported. An advantage of using the controlled drying method is its simplicity: both 2-D and 3-D assemblies can be prepared simply by placing a drop of a colloidal dispersion of monodisperse particles on a flat substrate and allowing the solvent to evaporate slowly. Ohara *et al.*[112] reported observations of reversible 2-D assembly of polydisperse gold nanocrystals from suspension. The structures are characterized by hexagonal domains of large particles at the centre, surrounded radially by successively smaller particles. Antonietti *et al.*[177] studied the 2-D assembly during drying under controlled conditions of dilute suspensions of polydisperse polystyrene microgel particles (5–50 nm) in organic solvents. The resulting "Zenon"-packed structures can be described as a special form of a regular lattice. A systematic variation of the solvent quality and the surface tension of the suspension showed that the ordering is driven by size-dispersive van der Waals attraction between the particles.

The formation of small 2-D binary crystals was originally observed in a mixture of alkanethiol-derivatized gold nanoparticles (Figure 7.8(a)).[170,178] Three different types of self-organization are recognized. First, in the mixture of particles with size ratio equal to 0.58, binary 2-D crystals with a local stoichiometry corresponding to LS_2 formed. For a size ratio <0.48, segregation into two hexagonal closed packed 2-D crystals was observed. Large particles with similar radii of size ratio ~0.87 form a random 2-D alloy. These results are consistent with the results obtained for bulk crystallization of binary colloidal crystals and for inter-metallic alloys. Subsequently, 2-D binary crystals of LS stoichiometry were observed in the same system (Figure 7.8(b)).[179]

The formation of binary 2-D crystals from nanoparticles, in addition to entropic factors, is a result of the balance of attractive van der Waals and repulsive steric forces. Capillary forces operating in thin liquid films may also lead to size segregation.[180] Some mixtures of bidisperse hard spheres organize into 3-D super-lattices driven entirely by the entropic gain,[181] and it has been speculated that the same was true for the 2-D lattices.[178,182] However, it was shown that ordered lattices of bidisperse particles cannot form with hard-sphere interactions[183] and only LS (but not LS_2) crystals form with soft-sphere interactions.[184]

Shevchenko *et al.*[185] reported a trilayer super-lattice of magnetic $CoPt_3$ nanoparticles of two different sizes (4.5 and 2.6 nm) consistent with the LS_5 structure (Figure 7.8(c)). Redl *et al.*[186] reported the next logical step in the development of nanoparticle-based binary crystals – the preparation of 3-D superstructures from materials with distinctly different properties (PbSe and Fe_2O_3 nanocrystals). Precisely ordered, large single domains (up to $2\,mm^2$) of LS_2 and LS_{13} super-lattices were obtained by simply allowing the solvent (dibutyl ether) to evaporate overnight. Co-crystallized assemblies of PbSe and γ-Fe_2O_3 nanocrystals were prepared following the examples of binary crystals from colloidal particles.[172,173,187,188] The LS_2 and LS_{13} super-lattices of best quality have been formed at a size ratio = 0.58, which is in good agreement with the calculated value for the assembly of micrometre-sized colloidal hard spheres.[176] A projection of the (001) plane of 3-D LS_2 binary

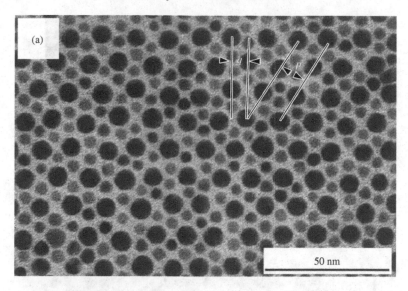

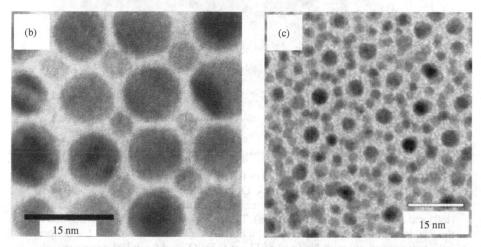

Figure 7.8 TEM micrographs showing examples of binary crystals from nanoparticles. (a) LS_2 2-D crystals from Au nanoparticles of two distinct sizes with size ratio < 0.58. The super-lattice planes marked as **p** and **q** have a measured interplanar angle of 30° and average periodicities of 13 and 8 nm, respectively. Taken from Ref. [170]; with permission of the Nature Publishing Group. (b) Section of a 2-D LS (NaCl-type) super-lattice of two different size silver nanoparticles stabilized by decanethiol. Taken from Ref. [179]; with permission of Wiley-VCH. (c) Trilayer LS_5 super-lattice (isostructural with $CaCu_5$) of $CoPt_3$ nanoparticles of two different sizes (4.5 and 2.6 nm). Taken from Ref. [185]; with permission of the American Chemical Society.

Figure 7.8 (*continued*) (d) Projection of the (001) plane of 3-D LS_2 binary crystals (isostructural with AlB_2) of 11 nm γ-Fe_2O_3 and 6 nm PbSe nanocrystals built up of alternating layers of large and small particles. The insets show a high magnification (left) and a fast Fourier transform (right). Taken from Ref. [186]; with permission of the Nature Publishing Group. (e) Depiction of a (100) plane on an LS_{13} super-lattices (isostructural with $NaZn_{13}$) of 11 nm γ-Fe_2O_3 and 6 nm PbSe nanocrystals. Taken from Ref. [186]; with permission of the Nature Publishing Group.

crystals of 11 nm γ-Fe_2O_3 and 6 nm PbSe nanoparticles is shown in Figure 7.8(d). A transmission electron microscope (TEM) image of the (100) plane projection of the LS_{13} super-lattice, with each large Fe_2O_3 nanoparticle surrounded by eight small PbSe nanocrystals to form a square array, is shown in Figure 7.8(e).

Velikov *et al.*[88] reported the first fabrication of binary colloidal crystals using a controlled drying process on a vertical substrate in a layer-by-layer process. A 2-D hexagonal close packed (hcp) crystal of large silica spheres (406 nm) was used as a template, on which small silica (202 and 220 nm) or polystyrene (394 nm) particles were deposited (Figure 7.9). They studied a narrow range of size ratios, 0.48–0.54, where the formation of the LS_2 structure was expected on the basis of experimental observations in bulk crystallization[173,174] and on packing arguments.[172,187]

For a size ratio = 0.54 and a relatively high volume fraction ($\phi = 4.3 \times 10^{-4}$), a slightly corrugated but "complete" close packed layer of small particles on top of the large ones was observed. Decrease of the volume fraction (to 2.1×10^{-4}) led to the formation of large ($\sim 200 \, \mu m^2$) areas still with hexagonal but more open packing with three neighbours (Figure 7.9(b)); the small spheres filled in all the hexagonally arranged crevices made by the first layer of larger spheres. The next layer of large particles deposited itself exactly on top of the first layer, which leads to the formation of the LS_2 binary colloidal crystal. Continuation in a layer-by-layer fashion leads to an LS_2 binary colloidal crystal of desired thickness. Further decrease of the volume fraction led to the formation of another open structure with hexagonal symmetry. This structure, with stoichiometry LS, is similar to LS_2 but instead of six particles there are now three small particles touching each large one (Figure 7.9(a)). Binary colloidal crystals of the NaCl type have been predicted to be a stable phase in bulk crystallization,[189] however small crystallites of binary crystal of stoichiometry LS were only observed recently for the first time.[188]

A completely different arrangement of the small spheres, in addition to the already described LS and LS_2 structures, for size ratios between 0.48 and 0.5 was observed.[88] At volume fractions ($\phi = 2.1 \times 10^{-4}$) lower than the one required for a "complete" coverage on the top of the big particles, each small sphere had four neighbours (Figure 7.9(c)). Each large sphere was surrounded in a ring by six small spheres (similar to the LS_2 binary crystal), but now the rings were rotated in such a way that the crevices formed in between the large spheres were filled with three small spheres arranged in a (planar) triangle or kagomé lattice. So far, the formation of LS_3 binary crystals has not been observed in bulk crystallization.

The complex structure formation in the second layer is the result of the interplay of geometrical packing constraints, minimization of the surface free energy of the drying liquid film and surface forces due to the curved menisci. During the drying process, the partially immersed particles experience attractive capillary forces. The specific packing that is obtained is determined by the local particle concentration, the symmetry and orientation of the already deposited layers and local hydrodynamic fluxes.[69] A schematic representation of the process is given in Figure 7.9(d). At sufficiently high volume fraction the smaller particles form a 2-D hexagonal but not flat layer with lines of touching particles parallel with the drying front. At slightly lower volume fraction, ϕ_3, complete coverage by the second layer is not possible so that the dark particles (see Figure 7.9(d)), which are sticking out of the drying film the most, are lost first. This leads to the formation of a kagomé net (Figure 7.9(c)). At yet lower volume fraction, ϕ_2, the structures formed are again explained by having fewer particles in the drying

Figure 7.9 Binary colloidal crystals grown using a layer-by-layer deposition of particles. (a) SEM micrograph of a colloidal crystal (size ratio = 0.48) with a stoichiometry *LS*. Small spheres ordered in an open hexagonal lattice on top of a 2-D crystal of large spheres. Scale bars = 2 μm. (b) SEM micrograph showing binary colloidal crystals from silica particles (size ratio = 0.54) with a stoichiometry of *LS*$_2$. Open packing with hexagonal symmetry occurs where each small sphere has three neighbours on top of a 2-D crystal of large spheres. Scale bar = 2 μm.

front and placing them at the lowest available points; the crevices formed by the hexagonally arranged first layer are guiding the formation of *LS*$_2$ structure (Figure 7.9(b)). At ϕ_1 there are even fewer particles in the drying front and now filling up lowest points in a line parallel to the drying film results in the *LS* structure shown in Figure 7.9(a).

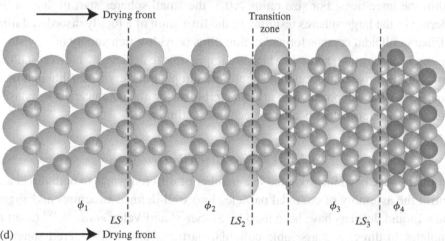

Figure 7.9 (*continued*) (c) SEM image of colloidal crystals (size ratio = 0.50) with stoichiometry LS_3. This is a magnified region with a kagomé net of small spheres on top of a 2-D crystal of large spheres. Scale bar = 1 μm. (d) Schematic representation of the structure of small spheres arranged onto a 2-D hexagonal layer of larger spheres with size ratio = 0.5 as a function of the particle volume fraction. Particles are lined up parallel to the drying front and go into sites favoured by the templating effect of the first layer and the forces resulting from the particle–liquid interfacial tension. The volume fraction, ϕ, determines how many of the available low lying sites can be occupied. The next layer of large particles will form LS, LS_2 and LS_3 binary crystals respectively. Images (a)–(c) taken from Ref. [88]; with permission of the American Institute for the Advancement of Science.

Velikov *et al.*[88] also demonstrated the fabrication of non-close packed crystals with a low packing fraction of $\phi = 0.6046$ from single size spheres using binary crystals of two types of particles. In the first step, a composite LS_2 binary crystal of alternating layers of inorganic (*e.g.* silica) and organic (*e.g.* polystyrene) particles was prepared. Then, the organic particles were removed by burning. This non-close packed structure is yet another example of how the layer-by-layer method can be used to create new crystal structures that cannot be grown in bulk. Later on, Wang and Möhwald[190] extended this layer-by-layer approach using spin-coating. This method provides significant acceleration of the growth process.

Kitaev and Ozin[93] used accelerated evaporation-induced co-assembly of binary dispersions of monodisperse particles with large size ratios to produce hexagonally closed packed monolayers of large spheres with superimposed patterns of small spheres confined within the natural surface relief of interstitial spaces in the large spheres' lattice (Figure 7.10). In this approach, binary assembly has been possible for a size ratio <0.225. For larger ratios, smaller spheres do not arrange well within the interstices. For size ratios >0.3, the small spheres start to disrupt the ordering of the large spheres resulting in the formation of a totally disordered film.

Binary colloidal crystals formed in thin films provide a rich variety of new structures and composition, some of which are impossible to achieve by bulk crystallization.[88,93] These materials could find application in photonic crystals. Although in an initial stage of development, they also hold promise for making materials responsive to magnetic, electrical, optical and mechanical stimuli.[106]

7.3.2.5 Template-directed assembly

Templates have been incorporated in a number of different ways to both direct and control the assembly of colloidal particles into well-defined structures and aggregates. Liquid droplets have been used by Stöber[191] and Velev *et al.*[192–194] to act as templates to direct and assemble colloidal particles into well-ordered spherical aggregates. The methods using droplets as a template will be considered in detail later in this chapter. The use of physical templates to direct the direction of crystal growth was demonstrated by van Blaaderen *et al.*[195] Two major techniques were used to create templates: (i) physical and (ii) chemical patterning of the substrate (see Figure 7.11). A comprehensive review of various unconventional techniques for fabricating complex nanostructures was recently published by Xia *et al.*[196]

(i) Physically patterned substrates

Physical templates based on a pre-fabricated relief pattern on the surface of a substrate that dictates and guides the nucleation and growth of colloidal crystals have been used by many groups.[4,7,8,195,197,198] Templates consisting of 2-D arrays of holes

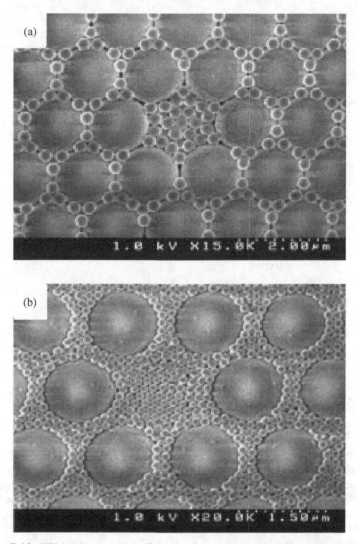

Figure 7.10 SEM micrographs of the surface patterns produced by accelerated evaporation co-assembly of large (diameter 1280 nm) and small colloidal particles of different sizes and different volume fractions. (a) size ratio = 0.203, ϕ_L (large) = 0.014, ϕ_S (small) = 2.5×10^{-4}. (b) size ratio = 0.113, ϕ_L = 0.017, ϕ_S = 4.1×10^{-4}. Taken from Ref. [93]; with permission of Wiley-VCH.

or trenches can be fabricated using lithography techniques (for an overview of lithography, see *e.g.* Ref. [196]). As the suspension dries and slowly dewets the template, the particles are pushed into the holes and/or ditches to form densely packed structures (Figure 7.11). During the liquid dewetting process, capillary and hydrodynamic forces are exerted on each particle and also lead to attraction between the

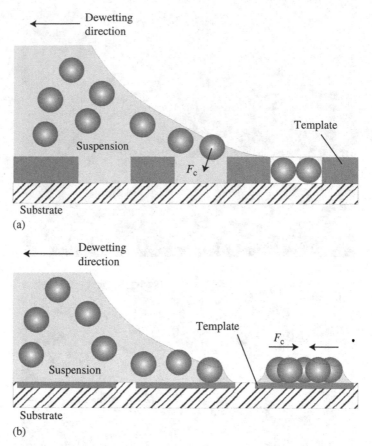

Figure 7.11 Schematics of the mechanism of template-assisted assembly in drying liquid films on (a) physically and (b) chemically patterned substrates. (a) A cross-sectional view of the process of crystal formation, illustrating the action of the capillary force (F_c) on the colloidal particles next to the rear edge of the liquid meniscus. (b) Ordering and focusing of particles on patterned solid surfaces of different wettability due to the action of the lateral capillary force.

particles and between the particle and the substrate (Figure 7.11(a)). The capillary force is partially directed toward the template, and thus capable of pushing particles into template holes. The excess spheres are moved along with the dewetting liquid with the help of hydrodynamic forces. These observations are consistent with the results of Velikov *et al.*[88] for layer-by-layer templating, starting from a 2-D layer of colloidal particles. In their study, they noticed that particles in the second layer were carried by the capillary forces and liquid flow along the edge of the liquid layer and placed at the crevices between the hexagonal packed big particles from the first layer. The maximum number of particles that each template hole can

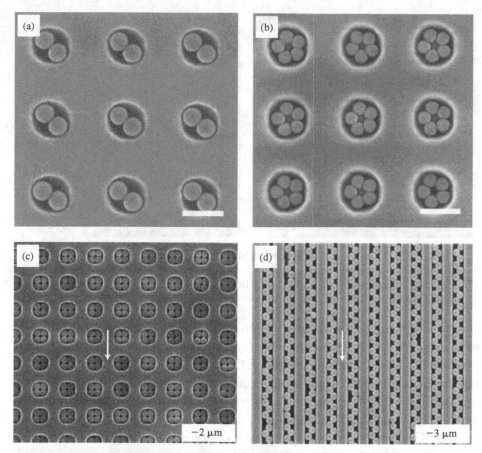

Figure 7.12 SEM micrographs of aggregates with complex shapes formed by templating polystyrene spheres against physical templates with different shapes and dimensions. (a) and (b) 2-D arrays of colloidal aggregates assembled by confinement in templates with cylindrical holes. Scale bars = 2 μm. Taken from Ref. [199]; with permission of the American Chemical Society. (c) 2-D arrays of aggregates assembled using a template with square holes. Taken from Ref. [199]; with permission of the American Chemical Society. (d) Arrays of colloidal aggregates assembled in trenches. The arrows in (c) and (d) indicate the flow direction for the liquid slug. Taken from Ref. [199]; with permission of the American Chemical Society.

accommodate and the structural arrangement are determined by the ratio of the particle size and template dimensions.

Xia *et al.* demonstrated the use of physically patterned templates in the assembly of spherical colloidal particles into 1-D and 2-D aggregates with well-defined shapes.[4,199] They showed how to make a rich variety of polygonal, polyhedral, spiral and hybrid aggregates that are difficult (or impossible) to fabricate using other methods (see Figure 7.12). The aggregates have been assembled from polystyrene spheres using

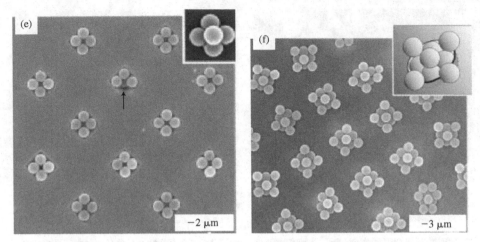

Figure 7.12 (*continued*) (e) Double layer array of square pyramidal clusters in pyramidal cavities. Taken from Ref. [4]; with permission of Wiley-VCH. (f) Complex aggregates assembled in cylindrical holes at high particle concentration. Taken from Ref. [4]; with permission of Wiley-VCH.

templates with different geometries.[196] The depth of the template plays an important role, leading to the formation of double layer structures. An example of the square pyramidal clusters is presented in Figure 7.12(e). These micro-clusters can be sintered together by heating slightly above the temperature of glass transition of the polymer material. The "welded" aggregates can then be released from the substrate by dissolving the (photoresist) template, followed by sonication in a liquid bath. The method is capable of generating complex aggregates with potential use in photonics and electronics and as model systems in condensed matter physics. However, a noticeable drawback of this process is the significantly low yield of aggregates.

Two-step deposition allows the formation of even more complex structures. In the second deposition step, the assemblies resulting from the first deposition step serve as templates with newly defined geometrical parameters. Increasing the concentration of colloidal particles also affects the assembly process and leads to more complex structures. Examples of aggregates of spherical particles assembled at relatively high particle concentrations are shown in Figure 7.12(f). Physical templates have been employed by several groups for directing the orientation of colloidal crystals obtained by convective self-assembly.[197,200–203] Yin *et al.*[203] employed 2-D regular arrays of square pyramidal pits or V-shaped grooves etched into the surfaces of Si (100) substrates for the fabrication of large colloidal crystals with their (100) planes oriented parallel to the surfaces of the supporting substrates. The capability and feasibility of this approach has been demonstrated by crystallizing spherical colloids (>250 nm) into (100)-oriented crystals over areas as large as several square centimetres (see Figure 7.13(a)).

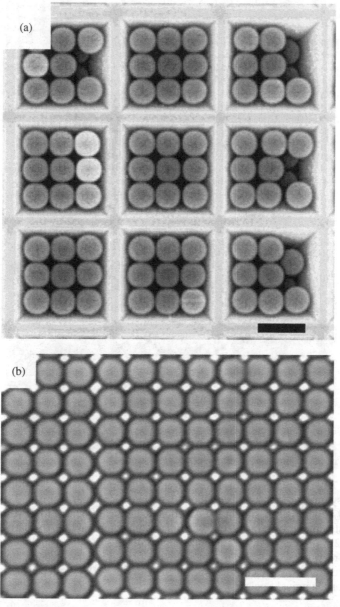

Figure 7.13 SEM micrographs showing examples of the use of physical templates used to grow colloidal crystals. (a) Three layers of a (100)-oriented colloidal crystal arranged on square pyramidal pits that were patterned in the surfaces of Si(100). Scale bar = 2 μm. Taken from Ref. [203]; with permission of the American Chemical Society. (b) One layer of colloidal particles arranged on a square symmetric pillar-shaped silicon template. Scale bar = 500 nm. Taken from Ref. [197]; with permission of the American Chemical Society.

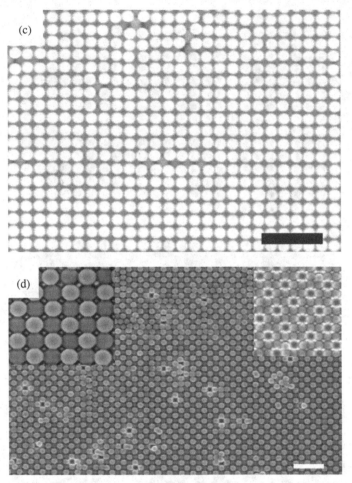

Figure 7.13 (*continued*) (c) Three layers of colloidal particles arranged on a square symmetric pillar-shaped silicon template. Scale bar = 1 μm. Taken from Ref. [197]; with permission of the American Chemical Society. (d) Two-layer bcc square structure grown on a template with a pitch comparable to the particle diameter (202 nm). The left inset is an enlargement of the structure whilst the right inset shows ordering of vacancies in a monolayer of spheres deposited at a lower particle concentration. Scale bar = 1 μm. Taken from Ref. [197]; with permission of the American Chemical Society.

Hoogenboom *et al.*[197] investigated the influence of the parameters that determine how the surface topography of templates affects colloidal crystal structures made by the vertical deposition method. They demonstrated that with the use of pillar-shaped templates, large defect-free square symmetric monolayers and body-centred cubic (bcc) and simple cubic (sc) colloidal crystals could be grown (Figure 7.13). Their results indicate that the direction of gravity plays a crucial role in the template-directed convective assembly. A two-layer bcc square structure made by "controlled drying" on a template with a pitch comparable to the particle diameter is shown in Figure 7.13(d). The left inset is a zoomed-in enlargement of the structure; the right inset shows ordering of vacancies in a monolayer of spheres deposited at a lower particle concentration.[8,197]

Schaak *et al.*[204] used a similar approach to assemble crystal arrays on lithographically templated surfaces. They further extended the method to assemble 2-D colloidal crystal binary super-lattices that can adopt a variety of structures. Graphite, kagomé, bcc, open hexagonal, tetragonal and linear chain structures can be formed by adjusting the ratio of the sphere diameter to the template diameter. Colloidal templating was also demonstrated in micro-capillaries leading to the formation of colloidal crystals of cylindrical shape.[205]

(ii) Chemically patterned substrates

Templates for particle assembly can also be made by using solid substrates whose surface is chemically patterned into domains with different wettability or surface charge of desired shape and size. Areas with positive or negative surface charges selectively attract particles of opposite charge. During the dewetting process, the deposited particles will be rearranged under the action of the capillary forces, Figure 7.11(b). When the templates are created with domains of different wettability, the shape of the withdrawing meniscus changes. This will give rise to selective deposition of particles only in the regions with concave meniscus.

Whitesides *et al.* pioneered the used of patterned molecular monolayers as templates directing the deposition and assembly of charged objects on a solid substrate.[196,206–208] This approach was developed further by Aizenberg *et al.*[209] and Hammond *et al.*[210,211] Aizenberg *et al.* used micro-patterned substrates with cationic and anionic regions. After dewetting, the particles attached are entrapped in residual water droplets on the surface. During the evaporation of water, the capillary force "focuses" the particles in the domain centre leaving an ordered cluster after drying is complete (Figure 7.11(b)). Very well-ordered 2-D particle arrays of desired symmetry and lattice constant much larger than the particle diameter were produced by this procedure.[209] It was demonstrated that other types of colloidal interactions can be used to deposit particles onto patterned substrates.[198,210] Hammond *et al.* demonstrated how to fabricate 2-D arrays of particles (Figure 7.14(a)–(d)), on polyelectrolyte surfaces using patterned layer-by-layer polyelectrolyte films as functional templates.[210,212] The particle position is determined by the electrostatic interaction.

Masuda *et al.*[213] developed a process for fabrication of particle wires through self-assembly onto hydrophilic regions of self-assembled molecular layers. Layers of octadecyltrichlorosilane were formed on a silicon substrate and modified by

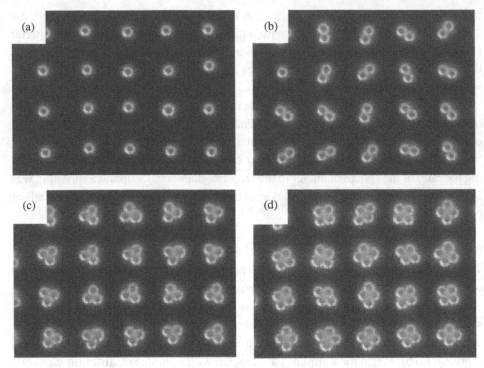

Figure 7.14 Examples of using chemically patterned templates for particle assembly. (a)–(d) Optical images of 2-D colloidal arrays with a different number of particles formed on circularly patterned polyelectrolyte templates. Particle diameter = 4.34 μm. Taken from Ref. [212]; with permission of Wiley-VCH.

UV irradiation to create a pattern of hydrophobic octadecyl and hydrophilic silanol groups. Suspensions of particles (550 or 800 nm) in ethanol or water were dropped onto the patterns. Due to selective wetting of the hydrophilic areas, particle wires formed between two droplets and colloid crystals formed at both ends of the particle wire after complete drying of the solution.

Fustin *et al.*[214] demonstrated the growth of microscopically structured colloidal crystals *via* convective self-assembly onto chemically patterned substrates. The authors deposited colloidal crystals on chemically patterned surfaces with hydrophilic and hydrophobic areas (Figure 7.14(e)–(h)). They used pattern dimensions much larger than the individual particle size to control the microscopic crystal shape rather than influence the crystal lattice geometry (as achieved in colloidal epitaxy). Better results were obtained when the lines of a stripe pattern are oriented parallel to the withdrawing direction rather than perpendicular to it. The deposition resolution (defined as the minimum feature size on which particles can be deposited) was found to depend on the wetting contrast and increases with lower average hydrophobicity of the substrate. In contrast to nanopatterned substrates, they found a non-linear

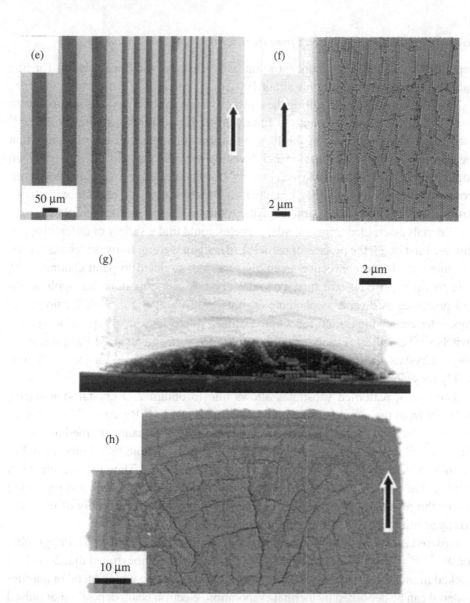

Figure 7.14 (*continued*) (e) SEM image showing vertical stripes of colloidal crystals of varying width grown on a perfluorodecyltrichlorosilane-SiOH pattern. The arrows indicate the direction in which the suspension was withdrawn. The dark areas correspond to the silica surface (SiOH) covered by latex particles, while the bright regions correspond to the hydrophobic perfluorodecyltrichlorosilane monolayer. Taken from Ref. [214]; with permission of the American Chemical Society. (f) SEM image showing the stepwise increase of the thickness from the edge to the centre of the stripe (top view stripe edge). The arrows indicate the withdrawing direction from the latex suspension. Taken from Ref. [214]; with permission of the American Chemical Society. (g) Cross-section through a stripe (after breaking the substrate). Taken from Ref. [214]; with permission of the American Chemical Society. (h) Edge at the beginning of a stripe (top view). Taken from Ref. [214]; with permission of the American Chemical Society.

dependence of the colloid concentration and substrate withdrawal speed for line patterns with dimensions below about 100 μm.

The formation of stripes during the convective assembly of crystals occurs due to instabilities in the hydrodynamics of the wetting liquid films. Kondic and Diez[215] used numerical simulation of the flow on a homogeneous substrate to show that the contact line becomes unstable and develops periodic patterns. This mechanism could explain the results of Masuda *et al.*[216] Using convective assembly under controlled drying conditions, they obtained an ordered array of particle wires constructed from close packed particles without the need for patterned templates.

Materials assembled from colloidal particles could find a variety of different applications. First of all, the process of particle ordering in wetting films, which resembles in some aspects the convective assembly process, is related to paint coating.[217,218] Due to their periodic structure, colloidal crystals are important for applications and processes as diverse as photonic crystals,[3,6,9,60,85,87,90,91,219–225] diffraction gratings, interference filters, anti-reflection coatings, micro-lenses,[80,84] optical filters and switches,[226] chemical sensors[227] and substrates for surface-enhanced Raman scattering.[228] Crystals assembled from particles can serve as templates for the creation of highly mesoporous and macroporous materials.[5,228–233]

The use of patterned substrates allows one to obtain 2-D crystal symmetries different from the simple close packed hexagonal array.[234] Recently, 3-D colloidal crystals were also utilized to produce a surface with 2-D nanopatterned arrays.[235] Tan *et al.*[236] reported the use of convective assembly for the fabrication of a 2-D periodic non-close packed array of non-spherical particles. They used reactive ion etching that converts the spheres into non-spherical particles. They also report the use of the resultant patterned Si substrates as templates for assembly of uniquely arranged microspheres of different diameters.

Ordered 2-D particle monolayers have been successfully applied as lithographic masks.[79,200,234,237,238] Deckman *et al.*[37,76–78,239,240] first demonstrated that 2-D close packed monolayers of particles could be used as masks into which metal or another material can be deposited by thermal evaporation, electron beam deposition or pulsed laser deposition from a source normal to the substrate (see Figure 7.15). After metal deposition, the nanosphere mask can be removed by sonicating the sample in a solvent, leaving behind the material deposited through the nanosphere mask on the substrate. Later, Van Duyne *et al.* extended the method for the development of double layer nanosphere masks.[237,241–245] This method is now recognized as an inexpensive and general materials nanofabrication technique.[79,241–249] Ingert and Pileni[250] later demonstrated colloidal lithography based on 10 nm ferrite nanocrystals.

In biological studies, ordered arrays serve as convenient micro-patterned surfaces for studying processes such as cell adhesion.[251] The wide range of available particle sizes and compositions, along with the possibility to graft various

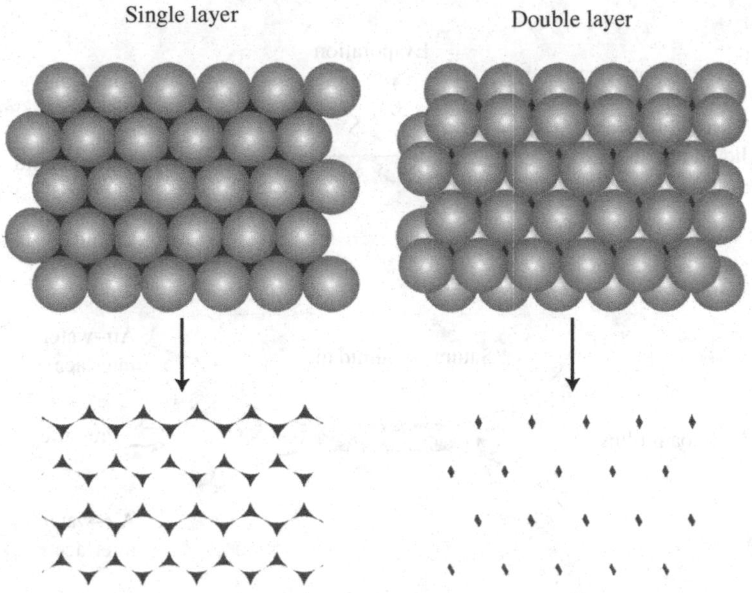

Single layer   Double layer

Figure 7.15 Schematic illustration of the use of single layer and double layer colloidal crystals in nanosphere lithography.[237] The black areas represent the areas where the metal is deposited.

bio-active molecules onto the particle surface, imply that these studies will expand in the future.

7.4 Materials from Colloidal Structures Assembled between Two Liquid Interfaces

7.4.1 Assembly in wetting films on a heavier liquid

In this method, schematically illustrated in Figure 7.16(a), colloidal structures are formed during the evaporation of a colloidal suspension on top of an inert heavier liquid. The fluidity of the two interfaces provides specific advantages over the assembly on a solid substrate. The method can, for example, be used to assemble small particles and biomolecules, as both interfaces of the fluid film are molecularly smooth. In addition, the tangential mobility of the surfaces provides a "non-sticky" substrate, where the particles are free to move, rotate and rearrange into defect-free lattices. The lack of friction between the particles and the substrate also enhances the action of the strongly attractive capillary forces that may facilitate the assembly process in films thinner than the particle diameter. However, one needs appropriate liquid substrates that are more dense than the particle dispersion and immiscible with it. Up to now, only two liquids have been used as substrates – mercury and fluorinated oil.

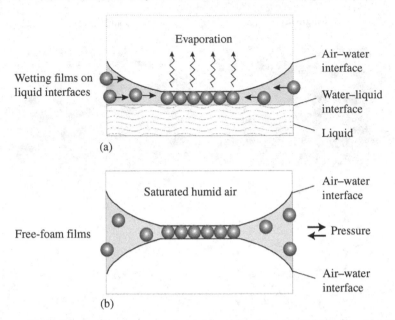

Figure 7.16 Schematic representation of particle assembly in (a) a wetting film on top of a heavier liquid and (b) a free standing thin liquid film.

The first convenient liquid substrate that can be used for 2-D assembly is perfluorinated oil which is an inert dense liquid with low volatility. Lazarov *et al.*[63] were the first to characterize the mechanism of latex microsphere assembly on perfluoromethyldecalin (PFMD). To ensure good and even spreading of the film, the authors used a mixture of fluorinated surfactants. The dynamics of the assembly process in wetting films on a heavier liquid was similar in many aspects to the assembly process in wetting films on solid surfaces. When the film thickness becomes approximately equal to the particle diameter, the particles in the film centre exhibit extraordinarily long-ranged attraction from tens of micrometres away caused by capillary forces. The particles organize into small well-ordered islands. If the rate of evaporation is high, the water film around the ordered aggregates may dry completely and the aggregated islands are left behind as separate discontinuous domains. At moderate and slow evaporation rates, however, the aggregates attract each other, merging and rearranging into larger ordered clusters. Eventually, at higher concentrations of particles, "2-D foams" are formed, where the ordered particles are packed into a continuous structure with large areas void of particles surrounded by particle bridges.

The growth of large crystalline areas is brought about *via* convective assembly by flux resulting from the evaporation of water, similar to the wetting films on solid substrates. Bilayers and trilayers can be formed at a decreased rate of water evaporation and increased particle concentration. Notably, the multilayers formed show an absence of square lattice bands in the transition regions, observed when at least one of the interfaces is solid. The reason is that as both of the interfaces are fluid, they can eventually bend and accommodate a sharper transition between the stable close packed hexagonal multilayers. The 2-D crystals formed can be mounted onto solid surfaces by evaporation of the fluorinated oil that leads to gentle deposition of the arrays on the substrate below.[63]

2-D crystallization of proteins on a mercury surface in a closed trough was reported by Nagayama *et al.*[252–255] The wettability of the mercury surface was increased by keeping the surface clean under an environment of pure oxygen. The spreading and drying process is relatively quick and after a period of 10–60 s the protein arrays formed are picked up on a carbon film-coated electron microscopy grid dropped from above. The method was successful in crystallization in more than 1/3 of the proteins studied (including ferritin, ATP-ase and membrane proteins) and appears particularly promising to the study of membrane proteins that are difficult to crystallize in 3-D.[252] The crystallization process is enhanced in the presence of glycerol and glucose, possibly through slower water evaporation and osmotic resistance to film rupture.[180,255] Mercury was used as a substrate for the crystallization of micrometre-sized latex particles.[61]

7.4.2 Assembly in free-standing liquid films

Solid particles are often present in the free liquid films formed between bubbles of foams or droplets of emulsions. Systems containing solid particles in thin liquid films are, *e.g.*, the "Pickering" emulsions stabilized by solid particles. Naturally, foam and emulsion liquid films could also serve as sites for 2-D assembly of colloidal arrays. Model free standing thin liquid films can be formed by sucking liquid out of a biconcave meniscus held in a capillary (Figure 7.16(b)). The microscopic observation is carried out in reflected monochromatic illumination that allows measuring the film thickness *via* interferometry. The first system of interest studied by this method was the stratification of micelles inside foam films.[256–258] It was shown that the stratification can be a major factor in the stabilization of liquid films in foams and emulsions. This phenomenon can also be observed with nanometre-sized latex particles and protein molecules.[259–261]

We have investigated the dynamics of structure formation in foam films containing negatively charged sulphate latex microspheres and different molecular stabilizers (surfactants) of the film.[262,263] The major factor affecting the particle dynamics and the resulting structures was the type of the surface active film stabilizer used. Foam films from solutions containing sodium dodecyl sulphate (SDS), a strong anionic surfactant, were very stable but all entrapped latex particles were immediately pushed out of the film area during the first few seconds of film thinning and before finishing the thinning stage of film evolution. In the case of films stabilized by protein (bovine serum albumin, BSA) entrapment of a few particles inside the film area for short periods of time was achieved, due to the high interfacial viscoelasticity of the protein monolayers. The entrapped particles were typically collected inside the thicker "dimple" in the film centre and formed small 2-D arrays, attracting each other by the lateral capillary forces from distances that could reach up to 100 μm. The small ordered clusters were "spit" out at a later stage of the film thinning. At higher particle coverage of the surfaces (~50% by area) the density of the particle aggregates changed from single clusters to "bridges" and 2-D foam-like metastable structures similar to those observed on the surface of fluorinated oil were formed.

A major change in the particle dynamics was observed in films stabilized by dodecyltrimethylammonium bromide (DTAB) or hexadecyltrimethylammonium bromide (HTAB) cationic surfactants which are charged oppositely to the particles. These surfactants modify the surfaces of the polystyrene microspheres by coupling to the intrinsic negative charges. This results in particle hydrophobization. The modified particles tend to adsorb directly on the air–water surfaces of the film because of their increased hydrophobicity.

Micrographs of the basic stages of film evolution in the case of cationic surfactants are shown in Figure 7.17.[262,263] After the first formation of the film, the particles were pushed out into the surrounding meniscus and the film area was void of particles. During the expansion of the first film, however, the particles in the ring closest to the periphery got in contact with the upper surface and stuck to it, linking together the lower and the upper surfaces (Figure 7.17(a)). After separating the surfaces by closing the film, these particles got sucked into the middle (Figure 7.17(b) and (c)). An ordered cluster of particles formed binding the two film surfaces together. Each subsequent opening and closing of the film added new particles to the ones already stuck in between the interfaces, leading to the formation of high-quality 2-D particle arrays with almost no lattice defects (Figure 7.17(d)–(f)). The size of the single crystalline domains obtained was up to $3 \times 10^4 \, \mu m^2$. Recently, detailed observations of the interesting dynamics of particles between two approaching oil–water interfaces with adsorbed colloidal particles were made by Horozov *et al.*,[264] discussed already in Chapter 1.

These results prove that the major factor for 2-D assembly in free-standing liquid films is the capillary interaction between the particles. Capillary interaction could occur both above and below a film thickness equal to the particle diameter, but the menisci around the particles will have different curvatures. Notably, both of these interactions are attractive and the interaction energy can exceed the thermal energy kT by many orders of magnitude for distances as large as 100 times the

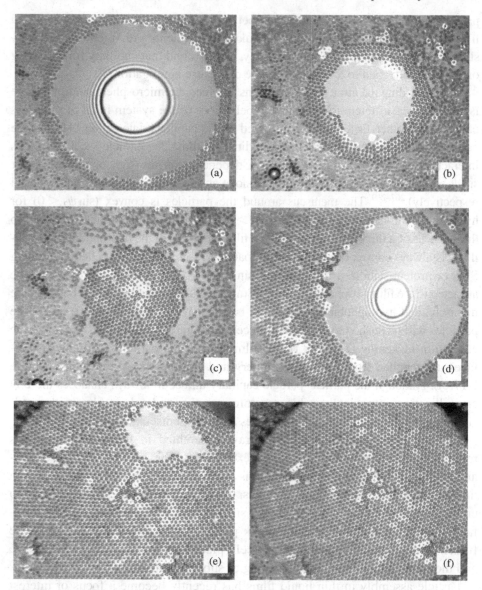

Figure 7.17 Micrographs of the formation of 2-D latex crystals in a free-standing foam film. Successive opening, (a) and (d), and closing, (c) and (f), of the film leads to particle structuring *via* the "zipping" mechanism – a particle bridges the two liquid surfaces.[262,263] The microspheres are 7 μm in diameter. Taken from Ref. [262]; with permission of the American Chemical Society.

particle diameter.[262] Strong interaction between particles in a free-standing liquid film has also been observed by Sur and Pak in experiments with micron-sized polystyrene latex particles.[265] In another study, it was found that capillary forces acting on particles attached to a lipid membrane can cause aggregation.[266]

Understanding the attractive interactions between the microspheres in films alone is, however, insufficient to describe the behaviour of the system in all of the cases observed. Another factor to be considered is the capillary interaction between the particles and the curved menisci surrounding the film. The particle – cell meniscus interaction can be attractive or repulsive depending on the sign of the parameter $\sin \psi_1 \sin \psi_2$ (where ψ_1 and ψ_2 denote the slope angles at the cell and particle menisci respectively).[267–269] The meniscus around the particles is convex ($\sin \psi_2 < 0$) for hydrophilic particles, and concave ($\sin \psi_2 > 0$) for hydrophobic particles attached to the surfaces of a thicker film, whereas the film meniscus in the hydrophilic glass capillary is always convex ($\sin \psi_1 < 0$). 2-D particle assemblies are expelled in thinning SDS or protein-stabilized films because $\sin \psi_1 \sin \psi_2 > 0$, *i.e.* these particles will be attracted to the film periphery (Figure 7.18(a)). Adversely, the interaction between the meniscus and hydrophobic particles will be repulsive ($\sin \psi_1 \sin \psi_2 < 0$) and these particles will be pushed towards the film centre (Figure 7.18(b)). This is exactly the case with thickening DTAB or HTAB films, where the particles bridging the surfaces are repelled from the film periphery and compressed in the centre (Figure 7.17(c)). Although the particle–particle capillary attraction always enhances the 2-D crystallization, intrinsically stable structures are obtained only in films with repulsive capillary forces between the particles and the meniscus.

The practical application of 2-D arrays assembled in freely standing films requires development of procedures for extraction of the assembled layers as separate solid state materials. This method is technologically more complicated than the ones reported in the previous sections, though it has the potential advantage to create single crystalline arrays of bigger size. One way to achieve this is the ultrafast freezing of the foam films by quick plunging in liquefied gasses.[261,270] The frozen films can be detached from the cell, placed upon an appropriate substrate and the water can be removed by melting or sublimation.

Particle assembly in thin liquid films has recently become a focus of interest from both academia and industry due to the effective stabilization of foams and emulsions by solid particles, and additional possibilities for making novel macroporous material and solid foams.[271] Alargova *et al.*[272] recently developed a new class of polymer microrods that can be used as "super-stabilizers" of foams.[273] The intertwined cylindrical particles adsorbed on the film surfaces form a thick barrier that prevents any drying and breakdown. Foams of extreme stability were reported, where the liquid can be all dried out and the structure is still preserved by the entangled microrods. In contrast, Binks and Horozov[274] showed how, by optimizing

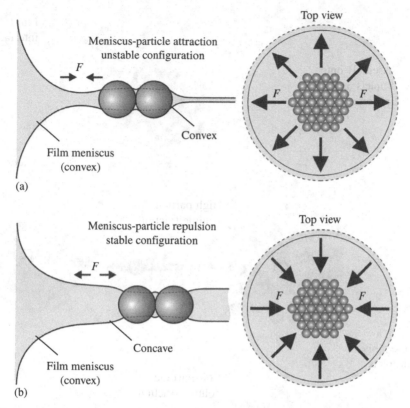

Figure 7.18 Schematic of the capillary interactions acting between particle arrays in a free-foam film and in the surrounding meniscus for (a) an unstable and (b) a stable configuration. The structure within the film is compressed by the lateral capillary forces and is in a stable configuration (b) only when the particles are partially hydrophobized in the presence of cationic surfactant due to the repulsion between the convex and concave menisci.

particle hydrophobicity, nanoparticles of silica can be effective foam stabilizers alone, discussed in Chapter 1.

7.5 Materials from Colloidal Structures Assembled Around and Inside Droplets

The preceding sections demonstrate that assembly of colloidal particles on a wide variety of flat solid or fluid surfaces can be achieved by appropriate design of interactions and processes. Interesting materials, however, can also be obtained when colloids are assembled at spherical instead of flat surfaces. Examples of systems where the surfaces are parts of self-contained colloidal objects are emulsion droplets and

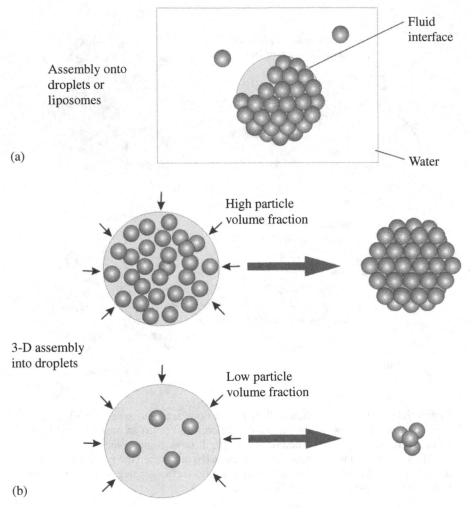

Figure 7.19 Schematic of the two modes of particle assembly (a) around and (b) inside liquid droplets.

lipid vesicles. The assembly process driven by the liquid–liquid interfaces of a droplet can take place both at the interface and inside the droplet. A schematic of the possible assembly processes around and inside a spherical liquid–liquid interface is shown in Figure 7.19.

7.5.1 Structured shell assembly around droplets

The assembly of colloidal particles at spherical interfaces was first studied in the context of emulsion stabilization. Beginning with the pioneering work of Ramsden[275] and Pickering,[276] shells of colloidal particles adsorbed around droplets have been used as an alternative to surfactant molecules to stabilize emulsions. Extensive reviews have been written on this subject.[277,278]

The first extensive demonstration of the use of emulsion droplets as templates for the assembly of monodisperse latex particles was reported by Velev *et al.*[192–194] Latex microspheres were collected, ordered and fixed together on the surfaces of micron-size octanol droplets in water. Subsequently, the droplets were dissolved in the surrounding environment to obtain a suspension of microspheres in the form of ordered hollow "supra-particles" (*i.e.* capsules with surfaces that are composed of a close packed layer of colloidal particles linked together to form a solid shell). The stages of the assembly process are schematically presented in Figure 7.20.

In the first step, the surface of negatively charged latex particles was modified by adding the amino acid lysine at neutral pH. As a result, the particles spontaneously adsorb and assemble on the droplet surfaces due to screening of the surface charges. The next step in the assembly was to add octanol and stir the oil and sensitized latex together. During this process the oil is dispersed into droplets and the latex microspheres adsorb around these droplets and form close packed ordered shells. Casein that adsorbs and forms a protective barrier by steric repulsion is added at the next stage to stabilize the latex-covered emulsion droplets. The droplets covered by an ordered layer of particles remain dispersed and stable after being sterically protected with casein. To bind the particles in the shells, a mixture of HCl and $CaCl_2$ was added in the next step. The particles around the droplet stick to each other to form a fixed rigid structure. In the next step of the process the template octanol droplets were removed by adding 50 vol.% of ethanol to the water phase. The process results in hollow "supra-particles" from microstructured latex particles (see Figure 7.21).

The droplet templating concept can be modified in several ways by using different particles and binding techniques. On the macroscopic scale, Huck *et al.*[279] employed emulsion droplets to assemble balls out of complex hexagonal objects of sub-millimetre size. Hexagons with one hydrophobic face are assembled around oil droplets in water, while hexagons with one hydrophilic face are assembled around water droplets in oil.

A similar class of supra-particles was prepared by assembling particles around droplet as templates in water-in-oil emulsions (Figure 7.22).[280] This technique for the preparation of hollow, elastic capsules with sizes ranging from micrometres to millimetres was developed by Dinsmore *et al.* The fabrication process is based on a general three-step assembly (Figure 7.22(a)). First, a suspension of the material to be encapsulated is emulsified in an immiscible fluid containing colloidal particles that adsorb on the surface of the emulsion droplets. Second, a shell is formed by locking the particles together around the droplets by controlled heating. Third, if

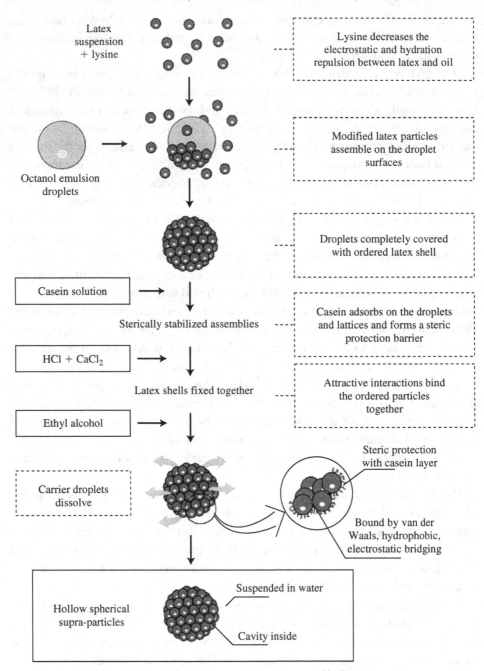

Figure 7.20 Schematic of the approach of Velev *et al.*[192–194] for the assembly of ordered supra-particles using emulsion droplets as templates.

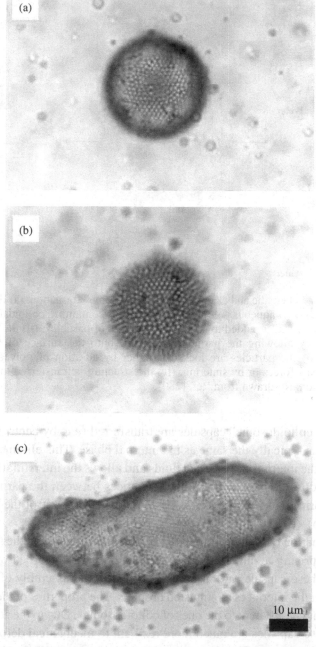

Figure 7.21 Optical microscopy images of supra-particles assembled from octanol-in-water droplets stabilized by 1 μm latex microspheres. (a) and (b) Top and bottom images of the same supra-particle. (c) An elongated particle formed when some octanol is ejected from the carrier droplet. Taken from Ref. [192]; with permission of the American Chemical Society.

(a)

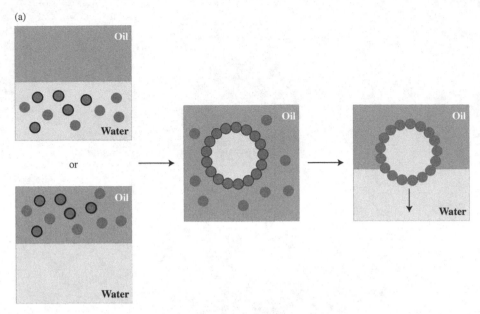

Figure 7.22 Schematic and examples of "colloidosome" formation. (a) Illustration of the process: (i) aqueous solution is added to oil containing colloidal particles or, alternatively, oil is added to aqueous dispersion of particles; (ii) the system is emulsified by allowing the particles to adsorb onto the surface of the water droplets; (iii) the particles are locked together by addition of polycations, by van der Waals forces or by sintering; (iv) the structure is transferred to water by centrifugation. Re-drawn from Ref. [280].

required, the "colloidosome" capsules are transferred (*e.g.* by centrifugation) into a solvent that is typically the same as the internal phase. This eliminates the interface between the internal and external fluids and allows the interstitial holes to control the colloidosome permeability. The interstices between the particles form an array of uniform pores, whose size is easily adjusted to control the permeability (Figure 7.22(b)–(d)).

In a continuation of their work, Weitz *et al.* have demonstrated the fabrication of capsules comprising a single layer network of polymer adsorbed onto colloidal particles.[281] These capsules are made using directed self-assembly of the polymer and particles at the interfaces of emulsion droplets. The properties of these structures are reminiscent but different to the ones of colloidosomes. Due to the adsorbed polymer layer, these capsules are much more resilient to mechanical deformation than sintered colloidosomes. The fabrication of colloidosome microcapsules with a shell of polymeric microrods (Figure 7.22(e)), was also demonstrated recently.[282] The underlying principle of the method is similar to the one illustrated in Figure 7.22(a). The synthesis involves three stages: (i) A hot aqueous solution of agarose

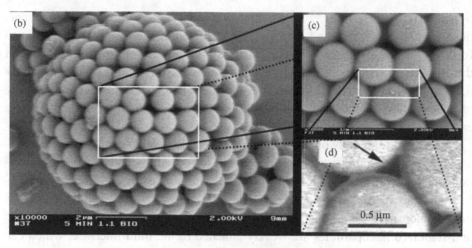

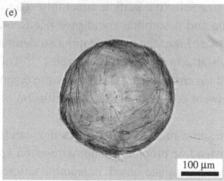

Figure 7.22 (*continued*) (b) SEM image of a dried, 10 μm diameter colloidosome composed of 0.9 μm diameter polystyrene spheres. (c) and (d) Close-ups of (b) and (c) respectively. The arrow points to one of the 0.15 μm holes that define the permeability. (e) "Hairy" colloidosome microcapsule produced by transferring micro-rod-coated agarose beads into water. (b)–(d) taken from Ref. [280]; with permission of the American Institute for the Advancement of Science and (e) from Ref. [282]; with permission of the American Chemical Society.

is emulsified in oil in the presence of rod-like polymeric particles to produce a water-in-oil emulsion stabilized by the solid particles, and the system is cooled to set the agarose gel. The function of the gel cores is to support the particle shell and to give the microcapsules enough stiffness to be separated from the oil phase by centrifugation. (ii) The suspension of aqueous gel microcapsules obtained is diluted with ethanol and centrifuged to separate them from the supernatant. (iii) The microcapsules are washed with ethanol and water and re-dispersed into water. This technique allows preparation of colloidosome capsules of diameters ranging from several tens to several hundreds of micrometres.

Recently, Croll and Stöver have demonstrated that soft, swellable microgel particles can be assembled and then covalently linked at the oil–water interface to form gel "tectocapsules".[283,284] The particles are first allowed to self-assemble at the oil–water interfaces of a propyl acetate–water mixture and then are covalently fixed in place to form flexible microgel capsules. The wall thickness and stability of the capsules can be controlled by varying the particle loading in the oil phase and the solvent composition.

Nanoparticle assembly on droplet interfaces can also lead to a multitude of materials with interesting properties. Size-dependent self-assembly and 2-D phase separation on liquid–liquid interfaces has been shown by Russell and collaborators.[285] In contrast to micron-sized particles, the thermal energy causing spatial fluctuations of the nanoparticles is comparable to the interfacial energy in their system. The ligand-stabilized nanoparticles are weakly attached to the fluid interface, which can be used to induce size-selective particle assembly. The authors show that size-dependent adsorption and desorption could give rise to a 2-D phase separation at the fluid–fluid interface. Later the same group also demonstrated chemical cross-linking of the ligands attached to the nanoparticles.[286] This allows stabilization of the nanoparticle shell *via* cross-linking. The composite organic–inorganic, nanometer thick membranes prevent convection but allow diffusion of small molecules across the interface.

Lipid vesicles are supra-molecular structures that can be used as a template for colloid assembly instead of droplets. Nanometre-sized species such as biomolecules can be a subject of template-based, interaction tailored assembly. A variety of membrane-specific proteins have been assembled into ordered structures onto reconstituted or specially prepared liposomes. A few of the best-known examples are bacteriorhodopsin,[287–290] cholera toxin, human coagulation factor[291–293] and the S-proteins presented in the previous sub-section. The possibility for incorporation of large micron-sized particles in a spherical lipid bilayer has also been demonstrated.[294,295] Liposomes have been used for the binding and assembly of ferritin using electrostatic attraction.[296] The liposomes are positively charged by a small fraction of the cationic surfactant HTAB. As the environment is sustained at pH = 6, the ferritin molecules are above their isoelectric point of 4.8 and possess a negative charge. Vesicles covered with ordered ferritin shells have been obtained (Figure 7.23). The ferritin shells can be fixed by glutaraldehyde, a common agent for protein molecule cross-linking. Moreover, the liposomes inside the cross-linked ferritin structures can be extracted by solubilization with nonionic surfactant, leaving behind ordered aggregates of cross-bound protein that are analogues of the latex supra-particles described above.

Russell and collaborators have demonstrated hierarchically structured nanoparticle arrays fabricated by a combination of two self-assembly processes on different length

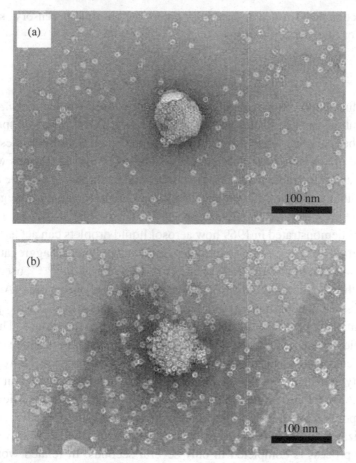

Figure 7.23 TEM micrographs of supra-particle assemblies from ferritin around liposome templates. (a) Vesicle with adherent non-fixed ferritin array. The liposomes are positively charged by incorporating some amount of HTAB. (b) Ferritin shell cross-linked by glutaraldehyde around a liposome. Courtesy of O.D. Velev.

scales.[297] They used the condensation of micrometre-sized water droplets on the surface of a polymer solution as spherical cavities that self-assemble into a well-ordered hexagonal array ("breath figures"), in combination with self-assembly of CdSe nanoparticles at the polymer solution–water droplet interface. Complete evaporation of the solvent and water leaves the particles assembled into an array of spherical cavities and allows for *ex situ* investigation. Fluorescence, confocal, transmission electron and scanning electron microscope images show that the CdSe nanoparticles are confined at the polymer solution–water interface, where they form a 5–7 nm thick layer. The wide range of assembly techniques of particles around droplets overviewed above opens new routes to encapsulation and to fabricating highly functionalized ordered materials, potentially useful in sensory, separation membrane or catalytic applications.

7.5.2 3-D assembly inside droplets

A general limitation of the methods described in the previous section (*i.e.* Section 7.5.1) is that the final geometry of the structure assembled is always a spherical or close to spherical shell. Therefore, methods for the assembly of particles into 3-D spheres and in non-spherical close packed structures or into aggregates with complex shape are of technological importance. Several examples of particle assembly in droplets that are dispersed in a gas or liquid or are suspended at the gas–liquid interface are presented in this sub-section.

Stöber[191] demonstrated in 1969 how aerosol liquid droplets can act as templates for the assembly of latex particles into polygonal and polyhedral aggregates of various sizes and shapes (Figure 7.24(a) and (b)). Iskandar *et al.*[298–300] used a spray-drying method to produce silica particles with ordered mesopores. A colloidal mixture of silica nanoparticles and polystyrene latex nanoparticles was sprayed as droplets into a vertical reactor separated into two temperature zones. The solvent in the droplets had evaporated at the front part of the reactor to produce composite assemblies consisting of silica and polystyrene nanoparticles (Figure 7.24(c) and (d)). The polystyrene nanoparticles in the droplets had then evaporated in the back portion of the reactor to produce a powder consisting of mesoporous silica spheres. The mesopores had been arranged into a hexagonal pattern, indicating that a self-organization process occurred spontaneously during the solvent evaporation. The entire process is completed in only several seconds. In related work, Moon *et al.* have demonstrated how electrospraying of an aqueous colloidal suspension can be used to make uniform 3-D assemblies and their inverse structures – "photonic balls".[301] The size of the balls can be controlled through the strength and frequency of the electric field and particle loading in the suspension. The photonic balls display angle independent colour and could be used as pigments for reflection-based coatings.

Velev *et al.*[229] have developed an "outside-in" templating technique to make 3-D organized balls inside emulsion droplets suspended at liquid interfaces. The particles are assembled inside suspended droplets that act as well-defined confinement templates for the growth of 3-D supra-particles of defined size and shape. The template water droplets float on the surface of PFMD, a heavier liquid immiscible with water. The colloidal particles inside the droplets are gradually concentrated by drying to form 3-D crystals, yielding a variety of novel microstructured symmetric supra-particles. The structures are to a certain extent the 3-D analogues of the supra-particles described in Section 7.5.1 and in Refs. [192–194].

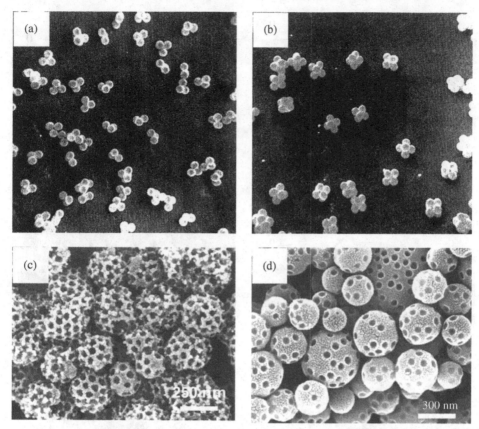

Figure 7.24 SEM micrographs of structures assembled in evaporating water droplets. (a) and (b) Silica particle aggregates with complex shape (particle diameter = 790 nm). Taken from Ref. [191]; with permission of Elsevier. (c) and (d) Spherical silica particles produced by spray drying and pyrolysis of a suspension containing different amounts of polystyrene latex (diameter 79 nm) and silica particles (diameter 5 nm). Taken from Ref. [300]; with permission of the American Chemical Society.

A major issue in the practical implementation of this method is the control of the shapes of the obtained particles *via* the initial shapes of the suspended template droplets. The shape of a floating droplet (or lens) is controlled by the interplay of the gravity and interfacial tension, which is captured by the Bond number β. The droplet shape is close to spherical at values of $\beta \rightarrow 0$ and flattens as the Bond number increases. Examples of the theoretically attainable shapes of the template droplets are displayed in Figure 7.25. The shape of the periphery of the droplet at the three-phase contact line is determined by the balance of the three interfacial tensions acting there. For water droplets floating on PFMD, the tension at the lens-air boundary is approximately equal to the sum of the tensions at the other two

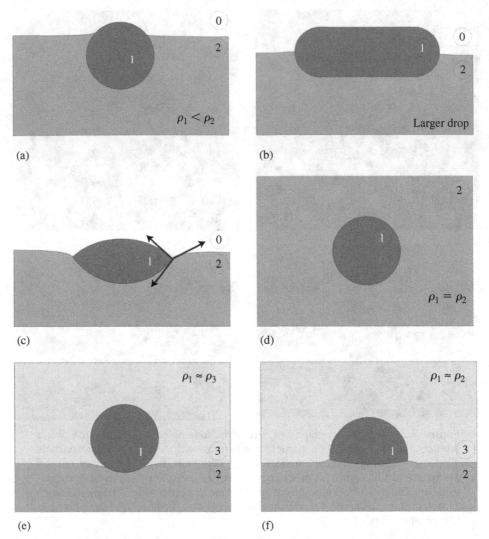

Figure 7.25 Examples of different initial shapes of suspended template droplets that can be realized by varying the effects of gravity and the interfacial tensions. (a) Small, nearly spherical droplet at an air–liquid surface. (b) Flat discoidal or ellipsoidal droplet. (c) Lens. (d) Spherical droplet suspended under microgravity. (e) Nearly spherical droplet. (f) Nearly hemi-spherical droplet.

interfaces. These droplets will have a smooth boundary at the three-phase contact line rather than a sharp contact edge.

For 1 μl template droplets from pure liquids, $\beta < 0.05$, and thus the droplet shape is essentially spherical. The assembly resulted in smooth symmetric balls < 0.5 mm in size and of low density. In low-magnification optical microscopy, the particles are

Figure 7.26 Examples of different crystalline assemblies obtained by outside-in templating – assembly inside droplets suspended at a gas–liquid surface. (a) Balls from crystals of 270 nm diameter latex particles. (b) Discs with different thicknesses from flattened droplets. (c) Doughnut-like toroidal assemblies formed in the presence of fluorinated surfactant. (d) Anisotropic spherical assemblies from separated regular (white) and magnetic (grey) latex particles. Scale bar = 500 μm. Taken from Ref. [229]; with permission of the American Institute for the Advancement of Science.

seen as composed of brightly reflecting coloured patches, *e.g.* green and red for assemblies made of 300 nm latexes – Figure 7.26(a). The colours observed give a clear indication of the long-range ordering of the latexes in the large supra-particles. The direct visualization of the structure by SEM showed large domains of hexagonally packed particles on the supra-particle surface (Figure 7.27). The internal structure of the balls, studied along the edges of broken particles (Figure 7.27(b)), showed that the bulk is made up of 3-D long-range ordered microspheres.

The most interesting feature of this method is the easy control of the shape of the assemblies *via* the droplet size and interfacial tension. Different values of the Bond number at the lower and upper droplet interfaces can be adjusted experimentally by varying the relative densities of the fluids or by adding appropriate surfactants (Figure 7.28). The fluorocarbon–water–air system provides a great degree of flexibility in independently controlling the interfacial tensions by use of surfactants that adsorb exclusively on the different interfaces.[302] For larger droplets in the

Figure 7.27 SEM micrographs of the structure of 3-D supra-particles. (a) A typical area on the surface. (b) A view of the bulk along the edge of a broken particle. Courtesy of O.D. Velev.

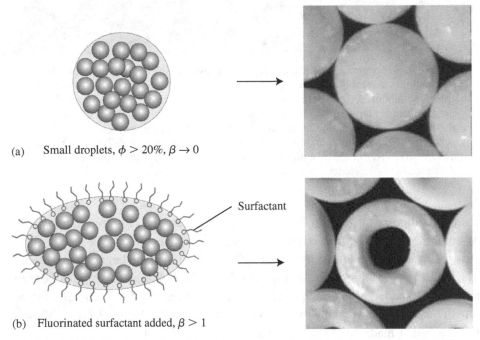

(a)   Small droplets, $\phi > 20\%$, $\beta \to 0$

Surfactant

(b)   Fluorinated surfactant added, $\beta > 1$

Figure 7.28 Two examples of how the process of outside-in templating can lead to assemblies of different shape and structure. (a) Smooth, symmetric spheres and (b) doughnut-like toroids.

presence of surfactants, values of $\beta > 1$ are accessible; thus, lens-like, discoidal and ellipsoidal templates with various three-phase contact angles can be obtained. The shape of the supra-particles formed, however, is also strongly impacted by the coupling of hydrodynamics and mass accumulation during the drying/crystallization process. If the initial volume fraction of the microspheres in the droplets is close to the crystallization threshold of *ca.* 50–60 vol.%, the assemblies obtained are close in size and shape to those of the original droplets. However, if the latex concentration is below about 20%, the particles accumulate at the droplet periphery during water evaporation, which leads to the formation of flattened discs or ellipsoids (Figure 7.28(b)).

Further change in the supra-particle shape can be achieved by adding a fluorinated surfactant, sodium perfluorooctanoate. Droplets containing surfactant first flatten and then develop a dimple during drying, which at higher surfactant concentrations forms a hole in the middle. These dried assemblies are sub-millimetre toroids or crystalline "doughnuts" (Figure 7.26(c)). Similar behaviour has been also observed in spray-dried nanoparticle suspensions.[298] Another important practical feature of this method is that complex composite and aniosotropic assemblies can be fabricated when the original drop contains a mixture of different types of particles. Magnetic particles could be segregated by applying constant magnetic field during the drying/crystallization process. Complete separation of regular and magnetic microparticles into two hemispheres of dried assemblies is demonstrated in Figure 7.26(d). A valuable property of such anisotropic particles is that they can be aligned and manipulated by a magnetic field. The arrays flip and change the displayed face from white to brown and back as the gradient of the field is reversed. As the wetting of the particles plays an important role in this assembly, the method can be modified to make particles of anisotropic wettability, similar to the so-called "Janus" beads.[303,304]

Similarly to the approach of Stöber,[191] Pine and collaborators demonstrated the fabrication of supra-particles with complex geometry using the confinement into emulsion droplets to organize particles inside into close packed structures.[305] To produce these complex structures, they used the method of removing the liquid from emulsion droplets, originally demonstrated by Velev and Nagayama.[194] The process begins with a suspension of latex particles in an organic solvent such as toluene. The swollen particles interact only through short-ranged steric (entropic) repulsion and to a good approximation can be considered as hard spheres. After the spheres are dispersed in toluene, water was added and mixed to create an oil-in-water emulsion consisting of small droplets of toluene ranging from 1 to 10 μm in diameter. The toluene was then evaporated from the system, forcing the hard sphere–like particles inside each droplet to pack together. The critical step of the evaporation process is a mechanically stable intermediate stage with spherical packing, formed when the particles touch one another on the surface of the droplet. Removing more oil at this stage causes the droplet to deform, generating capillary forces that ultimately lead to very quick re-arrangement of the particles.

After the toluene is completely evaporated, the particles de-swell, at which point the inter-particle van der Waals attractions increase and the particles stick to one

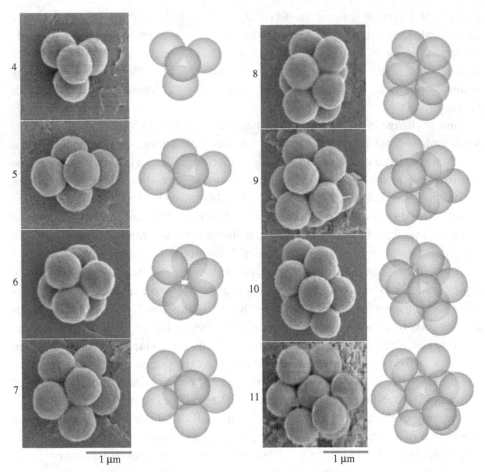

Figure 7.29 SEM micrographs exemplifying some of the complex particle "molecules" containing different numbers of spheres (given) assembled in droplets. Taken from Ref. [305]; with permission of the American Institute for the Advancement of Science.

another forming small colloidal aggregates. The surface charges on the outer particles prevent the clusters from aggregating with other clusters in the water phase. The clusters formed range from doublets, triangles and tetrahedra to exotic polyhedra not found in infinite lattices (Figure 7.29). Each cluster originates from a single droplet containing n spheres, but n varies from cluster to cluster because the initial droplets are not uniform in size. If the centre of each sphere is treated as a vertex of a polyhedron, the first few observed configurations are a line segment ($n = 2$), triangle ($n = 3$), tetrahedron ($n = 4$), triangular dipyramid ($n = 5$) and octahedron ($n = 6$). Similar structures occur in many common molecules or coordination complexes. Clusters containing 7–10 particles have been identified as members of a set of highly symmetric polyhedra known as convex deltahedra. It seems that a key factor responsible for the observed configuration is the minimization of the total surface area of the drying liquid between the spheres. A similar case of unusual arrangement of the particles because of surface area minimization was observed in earlier experiments on 2-D assembly of colloidal particles.[85]

The clusters assembled in droplets can be separated into fractions of single size by gradient centrifugation. The controlled structure of the separated assemblies offers a variety of new areas for further development and potential application. For example, aggregates of complex shape can serve as a tool to study the interaction and crystallization of "colloid molecules" with non-spherical interaction potential, tuning the symmetry of colloidal sphere crystals or controlling nucleation in colloidal crystals. Yin and Xia have shown that adding a small amount of square ($n = 4$) clusters to a dense suspension of spheres influences the epitaxial growth direction of colloidal crystals on flat substrates.[306]

Mesoporous particles with complex shape and composition can find many useful applications. For instance, the "doughnuts" (Figure 7.26(c)) may offer advantages in chromatography over the usual spherical particles by virtue of their combination of reduced mass transfer length scale, reduced pressure drop and uniform porosity. Particles with magnetic and colour anisotropy (Figure 7.26(d)) can find application in switchable coatings, electronic paper, magneto-optical devices or in electrorheological or magnetorheological fluids. In addition to being anisotropic and magnetic, these particles are also of uniform porosity, monodisperse in size and highly structured, and can be produced in different shapes. These methods can be modified to the assembly of inorganic particles from silica, TiO_2, ZrO_2, metals or semi-conductors or particles where the structure of the crystal is replicated in a solid matrix. Millman *et al.*[307] have recently demonstrated how droplet-based microfluidic chips controlled by electric fields can serve as microscopic factories, where a wide variety of particles and capsules can be assembled or synthesized from polymer precursors.

7.6 Concluding Remarks

In concluding this chapter we look back to see a breathtaking variety of nano- and microstructured materials that have been assembled with the use of fluid–fluid interfaces. Many of these structures modify the properties of solid surfaces where they are deposited as coatings (*e.g.* Sections 7.2 and 7.3). Others exist as self-sustained objects of unique structure and properties, *e.g.* Sections 7.4 and 7.5. Many structures act as stabilizers in colloidal systems such as foams and emulsions. The assembly of various biological molecules adds another level of material functionality. The surface-based or surface-guided processes appear to be the key to the future fabrication of functional nanostructures. Many new applications, especially in areas bridging colloidal assembly and high technologies, are still forthcoming possibly including various micro-sensors, nanodevices, media for ultrahigh density memory storage, photonic materials, better catalysts and drastically improved consumer products.

References

1. Y.N. Xia, P.D. Yang, Y.G. Sun, Y.Y. Wu, B. Mayers, B. Gates, Y.D. Yin, F. Kim and Y.Q. Yan, *Adv. Mater.*, **15** (2003), 353.
2. Y.N. Xia, B. Gates, Y.D. Yin and Y. Lu, *Adv. Mater.*, **12** (2000), 693.
3. Y.N. Xia, B. Gates and Z.Y. Li, *Adv. Mater.*, **13** (2001), 409.
4. Y.N. Xia, Y.D. Yin, Y. Lu and J. McLellan, *Adv. Funct. Mater.*, **13** (2003), 907.
5. O.D. Velev and E.W. Kaler, *Adv. Mater.*, **12** (2000), 531.
6. A. van Blaaderen, *Mat. Res. Soc. Bull.*, **23** (1998), 39.
7. A. van Blaaderen, J.P. Hoogenboom, D.L.J. Vossen, A. Yethiraj, A. van der Horst, K. Visscher and M. Dogterom, *Farad. Disc.*, **123** (2003), 107.
8. A. van Blaaderen, *Mat. Res. Soc. Bull.*, **29** (2004), 85.
9. A. Stein and R.C. Schroden, *Curr. Opin. Solid State Mater. Sci.*, **5** (2001), 553.
10. C. Murray, *Mat. Res. Soc. Bull.*, **23** (1998), 33.
11. A.K. Arora and B.V.R. Tata, *Adv. Colloid Interf. Sci.*, **78** (1998), 49.
12. A.D. Dinsmore, J.C. Crocker and A.G. Yodh, *Curr. Opin. Colloid Interf. Sci.*, **3** (1998), 5.
13. P. Pieranski, *Contemp. Phys.*, **24** (1983), 25.
14. A.G. Yodh, K.H. Lin, J.C. Crocker, A.D. Dinsmore, R. Verma and P.D. Kaplan, *Phil. Trans. Roy. Soc. Lon. Ser. A*, **359** (2001), 921.
15. J.N. Israelachvili, *Intermolecular and Surface Forces*, 2nd edn., Academic Press, London, 1992.
16. D.F. Evans and H. Wennerström, *The Colloidal Domain: Where Physics, Chemistry, Biology and Technology Meet*, Wiley-VCH, New York, 1999.
17. S. Asakura and F. Oosawa, *J. Polym. Sci.*, **33** (1958), 183.
18. P.A. Kralchevsky and K. Nagayama, *Adv. Colloid Interf. Sci.*, **85** (2000), 145.
19. P.A. Kralchevsky and N.D. Denkov, *Curr. Opin. Colloid Interf. Sci.*, **6** (2001), 383.
20. W.B. Russell, D.A. Saville and W.R. Schowalter, *Colloidal Dispersions*, Cambridge University Press, Cambridge, 1999.
21. D. Leckband, *Ann. Rev. Biophys. Biomol. Str.*, **29** (2000), 1.
22. M. Ahlers, W. Muller, A. Reichert, H. Ringsdorf and J. Venzmer, *Angew. Chem. Int. Ed.*, **29** (1990), 1269.
23. N. Bowden, A. Terfort, J. Carbeck and G.M. Whitesides, *Science*, **276** (1997), 233.
24. N. Bowden, I.S. Choi, B.A. Grzybowski and G.M. Whitesides, *J. Am. Chem. Soc.*, **121** (1999), 5373.
25. I.S. Choi, N. Bowden and G.M. Whitesides, *J. Am. Chem. Soc.*, **121** (1999), 1754.
26. I.S. Choi, N. Bowden and G.M. Whitesides, *Angew. Chem. Int. Ed.*, **38** (1999), 3078.
27. I.S. Choi, M. Weck, N.L. Jeon and G.M. Whitesides, *J. Am. Chem. Soc.*, **122** (2000), 11997.
28. M. Weck, I.S. Choi, N.L. Jeon and G.M. Whitesides, *J. Am. Chem. Soc.*, **122** (2000), 3546.
29. J.W. Goodwin, R.H. Ottewill and A. Parentich, *J. Phys. Chem.*, **84** (1980), 1580.

30. F.C. Meldrum, N.A. Kotov and J.H. Fendler, *J. Phys. Chem.*, **98** (1994), 4506.
31. N.A. Kotov, F.C. Meldrum, C. Wu and J.H. Fendler, *J. Phys. Chem.*, **98** (1994), 2735.
32. F.C. Meldrum, N.A. Kotov and J.H. Fendler, *Mater. Sci. Eng. C*, **3** (1995), 149.
33. F.C. Meldrum, N.A. Kotov and J.H. Fendler, *Langmuir*, **10** (1994), 2035.
34. F.C. Meldrum, N.A. Kotov and J.H. Fendler, *Chem. Mater.*, **7** (1995), 1112.
35. N.A. Kotov, G. Zavala and J.H. Fendler, *J. Phys. Chem.*, **99** (1995), 12375.
36. H. Yoshimura, T. Scheybani, W. Baumeister and K. Nagayama, *Langmuir*, **10** (1994), 3290.
37. J.H. Dunsmuir, H.W. Deckman and J.A. Mchenry, US 4 801 476, (1989).
38. K.U. Fulda and B. Tieke, *Adv. Mater.*, **6** (1994), 288.
39. P. Pieranski, *Phys. Rev. Lett.*, **45** (1980), 569.
40. R. Aveyard, J.H. Clint, D. Nees and V.N. Paunov, *Langmuir*, **16** (2000), 1969.
41. G. Picard, I. Nevernov, D. Alliata and L. Pazdernik, *Langmuir*, **13** (1997), 264.
42. G. Picard, *Langmuir*, **14** (1998), 3710.
43. S. Reculusa and S. Ravaine, *Chem. Mater.*, **15** (2003), 598.
44. G.Y. Onoda, *Phys. Rev. Lett.*, **55** (1985), 226.
45. M. Kondo, K. Shinozaki, L. Bergstrom and N. Mizutani, *Langmuir*, **11** (1995), 394.
46. A. George and W.W. Wilson, *Acta Cryst. D*, **50** (1994), 361.
47. B.L. Neal, D. Asthagiri, O.D. Velev, A.M. Lenhoff and E.W. Kaler, *J. Cryst. Gr.*, **196** (1999), 377.
48. D. Rosenbaum, P.C. Zamora and C.F. Zukoski, *Phys. Rev. Lett.*, **76** (1996), 150.
49. O.D. Velev, E.W. Kaler and A.M. Lenhoff, *Biophys. J.*, **75** (1998), 2682.
50. A.S. Dimitrov, T. Takahashi, K. Furusawa and K. Nagayama, *J. Phys. Chem.*, **100** (1996), 3163.
51. M. Golosovsky, Y. Saado and D. Davidov, *Appl. Phys. Lett.*, **75** (1999), 4168.
52. T. Alfrey Jr., E.B. Bradford, J.W. Vanderhof and G. Oster, *J. Opt. Soc. Am.*, **44** (1954), 603.
53. W.C. Price, R.C. Williams and R.W.G. Wyckoff, *Science*, **102** (1945), 277.
54. R.W.G. Wyckoff, *Acta Cryst.*, **1** (1948), 292.
55. C.B. Murray, C.R. Kagan and M.G. Bawendi, *Science*, **270** (1995), 1335.
56. L. Motte, F. Billoudet and M.P. Pileni, *J. Phys. Chem.*, **99** (1995), 16425.
57. N.D. Denkov, O.D. Velev, P.A. Kralchevsky, I.B. Ivanov, H. Yoshimura and K. Nagayama, *Nature*, **361** (1993), 26.
58. N.D. Denkov, O.D. Velev, P.A. Kralchevsky, I.B. Ivanov, H. Yoshimura and K. Nagayama, *Langmuir*, **8** (1992), 3183.
59. B.G. Prevo and O.D. Velev, *Langmuir*, **20** (2004), 2099.
60. A.A. Chabanov, Y. Jun and D.J. Norris, *Appl. Phys. Lett.*, **84** (2004), 3573.
61. A.S. Dimitrov, C.D. Dushkin, H. Yoshimura and K. Nagayama, *Langmuir*, **10** (1994), 432.
62. L.E. Helseth and T.M. Fischer, *Phys. Rev. E*, **68** (2003), 042601.
63. G.S. Lazarov, N.D. Denkov, O.D. Velev, P.A. Kralchevsky and K. Nagayama, *J. Chem. Soc. Farad. Trans.*, **90** (1994), 2077.
64. A.S. Dimitrov and K. Nagayama, *Langmuir*, **12** (1996), 1303.
65. A.S. Dimitrov and K. Nagayama, *Chem. Phys. Lett.*, **243** (1995), 462.
66. S. Maenosono, C.D. Dushkin, Y. Yamaguchi, K. Nagayama and Y. Tsuji, *Colloid Polym. Sci.*, **277** (1999), 1152.
67. C.D. Dushkin, G.S. Lazarov, S.N. Kotsev, H. Yoshimura and K. Nagayama, *Colloid Polym. Sci.*, **277** (1999), 914.
68. A.F. Routh and W.B. Russel, *Am. Inst. Chem. Eng. J.*, **44** (1998), 2088.
69. D.J. Norris, E.G. Arlinghaus, L.L. Meng, R. Heiny and L.E. Scriven, *Adv. Mater.*, **16** (2004), 1393.

70. E. Adachi, A.S. Dimitrov and K. Nagayama, *Langmuir*, **11** (1995), 1057.
71. R.D. Deegan, O. Bakajin, T.F. Dupont, G. Huber, S.R. Nagel and T.A. Witten, *Nature*, **389** (1997), 827.
72. S. Maenosono, C.D. Dushkin, S. Saita and Y. Yamaguchi, *Langmuir*, **15** (1999), 957.
73. P.C. Ohara and W.M. Gelbart, *Langmuir*, **14** (1998), 3418.
74. J.A. Davidson and E.A. Collins, *J. Colloid Interf. Sci.*, **40** (1972), 437.
75. Y. Wang, D. Juhue, M.A. Winnik, O.M. Leung and M.C. Goh, *Langmuir*, **8** (1992), 760.
76. H.W. Deckman, J.H. Dunsmuir, S. Garoff, J.A. Mchenry and D.G. Peiffer, *J. Vac. Sci. Technol. B*, **6** (1988), 333.
77. H.W. Deckman and J.H. Dunsmuir, *J. Vac. Sci. Technol. B*, **1** (1983), 1109.
78. H.W. Deckman and J.H. Dunsmuir, *Appl. Phys. Lett.*, **41** (1982), 377.
79. J.C. Hulteen, D.A. Treichel, M.T. Smith, M.L. Duval, T.R. Jensen and R.P. Van Duyne, *J. Phys. Chem. B*, **103** (1999), 3854.
80. C.D. Dushkin, K. Nagayama, T. Miwa and P.A. Kralchevsky, *Langmuir*, **9** (1993), 3695.
81. C.D. Dushkin, H. Yoshimura and K. Nagayama, *Chem. Phys. Lett.*, **204** (1993), 455.
82. R. Micheletto, H. Fukuda and M. Ohtsu, *Langmuir*, **11** (1995), 3333.
83. S. Rakers, L.F. Chi and H. Fuchs, *Langmuir*, **13** (1997), 7121.
84. P. Jiang, J.F. Bertone, K.S. Hwang and V.L. Colvin, *Chem. Mater.*, **11** (1999), 2132.
85. K.P. Velikov, A. Moroz and A. van Blaaderen, *Appl. Phys. Lett.*, **80** (2002), 49.
86. R. Rengarajan, P. Jiang, D. Larrabee, V.L. Colvin and D.M. Mittleman, *Phys. Rev. B*, **64** (2001), 205103.
87. P. Jiang, G.N. Ostojic, R. Narat, D.M. Mittleman and V.L. Colvin, *Adv. Mater.*, **13** (2001), 389.
88. K.P. Velikov, C.G. Christova, R.P.A. Dullens and A. van Blaaderen, *Science*, **295** (2002), 106.
89. S.M. Yan, H. Miguez and G.A. Ozin, *Adv. Funct. Mater.*, **12** (2002), 425.
90. Y.A. Vlasov, X.Z. Bo, J.C. Sturm and D.J. Norris, *Nature*, **414** (2001), 289.
91. S. Wong, V. Kitaev and G.A. Ozin, *J. Am. Chem. Soc.*, **125** (2003), 15589.
92. H. Miguez, S.M. Yang and G.A. Ozin, *Appl. Phys. Lett.*, **81** (2002), 2493.
93. V. Kitaev and G.A. Ozin, *Adv. Mater.*, **15** (2003), 75.
94. R.M. Amos, J.G. Rarity, P.R. Tapster, T.J. Shepherd and S.C. Kitson, *Phys. Rev. E*, **61** (2000), 2929.
95. L.M. Goldenberg, J. Wagner, J. Stumpe, B.R. Paulke and E. Gornitz, *Langmuir*, **18** (2002), 3319.
96. F. Garcia-Santamaria, M. Ibisate, I. Rodriguez, F. Meseguer and C. Lopez, *Adv. Mater.*, **15** (2003), 788.
97. Y.H. Ye, F. LeBlanc, A. Hache and V.V. Truong, *Appl. Phys. Lett.*, **78** (2001), 52.
98. B.G. Prevo, J.C. Fuller and O.D. Velev, *Chem. Mater.*, **17** (2005), 28.
99. P. Jiang and M.J. McFarland, *J. Am. Chem. Soc.*, **126** (2004), 13778.
100. M.P. Pileni, *Langmuir*, **13** (1997), 3266.
101. M.P. Pileni, *New J. Chem.*, **22** (1998), 693.
102. M.P. Pileni, *Supramol. Sci.*, **5** (1998), 321.
103. M.P. Pileni, *Pure Appl. Chem.*, **72** (2000), 53.
104. M.P. Pileni, *Comp. Rend. Chim.*, **6** (2003), 965.
105. C.B. Murray, C.R. Kagan and M.G. Bawendi, *Ann. Rev. Mater. Sci.*, **30** (2000), 545.
106. C.P. Collier, T. Vossmeyer and J.R. Heath, *Ann. Rev. Phys. Chem.*, **49** (1998), 371.
107. M.P. Pileni, Y. Lalatonne, D. Ingert, I. Lisiecki and A. Courty, *Farad. Disc.*, **125** (2004), 251.

108. A.L. Rogach, *Angew. Chem. Int. Ed.*, **43** (2004), 148.
109. C.B. Murray, S.H. Sun, W. Gaschler, H. Doyle, T.A. Betley and C.R. Kagan, *IBM J. Res. Dev.*, **45** (2001), 47.
110. L. Motte, F. Billoudet, E. Lacaze and M.P. Pileni, *Adv. Mater.*, **8** (1996), 1018.
111. L. Motte, F. Billoudet, E. Lacaze, J. Douin and M.P. Pileni, *J. Phys. Chem. B*, **101** (1997), 138.
112. P.C. Ohara, D.V. Leff, J.R. Heath and W.M. Gelbart, *Phys. Rev. Lett.*, **75** (1995), 3466.
113. S. Narayanan, J. Wang, X.M. Lin, *Phys. Rev. Lett.*, **93** (2004), 135503.
114. D.R. Talham, *Chem. Rev.*, **104** (2004), 5479.
115. V. Santhanam, J. Liu, R. Agarwal and R.P. Andres, *Langmuir*, **19** (2003), 7881.
116. P.C. Ohara, J.R. Heath and W.M. Gelbart, *Angew. Chem. Int. Ed.*, **36** (1997), 1078.
117. T. Vossmeyer, S.W. Chung, W.M. Gelbart and J.R. Heath, *Adv. Mater.*, **10** (1998), 351.
118. C. Petit, T. Cren, D. Roditchev, W. Sacks, J. Klein and M.P. Pileni, *Adv. Mater.*, **11** (1999), 1198.
119. K.V.P.M. Shafi, I. Felner, Y. Mastai and A. Gedanken, *J. Phys. Chem. B*, **103** (1999), 3358.
120. M. Maillard, L. Motte, A.T. Ngo and M.P. Pileni, *J. Phys. Chem. B*, **104** (2000), 11871.
121. M. Maillard, L. Motte and M.P. Pileni, *Adv. Mater.*, **13** (2001), 200.
122. S.L. Tripp, S.V. Pusztay, A.E. Ribbe and A. Wei, *J. Am. Chem. Soc.*, **124** (2002), 7914.
123. L.V. Govor, G.H. Bauer, G. Reiter, E. Shevchenko, H. Weller and A. Parisi, *Langmuir*, **19** (2003), 9573.
124. D. Wyrwa, N. Beyer and G. Schmid, *Nano Lett.*, **2** (2002), 419.
125. S.W. Chung, G. Markovich and J.R. Heath, *J. Phys. Chem. B*, **102** (1998), 6685.
126. L. Motte and M.P. Pileni, *J. Phys. Chem. B*, **102** (1998), 4104.
127. C.A. Stowell and B.A. Korgel, *Nano Lett.*, **1** (2001), 595.
128. A. Taleb, C. Petit and M.P. Pileni, *Chem. Mater.*, **9** (1997), 950.
129. A. Taleb, C. Petit and M.P. Pileni, *J. Phys. Chem. B*, **102** (1998), 2214.
130. W.D. Luedtke and U. Landman, *J. Phys. Chem.*, **100** (1996), 13323.
131. S.A. Harfenist, Z.L. Wang, R.L. Whetten, I. Vezmar and M.M. Alvarez, *Adv. Mater.*, **9** (1997), 817.
132. A. Taleb, V. Russier, A. Courty and M.P. Pileni, *Phys. Rev. B*, **59** (1999), 13350.
133. C. Petit, A. Taleb and M.P. Pileni, *Adv. Mater.*, **10** (1998), 259.
134. V. Russier and M.P. Pileni, *Surf. Sci.*, **425** (1999), 313.
135. V. Russier, C. Petit, J. Legrand and M.P. Pileni, *Appl. Surf. Sci.*, **164** (2000), 193.
136. A.T. Ngo and M.P. Pileni, *Colloids Surf. A*, **228** (2003), 107.
137. J. Legrand, C. Petit, D. Bazin and M.P. Pileni, *Appl. Surf. Sci.*, **164** (2000), 186.
138. C. Petit and M.P. Pileni, *Appl. Surf. Sci.*, **162** (2000), 519.
139. A. Taleb, F. Silly, A.O. Gusev, F. Charra and M.P. Pileni, *Adv. Mater.*, **12** (2000), 633.
140. E. Rabani, D.R. Reichman, P.L. Geissler and L.E. Brus, *Nature*, **426** (2003), 271.
141. P.S. Shah, B.J. Novick, H.S. Hwang, K.T. Lim, R.G. Carbonell, K.P. Johnston and B.A. Korgel, *Nano Lett.*, **3** (2003), 1671.
142. V.E. Cosslett and R. Markham, *Nature*, **161** (1948), 250.
143. R. Markham, K.M. Smith and R.W.G. Wyckoff, *Nature*, **161** (1948), 760.
144. R.W. Horne and I.P. Ronchett, *J. Ultrastr. Res.*, **47** (1974), 361.
145. B. Wells, R.W. Horne and P.J. Shaw, *Micron*, **12** (1981), 37.

146. J.R. Harris, *Micron Microsc. Acta*, **22** (1991), 341.
147. J.R. Harris, *Micron*, **13** (1982), 169.
148. J.R. Harris, Z. Cejka, A. Wegenerstrake, W. Gebauer and J. Markl, *Micron Microsc. Acta*, **23** (1992), 287.
149. W.H. Massover, *Micron*, **24** (1993), 389.
150. P. Fromherz, *Nature*, **231** (1971), 267.
151. H. Yoshimura, *Adv. Biophys.*, **34** (1997), 93.
152. K. Nagayama, *Adv. Biophys.*, **34** (1997), 3.
153. S. Takeda, H. Yoshimura, S. Endo, T. Takahashi and K. Nagayama, *Protein. Str. Funct. Gen.*, **23** (1995), 548.
154. E. Adachi and K. Nagayama, *Adv. Biophys.*, **34** (1996), 81.
155. T. Scheybani, H. Yoshimura, W. Baumeister and K. Nagayama, *Langmuir*, **12** (1996), 431.
156. N.M. Green, *Meth. Enzym.*, **184** (1990), 51.
157. R. Blankenburg, P. Meller, H. Ringsdorf and C. Salesse, *Biochem.*, **28** (1989), 8214.
158. S.A. Darst, M. Ahlers, P.H. Meller, E.W. Kubalek, R. Blankenburg, H.O. Ribi, H. Ringsdorf and R.D. Kornberg, *Biophys. J.*, **59** (1991), 387.
159. W. Muller, H. Ringsdorf, E. Rump, G. Wildburg, X. Zhang, L. Angermaier, W. Knoll, M. Liley and J. Spinke, *Science*, **262** (1993), 1706.
160. U.B. Sleytr and T.J. Beveridge, *Tr. Microbiol.*, **7** (1999), 253.
161. U.B. Sleytr, P. Messner, D. Pum and M. Sara, *Angew. Chem. Int. Ed.*, **38** (1999), 1035.
162. U.B. Sleytr, E. Gyorvary and D. Pum, *Progr. Org. Coat.*, **47** (2003), 279.
163. U.B. Sleytr, B. Schuster and D. Pum, *IEEE Eng. Med. Biol. Mag.*, **22** (2003), 140.
164. M. Sara and U.B. Sleytr, *J. Bacteriol.*, **182** (2000), 859.
165. D. Pum and U.B. Sleytr, *Tr. Biotechnol.*, **17** (1999), 8.
166. B.K. Jap, M. Zulauf, T. Scheybani, A. Hefti, W. Baumeister, U. Aebi and A. Engel, *Ultramicr.*, **46** (1992), 45.
167. W. Chiu, A.J. AvilaSakar and M.F. Schmid, *Adv. Biophys.*, **34** (1997), 161.
168. E. Sackmann, *Science*, **271** (1996), 43.
169. E.E. Uzgiris and R.D. Kornberg, *Nature*, **301** (1983), 125.
170. C.J. Kiely, J. Fink, M. Brust, D. Bethell and D.J. Schiffrin, *Nature*, **396** (1998), 444.
171. M.J. Murray and J.V. Sanders, *Nature*, **275** (1978), 201.
172. M.J. Murray and J.V. Sanders, *Phil. Mag. A*, **42** (1980), 721.
173. S. Hachisu and Sh. Yoshimura, *Nature*, **283** (1980), 188.
174. P. Bartlett, R.H. Ottewill and P.N. Pusey, *Phys. Rev. Lett.*, **68** (1992), 3801.
175. B.A. Grzybowski, A. Winkleman, J.A. Wiles, Y. Brumer and G.M. Whitesides, *Nat. Mater.*, **2** (2003), 241.
176. M.D. Eldridge, P.A. Madden and D. Frenkel, *Nature*, **365** (1993), 35.
177. M. Antonietti, J. Hartmann, M. Neese and U. Seifert, *Langmuir*, **16** (2000), 7634.
178. M. Brust and C.J. Kiely, *Colloids Surf. A*, **202** (2002), 175.
179. C.J. Kiely, J. Fink, J.G. Zheng, M. Brust, D. Bethell and D.J. Schiffrin, *Adv. Mater.*, **12** (2000), 640.
180. M. Yamaki, J. Higo and K. Nagayama, *Langmuir*, **11** (1995), 2975.
181. M.D. Eldridge, P.A. Madden and D. Frenkel, *Mol. Phys.*, **80** (1993), 987.
182. C.N. Likos and C.L. Henley, *Phil. Mag. B*, **68** (1993), 85.
183. B.D. Rabideau and R.T. Bonnecaze, *Langmuir*, **20** (2004), 9408.
184. R.D. Doty, R.T. Bonnecaze and B.A. Korgel, *Phys. Rev. E*, **65** (2002), 061503.
185. E.V. Shevchenko, D.V. Talapin, A.L. Rogach, A. Kornowski, M. Haase and H. Weller, *J. Am. Chem. Soc.*, **124** (2002), 11480.
186. F.X. Redl, K.S. Cho, C.B. Murray and S. O'Brien, *Nature*, **423** (2003), 968.

187. J.V. Sanders, *Phil. Mag. A*, **42** (1980), 705.
188. N. Hunt, R. Jardine and P. Bartlett, *Phys. Rev. E*, **62** (2000), 900.
189. A.R. Denton and N.W. Ashcroft, *Phys. Rev. A*, **42** (1990), 7312.
190. D.Y. Wang and H. Möhwald, *Adv. Mater.*, **16** (2004), 244.
191. W. Stöber, *J. Colloid Interf. Sci.*, **29** (1969), 710.
192. O.D. Velev, K. Furusawa and K. Nagayama, *Langmuir*, **12** (1996), 2374.
193. O.D. Velev, K. Furusawa and K. Nagayama, *Langmuir*, **12** (1996), 2385.
194. O.D. Velev and K. Nagayama, *Langmuir*, **13** (1997), 1856.
195. A. van Blaaderen, R. Ruel and P. Wiltzius, *Nature*, **385** (1997), 321.
196. Y.N. Xia, J.A. Rogers, K.E. Paul and G.M. Whitesides, *Chem. Rev.*, **99** (1999), 1823.
197. J.P. Hoogenboom, C. Retif, E. de Bres, M.V. de Boer, A.K. Langen-Suurling, J. Romijn and A. van Blaaderen, *Nano Lett.*, **4** (2004), 205.
198. K.H. Lin, J.C. Crocker, V. Prasad, A. Schofield, D.A. Weitz, T.C. Lubensky, and A.G. Yodh, *Phys. Rev. Lett.*, **85** (2000), 1770.
199. Y.D. Yin, Y. Lu, B. Gates and Y.N. Xia, *J. Am. Chem. Soc.*, **123** (2001), 8718.
200. F. Burmeister, C. Schafle, B. Keilhofer, C. Bechinger, J. Boneberg and P. Leiderer, *Adv. Mater.*, **10** (1998), 495.
201. Y.H. Ye, S. Badilescu, V.V. Truong, P. Rochon and A. Natansohn, *Appl. Phys. Lett.*, **79** (2001), 872.
202. D.K. Yi, E.M. Seo and D.Y. Kim, *Appl. Phys. Lett.*, **80** (2002), 225.
203. Y. Yin, Z.Y. Li and Y. Xia, *Langmuir*, **19** (2003), 622.
204. R.E. Schaak, R.E. Cable, B.M. Leonar and B.C. Norris, *Langmuir*, **20** (2004), 7293.
205. J.H. Moon, S. Kim, G.R. Yi, Y.H. Lee and S.M. Yang, *Langmuir*, **20** (2004), 2033.
206. Y.N. Xia and G.M. Whitesides, *Angew. Chem. Int. Ed.*, **37** (1998), 551.
207. Y.N. Xia and G.M. Whitesides, *Ann. Rev. Mater. Sci.*, **28** (1998), 153.
208. Y.N. Xia, J. Tien, D. Qin and G.M. Whitesides, *Langmuir*, **12** (1996), 4033.
209. J. Aizenberg, P.V. Braun and P. Wiltzius, *Phys. Rev. Lett.*, **84** (2000), 2997.
210. K.M. Chen, X.P. Jiang, L.C. Kimerling and P.T. Hammond, *Langmuir*, **16** (2000), 7825.
211. B.F. Lyles, M.S. Terrot, P.T. Hammond and A.P. Gast, *Langmuir*, **20** (2004), 3028.
212. I. Lee, H. Zheng, M.F. Rubner and P.T. Hammond, *Adv. Mater.*, **14** (2002), 572.
213. Y. Masuda, K. Tomimoto and K. Koumoto, *Langmuir*, **19** (2003), 5179.
214. C.A. Fustin, G. Glasser, H.W. Spiess and U. Jonas, *Langmuir*, **20** (2004), 9114.
215. L. Kondic and J. Diez, *Colloids Surf. A*, **214** (2003), 1.
216. Y. Masuda, T. Itoh, M. Itoh and K. Koumoto, *Langmuir*, **20** (2004), 5588.
217. M.A. Winnik, *Curr. Opin. Colloid Interf. Sci.*, **2** (1997), 192.
218. J.L. Keddie, *Mater. Sci. Eng. Res. Rep.*, **21** (1997), 101.
219. J.D. Joannopoulos, R.D. Meade and J.N. Winn, *Photonic Crystals*, Princeton University Press, Princeton, 1995.
220. C.M. Soukoulis, ed. *Photonic Crystals and Light Localization in the 21st Century*, NATO Science Series C563, Kluwer Academic Publishers, Dordrecht, 2001.
221. C.M. Soukoulis, ed. *Photonic Crystals and Light Localization*, NATO ASI Series E315, Kluwer Academic Publishers, Dordrecht, 2000.
222. Y. Zhao, I. Avrutsky and B. Li, *Appl. Phys. Lett.*, **75** (1999), 3596.
223. Y. Zhao and I. Avrutsky, *Opt. Lett.*, **24** (1999), 817.
224. S.I. Matsushita, Y. Yagi, T. Miwa, D.A. Tryk, T. Koda and A. Fujishima, *Langmuir*, **16** (2000), 636.
225. A. Blanco, E. Chomski, S. Grabtchak, M. Ibisate, S. John, S.W. Leonard, C. Lopez, F. Meseguer, H. Miguez, J.P. Mondia, G.A. Ozin, O. Toader and H.M. van Driel, *Nature*, **405** (2000), 437.

226. G.S. Pan, R. Kesavamoorthy and S.A. Asher, *Phys. Rev. Lett.*, **78** (1997), 3860.
227. J.H. Holtz and S.A. Asher, *Nature*, **389** (1997), 829.
228. P.M. Tessier, O.D. Velev, A.T. Kalambur, J.F. Rabolt, A.M. Lenhoff and E.W. Kaler, *J. Am. Chem. Soc.*, **122** (2000), 9554.
229. O.D. Velev, A.M. Lenhoff and E.W. Kaler, *Science*, **287** (2000), 2240.
230. B.T. Holland, C.F. Blanford, T. Do and A. Stein, *Chem. Mater.*, **11** (1999), 795.
231. B.T. Holland, C.F. Blanford and A. Stein, *Science*, **281** (1998), 538.
232. P. Jiang, J. Cizeron, J.F. Bertone and V.L. Colvin, *J. Am. Chem. Soc.*, **121** (1999), 7957.
233. K.M. Kulinowski, P. Jiang, H. Vaswani and V.L. Colvin, *Adv. Mater.*, **12** (2000), 833.
234. F. Burmeister, W. Badowsky, T. Braun, S. Wieprich, J. Boneberg and P. Leiderer, *Appl. Surf. Sci.*, **145** (1999), 461.
235. X. Chen, Z.M. Chen, N. Fu, G. Lu and B. Yang, *Adv. Mater.*, **15** (2003), 1417.
236. B.J.Y. Tan, C.H. Sow, K.Y. Lim, F.C. Cheong, G.L. Chong, A.T.S. Wee and C.K. Ong, *J. Phys. Chem. B*, **108** (2004), 18575.
237. C.L. Haynes and R.P. Van Duyne, *J. Phys. Chem. B*, **105** (2001), 5599.
238. J.C. Hulteen and R.P. Van Duyne, *J. Vac. Sci. Technol. A*, **13** (1995), 1553.
239. H.W. Deckman and J.H. Dunsmuir, *J. Vac. Sci. Technol. B*, **1** (1983), 1166.
240. H.W. Deckman and T.D. Moustakas, *J. Vac. Sci. Technol. B*, **6** (1988), 316.
241. T.R. Jensen, G.C. Schatz and R.P. Van Duyne, *J. Phys. Chem. B*, **103** (1999), 2394.
242. T.R. Jensen, M.L. Duval, K.L. Kelly, A.A. Lazarides, G.C. Schatz and R.P. Van Duyne, *J. Phys. Chem. B*, **103** (1999), 9846.
243. T.R. Jensen, M.D. Malinsky, C.L. Haynes and R.P. Van Duyne, *J. Phys. Chem. B*, **104** (2000), 10549.
244. M.D. Malinsky, K.L. Kelly, G.C. Schatz and R.P. Van Duyne, *J. Phys. Chem. B*, **105** (2001), 2343.
245. A.D. Ormonde, E.C.M. Hicks, J. Castillo and R.P. Van Duyne, *Langmuir*, **20** (2004), 6927.
246. J.C. Riboh, A.J. Haes, A.D. McFarland, C.R. Yonzon and R.P. Van Duyne, *J. Phys. Chem. B*, **107** (2003), 1772.
247. C.L. Haynes, A.D. McFarland, M.T. Smith, J.C. Hulteen and R.P. Van Duyne, *J. Phys. Chem. B*, **106** (2002), 1898.
248. A.J. Haes, S.L. Zou, G.C. Schatz and R.P. Van Duyne, *J. Phys. Chem. B*, **108** (2004), 6961.
249. C.L. Haynes and R.P. Van Duyne, *J. Phys. Chem. B*, **107** (2003), 7426.
250. D. Ingert and M.P. Pileni, *J. Phys. Chem. B*, **107** (2003), 9617.
251. M. Miyaki, K. Fujimoto and H. Kawaguchi, *Colloids Surf. A*, **153** (1999), 603.
252. K. Nagayama, S. Takeda, S. Endo and H. Yoshimura, *Jpn. J. Appl. Phys. Part 1*, **34** (1995), 947.
253. H. Yoshimura, M. Matsumoto, S. Endo and K. Nagayama, *Ultramicr.*, **32** (1990), 265.
254. H. Yoshimura, S. Endo, M. Matsumoto, K. Nagayama and Y. Kagawa, *J. Biochem.*, **106** (1989), 958.
255. M. Yamaki, K. Matsubara and K. Nagayama, *Langmuir*, **9** (1993), 3154.
256. D.T. Wasan, A.D. Nikolov, P.A. Kralchevsky and I.B. Ivanov, *Colloids Surf.*, **67** (1992), 139.
257. A.D. Nikolov, P.A. Kralchevsky, I.B. Ivanov and D.T. Wasan, *J. Colloid Interf. Sci.*, **133** (1989), 13.
258. V. Bergeron and C.J. Radke, *Langmuir*, **8** (1992), 3020.
259. A.D. Nikolov and D.T. Wasan, *Colloids Surf. A*, **128** (1997), 243.

260. K. Koczo, A.D. Nikolov, D.T. Wasan, R.P. Borwankar and A. Gonsalves, *J. Colloid Interf. Sci.*, **178** (1996), 694.
261. N.D. Denkov, H. Yoshimura, K. Nagayama and T. Kouyama, *Phys. Rev. Lett.*, **76** (1996), 2354.
262. K.P. Velikov, F. Durst and O.D. Velev, *Langmuir*, **14** (1998), 1148.
263. K.P. Velikov and O.D. Velev, in *Emulsions, Foams and Thin Films*, eds. K.L. Mittal and P. Kumar, Marcel Dekker, New York, 2000, p. 84.
264. T.S. Horozov, R. Aveyard, J.H. Clint and B. Neumann, *Langmuir*, **21** (2005), 2330.
265. J. Sur and H.K. Pak, *Phys. Rev. Lett.*, **86** (2001), 4326.
266. K.D. Danov, B. Pouligny and P.A. Kralchevsky, *Langmuir*, **17** (2001), 6599.
267. P.A. Kralchevsky, V.N. Paunov, I.B. Ivanov and K. Nagayama, *J. Colloid Interf. Sci.*, **151** (1992), 79.
268. P.A. Kralchevsky and K. Nagayama, *Langmuir*, **10** (1994), 23.
269. V.N. Paunov, P.A. Kralchevsky, N.D. Denkov and K. Nagayama, *J. Colloid Interf. Sci.*, **157** (1993), 100.
270. N.D. Denkov, H. Yoshimura and K. Nagayama, *Ultramicroscopy*, **65** (1996), 147.
271. B.P. Binks, *Adv. Mater.*, **14** (2002), 1824.
272. R.G. Alargova, K.H. Bhatt, V.N. Paunov and O.D. Velev, *Adv. Mater.*, **16** (2004), 1653.
273. R.G. Alargova, D.S. Warhadpande, V.N. Paunov and O.D. Velev, *Langmuir*, **20** (2004), 10371.
274. B.P. Binks and T.S. Horozov, *Angew. Chem. Int. Ed.*, **44** (2005), 3722.
275. W. Ramsden, *Proc. Roy. Soc.*, **72** (1903), 156.
276. S.U. Pickering, *J. Chem. Soc.*, **91** (1907), 2001.
277. B.P. Binks, *Curr. Opin. Colloid Interf. Sci.*, **7** (2002), 21.
278. R. Aveyard, B.P. Binks and J.H. Clint, *Adv. Colloid Interf. Sci.*, **100–102** (2003), 503.
279. W.T.S. Huck, J. Tien and G.M. Whitesides, *J. Am. Chem. Soc.*, **120** (1998), 8267.
280. A.D. Dinsmore, M.F. Hsu, M.G. Nikolaides, M. Marquez, A.R. Bausch and D.A. Weitz, *Science*, **298** (2002), 1006.
281. V.D. Gordon, C. Xi, J.W. Hutchinson, A.R. Bausch, M. Marquez and D.A. Weitz, *J. Am. Chem. Soc.*, **126** (2004), 14117.
282. P.F. Noble, O.J. Cayre, R.G. Alargova, O.D. Velev and V.N. Paunov, *J. Am. Chem. Soc.*, **126** (2004), 8092.
283. L.M. Croll and H.D.H. Stöver, *Langmuir*, **19** (2003), 5918.
284. L.M. Croll and H.D.H. Stöver, *Langmuir*, **19** (2003), 10077.
285. Y. Lin, H. Skaff, T. Emrick, A.D. Dinsmore and T.P. Russell, *Science*, **299** (2003), 226.
286. Y. Lin, H. Skaff, A. Boker, A.D. Dinsmore, T. Emrick and T.P. Russell, *J. Am. Chem. Soc.*, **125** (2003), 12690.
287. J.L. Rigaud, B. Pitar and D. Levy, *Biochim. Biophys. Acta Bioenerg.*, **1231** (1995), 223.
288. J.L. Rigaud, M.T. Paternostre and A. Bluzat, *Biochemistry*, **27** (1988), 2677.
289. G.D. Eytan, *Biochim. Biophys. Acta*, **694** (1982), 185.
290. A. Darszon, C.A. Vandenberg, M.H. Ellisman and M. Montal, *J. Cell Biol.*, **81** (1979), 446.
291. M. Arnold, P. Ringler and A. Brisson, *Biochim. Biophys. Acta-Biomembr.*, **1233** (1995), 198.
292. S. Stoylova, K.G. Mann and A. Brisson, *FEBS Lett.*, **351** (1994), 330.
293. C. Pigault, A. Folleniuswund, M. Schmutz, J.M. Freyssinet and A. Brisson, *J. Mol. Biol.*, **236** (1994), 199.

294. K. Velikov, C. Dietrich, A. Hadjiisky, K. Danov and B. Pouligny, *Europhys. Lett.*, **40** (1997), 405.
295. K. Velikov, K. Danov, M. Angelova, C. Dietrich and B. Pouligny, *Colloids Surf. A*, **149** (1999), 245.
296. O.D. Velev, *Adv. Biophys.*, **34** (1997), 139.
297. A. Boker, Y. Lin, K. Chiapperini, R. Horowitz, M. Thompson, V. Carreon, T. Xu, C. Abetz, H. Skaff, A.D. Dinsmore, T. Emrick and T.P. Russell, *Nat. Mater.*, **3** (2004), 302.
298. F. Iskandar, L. Gradon and K. Okuyama, *J. Colloid Interf. Sci.*, **265** (2003), 296.
299. F. Iskandar, A. Mikrajuddin and K. Okuyama, *Nano Lett.*, **2** (2002), 389.
300. F. Iskandar, A. Mikrajuddin and K. Okuyama, *Nano Lett.*, **1** (2001), 231.
301. J.H. Moon, G.R. Yi, S.M. Yang, D.J. Pine and S. Bin Park, *Adv. Mater.*, **16** (2004), 605.
302. M. Morita, M. Matsumoto, S. Usui, T. Abe, N. Denkov and O. Velev, *Colloids Surf.*, **67** (1992), 81.
303. C. Casagrande and M. Veyssie, *Comptes Rendus Academie Sciences Serie Ii*, **306** (1988), 1423.
304. V.N. Paunov and O.J. Cayre, *Adv. Mater.*, **16** (2004), 788.
305. V.N. Manoharan, M.T. Elsesser and D.J. Pine, *Science*, **301** (2003), 483.
306. Y. Yin and Y. Xia, *Adv. Mater.*, **13** (2001), 267.
307. J.R. Millman, K.H. Bhatt, B.G. Prevo and O.D. Velev, *Nat. Mater.*, **4** (2005), 98.

Krassimir Velikov (left) received an M.Sc. degree in Chemical Physics and Theoretical Chemistry (1994) and an M.Sc. degree in Chemical Engineering with speciality in Separation Techniques in Various Industries and Environmental Protection (1997), both from the University of Sofia. He received a Ph.D. in Physics from the University of Utrecht in 2002 under supervision of Prof. van Blaaderen. The topic of his thesis was the syntheses and convective assembly of (core-shell) colloidal particles into photonic colloidal crystals with different unit cells and stoichiometries. He is presently a staff scientist in the Unilever Research Division in Vlaardingen, The Netherlands. His research interests include colloids, colloidal assembly and the physics and chemistry of soft condensed matter and complex fluids.

Orlin Velev (right) is an expert in colloid science and nanoscale engineering. His group studies and develops structures with electrical and photonic functionality, biosensors and microfluidic devices. He was the first to synthesize "inverse opals", one of the most widely studied types of photonic materials today. He also pioneered principles for microscopic biosensors with direct electrical detection, discovered novel types of self-assembling microwires and designed new microfluidic chips. He received M.Sc. and Ph.D. degrees from the University of Sofia, while also spending 1 year as a researcher in the Nagayama Protein Array Project in Japan. After graduating in 1996, he accepted a postdoctoral position in the Department of Chemical Engineering, University of Delaware. He initiated there an innovative programme in colloidal assembly and was promoted to research faculty in 1998. In 2001, he formed his new research group as an Assistant Professor in the Department of Chemical Engineering, North Carolina State University. He has contributed more than 75 publications. Recent awards include the NSF Career, Sigma Xi, Ralph E. Powe and Camille and Henry Dreyfus foundation awards.

Tel.: +1-919-5134318; *Fax*: +1-919-5153465; *E-mail*: odvelev@unity.ncsu.edu

8

Interfacial Particles in Food Emulsions and Foams

Eric Dickinson

Procter Department of Food Science, University of Leeds,
Leeds LS2 9JT, UK

8.1 Introduction

Many food colloids are stabilized, at least in part, by the presence of particulate material that accumulates at oil–water or air–water interfaces. As applied to emulsion droplets, this type of stabilization mechanism is commonly referred to as Pickering stabilization.* Some examples of particles involved in Pickering stabilization in food emulsions are casein micelles (in homogenized milk), egg-yolk lipoprotein granules (in mayonnaise) and fat crystals (in spreads and margarine).[1,2] In addition, dairy-type foams such as whipped creams and toppings are stabilized by a protective layer of partially aggregated emulsion droplets (or fat crystals) which adhere to the air bubbles during whipping.[1,2]

This chapter reviews recent progress in the making and stabilization of food emulsions and foams using solid particles. To put the topic into context, we need to make direct comparison with the interfacial and stabilizing roles of the key *molecular* species – emulsifiers, proteins and hydrocolloids. It will be assumed that the reader is familiar with the basic physico-chemical principles of surfactant and polymer adsorption, with the meaning of terms commonly used in colloid science like "flocculation" and "coalescence", and with the essence of the established theories of stabilization (and destabilization) of food emulsions and foams, as described in existing texts.[1-4]

At their simplest, adsorbed particles at a liquid interface provide a mechanical barrier against instability. In keeping with this idea, an emulsion is better protected against flocculation or coalescence when the adsorbed particles are preferentially wetted by the continuous phase, and thus lie predominantly on the convex side of the oil–water interface. The angle θ (measured into the aqueous phase) defines the

---

*Though widely attributed to Pickering (1907), it seems that the phenomenon was actually described 4 years earlier by W. Ramsden (*Proc. Roy. Soc.*, **72** (1903), 156) as discussed in Chapter 1.

position of the particle with respect to that interface, as illustrated in Figure 8.1 for the case of a spherical particle that is preferentially wetted by water ($\theta < 90°$). Resolving the forces (per unit length) parallel to the solid surface at the junction of the three phases leads to Young's famous equation[5]

$$\cos \theta = \frac{(\gamma_{po} - \gamma_{pw})}{\gamma_{ow}} \tag{8.1}$$

where γ_{po}, γ_{pw} and γ_{ow} are the particle–oil, particle–water and oil–water surface energies, respectively.

The value of the parameter θ is the single most important indicator of the type of emulsion likely to be stabilized by a particular kind of particles – whether they be silica beads, polystyrene latices or even bacteria.[6] So, in the food context, adsorbed protein particles ($\theta < 90°$) can stabilize oil-in-water (o/w) emulsions by forming a steric barrier layer on the outside of dispersed oil droplets. Conversely, fat crystals ($\theta > 90°$) can stabilize water-in-oil (w/o) emulsions by coating the outside of dispersed water droplets.

Once a particle has become attached to a fluid interface with a finite contact angle, it can be regarded as being irreversibly adsorbed, since the free energy of spontaneous desorption, ΔG_d, is very high compared with the thermal energy. That is, for a spherical particle of radius r at an air–water (a–w) interface of tension γ_{aw}, we have

$$\Delta G_d = G(\theta) - G(0) = \pi r^2 \gamma_{aw} (1 - \cos \theta)^2 \tag{8.2}$$

Taking $\gamma_{aw} = 72\,\mathrm{mN\,m^{-1}}$ and $\theta = 45°$, equation (8.2) gives a desorption free energy having an incredible magnitude of $1.2 \times 10^4 kT$ for a particle of diameter

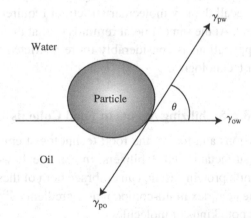

Figure 8.1 Representation of the location of a predominantly hydrophilic spherical particle at the oil–water interface. The contact angle θ, measured through water, is defined as the angle formed at the junction point of the phases – particle (p), oil (o) and water (w). The energies at the three interfaces are γ_{po}, γ_{pw} and γ_{ow}.

100 nm (r = 50 nm). Even though the energy is reduced to $\Delta G_d \approx 118\,kT$ for an equivalent 10 nm diameter particle, this still represents a negligible probability of desorption over normal experimental timescales. This simple calculation indicates that even a relatively small nanoparticle (down to the size of a large protein molecule) can be regarded as being essentially irreversibly adsorbed, so long as the contact angle is not too low.

In practice, the situation in a food emulsion or foam is complicated by the presence of other surface-active species present in addition to adsorbed particles. As well as reducing the tension at the oil–water or air–water interface, small-molecule surfactants (emulsifiers) may adsorb to the surface of the solid particles, thereby changing their wetting properties with respect to bulk oil and aqueous phases. Furthermore, where the particle is itself composed of lipid or protein, the sticking of the particle to the fluid surface may be followed immediately by spreading of lipid or protein molecules at the oil–water or air–water interface. Of particular importance in relation to the stabilization of aerated dairy emulsions is the spreading of emulsion droplets at the surface of newly formed bubbles during whipping. In such a complex situation, the stability properties of the system are influenced by the local dynamic interfacial properties as determined by the specific processing conditions and the particular combination of surface-active species present – particles, proteins and low-molecular-weight emulsifiers.

In order to understand the benefits, and also the disadvantages, of using solid particles in the formulation of food colloids, we need to be aware of the roles of the various surface-active species naturally present in foods, as well as those deliberately added during processing to facilitate dispersion, emulsification, fat-binding, thickening, foaming, gelation and water-holding.[7,8] Briefly reviewing this background allows us to recall the key molecular structural requirements for colloidal stabilization in general. At the same time, it reminds us that the range of chemicals available for food applications is considerably more restricted than that available for use in most other technologies.

8.2 Stabilizing Agents in Food Colloids

In formulating emulsions and foams, the food technologist employs two kinds of molecular species – surfactants and stabilizers. In practice, however, a single food ingredient, like the milk protein casein, can embrace both of these functional characteristics. And some complex multi-component ingredients, like whole egg yolk, contain a mixture of both kinds of molecules.

The role of the emulsifying (or foaming) agent is to promote emulsion (or foam) formation and short-term stabilization by means of interfacial action.[1,2] There are two broad classes of such species used in food processing – the small-molecule surfactants (lecithin, monoglycerides, polysorbates (Tweens), sugar esters, *etc.*) and the macromolecular surfactants, mainly proteins (from milk, egg, *etc.*), but also including some hydrocolloids (gum arabic, modified starch, propylene glycol alginate, *etc.*). The small-molecule surfactants used in foods are mainly nonionic (Spans, Tweens, monoglycerides) or zwitterionic (phosphatidylcholine). Some anionic surfactants are also used, such as the diacetyl tartaric esters of monoglycerides (DATEM), but cationic surfactants are not permitted. In practice, this means that emulsifiers with a wide range of hydrophile–lipophile balance (HLB) number are available, from anionic sodium stearoyl-2-lactylate, one of the most hydrophilic of food surfactants (HLB = 21), to oil-soluble propylene glycol stearate, one of the most lipophilic (HLB = 2). Although proteins (and some gums) commonly act in an emulsifying capacity in the food industry, the technical term "emulsifier" is confined almost exclusively to the low-molecular-weight amphiphiles. Somewhat confusingly, this same term also includes those small molecules that affect texture or shelf-life in ways other than by adsorbing at fluid interfaces, *e.g.* by interacting with starch or modifying fat crystallization.[7,8]

The role of the stabilizer in a food emulsion (or foam) is to enhance long-term physico-chemical stability (shelf-life). This may be achieved by adsorption, but not necessarily so.[9] Stabilizers are typically water-soluble macromolecules (proteins or polysaccharides) or dispersed particles composed of crystallized lipids (fat crystals) or insoluble biopolymers (casein micelles, starch granules). Being predominantly hydrophilic, the main stabilizing action of polysaccharides is to thicken and immobilize the aqueous continuous phase of foams and o/w emulsions *via* viscosity modification or gelation.[9,10] By contrast, the amphiphilic proteins have a strong tendency to adsorb at air–water and oil–water interfaces, forming stabilizing layers around gas bubbles and oil droplets. Therefore proteins are able to fulfil the foaming/emulsifying role as well as the stabilizing role.[1–3]

Small-molecule surfactants are good foaming/emulsifying agents because they diffuse rapidly to interfaces. This means that they are extremely efficient at rapidly lowering the interfacial free energy, and hence facilitating the formation of small bubbles/droplets under conditions of mechanical agitation.[11] In addition, prior to reaching full saturation coverage, surfactants are effective in generating interfacial tension gradients during close approach of newly formed colliding bubbles/droplets, thereby providing short-term dynamic stabilization *via* the Gibbs–Marangoni mechanism.[11] The lower diffusion coefficients of macromolecules and nanoparticles, and their less distinctly amphiphilic character, makes them potentially less suitable for operating in this role. Once the dispersed system has been produced, however, considerations of surface activity and/or surface tension gradients are of relatively little consequence as far as long-term stability with respect to aggregation and coalescence is concerned. The relevant physico-chemical factors then involved are the ones relating to the classical colloid stability mechanisms of electrostatic stabilization and, even more importantly, steric stabilization.[1,2]

Once adsorbed at the surface of a droplet or bubble, the good stabilizing agent must remain permanently attached to the air–water or oil–water interface. To confer long-term protection against flocculation or coalescence, the stabilizer should also be present at sufficient concentration to fully saturate the surface. Moreover, for effective steric stabilization, the protective polymeric layer should be of a thickness (at least a few nanometres) to provide surface-to-surface repulsion of sufficient range and strength to overcome the combined effects of fluctuating Brownian motion, externally applied forces, and the ubiquitous van der Waals attractive forces which inevitably drive any metastable colloidal system towards its final phase-separated equilibrium condition. For a polyelectrolyte stabilizing agent, or any copolymer adsorbing at a charged surface or in combination with other charged species (small ions or surfactants), the steric repulsion is supplemented by a repulsive electrostatic contribution. In reality, the separation of the steric and electrostatic components of the stabilization is not clear-cut, because the repulsion between the charged groups on the hydrophilic parts of stabilizing macromolecules is often an essential feature of the configurations of the adsorbate species providing the required layer thickness associated with good steric stabilization. This is the case, for instance, in the stabilization of o/w emulsions by adsorbed layers of β-casein or sodium caseinate.[12–14]

So long as they can be anchored permanently at full adsorbate surface coverage, polysaccharides are the ideal steric stabilizers[9] in aqueous media due to their high molecular weight and strongly hydrophilic character, features which together lead to highly expanded hydrated layers under "good solvent" conditions. For the most widely used hydrocolloid emulsifying agent, gum arabic (*Acacia senegal*), the strong anchoring to the surface of droplets of flavour oils (*e.g.* orange oil) has been attributed[15–17] to a small fraction (a few per cent) of covalently bound protein. While the presence of residual protein has also been implicated in the emulsion stabilizing properties of other hydrocolloids, such as certain fractions of pectin,[18] most of the commonly used polysaccharide stabilizers (xanthan, carrageenan, starch, *etc.*) lack the capacity for strong adsorption at fluid interfaces.[9] On the other hand, the covalent linking of a surface-active protein (*e.g.* β-lactoglobulin) to a non-surface-active polysaccharide (*e.g.* dextran) can be made to generate a hybrid biopolymer conjugate with excellent emulsifying properties.[19] This greatly enhanced stabilizing behaviour of the conjugate over the mixture of the separate individual biopolymers can now be readily explained[20] in terms of modern polymer adsorption theory.[21]

Many of the proteinaceous ingredients used for emulsification and foaming in food product formulation do actually exist as colloidal particles. In some cases the particulate form arises as a result of ingredient processing. For instance, the whey proteins in their native form in milk are highly soluble, but they become denatured and aggregated into particles of micron size and larger during the heating and drying processes used in the preparation of whey protein concentrates.[22] The other main protein component of milk, casein, occurs naturally in the form of self-assembled complexes with calcium phosphate. Although not as effective at low protein concentrations as the much less aggregated sodium caseinate, the micellar casein form is still a good emulsion stabilizing agent.[23,24] Together with the soluble milk proteins and some casein micellar fragments (sub-micelles), it is now well established[25–27] that polydisperse casein micelles (mean size a few hundred nm) are the main stabilizing entities in homogenized milk, as well as in recombined and synthetic o/w emulsions prepared with skim milk powder as emulsifier. Another ingredient containing polydisperse surface-active particles is egg yolk;[28] this is the traditional emulsifying agent used to make mayonnaise and salad dressings. The intact egg-yolk granules (mean size around 1 μm) are reasonably effective in stabilizing o/w emulsions,[29] although they are not so effective, apparently, as the mixture of dissociated lipoproteins and phosvitin released from the granules by extraction with salt.[30] Presumably the main reason for the greater stabilizing ability of dissociated casein micelles or dissociated egg yolk granules is that the monomeric proteins (or small aggregates thereof) can more readily give full saturated coverage of the surface of the oil droplets, which is an essential requirement for effective steric stabilization.

Foams are more difficult to stabilize. This is partly because the gas bubbles in an aerated solution of surfactant or protein are typically much larger and less dense than homogenized oil droplets, which leads to much higher rates of gravity creaming and coalescence.[1,2] But even more important is the significant solubility of the gas in the aqueous phase which allows relentless diffusive mass transport between bubbles of different sizes under the influence of Laplace pressure gradients. Unless bubbles are embedded in a solid matrix or surrounded by a rigid shell-like adsorbed layer, this disproportionation (Ostwald ripening) eventually leads to the loss of all but the largest bubbles in the system. There is current interest[31] in the potential of protein-based nanoparticles for inhibiting disproportionation in aerated food systems. This issue will be addressed later (see Section 8.5).

In contrast to o/w emulsions and foams, there are no equivalent oil-soluble polymers available to the food technologist for stabilizing w/o emulsions. The ubiquitous monoglycerides, and most of their common chemical derivatives, have lipophilic tails that are too short to provide long-term steric stabilization. Reasonably satisfactory in practice, however, for stabilizing the fine internal water droplets of w/o/w multiple emulsions,[32] are polyglycerol polyricinoleate (PGPR) or phosphatidylcholine-depleted soya lecithin. (PGPR is widely used in chocolate making as a viscosity modifier and dispersant of sugar granules.[33]) The most important w/o food emulsions are the fatty spreads. However, the long-term stability of margarine and many other yellow spreads is provided not by these small-molecule emulsifiers, but by the texture-controlling network of fat crystals dispersed in the continuous phase, some of which adhere to the surface of the dispersed water droplets.[34] As a result of their commercial importance in determining the texture of these semi-solid w/o emulsions, fat crystals are the class of particles that have been investigated in most detail in the context of food colloid stabilization. Accordingly, the behaviour of systems containing fat crystals is given substantial emphasis in this chapter (see Sections 8.3 and 8.4). First, though, we set the scene by reviewing briefly the classical theory of the origin of the wetting of solid particles by oil and aqueous phases.

8.3 Wettability and Contact Angles

The very existence of an interface between a pair of liquid phases is indicative of a strong imbalance of intermolecular forces. For an oil–water interface, the value of the tension depends on the characteristics of the molecular interaction potential between the unlike species (U_{ow}) relative to the characteristics of the potentials acting between the like species (U_{oo} and U_{ww}). To a good approximation, the oil–oil and oil–water pair potentials, U_{oo} and U_{ow}, can each be regarded as made up from a short-range repulsive part (the excluded volume interaction) and a long-range attractive part (the van der Waals interaction). Together with both these isotropic contributions, the more complex water–water potential, U_{ww}, has an extra medium-range contribution that is attractive and directional (the hydrogen bonding interaction). While the short-range repulsive interaction determines the interfacial thickness and structure,[1] it has relatively little influence on the actual numerical value of the interfacial tension.

It was argued semi-intuitively by Foulkes[35] that additivity of attractive energies on the molecular scale is consistent with independent contributions to the oil–water tension from hydrogen bonding and dispersion forces. Taking the oil–water dispersion term as the geometric mean of the oil–oil and water–water dispersion terms, γ_o^d and γ_w^d, by analogy with the geometric mean rule of molecular dispersion energies, the interfacial tension is given by

$$\gamma_{ow} = \gamma_{oa} + \gamma_{aw} - 2(\gamma_o^d \gamma_w^d)^{1/2} \tag{8.3}$$

where γ_{oa} ($= \gamma_o^d$) is the surface tension of the non-polar oil (with air) and γ_{aw} is the surface tension of water. Pragmatic support for the generality of applicability of the geometric mean rule comes from the success of the solubility parameter concept[36] for the prediction of the solubility of many non-electrolytes. Nevertheless, it was proposed[37] that an extra term I_{ow} be added to equation (8.3) to allow for the contribution from polar intermolecular forces

$$\gamma_{ow} = \gamma_{oa} + \gamma_{aw} - 2(\gamma_o^d \gamma_w^d)^{1/2} - I_{ow} \tag{8.4}$$

The value of γ_w^d in equations (8.3) and (8.4) has been estimated[35] to be $21.8 \, mN \, m^{-1}$ at 20°C, meaning that approximately 70% of the surface tension of pure water is attributable to hydrogen bonding interactions. The strongly preferred orientation of individual H_2O molecules at the air–water surface is reflected thermodynamically in the much lower surface entropy for liquid water as compared with a hydrocarbon or a triglyceride oil.[1]

Turning now to the case of a (solid) particle phase in contact with the two liquid phases, we can write down equivalent expressions to equation (8.4) for both γ_{po} and γ_{pw} as follows

$$\gamma_{po} = \gamma_p + \gamma_{oa} - 2(\gamma_p^d \gamma_o^d)^{1/2} - I_{po} \tag{8.5}$$

$$\gamma_{pw} = \gamma_p + \gamma_{aw} - 2(\gamma_p^d \gamma_w^d)^{1/2} - I_{pw} \tag{8.6}$$

By combining equations (8.4)–(8.6) with Young's equation (equation (8.1)), an expression can be obtained[38] for the three-phase contact angle in terms of the various interfacial energy parameters. Since typical liquid vegetable oils and solid triglyceride fat phases are both rather non-polar, we can assume that $I_{po} = 0$ and $\gamma_{oa} = \gamma_o^d$ for food systems. Therefore, the contact angle θ is related to the oil–water interfacial tension by

$$\cos \theta = (I_{pw} - I_{ow} + C)\gamma_{ow}^{-1} - 1 \tag{8.7}$$

where the coefficient C is defined by

$$C = 2[(\gamma_o^d)^{1/2} - (\gamma_w^d)^{1/2}][(\gamma_o^d)^{1/2} - (\gamma_p^d)^{1/2}] \tag{8.8}$$

If I_{ow} is insignificant compared with I_{pw}, equation (8.7) reduces to

$$\cos \theta = (I_{pw} + C)\gamma_{ow}^{-1} - 1 \tag{8.9}$$

Sets of equations similar to equations (8.7)–(8.9) have been presented by van Voorst Vader,[39] Lucassen-Reynders,[40] Campbell[38] and Johansson and Bergenståhl.[41]

In applying this theory to the case of triglyceride crystals at the oil–water interface, it has been recognized[38,42] that the constant term $(I_{pw} + C)$ in equation (8.9) is small compared with γ_{ow}. This means that, for solid particles composed of non-polar material and for typical values of γ_{ow} in the range 30–$50 \, mN \, m^{-1}$, the contact angle will tend to be high ($\theta \rightarrow 180°$), corresponding to nearly complete wetting of the particle by the oil. The contact angle becomes reduced for moderately polar monoglyceride particles at the interface between water and liquid vegetable oils. Only for low oil–water interfacial tensions, as produced for instance by adding emulsifiers to the system, does the contact angle become sensitive to γ_{ow}.

Campbell reported[38] the effect of various food emulsifiers on the contact angle in a system of solid fat (hardened palm oil) + soybean oil + water. The five commercial emulsifiers investigated were three types of monoglyceride sample (saturated, part-saturated, unsaturated), a sorbitan ester sample (HLB = 4.3) and a mixed phospholipid sample (lecithin). Table 8.1 shows that there is no significant effect on the measured contact angle of $\theta = 150 \pm 1°$ on adding 0.3 wt.% of any of these emulsifiers to the oil phase. However, with the addition of 1 wt.% sodium caseinate to the aqueous phase, it was found[38] that the measured contact angles were lower than 150° and dependent on the type of emulsifier. The zwitterionic lecithin, which is known to complex with proteins in solution and at interfaces, reduced the contact angle very substantially in the presence of sodium caseinate. It was suggested[38] that, as the measured γ_{ow} was not appreciably lower with protein than without it, the reason for the lowering of the contact angle in the presence of sodium caseinate is that the quantity I_{pw} in equation (8.7) is significantly affected by protein adsorption. Presumably the nature of the emulsifier influences the extent of protein adsorption at the solid fat surface.

Further experiments were carried out by Campbell[38] with the same emulsifiers added to the hardened palm oil phase. The results are shown in Table 8.2. We see that the measured contact angle falls as the fat phase becomes more polar due to the incorporation of the emulsifier. Considerable differences between the monoglyceride

Table 8.1 *Contact angle θ (measured through the aqueous phase) in the three-phase system of hardened palm oil (solid fat) + soybean oil (liquid) + water, with emulsifier (0.3% w/w) added to the oil phase, and with or without sodium caseinate (1% w/w) added to the aqueous phase. Taken from Ref. [38]; with permission of the Royal Society of Chemistry.*

	θ (°)	
Emulsifier	Without protein	With protein
Saturated monoglyceride	150	148
Partly saturated monoglyceride	151	142
Unsaturated monoglyceride	150	130
Sorbitan monooleate (Span 80)	151	130
Soybean lecithin	149	82
No emulsifier	150	–

samples are apparent, with the unsaturated one giving a much larger reduction in θ at the 1% level than does the saturated one at the 10% level. Surprisingly, however, when pure solid monoglyceride (100% emulsifier) is used instead of hardened palm oil, the low contact angle obtained ($\theta \sim 60\text{--}65°$) is nearly independent of the monoglyceride type. Thus it is clear that the monoglyceride crystals themselves can be regarded as being moderately polar in nature.

Johansson and Bergenståhl[41] used equation (8.7) to estimate the polar energy excess $I_{excess} = I_{pw} - I_{ow}$ for particle–water as compared with oil–water as influenced by various food emulsifiers. The quantity I_{excess} measures a particle's tendency to adsorb at the oil–water interface, whereas the quantity γ_{ow} represents the interfacial resistance to particle adsorption. The values of I_{excess} reported in Table 8.3 were separately estimated for palm stearin β' crystals approaching the soybean oil–water interface from the water side (advancing contact angles) and the oil side (receding contact angles). The polar excess energy is always positive, which means that fat crystals always have the tendency to adsorb at the triglyceride oil–water interface, even in the absence of emulsifiers. The excess energy is of the order of 10^{-14} J for a hypothetical particle of diameter 1 μm, corresponding to around $10^5\text{--}10^6$ more hydrogen bonds per particle than an oil–water interface of the same area.[41] The generally higher value of I_{excess} when crystals approach the oil–water interface from the water side can be attributed to the higher polarity of the surface of the particles when dispersed in water. The data in Table 8.3 show that the highest polar energy excess values are for the lactic acid esters of monoglycerides. In contrast, the lowest I_{excess} values are given by the hydrophobic lecithins, which have relatively little preference for the fat crystal–water interface as compared with the oil–water interface.

Table 8.2 *Contact angle* θ *(measured through the aqueous phase) in the three-phase system of hardened palm oil (solid)* + *soybean oil (liquid)* + *water, with various contents of emulsifier (1, 10 or 100 wt.%) in the solid fat phase. Taken from Ref. [38]; with permission of the Royal Society of Chemistry.*

Emulsifier	θ (°)		
	1%	10%	100%
Saturated monoglyceride	146	142	61
Partly saturated monoglyceride	138	81	63
Unsaturated monoglyceride	125	73	66
Sorbitan monooleate (Span 80)	142	–	–
Soybean lecithin	132	109	–

Table 8.3 *Effect of emulsifiers (1–2 wt.%) on the polar energy excess* $I_{excess} = I_{pw} - I_{ow}$ *for palm stearin* β' *crystals and soybean oil for advancing and receding contact angles. The receding contact angle corresponds to particles approaching the oil–water interface from the "normal" oil side, whereas the advancing contact angle refers to particles approaching from the water side, where they could be pushed by shear forces during mixing or homogenization. Taken from Ref. [41]; with permission of the American Oil Chemists' Society.*

Emulsifier	I_{excess} (mJ m^{-2})	
	Advancing θ	Receding θ
Lecithins	1.8–3.3	1.5–3.1
Monoolein	5.0	1.5
Lactic acid esters of monoglycerides	14.5	8.5
Ethoxylated alkyl ether	4.9	2.6
Ethoxylated sorbitan monostearate (Tween 60)	2.6	2.6
No emulsifier	7.2	2.5

8.4 Stabilization of w/o Emulsions by Fat Crystals

Dispersed fat crystals are the essential structural components of margarines and edible fatty spreads.[43] They crystallize from the partially saturated oil phase of w/o emulsions in the form of colloidal particles. The crystals interact and aggregate to form a three-dimensional network which provides long-term stability to the dispersed water droplets and a solid-like textural character to the food product.[44–46] Small-molecule emulsifiers are added to control the fat crystallization,[47,48] and they also influence the particle–particle interactions, as does the small amount of water present in the oil.[49] Emulsifiers also affect the wetting properties of the fat crystals at the oil–water interface.

Fat crystals contribute to the stability of margarine-type w/o emulsions[43] by attaching themselves to the surfaces of aqueous droplets and protecting them against coalescence *via* Pickering stabilization.[50] The effectiveness of the crystals in this role depends on their size, shape and morphology, as well as on the crystal surface wettability as influenced by the presence of other surface-active species such as emulsifiers and proteins.[38,40,51,52] The effect of the polarity of the crystal surface on the stability of w/o emulsions was demonstrated by Campbell.[38] Based on visual observations of emulsions (50 wt.% soybean oil) 1 h after preparation in a laboratory turbine-type mixer, it was found that the emulsion samples made with 1 wt.% monoglyceride crystals were stable, whereas those made with 1 wt.% triglyceride crystals (composed of tristearin or hardened palm oil) were not stable. This difference could be attributed to the high polarity of the monoglyceride crystals

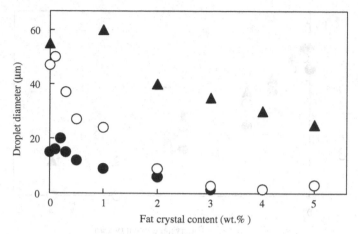

Figure 8.2 Influence of the concentration of triglyceride crystals (palm stearin β') in oil on the droplet diameter (mode value) of monoolein-stabilized water-in-soybean oil emulsions: ▲, 2% monoolein, 20% water; ○, 4% monoolein, 5% water; ●, 2% monoolein, 5% water. Taken from Ref. [52]; with permission of the American Oil Chemists' Society.

($\theta \approx 60°$) as compared with the triglyceride crystals ($\theta \approx 150°$), as illustrated by the data in Tables 8.1 and 8.2. Under the same conditions, it was observed[38] that 0.8 wt.% of the unsaturated monoglyceride was required for stability, whereas just 0.2 wt.% of the saturated monoglyceride was sufficient. This can be explained in terms of the difference in solubility of the monoglycerides in the oil phase; and it confirms the supposition that it is indeed the presence of dispersed fat crystals in the emulsion that is conferring stability, and not just the soluble emulsifier.

The influence of fat crystal concentration on w/o emulsion stability has been systematically investigated by Johansson *et al.*[52] The emulsions were prepared by Ultra-Turrax homogenization and droplet-size distributions were determined by static light scattering (Malvern Mastersizer) with a correction made for scattering from the fat crystals. Figure 8.2 shows a plot of the initial droplet size (mode value of the volume distribution) for monoolein-stabilized emulsions (95% or 80% soybean oil) as a function of the concentration of palm stearin β' crystals (specific surface area $\sim$ $9\,m^2\,g^{-1}$). In the absence of fat crystals the monoglyceride-stabilized emulsions were rather coarse, with an initial diameter in the range 15–55 μm, depending on the relative proportions of emulsifier and water. We can see from Figure 8.2 that the addition of fat crystals leads to a slightly increased droplet size at low concentrations followed by a substantial decrease at higher concentrations. Similar trends of behaviour were found for emulsions prepared with pure soybean phosphatidylcholine or a commercial lecithin as the emulsifying agent instead of monoolein.

Using sedimentation experiments, Johansson *et al.*[52] investigated the effect of fat crystals on the long-term stability of these w/o emulsions. Figure 8.3 shows the

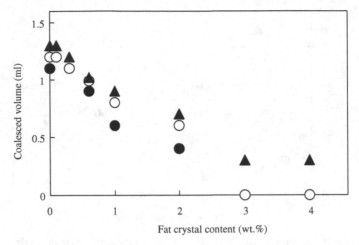

Figure 8.3 Influence of the concentration of fat crystals (palm stearin β') in oil on the volume of water separated from monoolein-stabilized water-in-soybean oil emulsions (4% emulsifier, 20% water, total emulsion volume 7.0 ml) due to coalescence. Water volume is plotted against the fat crystal content: ●, after 1 day; ○, after 6 days; ▲, after 4 weeks. Taken from Ref. [52]; with permission of the American Oil Chemists' Society.

sediment volume due to coalescence of monoolein-stabilized emulsions (4% emulsifier, 20% water) after 1, 6 and 28 days. We can see that the sample without fat crystals was rather unstable with respect to coalescence: 60–70% of the dispersed aqueous phase separated into a distinct lower layer after 1 day, and this same state was essentially the same after 1 month. The stability was gradually improved by addition of palm stearin β' crystals to the oil. A particle content of >0.5% was required to produce a significant lowering of the sediment volume. For a particle content of 3–4%, no phase separation was visible within 1 week, and only 10–20% of the water separated after 1 month. Figure 8.4 shows the corresponding sedimentation data for emulsions (25% water) stabilized with commercial lecithin (0.5%). In this case the reference emulsion (no fat crystals) with smaller water droplets (initial diameter $\approx$ 35 μm) was found to be substantially more stable than the monoolein emulsion: only ~25% of the dispersed aqueous phase separated after 4 days, and still only ~40% after 4 weeks. Inhibition of water separation from the lecithin-stabilized emulsion over a period of several days could be achieved with particle concentrations as low as 0.1–0.2%. For emulsion stabilization over a period of 1 month, the data in Figure 8.4 suggest an optimum fat crystal concentration of 1–2%, with higher fat contents leading to some loss of stability.

Mechanistically, the addition of fat crystals might be considered to enhance w/o emulsion stability in various ways. There are both bulk and interfacial rheological effects. The viscosity of the continuous phase increases,[53,54] which reduces the

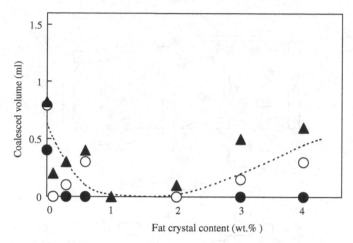

Figure 8.4 Influence of the concentration of fat crystals (palm stearin β') in oil on the volume of water separated from lecithin-stabilized water-in-soybean oil emulsions (0.5% emulsifier, 25% water, total emulsion volume 6.8 ml) due to coalescence. Water volume is plotted against the fat crystal content: ●, after 4 days; ○, after 6 days; ▲, after 4 weeks. The dashed line indicates maximum stability at 1–2 wt.% fat crystal content. Taken from Ref. [52]; with permission of the American Oil Chemists' Society.

rates of droplet sedimentation and flocculation. When the inter-droplet interactions are net attractive, and the droplet concentration is sufficient to form a gel-like network with a finite yield stress, the processes of sedimentation and flocculation may be inhibited completely. Association of fat crystals with the oil–water interface will typically tend to enhance the interfacial viscoelasticity, especially when the interactions between the adsorbed crystals are net attractive. This surface rheological effect will contribute towards stabilizing the emulsion against coalescence, in addition to the protection provided by the steric barriers of crystals surrounding the droplets. Another pertinent factor is that fat crystals are of higher density than the liquid oil. Therefore, the rate of sedimentation, as well as the associated gravity-driven flocculation and coalescence, is reduced in the presence of crystalline fat, because of the lower density difference between the dispersed phase and the continuous phase.

At low levels of addition, fat crystals may have a detrimental effect on emulsion stability due to bridging flocculation, as illustrated schematically in Figure 8.5(a). Fat crystal wetting by the aqueous phase, as determined by the value of the contact angle, is an important requirement for the formation of fat crystal bridges between dispersed water droplets. This bridging flocculation may lead on to enhanced coalescence, especially in the presence of a shear field. The increased droplet size at low fat contents observed in the monoolein-stabilized emulsion data in Figure 8.2 can probably be attributed to this mechanism. That this putative phenomenon

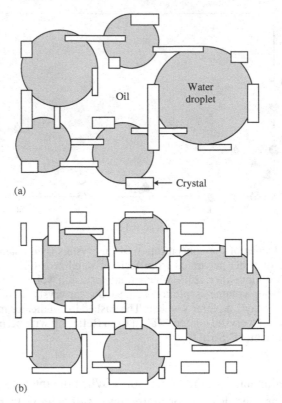

(a)

(b)

Figure 8.5 Highly schematic representation of the effect of the distribution of fat crystals on the stability of water droplets dispersed in oil. (a) Bridging flocculation at fat crystal contents well below full saturation coverage of the oil–water interface, and (b) inhibition of coalescence due to screening of droplets at fat crystal contents near full saturation coverage.

seems not to be also reflected in the sedimentation behaviour for the same system (Figure 8.3) has been explained[52] in terms of hindering effects dominating the consolidation process in these highly concentrated systems.

At higher levels of fat crystal addition, various stabilizing mechanisms take over: these are associated with the formation of a viscoelastic layer of fat crystals at the oil–water interface and with changes in rheology and density within the oil dispersion medium. Crystals inhibit coalescence by preventing droplets from touching each other, as illustrated in Figure 8.5(b). Johansson *et al.* have estimated[52] that the monoolein-based w/o emulsions (Figures 8.2 and 8.3) require thinner layers for long-term stabilization than do the lecithin-based w/o emulsions (Figure 8.4). This is attributed to the greater penetration of the crystals into the water droplets in the presence of commercial lecithin, *i.e.* the lower contact angles.[55] When an emulsion is prepared with a concentration of fat crystals below that corresponding to saturation coverage of the oil–water interface, destabilization occurs due to combined sedimentation, flocculation and coalescence. Once saturation coverage has been reached as a result of loss of interfacial area, further coalescence is inhibited by Pickering stabilization.

8.5 Stabilization of Air Bubbles in Water by Proteins and Nanoparticles

Compared to emulsions, it is generally an even greater challenge to control the stability of aerated systems. In many food foam formulations, the aqueous continuous phase is initially liquid-like, but it is commonly transformed into a solid-like matrix through subsequent cooking/processing, *e.g.* in the baking of a cake or the freezing of ice-cream.[1,2] In a wet foam, there is drainage of the intervening fluid between adjacent bubbles due to the combined effects of gravity and capillary action. This leads on to film thinning, bubble coalescence, structural collapse and finally the loss of foam volume, especially with accompanying mechanical or thermal disturbance. Furthermore, diffusion of gas occurs between bubbles of different sizes and Laplace pressures, leading to foam coarsening (disproportionation). For many wet foams, the kinetics of coarsening is well described by classical Lifshitz–Slyozov–Wagner (LSW) theory,[56] which uses a mean-field approach in a closed system, such that the concentration of gas is assumed to rise uniformly throughout the aqueous phase as smaller bubbles gradually dissolve and eventually disappear. This mass transport process is difficult to stop because even the most well-packed monolayer of small-molecule surfactants provides only a limited barrier to gas permeability.[57] And, of course, protein layers have even bigger holes through which gas molecules can readily diffuse.

Nevertheless, one potential way to retard or even prevent bubble shrinkage/growth is through the mechanical properties of the matrix or film surrounding the bubbles.[58] In order to become smaller, a bubble has to do work against the surface dilatational viscoelasticity. Hence, the rheological character of the material located around the bubble surface can provide a resistance to slow down the rate of bubble shrinkage, or may be even a barrier to prevent it altogether. At the level of LSW theory, it has been recently shown,[59] however, that the complete cessation of disproportionation is possible only for bubbles that possess a purely elastic adsorbed layer, or are contained within a purely elastic matrix, *i.e.* a solid foam. A highly viscous continuous phase can considerably slow down the shrinkage rate, but it cannot stop it completely.

It is well established[60,61] that irreversibly adsorbed globular proteins like β-lactoglobulin and ovalbumin form highly viscoelastic films at the air–water surface. It has also long been assumed by various authors – including this one[2] – that such globular proteins can be used as long-term foam stabilizing agents to prevent coarsening due to the formation of an elastic skin around expanding or shrinking bubbles. Recently, however, some direct experimental observations of single bubbles[62] and small clusters of bubbles[63] at a planar interface has provided unequivocal evidence that none of the common food proteins is actually capable of stopping the disproportionation process. Furthermore, in practice, none appears capable of slowing down the rate of bubble shrinkage by much more than a factor of, say, 2–3 times. This is illustrated by the data in Figure 8.6 which shows plots of the measured time-dependent radius $R_a(t)$ for bubbles stabilized by various food proteins,[64] together with the values of the surface dilatational modulus ε used as a single surface rheological parameter to fit the experimental data to the theory.[62] The *a posteriori* explanation for the modest influence of surface viscoelasticity on the coarsening rate is that, although the protein film is indeed predominantly elastic at short timescales (*i.e.* those typically measured in surface rheology experiments[60,61]), it is predominantly viscous at the relatively long timescales over which the surface area of a bubble changes during disproportionation. There may be some loss of interfacial material to the bulk phase as the bubble shrinks, although full protein desorption is usually taken to be an extremely slow process.[65] More likely, under the influence of the applied stresses, is partial displacement of the adsorbed protein layer, involving just a proportion of the originally adsorbed molecules, leading to buckling and wrinkling of the film, as recently predicted by computer

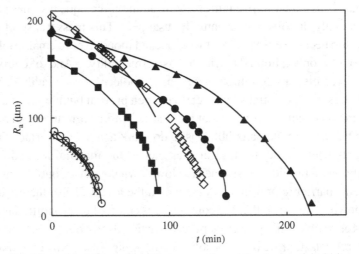

Figure 8.6 Plot of bubble radius R_a *versus* time t for isolated bubbles near a planar–air water surface stabilized at pH 7 by different proteins: ×, sodium caseinate; O, whey protein isolate (WPI); ▲, gelatin; ◇, ovalbumin; ●, β-lactoglobulin; ■, soy glycinin. The lines drawn through the points indicate the best fits of the theory to the experimental data, i.e. requiring the following values of the dilatational modulus, ε (mN m^{-1}): 0 (constant surface tension) for sodium caseinate and WPI; 2.3 for gelatin; 7.0 for β-lactoglobulin; 8.5 for soy glycinin; 12.5 for ovalbumin. Taken from Ref. [64]; with permission of the Royal Society of Chemistry.

simulation.[66] In the final stages of accelerating bubble shrinkage and film wrinkling, a rapidly disappearing gas bubble coated with β-lactoglobulin or soy glycinin leaves behind it an irregularly shaped particle of the aggregated denatured protein.[62,64] In the case of ovalbumin, which is well known[65] for its susceptibility to interfacial coagulation, we have observed,[64] even in the early stages of bubble shrinkage, the presence of stringy interfacial aggregates and some coagulated protein located between the surfaces of adjacent bubbles. (We have also found[64] ovalbumin to be exceptional in being the only protein that cannot be fitted well by our simple theory,[62] using ε as a single adjustable parameter; see Figure 8.6.)

Thus it would appear that protein films are generally capable only of slowing down bubble shrinkage/expansion, and not preventing it altogether. The question arises, then, whether there is any kind of interfacial film capable of arresting shrinkage completely. Small-molecule surfactants seem unlikely candidates, except perhaps for the unique case of gas micro-cells stabilized by sucrose esters at high-sugar contents, which have a special kind of crystalline polyhedral domain structure.[67] On the other hand, the use of partly hydrophobic nanoparticles would seem to be an idea worthy of consideration for this purpose, in the light of the recent success of Binks *et al.*[68–70] in using silica nanoparticles to stabilize emulsions based on various kinds of oil phases, including triglyceride oil.[69] Indeed, a preliminary study using the same kind of surface-active particles has shown[71] that very stable gas bubbles can be obtained. Moreover, it has been found[72] to be important to optimize the degree of hydrophobicity, in order to achieve a suitable balance between maximizing the extent of particle adsorption and minimizing the tendency for particle aggregation in the bulk phase prior to adsorption. For the case of silica particles (diameter 20 nm) with a third of the SiOH surface groups hydrophobically modified (67% SiOH), the addition of 3 M NaCl to the aqueous phase has been used to tune the hydrophilic–hydrophobic balance. It is essential that particles are not too hydrophobic, as otherwise they tend to bridge the surfaces of bubbles and cause coalescence;[73] in fact, such a bridging process is the basic instability mechanism underlying the use of hydrophobic particles as anti-foaming agents;[74] see Chapter 10 for a full discussion.

The level of bubble stability that can be achieved[64,72] with partly hydrophobic silica nanoparticles is very much greater than that obtainable with adsorbed proteins. Figure 8.7 shows that, whereas β-lactoglobulin-stabilized bubbles of size 50–100 μm shrink and disappear within a matter of several minutes, the equivalent nanoparticle-stabilized bubbles can be made to remain stable almost indefinitely (many hours or days). Microscopic observation would appear to indicate[71] that the formation of a rigid shell of strongly adsorbed nanoparticles around the surface is what prevents bubble collapse (see Figure 8.8). However, more recent images from confocal microscopy[64,72] suggest that a more complex structure-stabilizing mechanism may

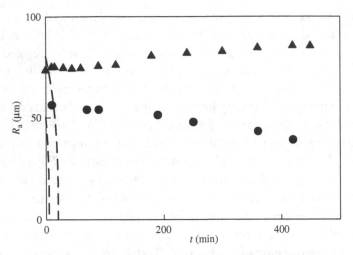

Figure 8.7 Plot of bubble radius R_a *versus* time t for two bubbles (▲, ●) near a planar air–solution surface. The bubbles are stabilized in a 3 M aqueous NaCl solution by partially hydrophobic silica particles. Shown for comparison (dashed lines) are experimental data for β-lactoglobulin-stabilized bubbles of similar size. Taken from Ref. [64]; with permission of the Royal Society of Chemistry.

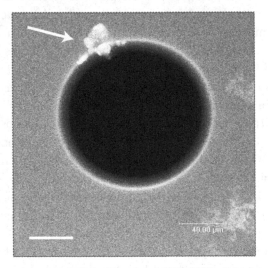

Figure 8.8 Confocal micrograph of a gas bubble stabilized by a layer of partially hydrophobic silica particles in 3 M aqueous NaCl solution. The particles were stained with the fluorescent dye Rhodamine B. The arrow shows the location of an aggregate of silica particles stuck to the bubble surface. Bar = 40 μm. Courtesy of Mr. Thomas Kostakis, University of Leeds.

be involved. The hydrophilic–hydrophobic balance ($\theta \approx 40°$) giving good interfacial stabilization is close to that also favouring nanoparticle aggregation, and hence much of the adsorbed particulate material is likely to be highly aggregated, and therefore unlikely to be present as a perfectly uniform close-packed monolayer around each bubble surface. Moreover, we have observed[64,72] that, whenever significant bubble stabilization by nanoparticles does takes place, these particles tend to form a weak gel-like network throughout the aqueous phase, with the particles adhering to the bubble surface also forming part of the same bulk gel network. What this implies is that it is the resistance to collapse of the whole network, as much as the resistance to collapse of the nanoparticle layer around an individual bubble, that actually determines the overall stability of the system.

The presence of protein particles is evident when aerated high-sugar systems of egg white (egg albumen) are observed by confocal microscopy.[75] We have speculated recently[64] that the special whipping and foam stabilizing properties of this familiar food ingredient might be attributable to stabilization by protein nanoparticles. Evidence for this comes from recent experiments[64] on rates of shrinkage of single bubbles stabilized by the major egg white protein, ovalbumin. Unlike results obtained with several other food proteins, those for ovalbumin were found not to be so reproducible, and occasionally some very long bubble lifetimes were recorded that did not fit the simple theory (see Figure 8.6). Moreover, visual observation under the confocal microscope has shown[31,64] that ovalbumin forms protein particles around individual bubbles as they shrink, and that some of the aggregates appear already to have formed part of a semi-soluble interfacial protein network before much shrinkage has actually taken place. We are tempted, therefore, to draw an analogy between the behaviour of whipped egg white protein and that of the system of bubbles stabilized by the partly hydrophobic silica nanoparticles. That is, we speculate that the well-known functionality of ovalbumin is associated with its tendency to form coagulated protein networks and surface-active stabilizing particles. This begs the question, of course, as to whether other kinds of food protein particles, with an even greater degree of effectiveness, could be directly engineered for this same purpose.

8.6 Spreading of Droplets and the Whipping of Cream

In a polyhedral foam, the presence of even a small quantity of particulate material is often highly detrimental to stability.[58,74] For a hydrophobic particle large enough to touch both surfaces of the thin liquid film between a pair of bubbles, the Laplace pressure in the film adjacent to the extraneous particle may become positive. This induces liquid flow away from the particle, causing the liquid to break contact with the particle, leading to film rupture. Another type of contaminating particle is one that spreads its contents at the air–water surface. The nearby film liquid is made to move in the same direction as the spreading particulate material, which induces a local thinning of the film and so enhances the probability of rupture.[2,58] Both of these mechanisms are probably involved in the destabilization of aqueous food foams by fatty particles. By way of example, we can refer to the considerably poorer foam stability of whole milk, which contains fat globules, as compared to (fat-free) skim milk. Another example – very familiar to the cook – is the detrimental effect of a trace of egg yolk on the foaming behaviour of egg white. Nevertheless, despite all this, the spreading of oil/fat at the air–water surface is actually a requirement for the making and stabilization of food colloids such as whipped cream, ice-cream and cake batter.

In commercial terms, the whipped dairy-based emulsion is a very important class of food colloid. To some extent it is a particle-stabilized foam, since the gas cells are stabilized by partly coalesced fat globules.[76–78] It has been demonstrated that the orthokinetic destabilization of the o/w emulsion is affected by shearing conditions,[79] the crystallization state of the fat,[51,80] and the presence of emulsifiers.[81–83] During shear-induced air incorporation there is controlled spreading of some liquid fat on the bubble surfaces,[51,77] which enhances droplet adsorption and aggregation. Hence the mechanism of spreading of individual fat globules at the air–water surface is highly relevant for understanding the factors controlling the whipping of dairy emulsions.

Whether an oil droplet will enter the air–water surface or remain in the bulk aqueous phase is thermodynamically determined by the so-called entering coefficient E defined by[84]

$$E = \gamma_{aw} + \gamma_{ow} - \gamma_{oa} \qquad (8.10)$$

The droplet enters the air–water surface under the condition $E > 0$. Once this has occurred, the droplet may either form a lens at the surface (partial wetting) or spread out at the surface (complete wetting). The tendency for a droplet to spread at the air–water surface is determined thermodynamically by the spreading coefficient S defined by[85]

$$S = \gamma_{aw} - \gamma_{ow} - \gamma_{oa} \qquad (8.11)$$

Spreading occurs under the condition $S \geq 0$. This condition is readily satisfied for many food-grade triglyceride oils on clean water, although not so for the less polar higher alkanes.

Figure 8.9 shows a schematic diagram of an oil lens resting at the air–water surface. The three-phase boundary disappears under three alternative conditions: (i) the oil droplet becomes expelled into the water ($E < 0$), (ii) the oil droplet becomes expelled into the air ($\gamma_{ow} + \gamma_{oa} < \gamma_{aw}$), or (iii) the oil spreads at the surface ($S > 0$). For a sunflower oil droplet, Hotrum *et al.*[86] have calculated the lens radius a_L as a function of the air–water surface tension γ_{aw}, assuming that the volumes of the upper and lower lens caps are given by[87,88]

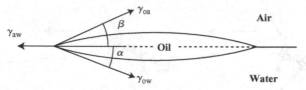

Figure 8.9 Representation of an oil lens resting at the air–water surface with contact angles α and β, respectively, on the water and air sides of the surface. The three tensions (γ_{aw}, γ_{ow} and γ_{oa}) satisfy the equation: $\gamma_{aw} = \gamma_{ow}\cos\alpha + \gamma_{oa}\cos\beta$.

$$V_{upper} = (\pi a_L^3/3\sin^3\beta)(\cos^3\beta - 3\cos\beta + 2) \tag{8.12}$$

$$V_{lower} = (\pi a_L^3/3\sin^3\alpha)(\cos^3\alpha - 3\cos\alpha + 2) \tag{8.13}$$

where the angles α and β are defined in Figure 8.9. The size of the lens increases with increasing γ_{aw} until the condition $S = 0$ is reached. Taking measured values of $\gamma_{ow} = 29\,\mathrm{mN\,m^{-1}}$ and $\gamma_{oa} = 28\,\mathrm{mN\,m^{-1}}$, this means that a_L grows asymptotically to infinity as $\gamma_{aw} \to 57\,\mathrm{mN\,m^{-1}}$.

While the above analysis is based on equilibrium thermodynamics, the processes of entering and spreading of oil droplets at air–water surfaces are kinetic processes. For stable emulsion droplets covered with a protein adsorbed layer, repulsive colloidal interactions provide an effective kinetic barrier, especially under quiescent conditions.[89] This means that, even when the condition $E > 0$ is satisfied, the entering of emulsion droplets may not be observed in practice.[90,91] Furthermore, the difference between the equilibrium and dynamic tensions has to be considered in real systems, where surfactants and/or proteins are present at the air–water surface, as well as at the oil–water interface. The tensions will be influenced by the rate of expansion of the interfaces, as well as by the concentrations of the various surface-active species present. Also, during spreading, the tensions may change due to the compression of the adsorbed layer by the expanding film.[86] So spreading may be transiently favoured when the dynamic air–water tension is high, but it may stop later (S < 0) following the lowering of γ_{aw} by adsorption of fresh surface-active material from the bulk.

The spreading behaviour of protein-stabilized oil droplets has been investigated experimentally recently[86,92] for solutions of milk proteins at various concentrations and surface expansion rates. Oil droplet spreading was observed for values of the dynamic steady-state air–water surface pressure Π_{aw} ($=72\,\mathrm{mN\,m^{-1}} - \gamma_{aw}$) of around $15\,\mathrm{mN\,m^{-1}}$ or lower, but no spreading was found for $\Pi_{aw} > 15\,\mathrm{mN\,m^{-1}}$. Hence the value of $15\,\mathrm{mN\,m^{-1}}$ represents the critical surface pressure Π_{cr} above which spreading is inhibited. This relatively high value of Π_{cr} for triglyceride oil

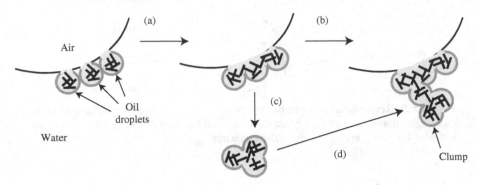

Figure 8.10 Representation of the process of surface-mediated partial coalescence. For $S > 0$, the fat globules attach to the protein-covered bubble surface to form interfacial flocs, which are subjected to partial coalescence (a). If a bubble bursts, a partially coalesced fat clump remains in the bulk (c). If a bubble does not burst (b), a fat clump from the bulk may partially coalesce with the adsorbed fat clump (d). Taken from Ref. [93]; with permission of the Royal Society of Chemistry.

spreading out of protein-stabilized oil droplets indicates that the air–water surface does not have to be completely void of adsorbed protein for spreading to occur.[93] This is relevant to the whipping of dairy emulsions, since the air bubbles incorporated during the early stages of whipping are stabilized by adsorbed protein,[94] which could potentially inhibit oil droplet spreading. Also relevant to emulsion whipping is the observation[93] that the presence of a competitively adsorbing small-molecule surfactant (Tween 20) can increase the critical surface pressure to $\Pi_{cr} \approx 22\,\mathrm{mN\,m^{-1}}$, and so can enhance the rate of spreading of liquid fat under less severe surface expansion conditions.

During the whipping of natural cream (fat content ~35 wt.%), the partially crystalline fat globules adhere to bubble surfaces, and they subsequently become clumped together by a process of surface-mediated partial coalescence,[93] as illustrated schematically in Figure 8.10. This process continues until a three-dimensional network of clumped fat globules is built up, which holds and stabilizes the incorporated bubbles, and gives the desired texture and mechanical strength (yield stress) to the final whipped cream (with a typical gas-to-liquid volume ratio of ~120%). Based on the results of the droplet spreading experiments described above, it seems reasonable to assume that fat droplet spreading and surface-mediated partial coalescence can only occur at bubble surfaces for which the dynamic surface pressure is below Π_{cr}. It has been suggested[93] that interfaces around (different) bubbles can be described by a surface pressure distribution, with newly formed bubble surfaces having low surface pressures, and bubble surfaces that are compressed, or at which protein is adsorbed, having higher surface pressures. The technological implication is that, by modifying the formulation and/or the whipping conditions, the fraction of the total surface area for which the condition $S > 0$ is satisfied can be manipulated, thereby influencing the rate of droplet adhesion and spreading, and hence optimizing the whipping process.

8.7 Some Concluding Remarks

It is clear that particles with sizes from the nanometre scale to the micrometre scale (and beyond) do play an important role in controlling the structure, rheology and stability of food colloids. The application of the underlying basic principles[95-98] to these systems is, however, still somewhat limited. This is due to a number of inter-related factors: the compositional complexity of food colloids, the effect of processing conditions (hydrodynamic, thermal, *etc.*) on the ingredient properties and on the system microstructure, and the lack of quantitative experimental information on the physico-chemical properties of particles (contact angles, size distributions, aggregation state, *etc.*) and their influence on the properties of fluid interfaces (adsorbed layer structure, surface rheology, *etc.*).

In order to make further significant progress, new techniques of contact angle measurement will have to be developed and exploited. Direct microscopic quantification of wettability is relatively straightforward for three-phase oil–water–fat systems,[38,55] since macroscopic solid fat surfaces can be reliably prepared with similar properties to those of small stabilizing particles. But, in the case of non-crystalline particles, which have to be compressed into a powder tablet or deposited on a surface, the directly measured contact angle is typically dependent on the particle packing density. And similar uncertainties exist with average contact angles determined indirectly from liquid penetration experiments using particle powder beds. For some protein-based particles, it may be feasible to determine contact angles at liquid interfaces by measuring "pull-off" forces of individual particles attached to the cantilever of an atomic force microscope.[99] Another promising novel technique[100] involves immobilizing the adsorbed particles at a macroscopic liquid interface by gelling the aqueous phase with a non-surface-active hydrocolloid (*e.g.* gellan gum), then fixing the liquid interface using a curable silicone elastomer, and imaging the relative penetration into the two bulk phases using scanning electron microscopy. The availability of more reliable data on contact angles for hydrophilic particles such as casein micelles and protein-coated droplets at various fluid interfaces would be an aid to the development of improved bio-nanoparticles for emulsion and foam stabilization.

There is also a need to design and exploit new techniques to determine the relationship between the state of aggregation and packing of particles at fluid interfaces and the properties of the corresponding emulsions and foams. One such promising technique is light backscattering,[101] which has already been used to study "fat structure formation" in emulsions containing food hydrocolloids.[83,102] Various types of light microscopy are already available for investigating particle structuring and dynamics on flat interfaces[103,104] and droplet surfaces.[105] As particle structuring at fluid interfaces influences their mechanical and viscoelastic properties, there is a requirement for information on the surface rheology of adsorbed particle monolayers in systems containing surface-active ingredients that can affect the inter-particle interactions as well as the contact angles (see Table 8.1). Steady-state surface shear viscometry, for instance, has been successfully used[106–108] to provide information on interactions of fat crystals at the oil–water interface in the presence of food proteins. When globular protein layers are subjected to large-scale deformation, non-linear fracture behaviour occurs,[109,110] with important implications for the bulk stability properties. The spreading behaviour of oil droplets at the air–water surface[111,112] is dependent on the large-deformation surface rheology of protein films, as illustrated schematically in Figure 8.11.[113] For a disordered protein (*e.g.* β-casein) forming a liquid-like adsorbed layer, the emulsion droplets can enter the interface at many different places, and it has been observed[113] that the oil spreads in a radial pattern. In contrast, for a globular protein (*e.g.* β-lactoglobulin or soy glycinin) forming a solid-like layer that fractures due to stress build-up during expansion, the

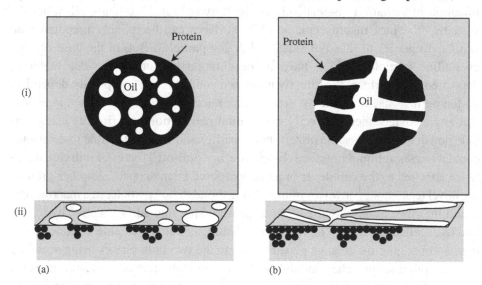

Figure 8.11 Influence on emulsion droplet spreading of the large-deformation mechanical behaviour of a protein film as viewed (i) from above the air–water surface and (ii) from the side. (a) For a *liquid-like* protein adsorbed layer, the droplets enter the interface at many points, and the oil (white) spreads *radially*. (b) For a *solid-like* protein adsorbed layer, the droplets enter the interface at cracks in the fractured film, and the oil spreads *erratically*. Taken from Ref. [113]; with permission of Elsevier.

droplet entering and spreading behaviour is initiated in the growing cracks of the fragmented protein film. This contrasting behaviour of the large-deformation interfacial rheology for different kinds of food proteins has been reproduced[66] in recent Brownian dynamics simulations of the expansion of gel-like adsorbed proteins layers. Further progress in the application of computer simulation in this field is anticipated. For instance, the same kind of Brownian dynamics model has also been successfully used[114] to simulate the evolving microstructure of a nanoparticle monolayer at a fluid interface subjected to uniaxial compression.

The author's intention in this chapter has been to convince the reader of the important role of solid particles at fluid interfaces in controlling the texture and shelf-life of traditional foods. In focusing on the colloid science aspects, of course, some other significant food science issues have had to be condensed or simplified, *e.g.* the effects of fat composition, temperature and emulsifiers on fat crystal formation and morphology in emulsions and foams.[4,43–48,115,116] Nevertheless, these related topics are already well covered in some of the references cited herein. The aim here has been to review those modern developments which seem to pertain most directly to some of the major scientific issues currently faced by the food industry. Looking ahead, there is good reason to believe that the intense current research interest in colloidal particles at liquid interfaces will generate exciting developments in the areas of improved food shelf-life, nutrient encapsulation technology and novel food formulations.

References

1. E. Dickinson and G. Stainsby, *Colloids in Food*, Applied Science, London, 1982.
2. E. Dickinson, *An Introduction to Food Colloids*, Oxford University Press, Oxford, 1992.
3. D.J. McClements, *Food Emulsions: Principles, Practice and Techniques*, CRC Press, Boca Raton, 1999.
4. P. Walstra, *Physical Chemistry of Foods*, Marcel Dekker, New York, 2003.
5. T. Young, *Phil. Trans.*, **95** (1805), 84.
6. L.S. Dorobantu, A.K.C. Yeung, J.M. Foght and M.R. Gray, *Appl. Environ. Microbiol.*, **70** (2004), 6333.
7. G. Charalambous and G. Doxastakis (eds.), *Food Emulsifiers*, Elsevier, Amsterdam, 1989.
8. G.L. Hasenhuettl and R.W. Hartel (eds.), *Food Emulsifiers and their Applications*, Chapman and Hall, New York, 1997.
9. E. Dickinson, *Food Hydrocoll.*, **17** (2003), 25.
10. E. Dickinson, in *Gums and Stabilisers for the Food Industry – 12*, eds. P.A. Williams and G.O. Phillips, Royal Society of Chemistry, Cambridge, 2004, p. 394.
11. E. Dickinson, in *Controlled Particle, Droplet and Bubble Formation*, ed. D.J. Wedlock, Butterworth–Heinemann, Oxford, 1994, p. 191.
12. E. Dickinson, *J. Dairy Sci.*, **80** (1997), 607.
13. E. Dickinson, *J. Chem. Soc. Faraday Trans.*, **94** (1998), 1657.

14. E. Dickinson, *Int. Dairy J.*, **9** (1999), 305.
15. R.C. Randall, G.O. Phillips and P.A. Williams, *Food Hydrocoll.*, **2** (1988), 131.
16. E. Dickinson, B.S. Murray, G. Stainsby and D.J. Anderson, *Food Hydrocoll.*, **2** (1988), 477.
17. A.K. Ray, P.B. Bird, G.A. Iacobucci and B.C. Clark Jr., *Food Hydrocoll.*, **9** (1995), 123.
18. M. Akhtar, E. Dickinson, J. Mazoyer and V. Langendorff, *Food Hydrocoll.*, **16** (2002), 249.
19. E. Dickinson and V.B. Galazka, *Food Hydrocoll.*, **5** (1991), 281.
20. R. Ettelaie, E. Dickinson and B.S. Murray, in *Food Colloids: Interactions, Microstructure and Processing*, ed. E. Dickinson, Royal Society of Chemistry, Cambridge, 2005, p. 74.
21. G.J. Fleer, M.A. Cohen Stuart, J.M.H.M. Scheutjens, T. Cosgrove and B. Vincent, *Polymers at Interfaces*, Chapman and Hall, London, 1993.
22. C.I. Onwulata, R.P. Konstance and P.M. Tomasula, *J. Dairy Sci.*, **87** (2004), 749.
23. E. Dickinson, in *Advanced Dairy Chemistry–1. Proteins*, 3rd edn., eds. P.F. Fox and P.L.H. McSweeney, Kluwer Academic/Plenum, New York, 2003, Part B, p. 1229.
24. H. Singh, in *Food Colloids: Interactions, Microstructure and Processing*, ed. E. Dickinson, Royal Society of Chemistry, Cambridge, 2005, p. 179.
25. P. Walstra and H. Oortwijn, *Neth. Milk Dairy J.*, **36** (1982), 103.
26. D.G. Dalgleish and E.W. Robson, *J. Dairy Res.*, **52** (1985), 539.
27. J.-L. Courthaudon, J.-M. Girardet, S. Campagne, L.-M. Rouhier, S. Campagna, G. Linden and D. Lorient, *Int. Dairy J.*, **9** (1999), 411.
28. P.K. Chang, W.D. Powrie and O. Fennema, *J. Food Sci.*, **42** (1977), 1193.
29. Y. Mine, *Food Hydrocoll.*, **12** (1998), 409.
30. M. Anton, V. Beaumal and G. Gandemer, *Food Hydrocoll.*, **14** (2000), 327.
31. B.S. Murray and R. Ettelaie, *Curr. Opin. Colloid Interf. Sci.*, **9** (2004), 314.
32. M. Akhtar and E. Dickinson, in *Food Colloids: Fundamentals of Formulation*, eds. E. Dickinson and R. Miller, Royal Society of Chemistry, Cambridge, 2001, p. 133.
33. J. Chevalley, in *Industrial Chocolate Manufacture and Use*, ed. S.T. Beckett, Blackwell, Oxford, 1999, p. 182.
34. D. Rousseau, *Food Res. Int.*, **33** (2000), 3.
35. F. N. Foulkes, *Ind. Eng. Chem.*, **56** (1964), 40.
36. J.H. Hildebrand and R.L. Scott, *Solubility of Non-Electrolytes*, 3rd edn., Reinhold, New York, 1950.
37. Y. Tamai, K. Makuuchi and M. Suzuki, *J. Phys. Chem.*, **71** (1967), 4176.
38. I.J. Campbell, in *Food Colloids*, eds. R.D. Bee, P. Richmond and J. Mingins, Royal Society of Chemistry, Cambridge, 1989, p. 272.
39. F. van Voorst Vader, *Chem. Eng. Technol.*, **49** (1977), 488.
40. E. H. Lucassen-Reynders, in *Scientific Basis of Flotation*, ed. K.J. Ives, NATO ASI Series E, Applied Sciences 75, Martinus Nijhoff, The Hague, 1984.
41. D. Johansson and B. Bergenståhl, *J. Am. Oil Chem. Soc.*, **72** (1995), 933.
42. D. Bargeman, *J. Colloid Interf. Sci.*, **40** (1972), 344.
43. D.P.J. Moran, in *Fats in Food Products*, eds. D.P.J. Moran and K.K. Rajah, Blackie, Glasgow, 1994, p. 155.
44. P. Walstra, in *Food Structure and Behaviour*, eds. J.M.V. Blanshard and P.J. Lillford, Academic Press, London, 1987, p. 67.
45. A.C. Juriaanse and I. Heertje, *Food Microstruct.*, **7** (1988), 181.
46. N. Garti and K. Sato, in *Crystallization and Polymorphism of Fats and Fatty Acids*, eds. N. Garti and K. Sato, Marcel Dekker, New York, 1988, p. 267.
47. O.J. Guth, J. Aronhime and N. Garti, *J. Am. Oil Chem. Soc.*, **66** (1989), 1606.
48. N. Krog and K. Larsson, *Fat Sci. Technol.*, **94** (1992), 55.

49. D. Johansson and B. Bergenståhl, *J. Am. Oil Chem. Soc.*, **69** (1992), 705, 718, 728.
50. S.U. Pickering, *J. Chem. Soc.*, **91** (1907), 2001.
51. D.F. Darling, *J. Dairy Res.*, **49** (1982), 695.
52. D. Johansson, B. Bergenståhl and E. Lundgren, *J. Am. Oil Chem. Soc.*, **72** (1995), 939.
53. J.M. Deman and A.M. Beers, *J. Texture Stud.*, **18** (1987), 303.
54. R. Vreeker, L.L. Hoekstra, D.C. ven Boer and W.G.M. Agterof, *Colloids Surf.*, **65** (1992), 185.
55. D. Johansson, B. Bergenståhl and E. Lundgren, *J. Am. Oil Chem. Soc.*, **72** (1995), 921.
56. A.S. Kabalnov and E.D. Schukin, *Adv. Colloid Interf. Sci.*, **38** (1992), 69.
57. E.A. Disalvo, *Adv. Colloid Interf. Sci.*, **29** (1988), 141.
58. A. Prins, in *Advances in Food Emulsions and Foams*, eds. E. Dickinson and G. Stainsby, Elsevier Applied Science, London, 1988, p. 91.
59. W. Kloek, T. van Vliet and M. Meinders, *J. Colloid Interf. Sci.*, **237** (2001), 158.
60. B.S. Murray and E. Dickinson, *Food Sci. Technol. Int. (Japan)*, **2** (1996), 131.
61. B.S. Murray, *Curr. Opin. Colloid Interf. Sci.*, **7** (2002), 426.
62. E. Dickinson, R. Ettelaie, B.S. Murray and Z. Du, *J. Colloid Interf. Sci.*, **252** (2002), 202.
63. R. Ettelaie, E. Dickinson, Z. Du and B.S. Murray, *J. Colloid Interf. Sci.*, **263** (2003), 47.
64. B.S. Murray, E. Dickinson, Z. Du, R. Ettelaie, T. Kostakis and J. Vallet, in *Food Colloids: Interactions, Microstructure and Processing*, ed. E. Dickinson, Royal Society of Chemistry, Cambridge, 2005, p. 259.
65. F. MacRitchie, *Adv. Colloid Interf. Sci.*, **25** (1986), 341.
66. L.A. Pugnaloni, R. Ettelaie and E. Dickinson, in *Food Colloids: Interactions, Microstructure and Processing*, ed. E. Dickinson, Royal Society of Chemistry, Cambridge, 2005, p. 131.
67. E. Dickinson, in *Food Macromolecules and Colloids*, eds. E. Dickinson and D. Lorient, Royal Society of Chemistry, Cambridge, 1995, p. 1.
68. B.P. Binks and S.O. Lumsdon, *Langmuir*, **16** (2000), 8622.
69. B.P. Binks and J.A. Rodrigues, *Langmuir*, **19** (2003), 4905.
70. B.P. Binks and C.P. Whitby, *Langmuir*, **20** (2004), 1130.
71. Z. Du, M.P. Bilbao-Montoya, B.P. Binks, E. Dickinson, R. Ettelaie and B.S. Murray, *Langmuir*, **19** (2003), 3106.
72. E. Dickinson, R. Ettelaie, T. Kostakis and B.S. Murray, *Langmuir*, **20** (2004), 8517.
73. R. Aveyard, B.P. Binks, P.D.I. Fletcher and C.E. Rutherford, *J. Disper. Sci. Technol.*, **15** (1994), 251.
74. P.R. Garrett, in *Food Colloids: Fundamentals of Formulation*, eds. E. Dickinson and R. Miller, Royal Society of Chemistry, Cambridge, 2001, p. 55.
75. C.K. Lau and E. Dickinson, *Food Hydrocoll.*, **19** (2005), 111.
76. W. Buchheim, N.M. Barfod and N. Krog, *Food Microstruct.*, **4** (1985), 221.
77. B.E. Brooker, M. Anderson and A.T. Andrews, *Food Microstruct.*, **5** (1985), 277.
78. H.D. Goff, *J. Dairy Sci.*, **80** (1997), 2620.
79. W. Xu, A. Nikolov and D.T. Wasan, *J. Food Eng.*, **66** (2005), 97.
80. K. Boode, C. Bisperink and P. Walstra, *Colloids Surf.*, **61** (1991), 55.
81. B.M.C. Pelan, K.M. Watts, I.J. Campbell and A. Lips, *J. Dairy Sci.*, **80** (1997), 2631.
82. E. Davies, E. Dickinson and R.D. Bee, *Int. Dairy J.*, **11** (2001), 827.
83. W. Xu, A. Nikolov and D.T. Wasan, *J. Food Eng.*, **66** (2005), 107.
84. J.V. Robinson and W.W. Woods, *J. Soc. Chem. Ind. (London)*, **67** (1948), 361.
85. W.D. Harkins and A. Feldman, *J. Am. Chem. Soc.*, **44** (1922), 2665.

86. N.E. Hotrum, T. van Vliet, M.A. Cohen Stuart and G.A. van Aken, *J. Colloid Interf. Sci.*, **247** (2002), 125.
87. R. Aveyard and J.H. Clint, *J. Chem. Soc. Faraday Trans.*, **93** (1997), 1397.
88. U. Retter and D. Vollhardt, *Langmuir*, **9** (1993), 2478.
89. E. Dickinson, B.S. Murray and G. Stainsby, *J. Chem. Soc. Faraday Trans. 1*, **84** (1988), 871.
90. K. Koczo, L.A. Lobo and D.T. Wasan, *J. Colloid Interf. Sci.*, **150** (1992), 492.
91. L.A. Lobo and D.T. Wasan, *Langmuir*, **9** (1993), 1668.
92. N.E. Hotrum, M.A. Cohen Stuart, T. van Vliet and G.A. van Aken, in *Food Colloids, Biopolymers and Materials*, eds. E. Dickinson and T. van Vliet, Royal Society of Chemistry, Cambridge, 2003, p. 192.
93. N.E. Hotrum, M.A. Cohen Stuart, T. van Vliet and G.A. van Aken, in *Food Colloids: Interactions, Microstructure and Processing*, ed. E. Dickinson, Royal Society of Chemistry, Cambridge, 2005, p. 317.
94. M. Anderson, B.E. Brooker and E.C. Needs, in *Food Emulsions and Foams*, ed. E. Dickinson, Royal Society of Chemistry, London, 1987, p. 100.
95. B.P. Binks, *Curr. Opin. Colloid Interf. Sci.*, **7** (2002), 21.
96. S.K. Bindal, G. Sethumadhavan, A.D. Nikolov and D.T. Wasan, *AIChEJ.*, **48** (2002), 2307.
97. G. Kaptay, *Colloids Surf. A*, **230** (2004), 67.
98. N.D. Denkov, *Langmuir*, **20** (2004), 9463.
99. S. Ecke, M. Preuss and H.J. Butt, *J. Adhes. Sci. Technol.*, **13** (1999), 1181.
100. V.N. Paunov, *Langmuir*, **19** (2003), 7970.
101. W. Xu, A. Nikolov, D.T. Wasan, A. Gonsalves and R. Borwankar, *J. Food Sci.*, **63** (1998), 183.
102. K. Koczo, D.T. Wasan, R.P. Borwankar and A. Gonsalves, *Food Hydrocoll.*, **12** (1997), 43.
103. S. Tarimala, S.R. Ranabothu, J.P. Vernetti and L.L. Dai, *Langmuir*, **20** (2004), 5171.
104. J.C. Fernandez-Toledano, A. Moncho-Jorda, F. Martinez-Lopez and R. Hidalgo-Alvarez, *Langmuir*, **20** (2004), 6977.
105. E. Vignati, R. Piazza and T.P. Lockhart, *Langmuir*, **19** (2003), 6650.
106. L.G. Ogden and A.J. Rosenthal, *J. Colloid Interf. Sci.*, **168** (1994), 539.
107. L.G. Ogden and A.J. Rosenthal, *J. Colloid Interf. Sci.*, **191** (1997), 38.
108. L.G. Ogden and A.J. Rosenthal, *J. Am. Oil Chem. Soc.*, **75** (1998), 1841.
109. A. Martin, M. Bos, M.A. Cohen Stuart and T. van Vliet, *Langmuir*, **18** (2002), 1238.
110. B.S. Murray, B. Cattin, E. Schuler and Z.O. Sonmez, *Langmuir*, **18** (2002), 9476.
111. N.E. Hotrum, M.A. Cohen Stuart, T. van Vliet and G.A. van Aken, *Langmuir*, **19** (2003), 10210.
112. M.A. Cohen Stuart, W. Norde, M. Kleijn and G.A. van Aken, in *Food Colloids: Interactions, Microstructure and Processing*, ed. E. Dickinson, Royal Society of Chemistry, Cambridge, 2005, p. 99.
113. G.A. van Aken, T.B.J. Blijdenstein and N.E. Hotrum, *Curr. Opin. Colloid Interf. Sci.*, **8** (2003), 371.
114. L.A. Pugnaloni, R. Ettelaie and E. Dickinson, *Langmuir*, **20** (2004), 6096.
115. E. Dickinson and D.J. McClements, *Advances in Food Colloids*, Blackie, Glasgow, 1995, p. 210.
116. S.M. Hodge and D. Rousseau, *J. Am. Oil Chem. Soc.*, **82** (2005), 159.

Eric Dickinson is Professor of Food Colloids in the Procter Department of Food Science at the University of Leeds. He is a physical chemist by training, with B.Sc. and Ph.D. degrees from the University of Sheffield, and postdoctoral experience at the Universities of Leeds, Oxford and California (Los Angeles). His research interests are in the stability, structure and rheology of food emulsions, foams and gels, the functional and interfacial properties of milk proteins, the interactions of proteins with surfactants and polysaccharides, and the statistical mechanics and computer simulation of adsorption, aggregation and gelation. He has authored the books "*Colloids in Food*" (1982, with Stainsby), "*An Introduction to Food Colloids*" (1992) and "*Advances in Food Colloids*" (1995, with McClements), and has edited several other books on the theme of food colloids. He is an Associate Editor of the journal *Food Hydrocolloids*.

Tel: +44-0113-3432956; *Fax*: +44-0113-3432982; *E-mail*: E.Dickinson@leeds.ac.uk

9

Collection and Attachment of Particles by Air Bubbles in Froth Flotation

Anh V. Nguyen,[a] Robert J. Pugh[b] and Graeme J. Jameson[a]

[a]Discipline of Chemical Engineering and Centre for Multiphase Processes, School of Engineering, The University of Newcastle, Callaghan, New South Wales 2308, Australia
[b]Institute for Surface Chemistry, Stockholm SE11486, Sweden

9.1 Introduction

Froth flotation has a long history (over 100 years) of development and widespread applications. Essentially, the process involves the attachment of finely dispersed hydrophobic particles to air bubbles to produce so-called "three-phase froths" on the surface of the flotation cell with the hydrophilic particles remaining dispersed in the suspension. In this way, the particles are separated, based on their differences in surface wettability. Although froth flotation includes several major elementary sub-processes, one of the most important operations involves the interaction of the selected suspended particles with a chemical reagent (a flotation collector) in order to make the surfaces sufficiently hydrophobic and become "targets" for bubbles generated in the cell. The "gangue" particles remain hydrophilic and do not interact with the collector reagent but remain dispersed in the suspension.

The following unit processes are also important:

(i) Generation and dispersion of gas bubbles in the presence of a surfactant (frother) in the pulp and the formation of the froth layer.
(ii) Collision of hydrophobic particles with gas bubbles.
(iii) Adhesion of hydrophobic particles to gas bubbles and the formation of particle–bubble aggregates.
(iv) Ascension of particle–bubble aggregates from the pulp into the three-phase froth.

Both the fundamental and practical aspects of froth flotation have been well studied and developed but the process still undergoes modification and advancement. In fact, froth flotation has been applied in many diverse industrial areas in addition to mineral processing. In recent years it has been adopted for the treatment and utilization of industrial wastewater, the recycling of plastics and bacteria separation

328

in bioengineering. Flotation has also been used to remove ink from used fibre (de-inking) enabling the fibre to be recycled and re-used in the manufacture of new paper. It has also been applied to the recovery of metallic silver from photographic residues and to separate heavy crude oil from tar sands. Flotation processes were reported to be used for the separation of biological materials, *e.g.* enzymes, albumin, penicillin, viruses and organic pollutants, *e.g.* phenols and chlorophenols.[1]

9.2 Mineral Flotation

This is by far the most important industrial application of surface chemistry. In recent years, one aspect which has received considerable interest involves both the role of the particle and the chemical frother on the stability of the bubbles. Overall, the dynamic behaviour of particles and the interaction with frothers are critical but poorly understood steps in industrial froth flotation and usually lead to loss of recovery across the froth phase. The frothers are usually nonionic molecules, which adsorb at the air–water surface and aid in the production of bubbles and stabilize the flotation froths. Frothers can be essentially divided into four chemical groups. The first group consists of aromatic alcohols such as cresol and 2,3-xylenol. A second group is the alkoxy types such as triethoxy butane (TEB). The third group consists of aliphatic alcohols such as 2-ethyl hexanol, diacetone and methyl isobutyl carbinol (MIBC). In recent years, a fourth important group of synthetic frothers consisting of polyethylene oxide (PEO), polypropylene oxide (PPO) and polybutylene oxide (PBO) groups has been introduced into the market.[2]

In addition to the chemical frother, mineral particles also play an important role in controlling the stability of the froth. Although it has been clearly established that foam stability can be increased or decreased by many different types of particles (see Chapter 10), to some extent the mechanism is complex, since frequently there are several different mechanisms operating in the same system. Usually, the particles have some critical degree of hydrophobicity, which plays a crucial role in the dynamics of the rupturing of thin foam (air–water–air) films. In fact, the use of particles as foam breakers is well known throughout industry, and hydrophobic particles are important ingredients in many foam-breaking formulations.[3] Both the particle size and shape have also been shown to play an important role in the flotation of minerals and systems have been studied with particles possessing a wide size range.[4,5]

9.3 Influence of Particles on the Structure and Stability of Froths

While foams (two-phase gas–liquid systems) are stabilized by adsorbed frother (surfactant) molecules (Figure 9.1(a)), froths (three-phase gas–liquid–solid systems) are usually stabilized by small particles with a critical degree of wetting, attached at the

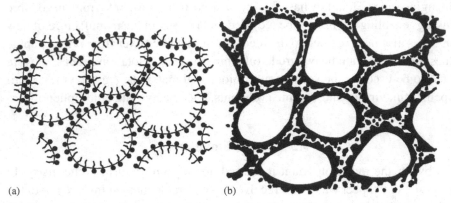

(a)   (b)

Figure 9.1 Schematic of the different types of stabilizer for air bubbles in water; (a) frother molecules and (b) solid particles.

(a)   (b)

Figure 9.2 (a) Photograph of froth heavily loaded with galena (PbS) particles, (b) froth stabilized by frother molecules during the flotation of fluorspar. Bubbles are in the mm–cm size range in both cases. Courtesy of Clariant, Germany.

gas–liquid surface and cause the bubbles to become "armoured" (Figure 9.1(b)). In the real world situation of mineral processing the bubbles are frequently stabilized by both particles and surfactants. In Figure 9.2(a) and (b) two extreme situations are shown corresponding to a heavily loaded particle-stabilized froth and a surfactant-stabilized foam.

In early studies, several ideas were developed based on the premise that coalescence of froths is prevented due to a steric interaction of particles attached to the interface.[6] Strongly adhering particles to the bubble generally produce more stable froths, and an increase in the contact angle that particles exhibit at the air–water surface to a certain critical value benefits froth stability. Also, as reported by many

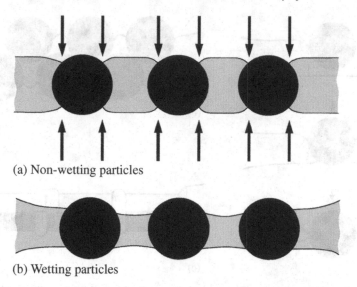

(a) Non-wetting particles

(b) Wetting particles

Figure 9.3 Influence of particle wettability on aqueous film stability as the film thickness approaches that of the particle diameter. (a) Non-wetting particles cause recession of the liquid around them and rupture of the film at the three-phase contact line as indicated by arrows. (b) Wetting particles retain liquid within the film delaying rupture. Taken from Ref. [7]; with permission of Elsevier.

flotation researchers, the stability and drainage of a three-phase froth (through the Plateau borders) depends on the hydrophobicity of the mineral particles present in the froth. Froths become stabilized and drainage of the liquid from a thin layer is restricted by hydrophobic solid particles. As the liquid film drains to a critical thickness, the non-wetted particles can reduce froth stability by inducing the liquid to dewet around the particle causing the liquid to recede from the particle at areas indicated by the arrows in Figure 9.3(a). This leads to rapid rupture. However, in the case of partial wetting, the particles trap the liquid making the film more stable, Figure 9.3(b). However, the influence of particle concentration, density and shape needs to be taken into consideration. As shown in Figure 9.4, the effect of low and high concentrations of non-wetting particles and the effect of plate-like particles, *e.g.* clay, on the thin film stability are suggested.

9.3.1 Capillary effects on froth stability

At least two alternative mechanisms have been suggested to explain the froth stabilization effects caused by hydrophobic particles adsorbed at the interface. The first effect results from a change in capillary pressure.[7] This is caused by the presence of adsorbed particles modifying the curvature of the gas–liquid surface,

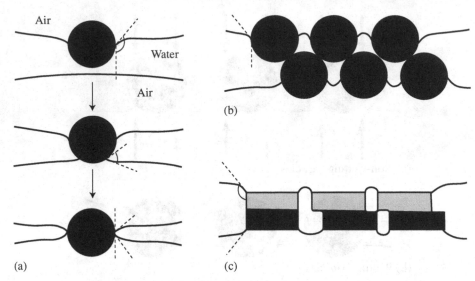

Figure 9.4 Effect of hydrophobic particles on bubble stability. (a) Poor mineralization of bubbles with very hydrophobic particles, (b) high mineralization of bubbles with bridging particle layers, (c) effect of plate-like particles.

which reduces the pressure difference between the Plateau borders and the three films associated with it. The situation is illustrated in Figure 9.5. In the case of the foam with no particles (Figure 9.5(a)), the liquid can flow from the film into the Plateau border, and then through the structure by gravity. The flow rate is proportional to the pressure difference ΔP expressed by

$$\Delta P = P_{\text{Film}} - P_{\text{PB}} = \gamma_{\text{aw}}/R_{\text{PB}} \tag{9.1}$$

where P_{Film} is the pressure in the films, P_{PB} is the pressure in the Plateau border, γ_{aw} is the surface tension of the liquid and R_{PB} is the radius of curvature of the gas–liquid surface (see Figures 9.5(b) and (c)). Therefore, when ΔP is high, the flow rate is increased which causes faster drainage, and the foam becomes less stable. If many hydrophobic particles are attached to the gas–liquid interface (see Figure 9.5(d)), the radius of curvature of the gas–liquid interface would be almost equal to that of the gas–Plateau border interface (see Figures 9.5(e) and (f)). This will cause the pressure difference to decrease leading to a more stable froth.

It is interesting to note that according to Lucassen,[8] capillary effects are especially important for small particles attached to interfaces where gravitational forces are negligible. This results from the fact that small floating solid particles at the fluid interface can interact with neighbouring particles because any solid particle will nearly always cause deformation of the interface. The extent of deformation becomes

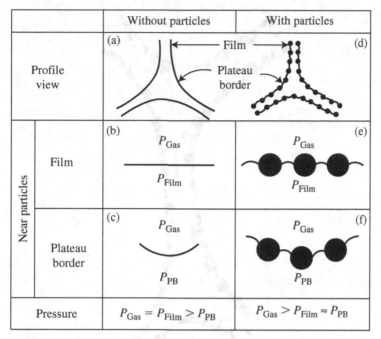

Figure 9.5 The effect of hydrophobic particles on the pressure difference between the foam film and the Plateau border. Taken from Ref. [7]; with permission of Elsevier.

altered by the approach of the neighbouring particle. This is especially important in cases where particles have an irregular wetting perimeter which disturbs the smoothness of the interface. Calculations made for a model particle with a sinusoidal edge indicate that the disturbances can become significant in magnitude.

9.3.2 Restricted drainage mechanism

It has also been suggested that particles attached to the thin film interface can hinder the overall drainage within the thin film causing the liquid passages to become constricted and tortuous. To some extent, the volume of the particle-stabilized froth must therefore be approximately proportional to the amount of hydrophobic solids present, but there is an upper limit and this will be discussed later in this chapter. In addition, particle size is important and if the size of the hydrophobic particle is small compared with the film thickness as discussed previously, then particles can arrange themselves at the gas–liquid interface and stabilize the films by the capillary mechanism described. If the particles are large, *i.e.* their diameter is larger than the film thickness, the particles can bridge and may rupture the film.

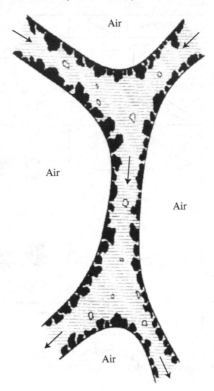

Figure 9.6 Schematic illustration of drainage in a particle-containing froth. Taken from Ref. [9]; with permission of Wiley.

9.3.3 Entrainment of particles in the froth phase and flotation

The process leading to the formation of three-phase froth layers involves not only the hydrophobic particles carried over in the froth but also the hydrophilic entrained particles in the gangue suspension.[9] These latter particles can only feebly adhere to the gas bubbles or are situated in the water film in the froth and are usually washed out during drainage, Figure 9.6. At the same time, the ascending air bubbles carry the hydrophobic particles to the top of the froth. Thus, the top of three-phase froth contains the more hydrophobic particles and has the higher grade; the grades of the floated material decrease from top to bottom of the froth. This phenomenon is referred to as a "secondary concentration effect," and is useful for upgrading the concentrate quality and has found its application in column flotation.

In most cases, complete separation between gangue and valuable mineral particles in the pulp zone is difficult to achieve, and nearly always some gangue minerals are transported into the froth with the entrained liquid. As the froth ages, some of the hydrophilic gangue returns back to the pulp due to drainage of slurry, while the remainder is carried up with the concentrate reducing the quality of the product.

It has been shown that the drainage of hydrophilic particles (or recovery of gangue in the concentrate) is largely affected by properties of the gangue minerals such as density and particle size. To date, the effect of the gangue characteristics on the drainage has been well established, but the influence of the froth characteristics on the gangue recovery has not been fully investigated.

9.3.4 Frothing studies with model quartz particles of well defined size and hydrophobicity

At the Institute for Surface Chemistry in Stockholm, we have studied the influence of size and hydrophobicity of mineral particles on the stability of froths using both the modified Bikermann column and the thin film balance[4] developed by Bulgarian researchers. In these experiments, surface modified quartz particles were used as models. The hydrophobicity of the quartz surface was controlled by reacting the dry surface with trimethylchlorosilane in cyclohexane under a dry environment following standard procedures. In order to evaluate the surface wetting, a quartz plate was also placed in the reaction vessel together with the particles. After the reaction, the plate and particles were removed and rinsed with cyclohexane and washed with water. Finally, the contact angle of a drop of water on the plate in air was determined using a Ramé-Hart goniometer. The hydrophobicity of the quartz particles was quantified from small-scale flotation experiments (which determine the per cent flotation yield) using a Hallimond tube apparatus.

The froths (containing quartz particles) were characterized by both dynamic and static frothing tests. The dynamic test (carried out during froth generation) quantified the equilibrium state of the froth whereas the static tests (after the gas flow has ceased) determined the rate of collapse of the froth. In this frothing study, a modified Bikermann test was used consisting of a glass column where the maximum equilibrium volume of the froth (H_{max}) was determined at a standard flow rate. A typical set of data is shown in Figure 9.7 for the dynamic frothing with four commercial mineral processing frothers (polypropylene glycol monomethyl ether (PPGMME), α-terpineol and MIBC). These results express the dynamic froth characteristics in terms of H_{max} versus the hydrophobicity of the particles expressed in terms of the flotation yield. From the results obtained with the small particle fraction (26–44 μm), there appears to be a distinct maximum corresponding to a flotation yield of about 70%. This corresponds to a critical degree of hydrophobicity with contact angle of 60° for the small particle fraction. This value seems to be reproducible in all the frother systems. Also, at higher flotation yield, the froth was found to collapse indicating the particles were acting as foam breakers, and it could be concluded that above this critical degree of hydrophobicity the particles appear to have a destabilizing effect on the system. These trends are observed at both high and low frother concentrations.

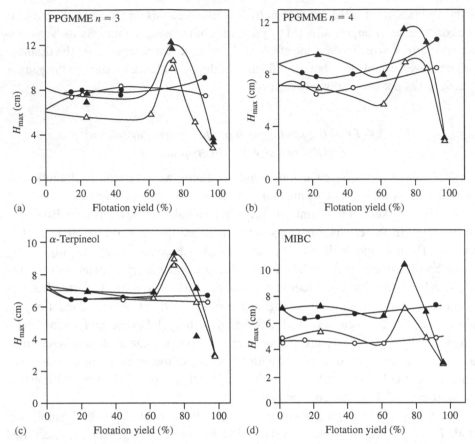

Figure 9.7 Relationship between the dynamic stability of froth, expressed as the maximum froth height at a flow rate of $60 \, l \, hr^{-1}$, and the hydrophobicity of the quartz particles expressed as the flotation yield for different frothers (given). Symbols refer to: $\triangle$ 26–44 μm size fraction and 20 mg l^{-1} froth concentration, $\blacktriangle$ 26–44 μm and 50 mg l^{-1}, $\bigcirc$ 74–106 μm and 20 mg l^{-1}, $\bullet$ 74–106 μm and 50 mg l^{-1}. Taken from Ref. [4]; with permission of Elsevier.

However, with the larger size fraction (74–106 μm), these effects were not observed. In fact, the particles do not appear to influence the stability of the system. Similar trends are observed at both low (20 mg l^{-1}) and high (50 mg l^{-1}) frother concentrations. Further experiments were carried out for a range of frother concentrations where a similar trend was observed for H_{max} values *versus* concentration.

9.4 Frothing and Flotation in the Absence of Collector and Frother

9.4.1 Flotation of hydrophilic colloidal particles

The frothing and flotation of hydrophilic metal hydroxides in the absence of frother and collector could be considered as an area of special interest and is sometimes referred to as contact-less flotation. There are several reviews on micro-flotation of solids which occurs in the presence of hydrolysing ions and in dissolved air flotation circuits. It has been shown that the region of floatability corresponds closely with the regions where precipitation of the metal ion occurs. In fact, maximum coagulation corresponds to maximum flotation. Hydrophilic solids, such as quartz, are usually readily coagulated by iron or aluminium ions, and various degrees of flotation occur with micro-bubbles. However, for flotation to occur, the particles need to be hydrophobic.

One theory suggests that naturally occurring organic compounds may be responsible for particle hydrophobicity, which causes frothing in some industrial dissolved air flotation operations.[10] It has been well documented that natural water frequently contains biologically derived surfactants which could stabilize micro-bubbles. It is interesting to note that in clean water systems, the flotation of a suspension of ferric hydroxide flocs can occur fairly readily in the presence of a few ppm of collector. Also, coagulation with hydroxyl ions increases the effective particle size and decreases the number of particles to be collected. General particle aggregation is needed, but in many cases a collector may also be needed to improve the process efficiency. Experiments of Kitchener and Gochin[11] have confirmed that the floatability of metal hydroxides is very sensitive to the presence of organic impurities in the system. They suggested that natural water contains surface-active impurities which are adsorbed onto precipitated metal hydroxides forming insoluble hydrophobic soaps that provide sites for bubble adhesion. Since the flocs have low density, then micro-bubble attachment to a few hydrophobic spots on the flocs would be sufficient to ensure flotation.

9.4.2 Flotation of naturally hydrophobic particles in aqueous solutions of inorganic electrolytes

The frothing and flotation of naturally hydrophobic particles in inorganic electrolyte solutions was first documented during the 1930's in the USSR. This work was mostly related to the flotation of coals in saline waters. Also, it was noted that the electrolytes play a role in stabilizing the froth. Several surface chemical mechanisms have been proposed to explain the flotation process. These range from the action of the electrolytes in (a) disruption of hydration layers surrounding the particles and enhancing bubble–particle capture, (b) reducing the electrostatic interactions and (c) increasing the charge on the surface of the bubbles to prevent primary bubble coalescence. However, none of these appear to satisfactorily explain the experimental observations.[12]

Craig *et al.*[13] assessed the inhibition of bubbles to coalescence in a series of electrolyte solutions by the application of a combining rule based on the nature of the cationic/anionic ion pair. This rule enables one to predict whether or not the electrolyte would inhibit coalescence of gas bubbles in the electrolyte solutions. Viscosity changes and electrostatic repulsion were ruled out as possible explanations. In fact, following conventional electrostatic double-layer theory, an increase in salt concentration would reduce the double-layer repulsion and should induce coalescence. It was also suggested that the coalescence in pure water was caused by a strong hydrophobic attractive force, which opposed the hydrodynamic repulsion existing between the colliding bubbles.

Paulson and Pugh[14] carried out flotation experiments with graphite (diameter ≈ 20 μm) using a series of different inorganic electrolytes at a range of concentrations (in the absence of an organic frother) in a small glass cylindrical column cell. The flotation of graphite (expressed as per cent recovery) as a function of the electrolyte concentration is shown in Figure 9.8. From this plot it can be seen that recovery generally increases with concentration and varies according to the cationic/anionic ion pair. In fact, it is possible to classify the electrolytes into three groups according to their flotation performance. Group A contains salts with divalent and trivalent cations or anions including $MgCl_2$, $CaCl_2$, Na_2SO_4, $MgSO_4$ and $LaCl_3$, and give high flotation response. In this group, flotation begins at about 0.02 M and reaches maximum recovery between 0.06 and 0.1 M. Group B includes NaCl, LiCl, KCl, CsCl and NH_4Cl which give medium flotation response with flotation

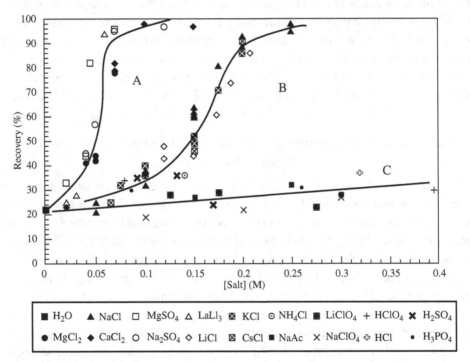

Figure 9.8 The flotation recovery of graphite particles *versus* the electrolyte concentration for a series of different types of electrolytes. Group A – high flotation performance. Group B – intermediate flotation performance. Group C – low flotation performance. Taken from Ref. [14]; with permission of the Americal Chemical Society.

beginning between 0.05 and 0.1 M electrolyte. Finally, Group C electrolytes (NaAc, NaClO$_4$, HClO$_4$, HCl, H$_2$SO$_4$, LiClO$_4$ and H$_3$PO$_4$) give a very low flotation response, even up to concentrations as high as 0.3 M. A plot of double-layer thickness *versus* flotation recovery showed that for Groups A and B electrolytes, a correlation exists between the flotation performance and the electrostatic double-layer thickness, which suggests that the electrostatic interaction plays a role in the process.

In this study, a relationship was also found between flotation recovery and the magnitude of the change in surface tension with respect to electrolyte concentration.[14] Also, a correlation showing a decrease in gas solubility occurred with increasing electrolyte concentration. Thus, the increased flotation performance of the hydrophobic graphite appears to be also linked to the increase in stability of the gas bubbles and froth caused by a decrease in dissolved gas concentration. Higher flotation recovery is attributed to an increase in the bubble–particle collision probability with higher concentration of smaller non-coalescing bubbles and to a reduction in the electrostatic interactions between particles and bubbles resulting in a more stable froth system.

9.5 The Flotation De-inking Process

The de-inking of fibres is an important step in waste paper recycling and the most common techniques used for de-inking of waste papers are washing and flotation. A typical industrial flotation cell is shown in Figure 9.9. The wash de-inking process is most efficient for removing the finer ink particles ($<20\,\mu$m) and, although it uses large volumes of water, the fibre and filler yields can be kept relatively low. Flotation de-inking is usually more effective in removing larger particles (20–300 μm) from newsprints and results in higher fibre and filler yields. Prior to washing and flotation steps, pulping is carried out where the ink is initially detached from the fibres. This process is enhanced by high shear conditions, chemical action and moderate temperatures in the range from 55 to 70°C.

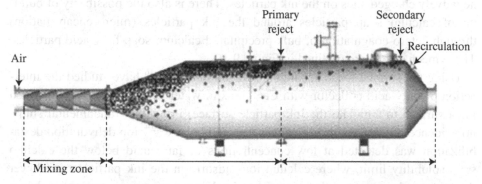

Figure 9.9 The Beloit PDM2 flotation cell for de-inking. Courtesy of Beloit, America.

9.5.1 Ink particles

Flotation mainly removes hydrophobic contaminant and ink particles in the size range 20–300 μm. However, the optimum sizes for flotation de-inking can be dependent on many different parameters such as type of ink and added chemicals.[15] The flotation rate for small particles was found to be rather low and they cannot be easily removed even by prolonged flotation. It was also found from the results of the ink particle distribution before and after flotation that a reduction of coarser particles occurred after flotation. This is important since it is the small ink particles which affect the brightness of the pulp most. It has been found that the main reason for the poor flotation ability of the small particles is the hydrodynamic effect that decreases the particle–bubble attachment efficiency. This will be discussed in Section 9.7.

9.5.2 Mechanism of calcium soap fatty acid interaction in de-inking

In Europe, long chain fatty acids in the presence of calcium ions are "the chemical workhorse" of the hot de-inking process and initially cause agglomeration of the ink particles. This causes the formation of relatively large hydrophobic agglomerates, which can be fairly easily removed by the stream of air bubbles in the flotation cell. The bubbles are carried to the froth and rupture, yielding good ink removal and minimal fibre loss. In general, this process has proved to be highly selective and economical. It is well known that the fatty acid renders the ink particles hydrophobic but the mechanism has been under dispute for several years. One mechanism suggests the Ca^{2+} ions induce a direct bridging mechanism with the anionic fatty acid species. It has also been suggested that the Ca^{2+} ions interact with the negatively charged surface groups on the oil-based ink particles. The adsorbed Ca^{2+} ions would reduce the negative charge and could lead to a carboxylic acid bridging mechanism with the Ca^{2+} ions, specifically adsorbing to the negatively charged sites on the ink particles. There is also the possibility of build-up of calcium soap particles around the ink particles (micro-encapsulation) through hetero-coagulation of bulk precipitated calcium soap fatty acid particles. These mechanisms are outlined in Figure 9.10.

Using the surface force apparatus, Rutland and Pugh[16] have studied the interaction of fatty acid collector with Ca^{2+} ions in water using a negatively charged mica surface to simulate the ink particle surface. From this fundamental study, no evidence of Ca^{2+} ion bridging was detected but a Ca^{2+} ion dehydration destabilization was detected at low concentrations of fatty acid below the calcium soap solubility limit, where calcium ions adsorb on the ink particles and lower the surface charge. At high fatty acid and calcium soap concentrations, calcium soap was precipitated in bulk solution. Overall, it was concluded that the main

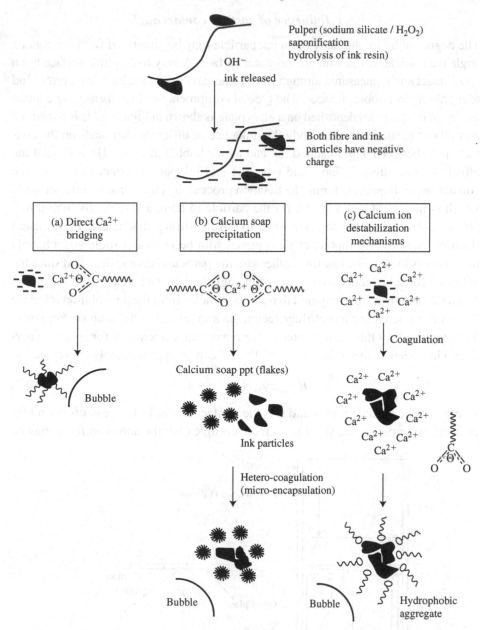

Figure 9.10 Possible mechanisms involved in calcium soap de-inking. Taken from Ref. [16]; with permission of Elsevier.

mechanism involves insoluble calcium soap formation and this becomes associated with the detached printing ink particles. The bubbles capture the aggregates and rise to the surface and the printing ink particles are skimmed off as froth.

9.5.3 Influence of particle contact angle

The degree of hydrophobicity of an ink particle may be quantified from the contact angle the particle makes with the air–water surface. A very hydrophilic surface has a low contact angle (measured through water) whereas contact angles $>90°$ correspond to highly hydrophobic surfaces. The type of equipment used for measuring contact angles of ink particles deposited on a glass plate is shown in Figure 9.11. Johansson[17] carried out measurements to study the influence of different chemicals on the contact angle between agglomerated ink and an air bubble in water. He evaluated the effect of ionic strength, fatty acid type, pH and ink particle concentration on the contact angle. However, during the flotation process, it is the dynamic contact angle which is important and, in order for the particle to float, a sufficiently high three-phase contact angle has to established to ensure strong attachment. For efficient flotation, the time for rupture of the aqueous film between a particle and a bubble must be less than the sliding time, otherwise the particles become detached since the adhesion force is a function of the contact angle and contact area.

The detachment force required to remove particles from the air–solution interface has been studied using a centrifuge technique and related to flotation performance. The magnitude of this critical force F (the maximum detachment for spherical particles) has been analysed in some detail[18] and can be approximately expressed by

$$F = \pi\gamma_{aw}r(1 - \cos\theta) \tag{9.2}$$

where r is the particle radius and θ is the contact angle. In the case of ensembles of irregular particles, the situation is more complex but the adhesion force may be

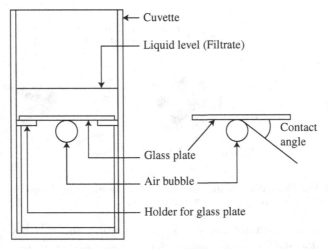

Figure 9.11 Setup for measuring the contact angle of an air bubble under water on an ink film. Taken from Ref. [17]; with permission of the Royal Institute of Technology, Stockholm.

estimated by using an average value of r. While the contact angle is a function of the adhesion force, it also plays an important role in the stability of the froth. For good particle recovery from the froth, the froth stability should not be too high for bubble–particle adhesion and the film elasticity should be fairly low. This indicates the clear advantage of having froth stabilized by particles instead of surfactants. Particles do not generally give high film elasticities of the air bubbles and do not therefore reduce the ink collection efficiency. This may explain why it is frequently favourable to include some coated paper grades together with old newsprint during de-inking. The coating and filler mineral particles can increase the foam stability without disturbing the ink particle–bubble adhesion.

Extensive flotation experimental studies were reported with de-inking flotation chemicals within the temperature range 20–70°C by Kaya and Oz[19] with a laboratory Wemco flotation cell with shredded black–white offset waste newspaper after slushing and pulping. Flotation was carried out with and without $CaCl_2$ but with Na-oleate and NaOH in the pulper. Without $CaCl_2$ and Na_2SiO_3, the highest brightness and lowest flotation recovery was obtained at 20°C. However, on adding these chemicals, the temperature was found to have a critical influence on flotation with brightness increasing but recovery decreasing as the temperature increased from 20 to 70°C. In practice, since flotation de-inking is usually carried out around 60°C rather than at a specific temperature, then many secondary parameters such as the solubility of the fatty acid, the bubble size and the froth structure may cause changes in the performance and further studies are needed to resolve some of these issues.

9.5.4 Influence of pH and surface charge of particles on de-inking flotation

Generally, flotation is carried out in the alkaline range up to pH = 10 with the efficiency decreasing at higher pH. This may be due to the production of a higher surface charge on the ink particles making them more highly dispersed and more difficult to attach to bubbles. In addition, pH can also affect the solubility of the fatty acids and other chemicals. Generally, a pH between 8 and 10 is reported to be the optimum for flotation de-inking. However, Dorris and Nguyen[20] reported that the highly negatively charged particles of black flexo news ink consisting of carbon black (with acrylic resin binder) did not float under alkaline conditions. On reducing the pH to the slightly acidic region, the flotation recovery improved and at the low pH of 3 the flotation rate became significant correlating with a low zeta potential of the ink particles. The relationship between flotation extent and particle zeta potential is shown in Figure 9.12.

This increase in recovery was explained by aggregation of small particles in the low pH range. Alzevedo et al.[21] have reported that neutral or acidic pulping conditions are most convenient for removal of toner particles by flotation de-inking. At higher pH, toner particles are released from fibres but poor flotation response was observed. Acidic conditions not only lead to an increase in the removal of toner but

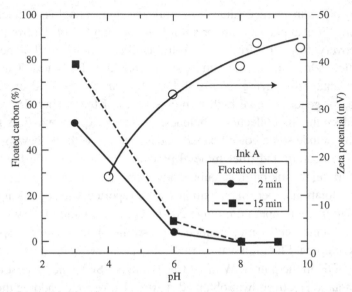

Figure 9.12 Influence of pH on the zeta potential of black flexo news ink particles consisting of carbon black in water (open points) and on flotation recovery at two flotation times given (filled points). Taken from Ref. [20]; with permission of Elsevier.

also to the removal of the mineral filler particles and it was suggested that in this case hetero-coagulation occurred between the smaller toner particles and filler.

9.6 Flotation of Plastic Wastes

Compared to mineral flotation, plastics flotation is just in its infancy. Plastics flotation research began in the 1970's, but there are few full scale applications of this technique in industry at present.[15] The surfaces of plastics are characterized by low surface energy and are hydrophobic in nature as shown in Table 9.1. However, some polymers contain polar groups with oxygen, nitrogen, chlorine or other atoms, which enable dipole–dipole interactions as well as Lewis acid–base interactions with the molecules or ions of reagents. Generally, there are three methods to realize selective plastics flotation:

(i)   undertaking flotation in a liquid medium (may be aqueous or non-aqueous mixture) with a specific value of surface tension,
(ii) selective surface modification (either by adsorption or hydrophobization) of plastics using chemical reagents,
(iii) selective wetting of plastic surfaces using physical modification techniques.

For plastics flotation, the particles are usually in the millimetre size range compared with the flotation of mineral particles which are usually in the micron size range. Our

Table 9.1 *Solid–vapour surface energy (γ_{sv}) and air–water–solid contact angle (θ_{aw}) for a range of plastics. Taken from Ref. [22]; with permission of Delft University of Technology.*

Plastic[a]	γ_{sv}/mN m^{-1}	θ_{aw}/°
GRPP	32.7	96.3
PS	43.0	86.3
ABS	42.7	83.7
PC	44.5	80.3
PA6	43.8	61.4
PMMA	43.8	72.9
PVC	42.3	84.6
POM	44.9	71.2

[a]GRPP: graft polypropylene, PS: polystyrene, ABS: acrylonitrile–butadiene–styrene, PC: polycarbonate, PA6: nylon 6, PMMA: polymethylmethacrylate, PVC: polyvinylchloride, POM: polyoxymethylene.

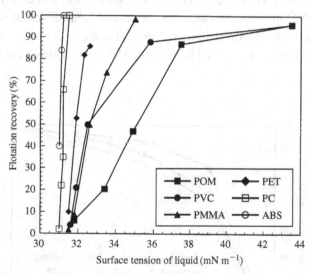

Figure 9.13 Effect of the liquid surface tension on the floatability performance of plastics using 4-methyl-2-pentanol as a frother (concentration 29.4×10^{-3} g l^{-1}, pH = 9.2, 25°C). Taken from Ref. [23]; with permission of Elsevier.

recent results[23] shown in Figure 9.13 clearly demonstrate the principles of the method where the difference in flotation performance of several different plastic particles is shown as a function of the solution surface tension (varied by adding surfactant). These results clearly show that polyethylene (PET), acrylonitrile–butadiene–styrene (ABS) and polycarbonate (PC) polymers give superior flotation compared to polyoxymethylene (POM) and polyvinylchloride (PVC) which float to a much lower extent. However, in cases where the critical surface energies of the plastics are similar, then the surfaces of specific types of plastics in the mixtures must be modified. This can be achieved by chemical conditioning using adsorption of wetting agents or polymers. In the flotation of naturally hydrophobic particles such as plastic, it is necessary to add a wetting agent or depressant such as tannic acid or sodium metasilicate to decrease the surface hydrophobicity and flotation recovery.[24] In Figure 9.14, the effect of particle size on the flotation performance of polymethylmethacrylate (PMMA) plastic particles in the presence of the flotation depressant tannic acid is shown. The flotation recovery is plotted *versus* time. Floatability of the coarse particles (-3.35 to $+2.83$ mm sieve size) increases slowly towards 100% flotation recovery at a low concentration of tannic acid but for the fine particles (-1.41 to $+1.00$ mm sieve size) the flotation recovery is higher but higher concentrations are needed. In addition, we have used the tannic acid depressant to control the flotation efficiency of several different particle size ranges of PMMA and PVC plastics using 4-methyl-2-pentanol frother.[24] It was demonstrated that by using a narrow size fraction and careful control of the depressant concentration, the successful separation of PMMA from PVC from a 50:50 wt.% mixture could be achieved. It was found that almost 100% of the PMMA plastic was recovered in the float compared with almost 100% of the PVC plastic remaining in the sink. This study showed the importance of

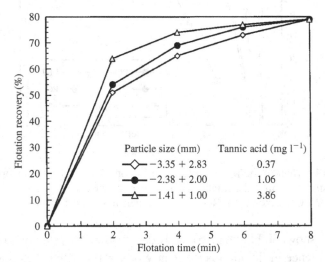

Figure 9.14 Flotation recovery of PMMA particles as a function of time and flotation depressant concentration for different particle size fractions at pH = 5.9 with 4-methyl-2-pentanol as frother (29 mg l^{-1}). Taken from Ref. [24]; with permission of Luleå University, Sweden.

particle size control and depressant concentration and demonstrates that it is an effective way to separate mixtures of plastic particles by flotation.

9.7 Attachment of Particles to Bubbles

One of the earliest applications was in the recovery of sphalerite (zinc sulphide, ZnS) minerals from finely ground ores at Broken Hill in Australia in 1905.[25,26] The flotation process depends on the ability of bubbles to collect particles from the suspension and carry them to the froth phase and the concentrate launder. The collection mechanism is one in which the hydrophobic particles attach to bubbles by the formation of a finite contact angle at the three-phase gas–liquid–solid contact (tpc). The bubble–particle collection involves a number of processes, which can be divided into impaction (collision), attachment and detachment.[27–32] Bubble–particle collision is determined by the physical properties of both the particle and the bubble and the hydrodynamics of the liquid flow, and has been studied extensively.[33–43] The modelling of the bubble–particle attachment and detachment interactions involves many unsolved complex problems such as hydrophobic attraction and de-wetting hysteresis on physically and chemically heterogeneous surfaces, and is not very well advanced.[44,45]

Attachment takes place when a bubble and particle approach each other closely. An intervening liquid film is formed in which interfacial forces become important, governing further stability of the liquid film between the vapour–liquid and solid–liquid interfaces. The net interfacial force between hydrophobic surfaces is attractive, resulting in destabilization of the liquid film. In this case, the liquid film becomes unstable and ruptures, leading to the formation of a three-phase contact line (TPCL) and attachment of the bubble. The bubble–particle contact line spreads further across the solid surface at a certain rate to form a stable wetting perimeter. The relaxation process initiated during rupture of the film leads to an equilibrium state, governed by the de-wetting dynamics and the thermodynamic properties of the gas–solid, liquid–solid and gas–liquid interfaces. The three steps of bubble–particle attachment are illustrated in Figure 9.15. This part of the chapter focuses on the physics governing bubble–particle attachment. In particular, the modelling and measurements of the attachment and contact times and the spreading of the gas–liquid–solid contact line will be reviewed. The interfacial forces, thinning of the intervening liquid film and deformation of the gas–liquid interface during the attachment interaction have been recently reviewed.[46] The attachment of nano- and submicron sized particles enhanced by Brownian diffusion will also be included.

We will consider here only the case where a bubble is rising in a quiescent liquid, which contains particles in suspension. The situation that exists in mechanical flotation cells, where contact is enhanced by shear and turbulence effects arising

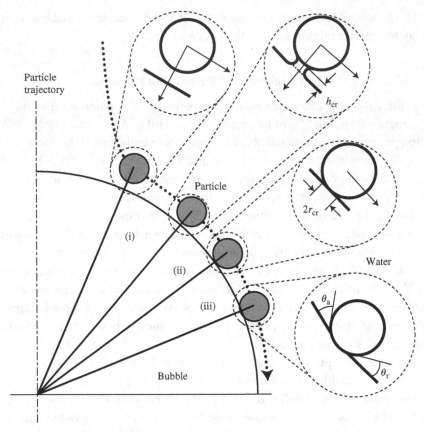

Figure 9.15 Bubble–particle attachment steps: (i) Thinning of the intervening liquid film to a critical film thickness, h_{cr}, (ii) rupture of the intervening liquid film formed when the bubble and the particle are very close together and formation of the three-phase contact nucleus with radius r_{cr}, and (iii) spreading of the TPCL from the critical radius to form a stable wetting perimeter with equilibrium contact angles (advancing, θ_a, and receding, θ_r).

from the rotating impeller, adds an additional range of complexity. However, the general principles considered here will still apply. The major assumption underlying the bubble–particle attachment is that, after meeting with an air bubble in a flotation cell, the particle must be sufficiently close to the bubble surface for thinning and rupture of the intervening liquid film to occur and the formation of a stable wetting perimeter to be established. The bubble–particle contact time, t_{con}, and attachment time, t_{att}, are introduced.[39,47,48] For attachment to occur the attachment time must not be longer than the contact time, *i.e.*

$$t_{att} \leqslant t_{con} \tag{9.3}$$

The attachment time is equal to the sum of the times of the bubble–particle attachment steps shown in Figure 9.15, *i.e.*

$$t_{att} = t_i + t_r + t_{tpc} \tag{9.4}$$

where t_i, the induction time, is the time required for the liquid film to thin to a critical film thickness, t_r is the time required for the film to rupture and form the three-phase contact nucleus and t_{tpc} is the time for the TPCL to expand from the critical nucleus radius to establish a stable wetting perimeter. Under normal condition, t_r is of the order of 1 ms which is significantly shorter than t_i and t_{tpc}, and is not considered in this review. The modelling and measurements of the contact time will be reviewed in Section 9.7.1. The induction time and TPCL expansion time will be reviewed in Sections 9.7.2 and 9.7.3, respectively. The attachment time measurements will be described in Section 9.7.4. The last section will provide a review on the attachment of nano- and sub-micron sized particles.

9.7.1 Bubble–particle contact time

Bubble–particle contact interaction can take place with different trajectories according to their respective directions of motion. If a particle approaches a bubble in a direction close to the local normal to the bubble surface, the momentum of the approach is high, causing strong deformation of the local gas–liquid interface (Figure 9.16). This bubble–particle interaction is referred to as collision contact. Another extreme case is the sliding contact when a bubble and a particle meet each other without any significant deformation of the local bubble surface. In this case, the particle usually slides on the bubble surface after the initial encounter. The experimental dependence of the particle velocity along the trajectories in the vicinity of the bubble surface on its polar position is shown in Figure 9.17 and indicates that the transition from the collision to sliding contact interaction occurs at the relative polar position $\varphi = 20°$.[49,50] Therefore, when the relative polar position of the bubble–particle contact is $\varphi < 20°$, the contact interaction is governed by collision and strong deformation of the local gas–liquid interface. If the relative polar position $\varphi > 20°$, the bubble–particle contact interaction is governed by sliding and the local deformation can be neglected.

9.7.1.1 Collision contact time

The collision time has been modelled based on the oscillation of an effective mass, m_{eff}, of the colliding particle under the action of the restoring force, F_{res}, of the deformation of the gas–liquid interface, Figure 9.18. The governing equation

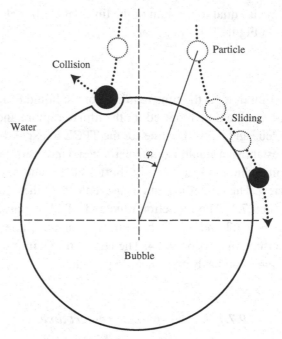

Figure 9.16 Schematic of the collision and sliding contact interactions of a particle moving relative to an air bubble in water. φ describes the polar position of the particle relative to the direction of the bubble motion.

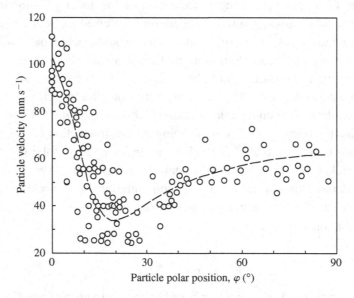

Figure 9.17 Velocity of glass spheres ($\approx 160 \, \mu m$ in diameter) along the trajectories in the vicinity of the bubble surface *versus* particle polar position. Taken from Ref. [29]; with permission of Elsevier.

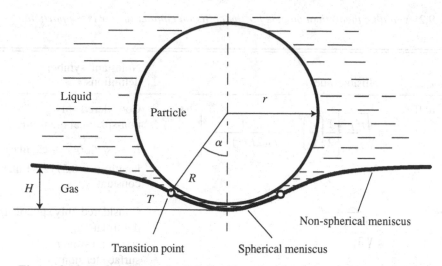

Figure 9.18 Deformation of a gas–liquid surface by contact with a particle. The deformed meniscus has a spherical part with a liquid film of radius of curvature, R, wrapping around the particle and a non-spherical part extending to the end of the local deformation.

for determining the collision contact time for the general case can be described as follows[51]

$$m_{eff} \frac{d^2 H}{dt^2} + F_d + F_{res} - m_p \left(1 - \frac{\delta}{\rho}\right) g = 0 \qquad (9.5)$$

where t is the reference time, H is the depth of the deformed gas–liquid interface at the apex, m_p is the mass of the particle, g is the acceleration due to gravity, δ and ρ are the liquid and particle densities, respectively, and F_d is the damping force which is a function of the oscillation velocity, dH/dt.

The non-linear equation (9.5) describes the damped oscillation, which can be solved for the collision contact time, t_c, being defined as half of the first period of the damped oscillation. The available models for collision time are summarized in Table 9.2. These models differ in the calculation of the restoring force and the effective mass. In some models, the gravitational and damped forces are not considered. In Figures 9.19 and 9.20, some of these models for the collision contact time are compared with the experimental data[57] obtained with both hydrophobic and hydrophilic glass spheres ($\rho = 2500 \, kg \, m^{-3}$) and Sn–Pb alloyed spheres ($\rho = 8500 \, kg \, m^{-3}$). The experimental data were obtained using light reflection microscopy. The experiments determined the local deformation of the air–water interface formed at the bottom of a sealed glass capillary. The intensity of the reflected light from the deformed interface was measured in terms of the grey level

Table 9.2 *Available theoretical models for the collision contact time. r is the particle radius.*

Author(s)	Collision time, t_c	Notes and symbol definitions
Philippoff[52]	$\pi\sqrt{\dfrac{r^3\rho}{\gamma_{aw}}}\left\{\dfrac{1}{2}\left(\dfrac{r}{L}\right)^2 + \dfrac{1}{\ln(2L/r)-\gamma}\right\}^{-0.5}$	Considered only non-spherical meniscus; $L = \sqrt{\gamma_{aw}/g\delta}$ – capillary length; $\gamma = 0.5772$ – Euler constant
Evans[53]	$\pi\sqrt{\dfrac{2r^3\rho}{3\gamma_{aw}}}$	Considered only spherical deformation; γ_{aw} = air–water surface tension
Scheludko et al.[54]	$\pi\sqrt{\dfrac{2r^3\rho}{3\gamma_{aw}}}\left\{\ln\dfrac{4L/r}{\sin\alpha(1+\cos\alpha)}\right\}^{0.5}$	Considered both menisci; α = angle of the transition point
Ye and Miller[55]	$\pi\sqrt{\dfrac{r^3}{3\gamma_{aw}}}\left[1 + \dfrac{2}{\pi}\times\arcsin\left[\left(1+\dfrac{3V_{rel}^3\gamma_{aw}\rho}{2r^3g^2(\rho-\delta)^2}\right)^{-0.5}\right]\right]$	Considered only spherical deformation; V_{rel} = bubble–particle approach speed
Schulze et al.[51]	$\pi\sqrt{\dfrac{r^3(\rho+1.5\delta)}{3\gamma_{aw}}}\left[\dfrac{\left\{\dfrac{1}{2}-\gamma+\ln\dfrac{2L}{r\alpha_m}\right\}\sqrt{2}}{\left\{\dfrac{1}{4}-\gamma+\ln\dfrac{2L}{r\alpha_m}\right\}^{0.5}}\right]$	Considered both menisci; α_m = maximum transition angle (measured in radians) to be determined by experiment
Nguyen et al.[56]	$\dfrac{\pi\sqrt{\dfrac{r^3(2\rho+\delta)}{6\gamma_{aw}}}}{0.506 + 0.04\ln\dfrac{r}{L} + \left\{0.041 + 0.03\ln\dfrac{r}{L}\right\}\ln We}$	Considered both menisci; $We = \{r^3(\rho+0.5\delta)V_{rel}^2/(3\gamma_{aw}L^2)\}^{0.5}$ – modified Weber number

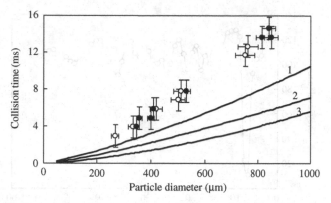

Figure 9.19 Comparison between the available models (lines) for the collision time as a function of the particle size and experimental data for hydrophobic (filled circles) and hydrophilic (unfilled circles) glass spheres. Lines: (1) Nguyen *et al.*,[56] (2) Philippoff,[52] (3) Ye and Miller[55] and Evans.[53] Re-drawn from Ref. [57]; with permission of Elsevier.

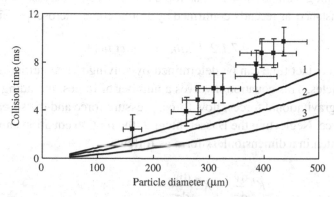

Figure 9.20 Comparison between the available models (lines) for the collision time *versus* particle size and experimental data (squares) for Sn–Pb alloyed spheres. Lines: (1) Nguyen *et al.*[56]; (2) Philippoff,[52]; (3) Ye and Miller[55] and Evans.[53] Re-drawn from Ref. [57]; with permission of Elsevier.

as a function of time (Figure 9.21) using a CCD (charge coupled device) high-speed camera and analysed using the Fourier transform. Recently, the experimental technique was improved using an oscilloscope to investigate both low and high frequency oscillations of the gas–liquid interface with an attaching particle.[58]

The available models predict lower collision times than the experimental values. This is thought to be due to the neglect of the drag force, F_d, acting on attaching particles in the near proximity of the gas–liquid interface in equation (9.5). The drag force on a particle is a function of the separation distance between the gas–liquid interface

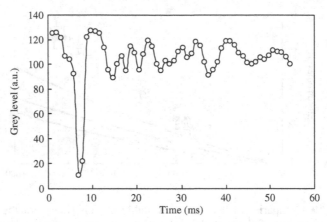

Figure 9.21 Oscillation (measured in terms of grey levels) of a deformed air–water surface by an attaching particle, as revealed by CCD high-speed light reflection microscopy. Re-drawn from Ref. [57]; with permission of Elsevier.

and the attaching particle surface. This force is expected to increase with decreasing separation distance, as recently confirmed by atomic force microscopy, Figure 9.22.[59]

9.7.1.2 Sliding contact time

The sliding contact time can be determined by solving the equation of motion for sliding particles. This equation involves a number of forces, including the viscous drag force, gravitational forces, flow force, pressure force and the so-called Basset (history) force. Neglecting the Basset integral, the motion equation can be conventionally written in a dimensionless form as follows

$$K' \frac{\mathrm{d}\vec{v}}{\mathrm{d}\tau} - K'' \frac{\mathrm{d}\vec{w}}{\mathrm{d}\tau} = (\vec{w} - \vec{v}) + \vec{v}_\mathrm{s} \tag{9.6}$$

where $K' = (1 + \delta/2\rho)St$, $K'' = (3\delta/2\rho)St$, $\tau = tU/R_\mathrm{a}$ in which U is the bubble slip (relative to the liquid) velocity, R_a is the bubble radius, $w = W/U$ is the dimensionless liquid velocity, $v = V/U$ is the dimensionless particle velocity and $v_\mathrm{s} = V_\mathrm{s}/U$ is the dimensionless terminal (settling) velocity of the particle. St is the particle Stokes number defined as

$$St = \frac{2r^2 U\rho}{9\mu R_\mathrm{a}} = \frac{1}{9} \frac{\rho}{\delta} \left\{ \frac{r}{R_\mathrm{a}} \right\}^2 Re \tag{9.7}$$

where μ is the liquid viscosity and r is the particle radius. The bubble Reynolds number, Re, in equation (9.7) is defined by $Re = 2R_\mathrm{a}\delta U/\mu$. Since the change in the

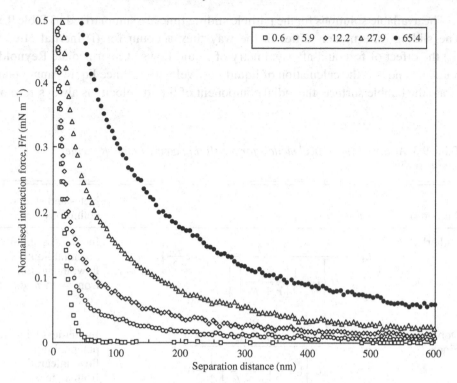

Figure 9.22 Interaction forces between a $600\,\mu\mathrm{m}$ diameter bubble and a $40\,\mu\mathrm{m}$ diameter hydrophilic glass sphere in aqueous $1\,\mathrm{mM}$ KCl solution *versus* the bubble–particle surface separation distance for different approach speeds (given in $\mu\mathrm{m\,s^{-1}}$), as determined by atomic force microscopy. At low approach speeds the surface forces dominate the interaction force, especially at close separation, while at higher velocities the hydrodynamic drag force becomes significant. Re-drawn from Ref. [59]; with permission of Elsevier.

particle polar position during the sliding interaction is significant, equation (9.6) can be further simplified employing the order-of-magnitude analysis, giving[60]

$$-K'\tilde{r}\left\{\frac{\mathrm{d}\varphi}{\mathrm{d}\tau}\right\}^2 = w_r - v_s \cos\varphi + \frac{K''}{\tilde{r}}\left\{w_\varphi \frac{\partial w_r}{\partial \varphi} - w_\varphi^2\right\} \quad (9.8)$$

$$K'\tilde{r}\frac{\mathrm{d}^2\varphi}{\mathrm{d}\tau^2} = \left\{w_\varphi - \tilde{r}\frac{\mathrm{d}\varphi}{\mathrm{d}\tau}\right\} + v_s \sin\varphi + \frac{K''}{\tilde{r}}\left\{w_\varphi \frac{\partial w_\varphi}{\partial \varphi} + w_\varphi w_r\right\} \quad (9.9)$$

where $\tilde{r} = 1 + r/R_a$. These two equations can be integrated for the sliding time using the initial condition $\varphi = \varphi_0$ at $\tau = 0$.

The available solutions for the particle sliding time are summarized in Table 9.3. The sliding time models differ in the way they account for (i) inertial effects, (ii) the effect of fore-and-aft asymmetry of liquid flows at intermediate Reynolds numbers and (iii) the calculation of liquid flow velocities. Since liquid cannot penetrate the bubble surface, the radial component of liquid velocity is always zero on

Table 9.3 *Available theoretical models for the sliding contact time. R_a is the bubble radius.*

Author(s)	Sliding time, t_{sl}	Notes and symbol definitions
Sutherland[39]	$$t_{sl} = \frac{2(r + R_a)}{U\left\{1 + \dfrac{1}{2}\left(\dfrac{R_a}{r + R_a}\right)^3\right\}} \ln\left\{\cot\frac{\varphi_0}{2}\right\}$$	Inertia-less sliding; potential liquid flow; integration from φ_0 to $\pi - \varphi_0$
Dobby and Finch[69]	$$t_{sl}(\varphi) = \frac{r + R_a}{U\left\{1 + \dfrac{1}{2}\left(\dfrac{R_a}{r + R_a}\right)^3\right\} + V_s} \cdot \ln\left\{\frac{\tan\dfrac{\varphi}{2}}{\tan\dfrac{\varphi_0}{2}}\right\}$$	Inertia-less sliding; potential liquid flow; integration from φ_0 to $\varphi \leqslant \pi/2$; included particle settling velocity
Dobby and Finch[69]	$$t_{sl}(\varphi) = (\varphi - \varphi_0)\frac{r + R_a}{U \cdot \overline{w}_\varphi + V_s \cdot \sin\varphi}$$	Inertia-less sliding; polar angles in radians; $0 < Re < 500$; immobile bubble surface; overbars describe the average values
Yoon and Luttrell[33]	$$t_{sl}(\varphi) = \frac{r + R_a}{U\left\{\dfrac{3}{2} + \dfrac{4}{15}Re^{0.72}\right\}\dfrac{r}{R_a}} \cdot \ln\left\{\frac{\tan\dfrac{\varphi}{2}}{\tan\dfrac{\varphi_0}{2}}\right\}$$	Inertia-less sliding; $0 < Re < 500$; immobile bubble surface; integration from φ_0 to $\varphi \leqslant \pi/2$

(*Continued*)

Table 9.3 *(Continued)*

Author(s)	Sliding time, t_{sl}	Notes and symbol definitions
Nguyen[42]	$$t_{sl} = \frac{r + R_a}{U(1 - B^2)A}$$ $$\ln\left\{ \left| \frac{\tan(\frac{\varphi}{2})}{\tan(\frac{\varphi_0}{2})} \left[\frac{\csc(\varphi) + B\cot(\varphi)}{\csc(\varphi_0) + B\cot(\varphi_0)} \right]^B \right| \right\}$$ $$A = \frac{V_s}{U} + \left\{ 1 + \frac{r}{R_a} \right\} \frac{X}{2}$$ $$+ \left\{ \frac{r}{R_a} \right\}^2 \frac{M}{2}$$ $$B = \left\{ 1 + \frac{r}{R_a} \right\} \frac{Y}{2A}$$ $$+ \left\{ \frac{r}{R_a} \right\}^2 \frac{N}{2A}$$	X, Y, M and N are functions of Re and the gas hold-up; inertia-less sliding; $0 < Re < 500$; immobile bubble surface; fore-and-aft asymmetry of liquid flows
Nguyen[42]	$$t_{sl} = \frac{r + R_a}{U(1 - B^2)A}$$ $$\ln\left\{ \left| \frac{\tan(\frac{\varphi}{2})}{\tan(\frac{\varphi_0}{2})} \left[\frac{\csc(\varphi) + B\cot(\varphi)}{\csc(\varphi_0) + B\cot(\varphi_0)} \right]^B \right| \right\}$$ A = as above $$B = \left\{ 1 + \frac{r}{R_a} \right\} \frac{Y}{2A} + \left\{ \frac{r}{R_a} \right\}^2 \frac{N}{2A} - St\left(1 - \frac{\delta}{\rho} \right) \frac{X^2}{4A}$$	X, Y, M and N are functions of Re and the gas hold-up; inertial sliding; $0 < Re < 500$; mobile bubble surface

the bubble surface. The tangential velocity component on the surface can be zero (if the bubble surface is immobile) or non-zero (if the bubble surface is mobile), depending on the contamination of the bubble surface as shown in Figure 9.23. As bubbles rise in a flotation cell, their surfaces becomes partially mobile because flotation surfactants and other contaminants are swept to the rear of the air bubble by the liquid resistance.[61–68] At present, there is only one model for the sliding time of particles on fully mobile bubble surfaces available in the literature (Ref. [42],

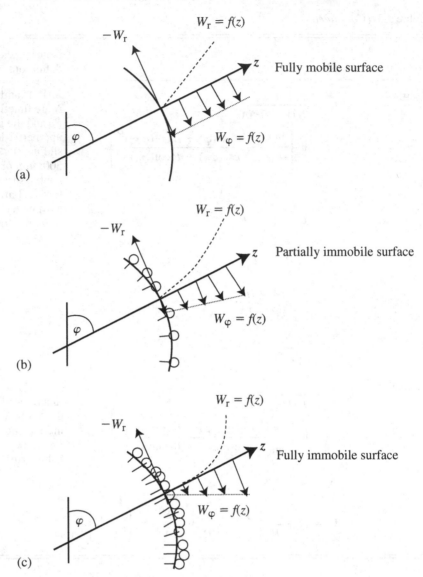

Figure 9.23 Effect of adsorbed surfactant (line + circle) and other surface contaminants on the liquid velocity components at the bubble surface. (a) A bubble
surface in clean (contaminant-free) water is fully mobile. (b) As it rises, the
bubble surface becomes partially immobile because the surface contaminants are
swept to the rear by the shear exerted by the liquid. The tangential liquid velocity
as a function of the radial distance, z, measured from the bubble surface is shown
by the parallel lines with the arrows. The profiles of the radial, W_r, and tangential,
W_φ, components of liquid velocity at the bubble surface are shown by the dashed
lines. (c) A bubble surface fully covered by surfactant and other contaminants is
immobile.

Table 9.4 *Available models for the maximum sliding contact position,* φ_m.

Author(s)	Maximum particle polar position, φ_m	Notes and symbol definitions
Sutherland[39]	$\varphi_m = \pi - \varphi_0$	Inertia-less sliding; potential liquid flow; the sliding begins at φ_0
Dobby and Finch[69]	$\varphi_m = \dfrac{\pi}{2}$	Inertia-less sliding; potential liquid flow
Finch and Dobby[32]	$\varphi_m = 9 + 8.1\rho + (0.9 - .09\rho)$ $(78.1 - 7.37 \log Re)$	Inertia-less sliding; polar angles in radians; $0 < Re < 500$; immobile bubble surface
Yoon and Luttrell[33]	$\varphi_m = \dfrac{\pi}{2}$	Inertia-less sliding; $0 < Re < 500$; immobile bubble surface
Nguyen[42]	$\varphi_m = \arccos \dfrac{\sqrt{(X + C)^2 + 3Y^2} - (X + C)}{3Y}$	X, Y are functions of Re and the gas hold-up; C is a function of Re and ρ; inertia-less sliding; $0 < Re < 500$; immobile bubble surface; fore-and-aft asymmetry of liquid flows
Nguyen[42]	$\varphi_m = \arccos \dfrac{\sqrt{(X + C)^2 + C_1^2 X^4} - (X + C)}{C_1 X^2}$	C_1 is a function of the particle Stokes number; $0 < Re < 500$ inertial sliding; mobile bubble surface

Table 9.3). The other models were developed using the immobile bubble surface. Solutions for partially mobile surfaces are not available yet.

How long can the sliding contact last? The answer to this question is important in the determination of the particle attachment. If a particle becomes attached to the bubble surface during the sliding interaction, it remains in contact with the bubble indefinitely. Non-attached particles can only remain in contact with the bubble for some time until some critical condition, *e.g.* the maximum particle polar position, φ_m, is reached. Dobby and Finch[69] showed that φ_m is the angle at which the radial component of the particle settling velocity (directed towards the bubble surface) is equal to the radial component of the liquid velocity calculated at the particle centre (directed away from the bubble surface). This solution for φ_m can be generally

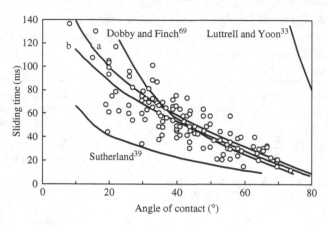

Figure 9.24 Model predictions (lines, $\varphi_m = 90°$) and experimental data (points) for the sliding time of hydrophobic glass spheres in water: $R_a = 1.41$ mm, $r = 74$ μm; $\rho = 2500$ kg m^{-3}, $V_s = 21.4$ mm s^{-1}, $U = 84.2$ mm s^{-1} and $\mu = 0.001$ kg m^{-1} s^{-1}. Curve a: inertia-less model with an immobile bubble surface.[42] Curve b: inertial solution of equations (9.8) and (9.9). Re-drawn from Ref. [42]; with permission of Elsevier.

determined from the condition when the radial component of the particle velocity at the bubble surface is zero.[42] Such an approach is useful for investigating the influence of the flow fore-and-aft asymmetry and inertia on φ_m. The available solutions are summarized in Table 9.4. Briefly, the potential and Stokes flows are fore-and-aft symmetrical, giving $\varphi_m = 90°$, whereas the liquid flow passing air bubbles with intermediate Reynolds number is fore-and-aft asymmetrical. The streamlines of such flows approach the bubble surface close in the leading half surface, while deviate from the surface in the back half surface, leading to the condition $\varphi_m < 90°$. Finally, φ_m can be smaller than 90° due to inertia of the sliding motion, in particular on the mobile bubble surface, on which the liquid tangential velocity is different to zero.

The available experimental data for the sliding time has been obtained with stroboscopy and high-speed video microscopy using captive (stationary) bubbles in flowing water.[35] Recently, the technique has also been used to study the sliding interactions in bubble-drop and drop–drop systems, as found in the flotation applications in bitumen recovery from oil sands.[70,71] Comparison between the available models and experimental data is shown in Figures 9.24 and 9.25. Although the models follow the experimental trend, fluctuation in the available data is not small and the experimental sliding time is difficult to analyse in order to obtain information about the sliding interaction with the liquid film thinning and rupture and three-phase contact spreading. Indeed, the new experiments[72] show that the sliding

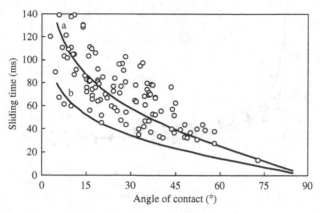

Figure 9.25 Sliding time of silanized ground glass particles in water *versus* angle of contact, φ_0. Points: experimental data with $R_a = 1.38$ mm, $r = 120\,\mu$m (mean), $\rho = 2500\,\text{kg}\,\text{m}^{-3}$, $V_s = 29\,\text{mm}\,\text{s}^{-1}$, $U = 125\,\text{mm}\,\text{s}^{-1}$, $\mu = 0.001\,\text{kg}\,\text{m}^{-1}\text{s}^{-1}$, $\varphi_m = 90°$. Lines a and b describe the inertia-less models with an immobile and mobile bubble surface, respectively. Re-drawn from Ref. [42]; with permission of Elsevier.

contact time with a TPCL formed between a bubble and a particle is significantly longer than predicted.

9.7.2 Induction time

Although there are several models available to describe the drainage rate of the liquid film between an air bubble and a solid surface,[73–81] these models are rarely applied to describe the induction time in mineral flotation systems because they are developed for drainage of films of large radius. The drainage of such films causes the formation of a dimple with non-uniform film thickness.[44,82,83] However, the drainage of small films between a bubble and a small particle in flotation does not create dimples and the film thickness remains approximately the same along the film radius.[84]

The liquid flow in a thin film is described by the continuity and Navier–Stokes equations. Considering that the film thickness is significantly smaller than the other two dimensions of the film, the equations can be simplified into a mathematically tractable expression, known as the lubrication approximation. The induction time required for the liquid film to thin to a critical film thickness can be determined, to a first approximation, from the lubrication theory using the Stefan–Reynolds equation for film thinning during the collision interaction or the Taylor equation for film thinning during the sliding interaction.[29] The Stefan–Reynolds equation for the collision interaction is only a first approximation applied locally to a small film area since the liquid film between a colliding particle and a deformed meniscus wrapping around the particle is not perfectly planar as assumed.

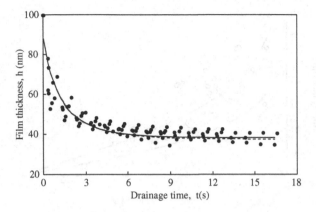

Figure 9.26 Stefan–Reynolds equation (lines) and experimental data (points) for the thickness of stable planar 1 mM KCl films on hydrophilic glass surfaces *versus* time. The disjoining pressure was calculated using the non-retarded van der Waals and electrical double-layer interactions. Both the latter interactions at constant surface charge (continuous line) and constant surface potential (dashed line) are considered but show little difference. Re-drawn from Ref. [85]; with permission of Elsevier.

The Stefan–Reynolds equation describes the correlation between the driving force, F, the film thickness, h, the film radius, R_f, and the approach speed of the film surfaces, V, by

$$F = -\frac{3\pi\mu V R_f^4}{2mh^3} \tag{9.10}$$

where m is equal to 1 or 4 for immobile or mobile bubble surfaces, respectively. The driving forces may be gravitational forces, applied forces or surface forces. If the gravitational forces can be neglected, the driving force is determined by $F = \pi R_f^2$ $(P_\sigma - \Pi)$ where P_σ is the capillary pressure and Π is the disjoining pressure. The Stefan–Reynolds equation (9.10) for film drainage can be rearranged to give

$$\frac{dh}{dt} = -\frac{2mh^3}{3\mu R_f^2}(P_\sigma - \Pi) \tag{9.11}$$

This equation can be integrated numerically to obtain the dependence of film thickness on time. Equation (9.11) has been confirmed by experiments. Shown in Figure 9.26 are the experimental and theoretical results for thinning aqueous films on a hydrophilic glass surface in 0.001 M KCl solutions. The experimental data were obtained by six measurements with the modified Derjaguin–Scheludko cell.[85] The average film radius was *ca.* $R_f = 90\,\mu\text{m}$. The measured capillary pressure was about 250 Pa and the numerical integration was carried out using a four-step Runge–Kutta algorithm.

The liquid films between a bubble and a hydrophobic solid surface are unstable and rupture at a critical thickness larger than ~ 40 nm, Figure 9.27. During the collision

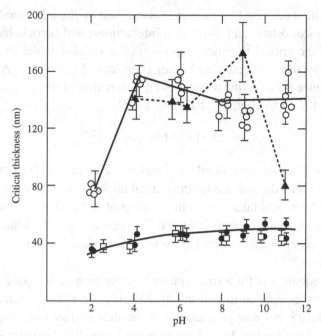

Figure 9.27 Critical thickness for rupture of aqueous dodecylamine hydrochloride (DCA) films between an air bubble and a silica surface *versus* pH and ionic strength. ▲: 10^{-6}M DCA and 10^{-4}M KCl; ○: 10^{-5}M DCA; ●: 10^{-5}M DCA and 10^{-2}M KCl; □: 10^{-5}M DCA and 10^{-1}M KCl. Taken from Ref. [29]; with permission of Elsevier.

interaction, the driving force for film thinning to a critical thickness, h_{cr}, is due to the particle inertia, which causes the strong gas–liquid interface deformation. The capillary pressure, $2\gamma_{aw}/r$, of the deformed meniscus is the major pressure component in equation (9.11), while the negative disjoining pressure determines h_{cr}. In the calculation of the induction time during the collision, the disjoining pressure can be neglected relative to the capillary pressure, and integration of equation (9.11) from $h = \infty$ to $h = h_{cr}$ gives

$$t_i = \frac{3\mu R_f^2 r}{2m\gamma_{aw}h_{cr}^2} \tag{9.12}$$

Schulze[35] obtained the following useful expression for the film radius, R_f, in terms of the particle and bubble physical properties

$$R_f = \frac{\pi r(656.9 - 87.4 \ln r)(V_{rel}t_c)^{0.62}}{180} \tag{9.13}$$

where the particle radius r is measured in μm, V_{rel} is the bubble–particle approach speed and t_c is the collision contact time.

The critical thickness for film rupture is controlled by the interfacial properties, especially the electrostatic and dispersion interactions, and particle hydrophobicity. Therefore, the critical thickness represents the variable which is affected by adsorption of flotation collectors and other chemicals, Figure 9.27. An empirical power dependence for the critical thickness as a function of the advancing contact angle, θ_a, and surface tension, γ_{aw}, has been used[35]

$$h_{cr} = 23.3\{\gamma_{aw}(1 - \cos \theta_a)\}^{0.16} \tag{9.14}$$

where γ_{aw} and h_{cr} have the units of mN m^{-1} and nm, respectively. The power dependence follows a similar dependence for the critical thickness of free foam films.[86] The dependence of the critical thickness on the product of surface tension and advancing contact angle is an analogue to the Frumkin–Derjaguin equation, which relates the product of surface tension and contact angle with the surface interaction energy at the primary minimum.[87]

The Taylor equation is for a solid sphere approaching an interface with "point contact" and can be applied to film thinning during the sliding interaction, when the bubble surface is not wrapping around the particle surface as during the colliding interaction. The classical Taylor equation can be modified to describe the influence of the bubble surface mobility, similar to the Stefan–Reynolds equation. Recently, the theory was improved by incorporating the slippage at the solid–liquid interface.[88] The major driving forces for the film thinning to critical thickness are gravitational and hydrodynamic forces, which are functions of the polar position of the sliding particle on the bubble surface. Consequently, a simple, explicit model for the film drainage and induction time as in the case of the collision process fails. The full balance of forces in both the radial and tangential directions is required and proves a further difficulty.

9.7.3 Spreading and relaxation of TPCL

After film rupture, the spreading of the TPCL on the particle surface takes place under the influence of the uncompensated air–water interfacial tension, γ_{aw}. During film pull-back, work must be done to de-wet the solid resulting in the expansion of the gas–liquid–solid contact until equilibrium is established. In flotation, the velocity of the TPCL was first estimated by Philippoff[52] who followed the method of calculating the rate of expansion of a hole in a soap bubble during bursting. The velocity of the film recession was determined from the balance between the work done by the surface tension force and the kinetic energy supplied to the receding film. The force per unit length required to de-wet the solid surface is given by $\gamma_{aw}(1 - \cos \theta_r)$, where θ_r is the receding contact angle and is smaller than the equilibrium contact angle, θ_0.

The calculation did not consider the influence of the solid–liquid interface on the film recession and the calculated velocity is about 10 times higher than the experimental data. Philippoff argued that the viscous effect is significant and the TPCL velocity must be dependent on the dynamic contact angle.

The fundamentals of the TPCL expansion relevant to flotation were established by Scheludko and collaborators.[54,89] It was shown that a three-phase contact arises from a hole in the intervening liquid film during the film rupture process. Any hole with a radius smaller than the critical radius collapses resulting in no rupture of the intervening liquid film. Only the three-phase contact with a radius larger than the (non-zero) critical hole radius can be formed and expand under the driving interfacial forces. Both molecular-kinetic[90] and hydrodynamic[91,92] approaches have been used to establish the relationship between the velocity of the TPCL motion and the dynamic contact angle. The hydrodynamic theory determines the dynamic contact angle by analysing the hydrodynamic bending of the gas–liquid meniscus due to TPCL motion. Recent experiments[93] show that the hydrodynamic theory does not correctly describe the dynamic TPCL with a small radius between a small bubble and a solid planar surface. Combination of the hydrodynamic and molecular-kinetic theories could provide a better prediction.[94] The simplified molecular-kinetic theory is described below.

The Blake and Haynes molecular-kinetic theory[90] is based on the statistical mechanics treatment of the transport processes of molecules and ions[95,96] and gives

$$V_{TPCL} = \vartheta \sinh \left\{ (\cos \theta - \cos \theta_0) \frac{\gamma_{aw}}{a\vartheta} \right\} \qquad (9.15)$$

where ϑ is the rate of the molecular jumping (back and forward) at equilibrium and a is the mobility of TPCL displacement ($\vartheta = 2\lambda\nu$, where λ is the mean molecular jumping distance and ν is the frequency of molecular jumping at equilibrium). Apart from the surface tension, the other two model parameters in equation (9.15) are not usually known and have to be determined by best fit to experimental data. Experiments of the spreading and relaxation of the TPCL between a bubble and a particle relevant to flotation are few. Important parameters of TPCL motion, such as the TPCL radius or the dynamic contact angle, are difficult to measure directly with small particles and air bubbles typically employed in flotation. Useful experimental data can be obtained by focussing on the area of the air bubble surface where the TPCL expansion takes place, where a particle-dropping technique has been used. In this technique, an air–water surface was formed at the bottom of a capillary, Figure 9.28. A single small solid particle ($\sim$100 μm diameter) was transferred into the liquid contained in the capillary and dropped onto the interface. The TPCL expansion process can be viewed and imaged from underneath the surface with high-speed video microscopy.

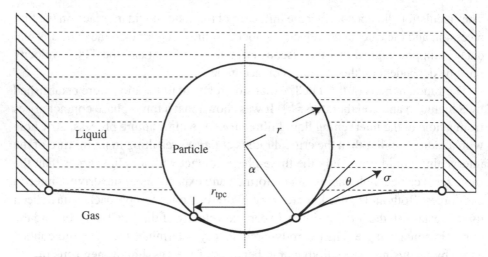

Figure 9.28 Three-phase contact of a small particle attached to an air–liquid surface formed at the bottom of a glass capillary in the particle-dropping experiments for studying TPCL expansion and relaxation. Re-drawn from Ref. [97]; with permission of Elsevier.

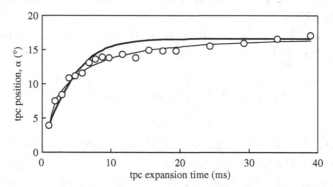

Figure 9.29 Experimental results (points) and theoretical predictions (lines) for the polar position of the TPCL on a hydrophobic glass sphere ($r = 172.3 \, \mu m$) *versus* time. Thick and thin lines describe the best fit predictions with (Ref. [97]) and without (equation (9.18)) the line tension term, respectively. Re-drawn from Ref. [97]; with permission of Elsevier.

Typical experimental results obtained with the particle-dropping technique are shown in Figure 9.29. For small particles, one obtains the following approximation for the velocity of the TPCL spreading on the particle surface and the dynamic contact angle,[97] respectively

$$V_{\text{TPCL}} = r \frac{d\alpha}{dt} \tag{9.16}$$

$$\sin \alpha \sin(\alpha - \theta) = \frac{2}{3}\left[\frac{r}{L}\right]^2 \left(\frac{\rho}{\delta} - \frac{2 + 3\cos\alpha - \cos^3\alpha}{4}\right) \tag{9.17}$$

where L is the capillary length (see Table 9.2). Equation (9.17) describes the simplified balance between gravitational and surface tension forces. Solving equations (9.15) and (9.16) gives

$$\frac{d\alpha}{dt} = \frac{\vartheta}{r}\sinh\left\{(\cos\theta - \cos\theta_0)\frac{a\gamma_{aw}}{\vartheta}\right\} \tag{9.18}$$

Equation (9.18) in conjunction with equation (9.17) can be integrated to obtain the TPCL position, α, on the particle surface *versus* time. The results are shown by the thick line in Figure 9.29. The model can be improved by including additional parameters, including line tension.[97]

Equations (9.16)–(9.18) can be further simplified for fine particles in flotation systems for which one has $r \ll L$. In this circumstance, equation (9.17) gives $\theta \cong \alpha$. The argument of the sinh function in equation (9.18) is small and can be linearized. Integration of the simplified equation leads to the following equation for the TPCL position, α, as a function of time, t

$$\tan\left(\frac{\alpha}{2}\right) = \tanh\left\{\frac{a\gamma_{aw}t\sin\theta_0}{2r}\right\}\frac{\sin\theta_0}{\cos\theta_0 + 1} \tag{9.19}$$

The particle polar position at equilibrium, α_e, is needed for the determination of the TPCL expansion time, t_{tpc}. For small particles, equation (9.17) can be approximately solved, leading to

$$\alpha_e = \theta_0 + B_o \tag{9.20}$$

where $B_o = \frac{2}{3}\left[\frac{r}{L}\right]^2\left(\frac{\rho}{\delta} - 1\right)$. Finally, the TPCL expansion time determined from equation (9.19) gives

$$t_{tpc} = \frac{2r}{a\sigma\sin\theta_0}\text{atanh}\left\{\tan\left(\frac{\alpha_e}{2}\right)\frac{\cos\theta_0 + 1}{\sin\theta_0}\right\} \tag{9.21}$$

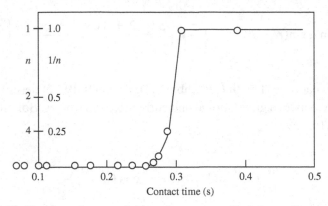

Figure 9.30 Number of successful attachments, n, of a 2.5 mm diameter air bubble onto a flat pyrite surface pre-treated with xanthate in water *versus* the contact time. The attachment time determined is $t_{att} = 290 \pm 10$ ms. Re-drawn from Ref. [47]; with permission of Springer.

Therefore, knowing the equilibrium contact angle, θ_0, the TPCL mobility, a, and other physical properties of bubbles and particles, the TPCL expansion time between a bubble and a particle can now be calculated.

9.7.4 Attachment time measurements

The usual experimental determination of the attachment time is based on the fact that after the collision stage the mineral particle must stay in contact with a bubble for a certain time, which is at least equal to the sum of the induction time for the liquid thin film drainage to a critical thickness and the TPCL expansion time, in order to be attached on the bubble surface. Sven-Nilsson[47] was among the first to recognize the importance of the attachment time in flotation kinetics. The author determined the attachment time by making direct contact between a captive bubble and a flat mineral surface for different times. The number of successful attachments was plotted against the contact time. The attachment was determined as the minimum possible contact time from the plot, Figure 9.30.

The Sven-Nilsson technique has been successfully applied in flotation research.[98–100] The research has established the importance of the attachment time as a kinetic measure of the flotation recovery *versus* pH and reagent concentration, while the contact angle failed to describe the flotation behaviour. The major advantage of the Sven-Nilsson technique is the relatively easy experimental procedure. However, it is quite understandable that the measurement does not truly represent the film thinning process between a bubble and a particle in an actual flotation system.

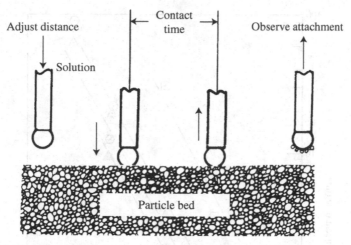

Figure 9.31 Schematic of the measurement of the attachment time using a particle bed. A fresh bubble attached to the bottom of a capillary is pushed downwards through the aqueous solution such that the tip of the bubble is kept in contact with the bed of particles for a controlled contact time. After that, the bubble and the capillary are returned to the original position. Particle attachment is directly observed through a microscope. Re-drawn from Ref. [55]; with permission of Elsevier.

An alternative method of measuring the attachment time is to use a bed of particles in place of a flat mineral surface. This type of technique was first developed by Glembotsky[101] and has been successfully used by many others.[48,55,71,102–105]

Figure 9.31 shows the principle involved in the determination of the attachment time with a particle bed. A fresh captive bubble with known size, held on a bubble tube, is pushed downward through the solution such that the tip of the bubble is kept in contact with the bed of particles for a controlled contact time. After that, the bubble together with the tube is returned to the original position. Then, the bubble is observed through a microscope to determine whether attachment of particles at the bubble surface has occurred during the controlled contact time. The experiment at this stage is repeated several times at different positions on the particle bed to obtain the percentage of observations which resulted in attachment at a given contact time. Next, the controlled contact time is changed to different levels, and the same procedure repeated at each contact time level, so that a distribution of per cent attachment *versus* controlled contact time is obtained. The attachment time is determined by the contact time at which 50% of the observations result in attachment. The dependence of the attachment time on the particle size can be determined with the particle bed technique. A typical result is shown in Figure 9.32 for coal particles. The attachment time increases with an increase in the particle size. This result confirms the theories described by equations (9.12) and (9.21).

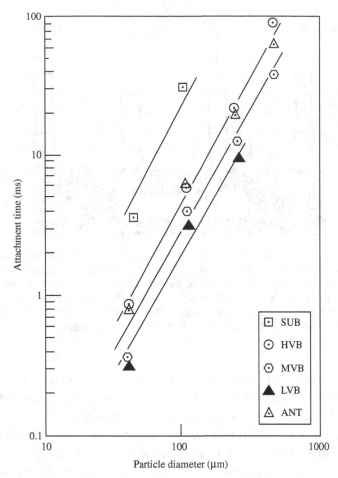

Figure 9.32 Attachment time *versus* particle diameter for five different coal types (given). Re-drawn from Ref. [55]; with permission of Elsevier.

The influence of flotation surfactant collector concentration and pH on the attachment time and flotation recovery of quartz particles is demonstrated in Figures 9.33 and 9.34, respectively.[104] As shown in Figure 9.33, the highest flotation recovery occurs for the shortest attachment time. Outside the optimum range of collector concentration, the flotation recovery is low and the attachment time is long. If the concentration of the dodecylammonium hydrochloride (DAH) collector is lower than the optimum, the collector adsorption on particles is low and their surfaces are less hydrophobic. If the DAH concentration is higher than the optimum, the formation of a second layer of collector makes the particle surface hydrophilic again. The correlation between the attachment time and the surface hydrophobicity can be explained by

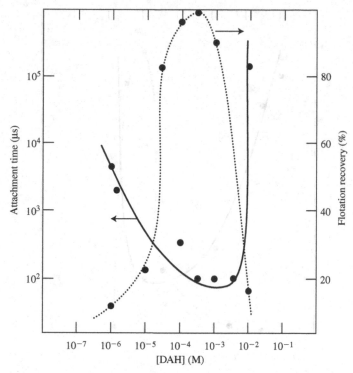

Figure 9.33 Effect of DAH concentration in water on the attachment time (given as time of electrical pulses which is proportional to the attachment time) and flotation recovery of quartz particles (140–203 μm) at pH = 6.6. Re-drawn from Ref. [104]; with permission of Elsevier.

considering the dependence of the induction time and critical film thickness on the contact angle, *viz*, equations (9.12) and (9.14), and the dependence of the TPCL spreading time on the contact angle, as described by equation (9.21). Similarly, outside the optimum range of pH shown in Figure 9.34, the adsorption density of DAH at particle surfaces decreases with decreasing pH and the precipitation of DAH to form neutral amine is significant at high pH.

9.7.5 Attachment of nano- and sub-micron sized particles

The attachment mechanisms described in the previous sections are governed by the particle inertia and hydrodynamic forces. The theories show that the collection efficiency of particles decreases with a decrease in the particle size. There also exists a second group of theories and models showing the opposite trend, namely, for very fine particles the collection efficiency increases with decreasing particle size. Therefore, the hydrodynamic and diffusion flotation and attachment régimes

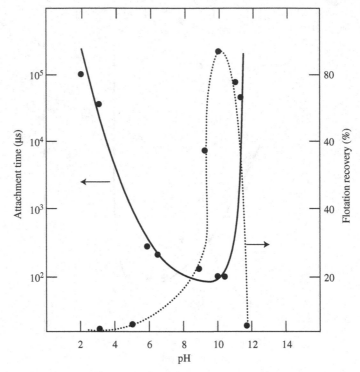

Figure 9.34 Effect of pH on attachment time and flotation recovery of quartz particles (140–203 μm) in 10^{-5}M DAH solutions. Re-drawn from Ref. [104]; with permission of Elsevier.

have been identified and are schematically shown in Figure 9.35. The results are interesting in that they predict a minimum in the collection efficiency at a particle diameter around 1 μm. Below this diameter, the collection of sub-micron and nano-sized particles is enhanced by Brownian diffusion, and for larger particles the hydrodynamic and inertial interactions are more favourable for collection.

Reay and Ratcliff[106] were among the first to assess the capture and attachment of fine particles by diffusion. The particle diffusivity, D, was determined by the Stokes–Einstein relation described by

$$D = \frac{k_B T}{6\pi \mu r} \tag{9.22}$$

where k_B is the Boltzmann constant, T is the absolute temperature and μ is the liquid viscosity. The particle capture was modelled using the mass transfer theory in the high Peclet number régime. The diffusive collection efficiency, E, was calculated by

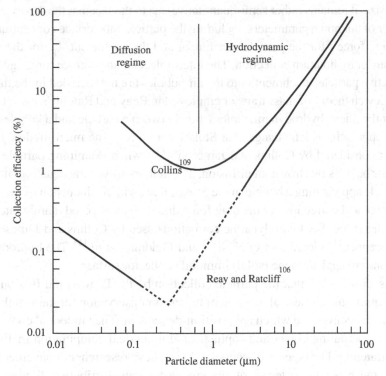

Figure 9.35 Collection efficiency as a function of particle diameter separating the Brownian diffusion from the hydrodynamic régime.

dividing the net flow of diffusive particles to the bubble surface per unit time by the number of particles swept out by the bubble per unit time, giving

$$E = f \frac{Sh}{Pe} = \frac{f}{(Pe)^{2/3}} \tag{9.23}$$

The Sherwood, Sh, and Peclet, Pe, numbers in equation (9.23) are, respectively, defined by $Sh = 2R_a k_p/D$ and $Pe = 2R_a U/D$, where R_a is the bubble radius, U is the bubble rise velocity and k_p is the particle mass transfer coefficient. In the high Peclet number régime of mass transfer, one obtains $Sh = Pe^{1/3}$.[107,108]

In equation (9.23), the function f is defined as $f = 1 - c_s/c$, where c and c_s are the particle concentrations in the bulk and near the bubble surface, respectively. The function f describes the dimensionless driving force for the diffusive mass transfer of particles towards the bubble surface and has the properties that $f \to 1$ for strong particle attachment onto the bubble surface and $f \to 0$ for weak particle attachment. Reay and Ratcliff[106] argued that f depends mainly on the chemical nature of the bubble and particle surfaces and the liquid phase. Although Reay and Ratcliff's theory predicts the correct trend for the collection of diffusive particles as a function of the particle size, it remains rather semi-quantitative due to the fact that the theory contains a number of unknown parameters, including the particle sub-surface concentration, c_s, the driving force f for the particle attachment and the coefficient, k_p, for the particle mass transfer by flotation collection. The intermolecular and surface forces governing the selective particle attachment onto the air bubbles are not included in the theory.

The Levich theory of mass transfer employed in Reay and Ratcliff's model did not consider the micro-hydrodynamic interaction between a particle and a bubble surface at close approach, which changes the Stokes drag force. The micro-hydrodynamics was later considered by Collins and Jameson[109,110] when examining particle collection by air bubbles by Brownian diffusion. The authors solved the equation of motion of a particle approaching a bubble in the Stokes flow, with the inclusion of the van der Waals force and corrections to the drag force due to micro-hydrodynamic interaction at short distances. The hydrodynamic corrections used by Collins and Jameson were due to Brenner,[111] Goren and O'Neill[112] and Goldman *et al.*[113] The hydrodynamic corrections are applied to the (solid) immobile collector surface.

It was shown[109,110] that the particle collection by the Brownian diffusion mechanism requires the numerical solution of the mass conservation for the simultaneous diffusion and convection which are position-dependent. The numerical computation needs the computing power and sophisticated numerical computation methods, as demonstrated by Davis and colleagues.[114–116] These researchers conveniently formulated the problem in terms of the stochastic pair-distribution function, $p(\vec{z})$, which represents the probability density of finding a particle with the centre lying at the radial position $\vec{z}$ (measured from the bubble centre). The pair-distribution function satisfies the quasi-steady Fokker–Planck equation described by[117]

$$\nabla \cdot [p(\vec{z}) \vec{V}(\vec{z})] = 0 \qquad (9.24)$$

where ∇ describes the divergence operator and $\vec{V}$ is the relative velocity between the bubble and the particle. The solution of equation (9.24) satisfies the boundary conditions that the function p is zero at the bubble–particle contact and is equal to unity if the particle is significantly far from the bubble surface. The relative velocity in equation (9.24) can be resolved in terms of the relative (micro-hydrodynamic) mobility functions (of the position vector, $\vec{z}$) into the bubble–particle inter-centre line and its perpendicular. The numerical solution of the Fokker–Planck equation is then integrated to obtain the collection efficiency described by

$$E = \frac{1}{\pi U (r + R_a)^2} \oint -p(\vec{z}) \vec{V}(\vec{z}) \frac{\vec{z}}{z} \, \mathrm{d}A \qquad (9.25)$$

The integral in equation (9.25) is carried out over the bubble surface at the radial distance of $(r + R_a)$. Typical exact numerical results obtained by Davis and colleagues are

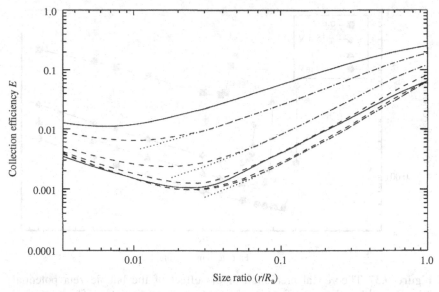

Figure 9.36 Numerical collection efficiency *versus* the particle-to-bubble size ratio at a fixed bubble Peclet number $Pe = 4 \times 10^6$. The upper solid line is for a bubble with a free interface and a scaled Hamaker constant $A = 0$ and the lower solid line is for a bubble with a rigid interface. The dashed lines are for $A = 5, 50, 500, 5000$ and ∞ (top to bottom). The dotted lines are the results of trajectory calculations in the absence of Brownian diffusion. Re-drawn from Ref. [116]; with permission of Elsevier.

shown in Figure 9.36. The unretarded component of the van der Waals force, determined by the Hamaker macroscopic approach (and the combined rules), was taken into consideration by the authors. Unfortunately, for asymmetric bubble–particle systems in flotation, the unretarded van der Waals force is repulsive and cannot be considered as the driving force for the selective attachment of particles onto the bubble surface. The retarded van der Waals force is influenced by the electrolyte concentration and can be attractive at large separation distances. The prediction of this attractive van der Waals force between a bubble and a particle has only recently become available.[118]

The effect of van der Waals, electrical double-layer and hydrophobic forces on the particle collection by diffusion was recently examined by Nguyen *et al.*[119] The modelling considers the particle trajectory approach and three predominant particle transport phenomena, including interception, gravity and Brownian diffusion in the mass balance equation, leading to

$$\frac{\partial c}{\partial t} + \nabla \cdot (c\vec{V} - \mathbf{D}\nabla c) = 0 \qquad (9.26)$$

where c is the particle concentration, t is the reference time, $\vec{V}$ is the non-Brownian component of the particle velocity and $\mathbf{D}$ is the tensor of the particle diffusivity

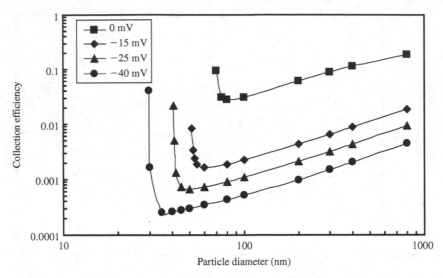

Figure 9.37 Theoretical prediction of the effect of the bubble zeta potential (given) on the collection efficiency at various particle diameters. The numerical constants were: bubble diameter (150 μm), particle refractive index (1.54), density (2600 kg m^{-3}), ζ-potential (-35 mV), Debye constant (50 nm^{-1}), hydrophobic force parameters (for the double exponential approximation): $K_1 = -7$ mN m^{-1}, $K_2 = -6$ mN m^{-1}, $\lambda_1 = 6$ nm and $\lambda_2 = 20$ nm. Re-drawn from Ref. [119]; with permission of Elsevier.

which is a function of the micro-hydrodynamic correction factors. The particle velocity, $\bar{V}$, is determined from the force balance by considering the intermolecular and surface forces, the liquid resistance and gravitational forces as well as the effect of the micro-hydrodynamics on the particle motion around the bubble. The numerical solution of equation (9.26) was then integrated to obtain the collection efficiency, described by

$$E = \frac{1}{c_\infty(U + V_s)} 2\int_0^\pi -J_r(\varphi) \sin \varphi d\varphi \qquad (9.27)$$

where J_r is the particle flux towards the bubble surface, c_∞ is the particle concentration in the bulk solution (far from the bubble surface), V_s is the particle terminal settling velocity, U is the bubble slip velocity and φ is the polar angle on the bubble surface (measured from the front stagnation point).

The recent theoretical results are shown in Figure 9.37.[119] The comparison between the model and experimental data obtained with silica nanoparticles from different sources is shown in Figure 9.38. The experimentally measured zeta potential for these particles was -35 mV. Furthermore, the non-DLVO colloidal forces between hydrophobic surfaces, referred to as hydrophobic forces, have not been precisely

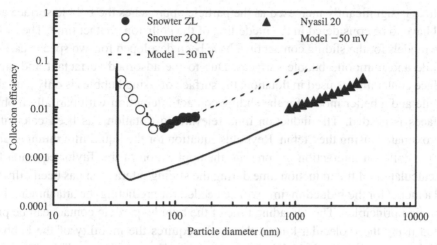

Figure 9.38 Comparison between model (lines) and experimental data (points) for collection efficiency of silica particles *versus* particle diameter. The numerical constants used in the calculation were the same as in Figure 9.37, except for $K_2 =$ $-20\,\text{mN}\,\text{m}^{-1}$ in the simulation with $-30\,\text{mV}$ for the bubble zeta potential. Re-drawn from Ref. [119]; with permission of Elsevier.

described and modelled and were therefore estimated in the modelling using the available experimental results obtained with the surface force apparatus and atomic force microscopy.[30] The hydrophobic forces are very long range and are stronger than the van der Waals attractive forces by many orders-of-magnitude.[120] Direct measurements with the surface force apparatus and atomic force microscopy now reveal, *e.g.* Refs. [121–123], that very small (nanometre and sub-micron) gas bubbles in solution and at hydrophobic surfaces[124–126] may influence the strong long range attraction between hydrophobic surfaces. The prediction of the forces between hydrophobic surfaces in the presence of the sub-microscopic gas bubbles has been a major challenge.[127] As a result, the inclusion of the hydrophobic forces in the modelling of the particle collection by Brownian diffusion still remains an unsolved problem.

9.8 Conclusions

In this chapter, both flotation and the attachment of mineral particles to air bubbles have been reviewed. In the latter, the emphasis has been on the determination of the collision and sliding contact times, the induction time required for the liquid film thinning to a critical thickness, the spreading time of the TPCL and the attachment time measurements. The available models for the collision contact time considered the importance of the local deformation of the bubble surface by the particle collision but still predict short collision times. The local drag force on the particle is significantly increased as the particle approaches the bubble surface and will have to be considered in the modelling of the collision contact time. The available models for the sliding contact time have been developed for two special cases – mobile and immobile bubble surfaces. Due to the adsorbed surfactant and other surface contaminants used in flotation, the surface of rising bubbles is only partially mobile and a better model for the sliding contact interaction with partially mobile surfaces is needed. The induction time relevant to flotation has been calculated approximately using the Stefan–Reynolds equation for the liquid film thinning during the collision interaction. At present, the application of the Taylor equation for the calculation of the induction time during the sliding interaction has been difficult and a model for the induction time will be needed for predicting the attachment time from first principles. The spreading time of the bubble–particle contact can be predicted using the molecular-kinetic theory. It requires the mobility of the bubble–particle contact spreading which has to be determined by experiments. The attachment time has been successfully measured using the contact time and has provided useful information about the influence of the particle size, collector concentration and pH on flotation recovery. These attachment phenomena are significantly influenced by the particle mass and hydrodynamic forces, and are therefore relevant to the flotation of particles with size greater than about $10\,\mu$m. For nano- and sub-micron sized particles, the particle attachment and flotation are enhanced by Brownian diffusion, and the role of the electrical double-layer force and the hydrophobic surface forces in the particle attachment is critical. The prediction for the attractive forces between hydrophobic surfaces remains a major challenge. Ideally, the experimental data for the direct force measurements between a particle and a gas bubble in the flotation aqueous solutions can be used for the prediction of the particle attachment. Hopefully, the colloid probe technique developed with micro-fabricated atomic force microscope cantilevers will be able to provide the required data for the hydrophobic forces between a particle and a bubble, in the near future.

Acknowledgement

The authors gratefully acknowledge the financial support provided by the Australian Research Council for the Centre for Multiphase Processes.

References

1. S. Lu, R.J. Pugh and E. Forssberg, *Interfacial Separation of Particles*, Studies in Interface Science Series, Vol. 20, Elsevier, Amsterdam, 2005.
2. R.J. Pugh, *Min. Eng.*, **13** (2000), 151.
3. R.J. Pugh, in *Handbook of Applied Colloid and Surface Chemistry*, ed. K. Holmberg, Wiley & Sons, New York, 2001, Chapter 8, p. 143.

4. G. Johansson and R.J. Pugh, *Int. J. Miner. Process.*, **34** (1992), 1.
5. A. Dippenaar, *Int. J. Miner. Process.*, **9** (1982), 1.
6. R.D. Crozier and R.R. Klimpel, in *Frothing in Flotation*, ed. J.S. Laskowski, Gordon and Breach, New York, 1989, p. 257.
7. S.W. Ip, Y. Wang and J.M. Toguri, *Can. Metall. Quart.*, **38** (1999), 84.
8. J. Lucassen, *Colloids Surf.*, **65** (1992), 131.
9. E.G. Kelly and D.J. Spottiswood, *Introduction to Mineral Processing*, Wiley-Interscience, New York, 1982.
10. J.A. Solari and R.J. Gochin in *Colloid Chemistry in Mineral Processing*, eds. J.S. Laskowski and J. Ralston, Elsevier, Amsterdam, 1992, p. 395.
11. J. Kitchener and R.J. Gochin, *Water Res.*, **15** (1981), 585.
12. P.K. Weissenborn and R.J. Pugh, *Proc. XIX Int. Min. Proc. Congr.*, Elsevier, Amsterdam, 1995, p. 125.
13. V.S.J. Craig, B. Ninham and R.M. Pashley, *J. Phys. Chem.*, **97** (1993), 10192.
14. O. Paulson and R.J. Pugh, *Langmuir*, **12** (1996), 4808.
15. K. Theander and R.J. Pugh, *Colloids Surf. A*, **240** (2004), 111.
16. M. Rutland and R.J. Pugh, *Colloids Surf. A*, **125** (1997), 33, 46.
17. B.M. Johansson, *Ph.D. Thesis*, Royal Institute of Technology, Stockholm, 1999.
18. I. Nishkov and R.J. Pugh, *Proc. XVII Int. Min. Proc. Congr.*, Vol. 2, Elsevier, Amsterdam, 1991, p. 339.
19. M. Kaya and D. Oz, *Proc. XIX Int. Min. Proc. Congr.*, Elsevier, Amsterdam, 2005, p. 167.
20. G.M. Dorris and N. Nguyen, *J. Pulp Pap. Sci.*, **21** (1995), 55.
21. M.A.D. Alzevedo, J. Drelich and J.D. Miller, *J. Pulp Pap. Sci.*, **25** (1999), 317.
22. N. Fraunholcz, *Ph.D. Thesis*, Delft University of Technology, The Netherlands, 1997.
23. H. Shen, R.J. Pugh and E. Forssberg, *Colloids Surf. A*, **196** (2002), 63.
24. H. Shen, *Licentiate thesis*, Luleå University, Sweden, 2000.
25. D.W. Fuerstenau, (ed.), *Froth Flotation: 50th Anniversary Volume*, American Institute of Mining Engineers, New York, 1962.
26. O.H. Woodward, *A Review of the Broken Hill Lead–Silver–Zinc Industry*, Australian Institute of Mining and Metallurgy, Melbourne, 1952.
27. G.J. Jameson, S. Nam and M.M. Young, *Miner. Sci. Eng.*, **9** (1977), 103.
28. K.L. Sutherland and I.W. Wark, *Principles of Flotation*, Australian Institute of Mining and Metallurgy, Melbourne, 1955.
29. H.J. Schulze, in *Developments in Mineral Processing*, Vol. 4, ed. D.W. Fuerstenau, Elsevier, Amsterdam, 1983.
30. A.V. Nguyen and H.J. Schulze, *Colloidal Science of Flotation*, Marcel Dekker, New York, 2004.
31. A.M. Gaudin, *Flotation*, McGraw-Hill, New York, 1957.
32. J.A. Finch and G.S. Dobby, *Column Flotation*, Pergamon, Oxford, 1990.
33. R.H. Yoon and G.H. Luttrell, *Miner. Process. Extract. Met. Rev.*, **5** (1989), 101.
34. G.S. Dobby and J.A. Finch, *Int. J. Miner. Process.*, **21** (1987), 241.
35. H.J. Schulze, *Miner. Process. Extract. Met. Rev.*, **5** (1989), 43.
36. B.V. Derjaguin and S.S. Dukhin, *Izv. Akad. Nauk SSSR, OTN: Metalurgiya i Toplivo (Div. Tech. Sci.: Metall. and Fuel)*, **1** (1959), 82.
37. J.F. Anfruns and J.A. Kitchener, *Trans. Inst. Min. Metall. (Sect. C)*, **86** (1977), 9.
38. M.E. Weber and D. Paddock, *J. Colloid Interf. Sci.*, **94** (1983), 328.
39. K. Sutherland, *J. Phys. Chem.*, **52** (1948), 394.
40. Z. Dai, S. Dukhin, D. Fornasiero and J. Ralston, *J. Colloid Interf. Sci.*, **197** (1998), 275.
41. J. Ralston, S.S. Dukhin and N.A. Mishchuk, *Int. J. Miner. Process.*, **56** (1999), 207.

42. A.V. Nguyen, *Int. J. Miner. Process.*, **56** (1999), 165.
43. Z. Dai, D. Fornasiero and J. Ralston, *Adv. Colloid Interf. Sci.*, **85** (2000), 231.
44. J. Ralston, D. Fornasiero and R. Hayes, *Int. J. Miner. Process.*, **56** (1999), 133.
45. R.H. Yoon, *Aufbereit.-Tech.*, **32** (1991), 474.
46. A.V. Nguyen, J. Drelich, M. Colic, J. Nalaskowski and J.D. Miller, in *Encyclopedia of Surface and Colloid Science*, ed. P. Somasundaran, Marcel Dekker, New York, 2006, in press.
47. I. Sven-Nilsson, *Kolloid-Zeit.*, **69** (1934), 230.
48. Y. Ye, S.M. Khandrika and J.D. Miller, *Int. J. Miner. Process.*, **25** (1989), 221.
49. H.J. Schulze and G. Gottschalk, *Aufbereit.-Tech.*, **22** (1981), 254.
50. H.J. Schulze and G. Gottschalk, in *Recent Advances in Mineral Processing, Proc. 13th Int. Miner. Proc. Congr.*, Vol. 2, ed. J.S. Laskowski, Elsevier, Amsterdam, 1981, p. 63.
51. H.J. Schulze, B. Radoev, T. Geidel, H. Stechemesser and E. Toepfer, *Int. J. Miner. Process.*, **27** (1989), 263.
52. W. Philippoff, *Trans. Am. Inst. Min. Eng.*, **193** (1952), 386.
53. L.F. Evans, *Ind. Eng. Chem.*, **46** (1954), 2420.
54. A. Scheludko, B.V. Toshev and D.T. Bojadjiev, *J. Chem. Soc. Faraday Trans. 1*, **72** (1976), 2815.
55. Y. Ye and J.D. Miller, *Int. J. Miner. Process.*, **25** (1989), 221.
56. A.V. Nguyen, H.J. Schulze, H. Stechemesser and G. Zobel, *Int. J. Miner. Process.*, **50** (1997), 97.
57. A.V. Nguyen, H.J. Schulze, H. Stechemesser and G. Zobel, *Int. J. Miner. Process.*, **50** (1997), 113.
58. A.D. Nikolov and D.T. Wasan, *Colloids Surf. A*, **250** (2004), 89.
59. A.V. Nguyen, G.M. Evans, J. Nalaskowski and J.D. Miller, *Exp. Therm. Fluid Sci.*, **28** (2004), 387.
60. A.V. Nguyen, *Int. J. Miner. Process.*, **37** (1993), 1.
61. M. Krzan, K. Lunkenheimer and K. Malysa, *Colloids Surf. A*, **250** (2004), 431.
62. Y. Wang, D.T. Papageorgiou and C. Maldarelli, *J. Fluid Mech.*, **390** (1999), 251.
63. R.L. Stefan and A.J. Szeri, *J. Colloid Interf. Sci.*, **212** (1999), 1.
64. R.B. Fdhila and P.C. Duineveld, *Phys. Fluids*, **8** (1996), 310.
65. R.M. Griffith, *Chem. Eng. Sci.*, **17** (1962), 1057.
66. G.J. Jameson, *NATO ASI Series E*, **75** (1984), 53.
67. E.K. Zholkovskij, V.I. Koval'chuk, S.S. Dukhin and R. Miller, *J. Colloid Interf. Sci.*, **226** (2000), 51.
68. A.V. Nguyen, *Am. Inst. Chem. Eng. J.*, **44** (1998), 226.
69. G.S. Dobby and J.A. Finch, *J. Colloid Interf. Sci.*, **109** (1986), 493.
70. G. Gu, R.S. Sanders, K. Nandakumar, Z. Xu and J.H. Masliyah, *Int. J. Miner. Process.*, **74** (2004), 15.
71. G. Gu, Z. Xu, K. Nandakumar and J. Masliyah, *Int. J. Miner. Process.*, **69** (2003), 235.
72. A.V. Nguyen and G.M. Evans, *J. Colloid Interf. Sci.*, **273** (2004), 271.
73. A. Scheludko, *Kolloid-Zeit. und Zeit. für Polymere*, **191** (1963), 52.
74. S. Hartland and J.D. Robinson, *J. Colloid Interf. Sci.*, **60** (1977), 72.
75. B.P. Radoev, D.S. Dimitrov and I.B. Ivanov, *Colloid Polym. Sci.*, **252** (1974), 50.
76. R.K. Jain and I.B. Ivanov, *J. Chem. Soc. Faraday Trans. 2*, **76** (1980), 250.
77. I.B. Ivanov (ed.), *Thin Liquid Films*, Marcel Dekker, New York, 1988.
78. C.Y. Lin and J.C. Slattery, *Am. Inst. Chem. Eng. J.*, **28** (1982), 147.
79. C.Y. Lin and J.C. Slattery, *Am. Inst. Chem. Eng. J.*, **28** (1982), 786.
80. J.D. Chen and J.C. Slattery, *Am. Inst. Chem. Eng. J.*, **28** (1982), 955.
81. J.D. Chen, P.S. Hahn and J.C. Slattery, *Am. Inst. Chem. Eng. J.*, **30** (1984), 622.

82. D. Platikanov, *J. Phys. Chem.*, **68** (1964), 3619.
83. D. Hewitt, D. Fornasiero, J. Ralston and L.R. Fisher, *J. Chem. Soc. Faraday Trans.*, **89** (1993), 817.
84. R. Tsekov, *Colloids Surf. A*, **141** (1998), 161.
85. H.J. Schulze, K.W. Stockelhuber and A. Wenger, *Colloids Surf. A*, **192** (2001), 61.
86. B. Radoev, A. Scheludko and E. Manev, *J. Colloid Interf. Sci.*, **95** (1983), 254.
87. N.V. Churaev and Z.M. Zorin, *Adv. Colloid Interf. Sci.*, **40** (1992), 109.
88. O.I. Vinogradova, *Int. J. Miner. Process.*, **56** (1999), 31.
89. H.J. Schulze, S. Tschaljowska, A. Scheludko and C. Cichos, *Freiberg Forschungsh. A*, **A568** (1977), 11.
90. T.D. Blake and J.M. Haynes, *J. Colloid Interf. Sci.*, **30** (1969), 421.
91. R.G. Cox, *J. Fluid Mech.*, **168** (1986), 169.
92. O.V. Voinov, *Mech. Liquid Gas*, **5** (1975), 76.
93. C.M. Phan, A.V. Nguyen and G.M. Evans, *Langmuir*, **19** (2003), 6796.
94. J.G. Petrov, J. Ralston, M. Schneemilch and R.A. Hayes, *J. Phys. Chem. B*, **107** (2003), 1634.
95. S. Glasstone, K.J. Laider and H.J. Eyring, *The Theory of Rate Processes*, McGraw-Hill, New York, 1941.
96. J.I. Frenkel, *Kinetic Theory of Liquids*, Oxford University Press, Oxford, 1946.
97. H. Stechemesser and A.V. Nguyen, *Int. J. Miner. Process.*, **56** (1999), 117.
98. J.S. Laskowski and J.D. Miller, *Reagents Minerals Industry*, Transactions of the Institute of Mining and Metallurgy, London, 1984, p. 145.
99. J.D. Miller, J.S. Laskowski and S.S. Chang, *Colloids Surf.*, **8** (1983), 137.
100. J.D. Miller, C.L. Lin and S.S. Chang, *Coal Prep.*, **1** (1984), 153.
101. V.A. Glembotsky, *Izv. Akad. Nauk SSSR (OTN)*, **11** (1953), 1524.
102. J. Laskowski and J. Iskra, *Inst. Mining Met. Trans. Sect. C*, **79** (1970), C6.
103. J.L. Yordan and R.H. Yoon, *Process Technol. Proc.*, **7** (1988), 333.
104. R.H. Yoon and J.L. Yordan, *J. Colloid Interf. Sci.*, **141** (1991), 374.
105. U. Bilsing, H.J. Schulze, H. Gruner and D. Menzer, *Freiberg Forschungsh. A*, **408** (1967), 55.
106. D. Reay and G.A. Ratcliff, *Can. J. Chem. Eng.*, **51** (1973), 178.
107. V.G. Levich, *Physicochemical Hydrodynamics*, Prentice-Hall, New Jersey, 1962.
108. J.T. Davies, *Am. Inst. Chem. Eng. J.*, **18** (1972), 169.
109. G.L. Collins, *Ph.D. thesis*, University of London, UK, 1975.
110. G.L. Collins and G.J. Jameson, *Chem. Eng. Sci.*, **32** (1977), 239.
111. H. Brenner, *Chem. Eng. Sci.*, **16** (1961), 242.
112. S.L. Goren and M.E. O'Neill, *Chem. Eng. Sci.*, **26** (1971), 325.
113. A.J. Goldman, R.G. Cox and H. Brenner, *Chem. Eng. Sci.*, **22** (1967), 653.
114. M. Loewenberg and R.H. Davis, *Chem. Eng. Sci.*, **49** (1994), 3923.
115. J.A. Ramirez, A. Zinchenko, M. Loewenberg and R.H. Davis, *Chem. Eng. Sci.*, **54** (1999), 149.
116. J.A. Ramirez, R.H. Davis and A.Z. Zinchenko, *Int. J. Multiphase Flow*, **26** (2000), 891.
117. G.K. Batchelor, *J. Fluid Mech.*, **52** (1972), 245.
118. A.V. Nguyen, G.M. Evans and H.J. Schulze, *Int. J. Miner. Process.*, **61** (2001), 155.
119. A.V. Nguyen, P. George and G.J. Jameson, *Chem. Eng. Sci.*, **61** (2006), 2494.
120. H.K. Christenson and P.M. Claesson, *Adv. Colloid Interf. Sci.*, **91** (2001), 391.
121. J. Ralston, S.S. Dukhin and N.A. Mishchuk, *Adv. Colloid Interf. Sci.*, **95** (2002), 145.
122. A.V. Nguyen, J. Nalaskowski, J.D. Miller and H.-J. Butt, *Int. J. Miner. Process.*, **72** (2003), 215.

123. H.K. Christenson and P.M. Claesson, *Science*, **239** (1988), 390.
124. J.D. Miller, Y. Hu, S. Veeramasuneni and Y. Lu, *Colloids Surf. A*, **154** (1999), 137.
125. J. Yang, J. Duan, D. Fornasiero and J. Ralston, *J. Phys. Chem. B*, **107** (2003), 6139.
126. N. Ishida, T. Inoue, M. Miyahara and K. Higashitani, *Langmuir*, **16** (2000), 6377.
127. N. Mishchuk, J. Ralston and D. Fornasiero, *J. Phys. Chem. A*, **106** (2002), 689.

Anh Nguyen (left) is Associate Professor at The University of Newcastle, New South Wales, Australia. He received the degrees of B.E. (1987) and Ph.D. (1992) in mineral processing from the Technical University of Kosice, Czechoslovakia. His current research interests include intermolecular and surface forces, interfacial rheology of adsorbed surfactants and particles in foam drainage and emulsification, and colloidal hydrodynamics of bubble–particle interactions. He has authored and co-authored over 100 publications, including one book on the colloidal science of flotation.

Robert Pugh (centre) holds positions as a Scientist, Institute for Surface Chemistry, Stockholm (1981-present), Adj. Professor, Luleå Technical University, and Associate Professor, Physical Chemistry Dept., Lund University. He has a Ph.D. from Imperial College, London (1972). Dr. Pugh has specialized in thin films and dispersed dynamic systems. He is a member of the American Chemical Society, the Royal Society of Chemistry and the American Institute of Metals and Mining. He has also held industrial research positions in Dow Chemicals, Zürich and Unilever Research. He has over 140 publications and books and consults in the area of surface and colloid chemistry for many different industries.

Graeme Jameson (right) has a B.Sc. from the University of New South Wales (1960) and a Ph.D. from the University of Cambridge (1963), both in chemical engineering. He is currently Laureate Professor and Director of the Centre for Multiphase Processes, University of Newcastle, Australia. His research interests are in the fluid dynamics of systems involving particles and bubbles. Recently he has been focussing on physical aspects of flotation, including particle–bubble interactions and the behaviour of particles in froths and foams. He is the inventor of the Jameson Cell, which is in worldwide use for the processing of minerals and fine coal, for the recovery of hydrocarbons from solvent extraction liquors and for flotation of algae from waste waters.

Tel: +612-49216189; *Fax*: +612-49216920; *E-mail*: Anh.Nguyen@newcastle.edu.au

10

Antifoam Effects of Solid Particles, Oil Drops and Oil–Solid Compounds in Aqueous Foams

Nikolai D. Denkov and Krastanka G. Marinova

Laboratory of Chemical Physics and Engineering, Faculty of Chemistry, Sofia University, 1164 Sofia, Bulgaria

10.1 Introduction

10.1.1 The antifoam effect

Foams appear as an integral part of various technological applications, such as ore and mineral flotation, tertiary oil recovery, production of porous insulating materials, fire fighting and many others. Foams are also encountered in certain types of consumer products, *e.g.* the mousses and ice-creams as food products, and shaving and styling foams in cosmetics.[1,2] It has been known for many years that the presence of oil droplets and/or hydrophobic solid particles in the aqueous foaming solutions can strongly reduce the foaminess and foam stability, which might be a problem in various applications.[1-9] For example, the fat particles in food products and the droplets of silicone oil used in personal care products (such as shampoos and hair/skin conditioners) have a strong antifoam effect, which should be suppressed to achieve an acceptable product quality from a consumer viewpoint.

On the other hand, excessive foaming might create serious problems in many industrial processes. Typical examples are during fermentation in drug and food manufacturing, the processing of drug emulsions and suspensions, pulp and paper production, industrial water purification, beverage production and packaging, textile dyeing, oil rectification and many others.[1,4,6] That is why special additives called "antifoams" or "defoamers" are widely used in these and other industrial applications to suppress foam formation or to destroy already formed foam.[1,3-9] The antifoams are also indispensable components of several everyday commercial products, such as washing machine detergents, paints and anti-flatulence drugs.[4]

We illustrate the antifoam effect in Figure 10.1 by showing the foam volume *versus* time, $V_F(t)$, for foam generated from a micellar solution of anionic surfactant Aerosol OT, AOT (sodium bis-2-ethylhexylsulphosuccinate) in the presence of

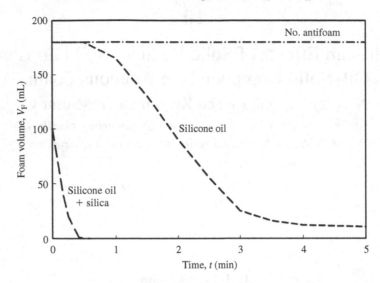

Figure 10.1 Foam volume V_F *versus* time t for foam generated by 10 hand-shakes (Bartsch test) from a 10 mM solution of the anionic surfactant AOT in the absence of antifoam and in the presence of 0.01 wt.% of two different antifoams – silicone oil or a compound of silicone oil + hydrophobized silica particles. Adapted from Ref. [22]; with permission of the American Chemical Society.

0.01 wt.% of two different antifoams — silicone oil and a compound of silicone oil + hydrophobized silica particles. One sees that the reference foam (in the absence of antifoam) is stable, whereas the two antifoams lead to relatively rapid foam decay; note the different manner of foam destruction by the two antifoams. The silicone oil does not affect the foaminess of the solution and the foam destruction starts only after an initial induction period, lasting for about 1 min after the foaming agitation was stopped. The main course of foam destruction continues for an additional 2 min and, afterwards, the residual foam remains stable for tens of minutes. In contrast, the oil + particle compound significantly reduces the foaminess of the AOT solution and destroys the foam completely in less than 20 s. These different patterns of foam destruction are related to two different modes of antifoam action,[8,10–25] which are explained in Section 10.2.3 and are discussed throughout this chapter.

Various terms are used to characterize the antifoam performance. The activity of antifoams characterizes their ability to prevent foam generation during agitation and/or to destroy rapidly pre-generated foams. Thus higher antifoam activity means less-generated foam and/or faster foam destruction.[3–9] The antifoam exhaustion (deactivation) is a process in which the antifoam loses its activity in the course of foam destruction.[5–8,12,19–21,24–28] The durability of an antifoam characterizes its ability to destroy a larger total amount of foam before exhaustion, or to maintain the

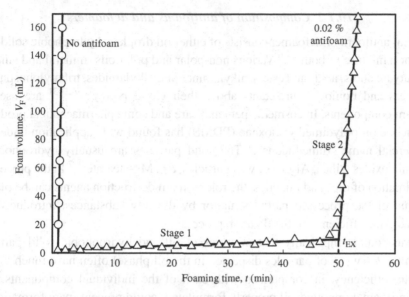

Figure 10.2 Foam volume V_F *versus* foaming time in the foam rise method (bubbling test) for a 10 mM AOT solution in the absence and in the presence of 0.02 wt.% silicone oil + silica particles compound (fast antifoam). Adapted from Ref. [20]; with permission of EDS.

instantaneous foam volume below a specified value (during continuous foaming) for a longer period of time.[8,12,19–21,24,25] Finally, the term antifoam efficiency is used to characterize the antifoam in a general sense, with respect to both activity and durability.

The meaning of these terms is illustrated in Figure 10.2, which shows the effect of an antifoam compound on the foam evolution in the foam rise test.[8,20,25] In this method, a controlled flux of nitrogen gas is continuously blown through a parallel set of glass capillaries and the dependence $V_F(t)$ is monitored. As seen from Figure 10.2, the increase in the foam volume is very fast in the absence of antifoam. In contrast, when only 0.02 wt.% of oil + silica particle compound was pre-dispersed in the foaming solution, the foam volume remained below 10 ml for about 50 min, as a result of the rapid rupture of the foam bubbles by the antifoam. After this initial stage of low foam volume being maintained (stage 1), a sudden, almost complete loss of the antifoam activity is observed; see the rapid foam growth during stage 2 in Figure 10.2. The sharp break in the curve $V_F(t)$, denoted by t_{EX}, indicates the moment of antifoam exhaustion, when the process of bubble destruction by the antifoam becomes too slow to compensate for the bubble generation. The total volume of the foam destroyed by the antifoam before its exhaustion is one possible measure of the antifoam durability.[8,20,25]

10.1.2 Composition of antifoams and defoamers

A typical antifoam or defoamer consists of either oil droplets, hydrophobic solid particles or a mixture of both.[3–10] Various non-polar and polar oils (mineral and silicone oils, fatty alcohols, acids and esters, alkylamines and alkylamides, tributylphosphates, thioethers and nonionic surfactants above their cloud point)[3,8,9,29–31] are used as antifoam components. In cosmetic, personal care and some pharmaceutical products, the silicone oil polydimethylsiloxane (PDMS) has found wide application under the commercial name "dimethicone".[4] The solid particles are usually hydrophobized inorganic oxides (silica, Al_2O_3) or wax particles, *e.g.* Mg-stearate.[3–7] In the process of froth flotation of ores and minerals, the role of foam destruction agents can be played by some of the processed particles and/or by the oily substances, introduced for enhancing the efficiency of the flotation process.

It was found empirically that mixtures of oil and hydrophobic solid particles (typically, 2–6 wt.% of particles dispersed in the oil phase) often have much higher antifoam efficiency, in comparison with each of the individual components.[3–8,23] Such antifoam compounds, if properly formulated, could prevent foam formation or destroy entirely the foam for seconds, at concentrations as low as 0.01–0.1 wt.% (see Figures 10.1 and 10.2). The mixture of PDMS and hydrophobized silica is a widely used antifoam compound in various technologies and consumer products (detergent powders, drugs) and it is sold under the commercial name "simethicone".[4] The reasons for the strong synergistic effect between oils and particles in the antifoam compounds are discussed in Section 10.5.

The commercial antifoams are usually sold in the form of oil-in-water emulsions with mean drop size between 3 and 30 μm. The size of the solid particles in the antifoam compounds is typically between 0.1 and several μm. Observations by optical and electron microscopy show that the solid particles in oil–solid compounds tend to adsorb on the surface of the oil drops (see Figure 10.3).[3,8,19,28,32] Since these particles are too hydrophobic to create steric stabilization of the compound globules (similar to that occurring in Pickering emulsions), appropriate surface-active polymers or additional, more hydrophilic solid particles are used to stabilize the commercial compound-in-water emulsions. The comparative studies of foam destruction by compounds and their emulsions showed that virtually the same mechanisms are operative for both forms (note that the compound is actually emulsified in the surfactant solution during foaming).[8,10,12,19] That is why, in the following consideration, we do not differentiate between a compound and its emulsion.

The terms antifoam and defoamer are usually used as synonyms. Sometimes, these terms distinguish two different ways of applying foam-destruction agents, which might have similar composition.[5,6] The antifoams are pre-dispersed in the foaming solutions with the major aim to prevent the formation of excessive foam upon solution agitation. In contrast, defoamers are sprayed or spread over an already

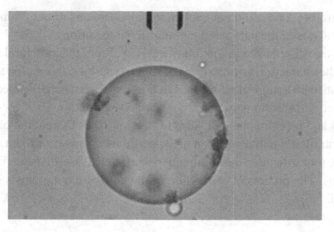

Figure 10.3 Optical microscopy image of antifoam globule containing silicone oil and silica particles (the dark objects of irregular shape adsorbed on the drop surface) dispersed in surfactant solution. Scale bar at top = 32 μm. Adapted from Ref. [8]; with permission of the American Chemical Society.

formed foam column, with the major aim to induce rapid foam collapse ("shock effect"). In this chapter we do not emphasize the differences between antifoams and defoamers, because the basic mechanisms of foam destruction are similar for both the substances. The main difference between antifoams and defoamers from a mechanistic viewpoint is the importance of the so-called "entry barrier", which characterizes how difficult it is for a pre-dispersed antifoam globule to enter the air–water surface. Since the defoamers are applied from the air phase, there is no entry barrier to prevent their appearance on the foam surface. The latter circumstance facilitates foam destruction by a defoamer, even when the entry barrier for pre-dispersed antifoam of the same composition is relatively high and, hence, the antifoam is not very efficient. All results presented here are obtained with surfactant solutions typical of detergent and personal care products, with antifoams that were pre-dispersed in the solutions prior to starting the foaming process.

10.1.3 Aim and structure of this chapter

The major aim is to present briefly our current understanding of the basic mechanisms of foam destruction by various types of antifoam entities – solid particles, oil drops and globules of mixed oil–solid compounds. The antifoam mechanisms always involve attachment of these entities to the fluid air–water surface, often followed by bridging of the surfaces of two neighboring bubbles and subsequent bubble coalescence. As explained in Section 10.5, the attachment of solid particles to the oil–water interface in antifoam compounds is very important for their high activity. Therefore, one could not explain the mechanisms of antifoaming without a detailed analysis of the particle–fluid interface interactions.

This chapter is organized in the following way. First, the structural elements of a foam and the characteristic timescales of foam dynamics, in relation to the mechanisms of antifoam action, are discussed in Section 10.2. Then, the mechanisms of foam destruction by solid particles are discussed in Section 10.3, by oil drops in Section 10.4 and by compound globules in Section 10.5. Depending on the particular type of antifoam entities and on the specific mechanism of foam destruction, the main factors affecting the antifoam performance are discussed throughout these sections. The process of exhaustion of the antifoam compounds is discussed in Section 10.5.3.

10.2 Foam Structure, Dynamic Timescales and General Modes of Antifoam Action

In this section we describe the general phenomena observed in the processes of foam formation and decay. The description is mostly phenomenological, without considering details of the specific mechanisms of antifoam action. These details depend strongly on the type of antifoam used (solid particles, oil drops or mixed oil–solid globules) and are discussed in the following Sections 10.3–10.5.

10.2.1 Foam structure

The bubble compaction in foams leads to the formation of several structural elements which have different dimensions and play different roles in the processes of foam destruction by antifoams. These elements[7,33,34] are the foam films intervening between two adjacent bubbles, the Plateau borders (PBs) and the nodes where four PBs meet with each other (see Figure 10.4). Since the characteristic dimensions and the capillary pressures of the PBs and their neighboring nodes are very similar, we discuss explicitly only the PBs, keeping in mind that the same phenomena (water drainage, oil drop entrapment and compression) occur in the nodes.

In the current section we discuss the characteristic dimensions of the foam films and PBs for foams, which are in mechanical equilibrium with underlying surfactant solution, under the action of gravity. The dynamics of approaching this equilibrium and the characteristic timescales of the respective processes are considered in the following Section 10.2.2.

10.2.1.1 Foam films

The foam films are characterized by their thickness, h, and radius, R_F (see Figure 10.4(e)). The film radius in dry foams (with volume fraction of air $\phi_A > 98\%$) is about twice as small as the bubble radius, $R_F \approx 0.5R_B$. The equilibrium film

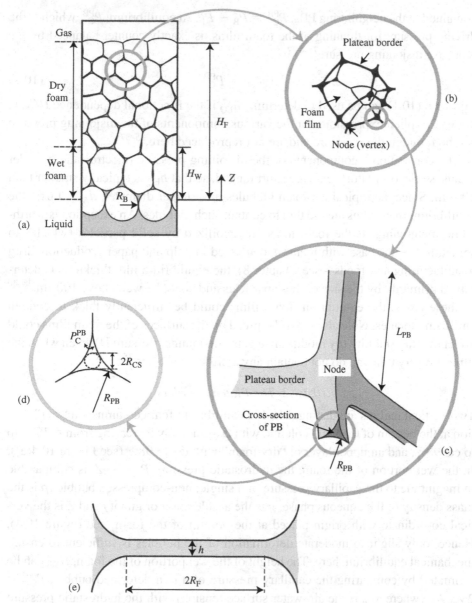

Figure 10.4 Schematic presentation of (a) foam column which is wet at the bottom and dry at the top and (b)–(e) the basic structural elements of the foam.

thickness, h_{EQ}, is determined by the balance of the capillary pressure of the PB walls, P_C^{PB}, and the disjoining pressure, $\Pi(h)$, which characterizes the forces acting between the two opposite surfaces of the foam film. The capillary pressure is defined as the difference between the pressure in the bubbles and the pressure in the liquid contained in the neighboring PBs, $P_C^{PB} = P_B - P_L$. At equilibrium, P_C^{PB}, which is the driving pressure for thinning of the foam films, is exactly counter-balanced by the repulsive disjoining pressure[7,34–37]

$$\Pi(h_{EQ}) = P_C^{PB} \tag{10.1}$$

Equation (10.1) can be used to determine h_{EQ} if the functional dependence $\Pi(h)$ is known. Explicit expressions for the various components of the disjoining pressure can be found in the literature and are not reproduced here.[35–37]

The conventional components of the disjoining pressure (electrostatic, van der Waals, steric, oscillatory, *etc.*) are short ranged, so that h_{EQ} is typically smaller than 100 nm. Since the typical antifoam globules have larger diameter, $d_A > 1 \, \mu m$, the equilibrium foam films are too thin to contain such globules. An exception is worthwhile mentioning. If the foam films are stabilized by solid particles (which can sometimes be the case with foams encountered in pulp and paper production, drug manufacturing and foods – see Chapter 8), the equilibrium film thickness is determined primarily by the size of these particles and could be well above 100 nm.[38–40] In these cases, the equilibrium foam films could be sufficiently thick to contain antifoam globules. Note also that oils spread on the surfaces of the foam films could affect the film stability (by modifying *e.g.* the disjoining pressure Π), even when the films are very thin and do not contain any antifoam globules.[15]

10.2.1.2 The Plateau borders

Two vertical regions are distinguished in equilibrium foam columns – a "wet" portion at the bottom of the foam column, with ϕ_A gradually increasing from $\approx 76\%$ up to ca. 98%, and an upper layer of "dry foam" with $\phi_A > 98\%$ (see Figure 10.4(a)). In the wet portion of the foam, the hydrostatic pressure, $P_g = \rho g Z$, is comparable in magnitude to the capillary pressure of a single, non-compressed bubble (ρ is the mass density of the aqueous phase, g is the acceleration of gravity and Z is the vertical co-ordinate with origin placed at the bottom of the foam (see Figure 10.4). Hence, only slight to moderate deformation of the bubbles is sufficient to ensure mechanical equilibrium here. The height of the wet portion of the foam, H_W, can be estimated by comparing the capillary pressure of a non-deformed bubble, $P_C^{NB} = 2\gamma_{aw}/R_B$ (where γ_{aw} is the air–water surface tension) with the hydrostatic pressure $P_g(H_W) = \rho g H_W$. This estimate leads to $H_W \sim 2\gamma_{aw}/(\rho g R_B)$, which is of the order of a centimeter for millimeter-sized bubbles. Therefore, foam columns with height above *ca.* several centimeters are prevailingly "dry" at equilibrium.

The bubbles in the upper, dry portion of the foam are strongly deformed to ensure sufficiently high capillary pressure, P_C^{PB}, which is able to compensate the increased hydrostatic pressure of the liquid in the PBs.[33,34] Important features of the dry foam are: (i) the main fraction of its liquid content is contained in the PBs and (ii) the length of the PBs, L_{PB}, is much larger than their cross-sectional dimension. For equilibrium dry foam, one can estimate the two characteristic dimensions of the PBs, namely the cross-sectional radius, R_{CS} and the radius of curvature of the PB wall, R_{PB}, which play important roles in the process of antifoam globule entry. R_{CS} determines whether antifoam globules of given radius get trapped or move freely in the interconnected network of PBs and nodes, whereas R_{PB} determines the capillary pressure, P_C^{PB}, which compresses the trapped antifoam globules, see Section 10.4.3.

To estimate R_{CS} and R_{PB} in the dry portion of the foam, one can start with the balance of the hydrostatic pressure, $P_g = \rho g Z$ (which is the driving force for water drainage from the foam column) and the capillary pressure, P_C^{PB} (which is the driving pressure for water suction from the surfactant solution into the foam)[7,8,14,16,34]

$$P_C^{PB}(Z) \approx \rho g Z \qquad (10.2)$$

Then, the radius of curvature of the wall of the Plateau channel, R_{PB}, can be estimated from P_C^{PB} by the expression[7,8,16,34]

$$R_{PB}(Z) = \frac{\gamma_{aw}}{P_C^{PB}(Z)} = \frac{\gamma_{aw}}{\rho g Z} \qquad (10.3)$$

Finally, the cross-sectional radius of the PB, R_{CS} (which is equal to the radius of a sphere inscribed in the Plateau channel, Figure 10.4(d)), can be found from geometrical considerations[8,14,16]

$$R_{CS}(Z) = \left(\frac{2\sqrt{3}}{3} - 1\right) R_{PB}(Z) \approx 0.155 \frac{\gamma_{aw}}{\rho g Z} \qquad (10.4)$$

These estimates are used in Section 10.4.3 to explain the effects of oil drops (as antifoams) and co-surfactants (as foam boosters) on foam stability.

10.2.2 Dynamics of foam evolution in the absence of antifoams

The foam evolution in the absence of antifoams is governed mainly by three inter-related processes: (i) thinning of the foam films, (ii) water drainage from the foam column and (iii) bubble coarsening due to gas diffusion from the smaller to the neighboring larger bubbles across the intervening foam films (analogous to Ostwald ripening of small crystallites and emulsion droplets).[7,15,33] The fourth possible process, namely bubble coalescence as a result of foam film rupture, is not considered in this sub-section, because we are interested mainly in foams which do not decay in the absence of antifoams.

The aforementioned processes (i)–(iii) are characterized by different timescales, which have an important impact on the modes of foam destruction by various antifoams. For this reason, the characteristic times of these processes are briefly described below and used in the subsequent Section 10.2.3 to explain the observed general modes of antifoam action.

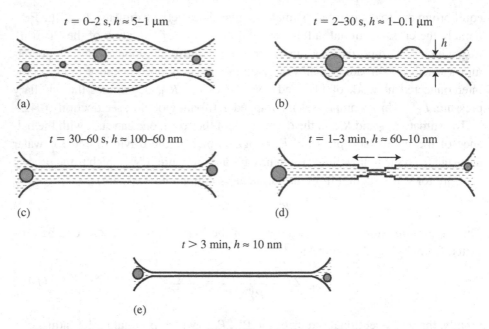

Figure 10.5 Main stages of foam film thinning: (a) Convex-lens-shaped "dimple", (b) planar film containing thicker channels, (c) plane-parallel film, (d) stratification – stepwise film thinning through formation and expansion of thinner spots and (e) thin film which is in equilibrium with the surrounding meniscus. The antifoam drops/particles are also shown schematically for comparison with the film thickness in the respective stage.

10.2.2.1 Dynamics of foam film thinning

Optical observations of millimeter-sized foam films, formed from solutions of low molecular mass surfactants like those used in detergency, showed that film thinning typically occurs in several consecutive stages (see Figure 10.5).[6,7,14,36] Dimple formation ($t \approx 0$–2 s, $h \approx 5$–1 μm) is where a convex-lens-shaped "dimple", with larger thickness in its center, is initially formed upon the mutual approach of two bubbles. This configuration is hydrodynamically unstable and an asymmetric outflow of liquid from the film leads to dimple disappearance within seconds after film formation. Drainage of a relatively thick planar film containing channels ($t \approx 2$–30 s, $h \approx$ 1–0.1 μm) follows in which the film contains several channels (dynamic regions with thickness 200–500 nm larger than the remaining planar portion of the film) and gradually thins down. Drainage of the thin plane-parallel film ($t \approx 30$–60 s, $h \approx 100$–60 nm) is the process in which the film has almost uniform thickness which gradually decreases with time. Film stratification ($t \approx 1$–3 min, $h \approx 60$–10 nm, see Figure 10.5(d)) is the stage realized in surfactant solutions with sufficiently high micelle volume fraction. The film thins in consecutive steps, which are due to oscillatory structural forces caused by "layering" of micelles in the film interior.[6,35,37,41,42] Each stepwise transition corresponds to a reduction of the number of micelle layers in the film (5 to 4, 4 to 3, *etc.*). Finally, the equilibrium black film ($t > 3$ min, $h \approx 5$–10 nm) appears in equilibrium with the surrounding meniscus.

The experiments with larger, centimeter-sized foam films also showed an initial stage of dimple formation, followed by hydrodynamic instability, which led to liquid outflow and reduction of the film thickness down to 1–2 μm within several seconds after film formation.[10] An important conclusion from the observations of foam film dynamics is that the first stages of film thinning are very short and the film thickness becomes smaller than the diameter of the typical antifoam globules, $d_A > 1$ μm, in less than 30 s. This fact implies that the antifoam globules, which are trapped in the foam films in the initial stage of film formation, should either break the films in less than 30 s or should leave them with the draining water. Thus one can estimate the characteristic time of foam film thinning, in relation to film rupture by antifoam globules, as $\tau_F \sim 30$ s.

The rate of foam film thinning is sometimes estimated by the Reynolds equation (see *e.g.* Ref. [43])

$$V_{RE} = -\frac{dh}{dt} = \frac{2}{3}\frac{h^3}{\eta R_F^2}\left[P_C - \Pi(h)\right] \qquad (10.5)$$

where V_{RE} is the rate of film thinning, P_C is the capillary pressure of the bubble, $\Pi(h)$ is the disjoining pressure, R_F is the film radius and η is the liquid viscosity. For h larger than *ca.* 50 nm, $\Pi(h)$ is usually negligible in comparison with P_C, and the driving pressure for film thinning can be estimated as $P_C \sim 2\gamma_{aw}/R_B$, where $\gamma_{aw} \approx 30$ mN m^{-1} and R_B is the bubble radius. The comparison of the theoretical predictions of equation (10.5) with the optical observations of foam films showed that Reynolds equation strongly under-estimates the rate of foam film thinning (*i.e.* the characteristic time of film thinning is strongly over-estimated), especially for foam films with diameter of the order of centimeters or millimeters. Theoretical and experimental studies showed[43–46] that this large discrepancy is due to the fact that the foam film surfaces in systems stabilized by low molecular mass surfactants are not usually plane-parallel or tangentially mobile. These two effects are not accounted for in the Reynolds equation and lead to faster film thinning, compared with the prediction of equation (10.5). Therefore, the Reynolds equation is rarely appropriate for estimating the rate of thinning of foam films with diameter above 1 mm, in relation to the antifoam effect.

10.2.2.2 Water drainage from quiescent foams

Recent theoretical and experimental studies[47–51] revealed three main periods in the water drainage from foam columns. The characteristic time, τ_{DR}, of the first two periods, which are governed by gravity, could be estimated by considering the water flow through the network of PBs and nodes. This network is considered as a porous medium with certain permeability, which depends on the liquid volume fraction, ϕ_L, and changes in the course of the drainage process.[47,49] The following estimate was derived theoretically[49] and confirmed experimentally for τ_{DR}

$$\tau_{DR} \sim \frac{H_F}{v_F} = \frac{H_F}{\left(K_m \rho g R_B^2/\eta\right)\phi_L^m}, \quad m = \frac{1}{2} \text{ or } 1 \tag{10.6}$$

where H_F is the foam height, v_F is the average velocity of the liquid through the network of PBs, K_m is a numerical constant ($K_1 \approx 6 \times 10^{-3}$ and $K_{1/2} \approx 2 \times 10^{-3}$) and m is a parameter which takes the value of 1 or 0.5 for tangentially immobile and tangentially mobile surfaces of the PBs, respectively. For the effects of tangential mobility, see refs. [49, 50]. Taking a typical value for a wet foam with $\phi_L \approx 0.25$, generated from an aqueous surfactant solution with $\rho = 1000\,\text{kg m}^{-3}$, one obtains the following estimate for the characteristic time of water drainage from the foam

$$\tau_{DR} \sim \frac{H_F \eta}{10 R_B^2} \tag{10.7}$$

Thus, for a foam with height $H_F = 20\,\text{cm}$ and mean bubble diameter $2R_B = 1\,\text{mm}$, made from aqueous solutions with viscosity $\eta = 1\,\text{mPa s}$, one obtains $\tau_{DR} \approx 80\,\text{s}$. As seen from equation (10.7), the characteristic drainage time strongly depends on bubble size and solution viscosity. The decrease of bubble diameter by a factor of 3 (down to $300\,\mu\text{m}$) leads to an increase of τ_{DR} up to $750\,\text{s}$.

The theoretical models[47,49] predict that the first stage of water drainage lasts for a period $\tau_{DR}/(m + 1)$, and during this stage about half of the total liquid contained in the initial foam, V_{L0}, drains linearly with time

$$\frac{V_{DR}(t)}{V_{L0}} = \frac{t}{\tau_{DR}}, \quad t \leq \frac{\tau_{DR}}{(m+1)} \tag{10.8}$$

The drainage of the remaining fraction of liquid during the second stage is somewhat slower and proceeds according to the expression

$$\frac{V_{DR}(t)}{V_{L0}} = 1 - \frac{m}{(1 + m)^{1+\frac{1}{m}}}\left(\frac{\tau_{DR}}{t}\right)^{\frac{1}{m}}, \quad t \geq \frac{\tau_{DR}}{(m+1)} \tag{10.9}$$

The latter equation shows that 95% of the liquid drains for 2–4 times τ_{DR}, depending on the tangential mobility of the PB walls (*i.e.* on the value of $m = 1$ or 0.5). As an illustration, Figure 10.6 presents the theoretically calculated volume of the liquid remaining in the foam during the process of water drainage, $V_L(t)/V_{L0} = [V_{L0} - V_{DR}(t)]/V_{L0}$, in accordance with equations (10.8) and (10.9). Note that water drainage can be slower in the presence of solid particles or oil drops in the

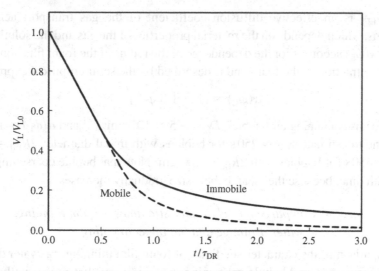

Figure 10.6 Change of the normalized liquid volume in the foam V_L/V_{L0} with the dimensionless time t/τ_{DR} as a result of water drainage. The theoretical predictions for bubbles with tangentially immobile (solid curve) or mobile (dashed curve) surfaces are plotted in accordance with equations (10.8) and (10.9).

foam, because these particles/drops obstruct the PBs and nodes, thus increasing the hydrodynamic resistance of the foam. This effect is not accounted for in equations (10.6–10.9).[5,6,14]

The theoretical modeling predicts that there is a third, much slower stage of liquid drainage, characterized by an exponential approach to the final equilibrium configuration of the foam at which the gravity force is counter-balanced by the capillary suction of the PBs.[48]

10.2.2.3 Characteristic time of bubble coarsening due to gas diffusion

Another process, which strongly affects the foam evolution, is the bubble coarsening due to gas diffusion across the foam films, from the smaller to the larger bubbles. The respective increase of the mean bubble radius is described by the expression[52,53]

$$R_B = R_{B0}\left(1 + \frac{t}{\tau_{CR}}\right)^{1/2} \tag{10.10}$$

where R_{B0} is the initial mean bubble radius. The characteristic time of bubble coarsening, τ_{CR}, can be estimated from the expression[53]

$$\tau_{CR} = \left[\frac{R_{B0}^2}{2D_{EFF}\alpha(\phi_L)}\right] \tag{10.11}$$

where D_{EFF} is an effective diffusion coefficient of the gas transport across the foam films, which depends on the material properties of the gas and the solution. The function $\alpha(\phi_L)$ accounts for the dependence of the radius of the foam films on the liquid volume fraction in the foam and is described by the semi-empirical expression[53]

$$\alpha(\phi_L) = [1 - 1.5\,\phi_L^{1/2}]^2 \qquad (10.12)$$

For dry foams containing air bubbles, $D_{EFF} \sim 5 \times 10^{-10}\,\mathrm{m^2 s^{-1}}$ and $\alpha(\phi_L) \approx 1$, which allows one to estimate $\tau_{CR} \approx 250\,\mathrm{s}$ for bubbles with initial diameter $2R_{B0} = 1\,\mathrm{mm}$, and $\tau_{CR} \approx 30\,\mathrm{s}$ for bubbles with $2R_{B0} = 300\,\mu\mathrm{m}$. Note that bubble coarsening decelerates with time, because the mean bubble size gradually increases.

10.2.2.4 Comparison of characteristic times for film thinning, water drainage and bubble coarsening

The comparison of the characteristic times of foam film thinning, τ_F, water drainage from the foam, τ_{DR}, and bubble coarsening, τ_{CR}, indicates that τ_F is usually shorter than both τ_{DR} and τ_{CR}. In other words, the foam films thin relatively rapidly, whereas the water drainage and bubble coarsening from the foam require longer time. For a typical foam with mean bubble diameter $2R_B = 1\,\mathrm{mm}$, one estimates $\tau_{DR} < \tau_{CR}$, *i.e.* drainage dominates the initial stage of foam evolution, as compared to bubble coarsening. In contrast, for foams containing smaller bubbles with $R_B < 400\,\mu\mathrm{m}$, τ_{CR} is shorter than τ_{DR} (*i.e.* the coarsening will be faster than drainage in the initial stage of foam evolution), due to the opposite dependences of τ_{DR} and τ_{CR} on R_B, see equations (10.7) and (10.11). At long times, $t \gg \tau_{DR}$, the processes of water drainage and bubble coarsening couple with each other and proceed in parallel.[51,53] The coarsening leads to a gradual increase of the mean bubble size, which is inevitably accompanied with a decrease of the number density of PBs and nodes per unit volume of the foam, and a slow drainage of the formed excess water.

Let us recall that the equilibrium PB cross-section, $R_{CS} \sim 5$–$50\,\mu\mathrm{m}$, is about two to three orders of magnitude larger than the equilibrium film thickness, $h_{EQ} < 50\,\mathrm{nm}$. The different magnitudes of R_{CS} and h, and the different timescales of the various processes, have important implications for the modes of antifoam action, which are discussed in the following sub-section.

10.2.3 Dynamics of foam destruction by antifoams

10.2.3.1 Location of the antifoam globule entry: fast and slow antifoams

The fact that the foam films become rapidly thinner than the antifoam globule size, whereas the PBs remain larger than these globules for longer time, means that there are two different scenarios for foam destruction by antifoams.[3,6,8,13] Foam film rupture by antifoam globules in the early stages of film thinning includes the formation of a

Figure 10.7 Image of a large (2 × 3 cm) foam film from a 10 mM AOT solution containing 0.01 wt.% of silicone oil + silica particles emulsion (fast antifoam) taken about 1 s after the film was formed and several milliseconds after the appearance of a hole in the film. The hole expands (illustrated by the black arrows) and eventually leads to film rupture. The observations are made by high-speed video camera. Taken from Ref. [11]; with permission of the American Chemical Society.

particle bridge (for solid particles) or an oil bridge (for oil drops or mixed oil–solid compounds) between the two opposite surfaces of the foam film – see Sections 10.3.1 and 10.4.1. If the formed bridge is unstable, it ruptures the foam film within seconds after its formation (see Figure 10.7). As a result, this mechanism of foam film rupture usually leads to complete foam destruction in a short period of time (seconds to tens of seconds); see the results for the oil–silica compound in Figure 10.1, for example. The term "fast antifoam" is used to denote substances which destroy the foam by this film-breaking mechanism.[8] The experiments showed[8,10–13,19–23] that mixed globules of appropriately formulated oil–solid compounds often (though not always) behave as fast antifoams, in which the solid particles ensure an ultra-low entry barrier, whereas the oil (if appropriately chosen) ensures unstable oil bridges.

By contrast, foam cell destruction after antifoam globule entry in the PBs and nodes can also take place. Here, the antifoam globules which are unable to break the foam films in the early stages of film thinning are first accumulated in the adjacent

PBs. They are afterwards compressed by the walls of the shrinking PBs (as a result of water drainage from the foam) and eventually enter the walls of the PBs, causing rupture of the neighboring foam films.[5,6,8,54] The foam destruction by this mechanism usually requires minutes or tens of minutes, because the water drainage from the foam (needed for compression of the trapped drops) is a slow process; see the results for silicone oil in Figure 10.1, for example. Furthermore, one usually observes residual foam which may remain stable for hours (the respective explanation is described in Section 10.4.3). The term "slow antifoam" is used to describe substances which destroy the foam by this mechanism.[8] The oil drops deprived of solid particles usually behave as slow antifoams.[5,6,8,14–18]

As explained in Section 10.4.3, one of the main factors determining whether a given antifoam would behave as fast or slow is the magnitude of the entry barrier of the antifoam globules. If the repulsive forces acting between the antifoam globules and the surfaces of the foam film are too strong, *i.e.* the entry barrier is high, the globules leave the foam film without forming bridges and the respective antifoam behaves as slow.

10.2.3.2 *Effect of antifoams on foaming: role of the kinetics of surfactant adsorption*

In general, the fast antifoams strongly reduce the foaminess of surfactant solutions, whereas the slow antifoams affect the foaminess only slightly.[8,14,15,17] To explain these effects, let us consider the process of foaming by agitation, *viz.* by shaking, stirring or pouring the solution from a nozzle. Such agitation leads to entrapment of air into the solution and to the formation of a wet foam with an air volume fraction below *ca.* 80%, in which the bubbles collide with each other. Foam films with thickness of several micrometers are formed in the zone of bubble–bubble contact. However, the intensive agitation rapidly separates the colliding bubbles from each other, so that the foam films have no time to thin. The film thickness remains larger than *ca.* 1 μm during the entire duration of the bubble encounters.

Therefore, an antifoam could reduce the solution foaminess only if its globules are able to destroy relatively thick foam films, *i.e.* if the antifoam behaves as fast. It is rather common to observe a weak effect, or even an increased foaminess of the surfactant solutions in the presence of dispersed oil droplets, whereas the same droplets may have a noticeable (slow) antifoam effect in quiescent foams generated from the same solutions (*e.g.* the results for silicone oil in Figure 10.1). The increased foaminess observed sometimes in the presence of oil-based antifoams is explained by reduced surface tension (as a result of oil spreading on the solution surface), which facilitates the expansion of the air–water surface and air entrapment.[17]

One particular feature of the antifoam effect during foaming and upon foam shear (*i.e.* under dynamic conditions) is that the antifoam activity can be affected strongly by the kinetics of surfactant adsorption.[8,19,32,55] The reason is that the bubble generation and deformation are related to the creation of new air–water surface, and it takes a certain time to cover this surface with a protective adsorption layer of surfactant. The kinetics of surfactant adsorption is particularly important for solutions of non-ionic surfactants, because the de-micellization rate and the monomer concentration are rather low in such solutions. As a result, the respective characteristic time of surfactant adsorption is often of the order of seconds and tens of seconds, which results in the formation of under-saturated adsorption layers upon bubble–bubble collision in dynamic foams. This effect facilitates the entry of the antifoam globules at the foam film surfaces and the subsequent foam film rupture in agitated foam, whereas the same antifoam could be rather inactive in quiescent foam of the same composition.

As an illustration of the effect of surfactant adsorption rate on the antifoam activity, we compare the foaminess and the foam stability of 10 mM AOT (anionic) and 0.6 mM APG (alkyl polyglucoside, nonionic) solutions (approximately 4 × critical micelle concentration, CMC) in the absence and in the presence of 0.01 wt.% silicone oil–silica compound. As seen from Figure 10.8(a), the initial foam volume generated by shaking AOT solutions in the presence of antifoam is several times larger than that generated from APG solutions under equivalent foaming conditions. However, after stopping the agitation, the AOT foam completely disappears within seconds, whereas the APG foam remains stable for hours. The reference samples of AOT and APG solutions (without antifoam) show rather good foaminess and no foam destruction in the timescale of Figure 10.8(a).

The results from related model experiments allowed us to explain the observed differences of the antifoam performance in AOT and APG solutions.[19] Measurements of the dynamic surface tension showed a very large difference in the adsorption kinetics of the two surfactants; more than 10 s were needed for saturation of the adsorption layer in the APG solution, whereas this process took less than 0.1 s in the AOT solution (see Figure 10.8(b)). Further, the entry barrier of the compound globules was much lower in AOT solutions (≈ 3 Pa) compared with APG solutions (>125 Pa) for saturated adsorption layers. The ultra-low entry barrier in AOT solutions explains why the antifoam globules are able to rapidly rupture the foam films in both agitated and quiescent AOT foams, whereas the high entry barrier in APG solutions precludes the foam film rupture in the APG-stabilized quiescent foams. Hence, the studied antifoam had significant activity in the foaming APG solution, mainly due to the slow APG adsorption. If the kinetics of APG adsorption was as fast as that of AOT, the antifoam would be much less efficient in suppressing the foaminess of APG solutions.

It is worthwhile noting that the results for APG-stabilized foams in Figure 10.8(a) illustrate the possible switch of a given antifoam from fast (during foaming) into

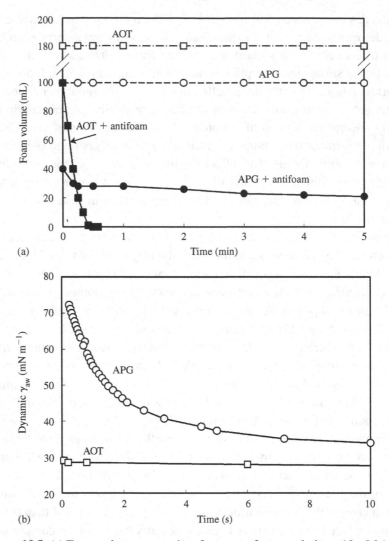

Figure 10.8 (a) Foam volume *versus* time for two surfactant solutions, 10 mM AOT and 0.6 mM APG, in the presence and in the absence of 0.01 wt.% silicone oil + silica compound (Bartsch test). (b) Dynamic surface tension of the same solutions measured by the maximum bubble pressure method in the absence of antifoam. Adapted from Ref. [19]; with permission of the American Chemical Society.

slow (in the quiescent foam). Indeed, the foaminess of the APG solution was strongly reduced in the presence of antifoam compound, whereas the evolution pattern of the quiescent APG foam was indicative of the action of a slow antifoam. As explained in the previous paragraph, such pattern reflects a low entry barrier of the antifoam globules during foaming due to formation of incomplete adsorption layers, and a high entry barrier when the adsorption layers are completed. Hence the antifoam behaves as fast (film breaking) during foaming and transforms into slow (acting through entry in the PBs) when the agitation is stopped and the surfactant is allowed to form complete adsorption layers on the foam film surfaces.

10.2.3.3 Effect of antifoams on the evolution of static (quiescent) foams

As already explained, the fast antifoams are able to destroy completely quiescent foams in a very short period of time (typically 3–30 s, depending on the antifoam activity and concentration), by rupturing the foam films in the early stages of their thinning. This means that the only important characteristic dimension in the foam is the film thickness, h, as compared to the diameter of the antifoam globules, d_A, and the relevant timescale is τ_F, which is imposed by the process of film thinning down to a thickness $h \sim d_A$. The other timescales discussed in Section 10.2.2 are too long to play an important role in the process of foam destruction.

In contrast, the dynamics of foam destruction by slow antifoams, *e.g.* oil drops in surfactant solutions of high concentration, is related to the rates of water drainage and bubble coarsening, because drop entrapment and compression in the PBs is needed for accomplishing the antifoam globule entry and foam cell destruction. In general, four distinct stages are observed in the evolution of quiescent foams destroyed by slow antifoams (see Figure 10.9):[8,14,15]

(i) During stage I, lasting for 1 to several minutes, the upper boundary of the foam does not change because no coalescence of the bubbles with the uppermost air phase takes place. The lower boundary of the foam rises with time, due to water drainage from the initially formed wet foam. During this stage the foam films rapidly thin down, the PBs and the nodes become much narrower (the duration of this stage is comparable to τ_{DR}) and the smallest bubbles shrink and disappear due to air diffusion across the foam films (characteristic time τ_C) (*cf.* Figures 10.10(a) and (b)).

(ii) During stage II ($t \gg \tau_{DR}$), the foam volume remains virtually constant because the water drainage is already very slow and no bubble coalescence occurs. However, optical observations evidence a significant restructuring of the foam cells during this period, due to bubble coarsening through air diffusion across the films. The latter process leads to a gradual, but significant decrease of the number density of PBs and nodes per unit volume of the foam. As a result, the antifoam globules gradually accumulate in the remaining nodes and PBs with time (see Figures 10.10(c) and (d)). In addition, the PBs and nodes slowly shrink with time due to water drainage from the foam, which leads to a decrease of the radius of curvature of the PB walls, R_{PB}, and to a gradual increase of the capillary pressure, P_C^{PB}, exerted by these walls on the trapped oil drops.

(iii) When a certain critical value of P_C^{PB} is reached, the foam destruction starts, primarily through rupture of the upper layer of bubbles where the compressing capillary pressure is the highest. This is the onset of stage III, denoted by t_{ON} in Figure 10.9(a). The rate of foam destruction, $v_D = -dH_F/dt$, is approximately constant during the main course of period III (see also the results for silicone oil in Figure 10.1).

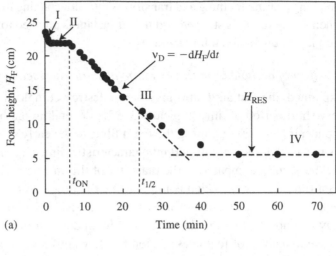

(a)

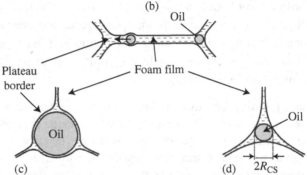

(b)

(c)   (d)

Figure 10.9 (a) Foam height, $H_F(t)$, *versus* time for a solution containing 0.1 M SDP3S (anionic surfactant), nonionic co-surfactant (foam booster) and 0.1 wt.% silicone oil as a slow antifoam. The time t_{ON} gives the onset of foam decay by bubble collapse and $t_{1/2}$ indicates the foam half-life. The roman numbers I–IV indicate different stages of foam evolution. Adapted from Ref. [15]; with permission of the American Chemical Society. (b) Schematic presentation of the oil drop migration from the foam film into the adjacent PB in the process of film thinning. (c) The walls of the shrinking PB compress the oil drop and asymmetric oil–water–air films are formed. (d) If the drop diameter is smaller than the cross-section of the PB, the drop is not compressed.

(iv) After a certain amount of foam is destroyed, the rate of foam decay gradually decreases and, eventually, stage IV is reached in which the foam volume remains almost constant for many minutes or even hours. Only large bubbles remain in the foam and the process of bubble coarsening is rather slow. The height of this residual, long-standing foam is denoted hereafter by H_{RES}. In Section 10.4.3 later, the process of foam destruction by slow antifoams is analyzed quantitatively, by relating the size of the antifoam globules and their entry barrier to the value of H_{RES} and to the stages of foam evolution described above.

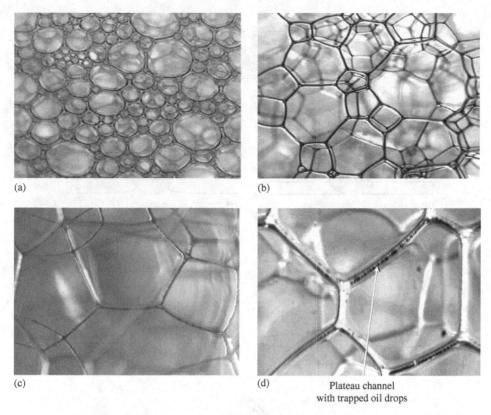

(a)

(b)

(c)

(d)   Plateau channel
with trapped oil drops

Figure 10.10 Photographs of foam cells just below the top of a foam column at different stages of the foam evolution (*cf.* Figure 10.9). (a) Wet foam in stage I. (b) Foam at the transition between stages I and II. (c) Air diffusion from the small bubbles towards the larger ones leads to the disappearance of the smallest bubbles and to the gradual accumulation of oil drops in the PBs during period II. When the capillary pressure at the top of the foam column exceeds the entry barrier of the oil drops, a destruction of the uppermost layers of bubbles is observed which is the beginning of stage III (not shown). (d) Enlarged view of Plateau channel in which trapped oil drops are seen. The image size is 2×1.5 cm in (a)–(c) and 1×0.75 cm in (d). Taken from Ref. [15]; with permission of the American Chemical Society.

10.3 Solid Particles as Antifoam Entities

10.3.1 Bridging–dewetting mechanism of foam film rupture: characteristic timescales

Experimental and theoretical studies showed that solid particles could rupture foam films by the so-called "bridging–dewetting" mechanism.[3,55–63] This mechanism implies that, first, the solid particle comes into contact with the two opposite surfaces of the foam film, forming a "solid bridge" between them (see Figure 10.11). Second, if the particle is sufficiently hydrophobic, the liquid dewets the particle surface so

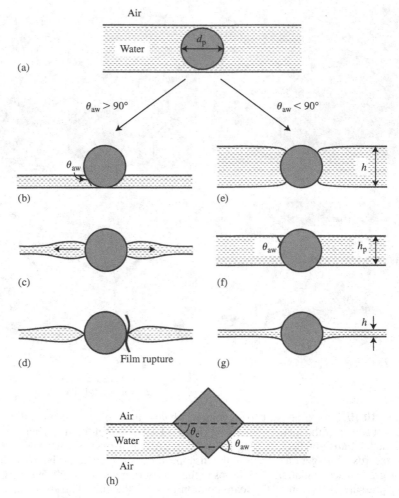

Figure 10.11 Schematic presentation of the bridging of foam film surfaces by a solid particle. (a)–(d) When the spherical particle has contact angle $\theta_{aw} > 90°$, it is dewetted by the liquid and the three-phase contact lines come into direct contact with each other perforating the foam film at the particle surface. (e)–(g) If the contact angle $\theta_{aw} < 90°$, the particle is not dewetted and the foam film remains stable. (h) Cone-shaped particle with slope angle θ_c can be dewetted if $\theta_{aw} > \theta_C$, when the particle is properly oriented to the film surfaces.

that the three-phase contact lines eventually come into direct contact with each other; the foam film gets perforated at the particle surface (see Figures 10.11(a)–(d)). The antifoam activity of solid particles is strongly related to their hydrophobicity, which is quantified by the three-phase contact angle air–water–solid, θ_{aw}, measured into water. Note that θ_{aw} depends not only on the particles used, but on the foaming solution as well.

Two timescales are important for the bridging–dewetting mechanism of foam film rupture by solid particles. First, the foam film should thin down to a thickness comparable with the particle diameter, d_P. This process is relatively fast, with a characteristic time $\tau_F(d_P) \leqslant 30\,s$, unless d_P is smaller than *ca.* $1\,\mu m$. Therefore, if the particles are with diameter $d_P > 1\,\mu m$, the rate of film thinning is not an obstacle for rapid foam destruction by this mechanism.

Another timescale is set by the moving contact lines along the particle surface, which push liquid from the vicinity of the particle into the neighboring film region (see Figure 10.11(c)). The characteristic time of this process can be estimated as, $\tau_{DEW} \sim (d_P/V_{DEW})$, where V_{DEW} is the sliding velocity of the contact lines along the particle surface. Direct optical observations by Dippenaar[58] showed that the motion of the contact lines (the particle dewetting) is very fast, $V_{DEW} \sim 1\,\mu m\,ms^{-1}$. Theoretical estimates, based on the hydrodynamic theory of contact line motion,[43] predict a similar value of V_{DEW}. By developing a detailed model of the dynamics of particle dewetting in foam films, Frye and Berg[55] calculated numerically τ_{DEW} and showed that, if the contact angle θ_{aw} is several degrees larger than the critical angle for dewetting (discussed in the next sub-section), τ_{DEW} is shorter than *ca.* $10\,ms$.

The above estimates show that the bridging–dewetting mechanism would lead to rapid foam destruction if solid particles with appropriate properties (see below) were present in the foaming solution. It is worthwhile noting that in detergency and many other applications where "strong" surfactants are typically used (*i.e.* very active surfactants with concentration above the CMC), solid particles are not very efficient foam breakers; the surfactant molecules adsorb on particle surfaces rendering them too hydrophilic to have a pronounced antifoam effect.[60–62] In such systems, oil-based antifoams are more efficient and have found wide application in practice.[3–5,60] On the other hand, foam control in the presence of solid particles is very important for the successful processing of ores and other minerals by the method of froth flotation, in which "mild" surfactants are typically used, see Chapter 9.

10.3.2 Influence of particle properties on antifoam effect

The two stages of the bridging–dewetting mechanism of foam film rupture by solid particles, Figure 10.11, imply two groups of important factors related to (i) bridge formation and (ii) particle dewetting. These factors are briefly discussed below.

With respect to bridge formation, the most important factors are the particle size and several other factors related to the particle entry barrier such as particle shape, surface charge, *etc.* As explained above, the particle size is not a problem for rapid bridge formation, unless $d_P < 1\,\mu m$. The factors related to the entry barrier are more difficult to quantify. As shown by Kulkarni *et al.*,[64,65] the surface charge could create significant electrostatic repulsion between the particle and the surfaces of the foam films (*e.g.* in the presence of adsorbed ionic surfactant), which might result in high entry barriers. An indirect proof for the importance of this effect was recently found in studies of Pickering emulsions stabilized by spherical solid particles of micrometer size.[66] Emulsification experiments showed that the particles enter the oil–water interface and stabilize the emulsions easier if the electrostatic repulsion between the particle and the oil–water interface is suppressed (*e.g.* at high NaCl concentration in the aqueous phase). Otherwise, the electrostatic barrier hampers particle adsorption and no stable emulsions are formed. Other types of surface forces caused by surfactant micelles (oscillatory surface forces[35,41,42]), adsorbed polymer molecules (steric repulsion), *etc.*, could create significant entry barriers which should be suppressed to effect bridge formation.

The experiments show that the entry barrier is strongly reduced and the bridges are easily formed when the particles have sharp edges. To explain this effect one can use Derjaguin's approximation,[67] which relates the force between a spherical particle and a planar surface, F_{PS}, with the disjoining pressure $\Pi_{PS}(H)$ between two planar surfaces, one of them being the foam film surface and the other (hypothetical) surface has the same properties as the particle surface

$$F_{PS}(\delta) = 2\pi R_P \int_{\delta}^{\infty} \Pi_{PS}(H)\mathrm{d}H \qquad (10.13)$$

Here R_P is the particle radius, δ is the distance between the particle forehead and the planar surface and H is a running variable. As seen from equation (10.13), the interaction force is proportional to the particle radius and is therefore lower in magnitude for smaller particles. If the solid particle is non-spherical, equation (10.13) can be used with R_P being replaced by the radius of curvature of the particle forehead, which is very small for sharp edges. Hence, the entry barrier of a solid particle with sharp edges (if properly oriented) would be much smaller, as compared to the barrier of a spherical particle having the same overall dimension. This "pin-effect" of the sharp edges strongly facilitates bridge formation for non-spherical particles.

With respect to dewetting, the most important factors are the particle hydrophobicity and shape. Theoretical and experimental studies[3,55,57–62] showed that the critical contact angle is 90° for complete dewetting of solid particles which have smooth convex surfaces, such as spheres, ellipsoids, disks and rods. Particles of contact angle $\theta_{aw} > 90°$ induce foam film rupture and foam collapse (see Figures 10.11(a)–(d)). Less hydrophobic smooth particles ($\theta_{aw} < 90°$) do not cause film rupture; they can even stabilize the foam by blocking the Plateau channels and reducing the rate of water drainage from the foam (see Figures 10.11(a), (e), (f) and (g)).

Various studies showed that foam films can be ruptured by less hydrophobic particles (θ_{aw} well below $90°$) if the latter have sharp edges and are properly oriented in the film.[3,55,57,58] An illustrative example of this possibility is the theoretical prediction for a cone-shaped particle, whose axis is oriented perpendicularly to the film surface (Figure 10.11(h)). Simple geometrical considerations show that cone-shaped particles with slope angle θ_c can be dewetted if

$$\theta_{aw} > \theta_c \tag{10.14}$$

Similar considerations imply that cubic particles with $\theta_{aw} > 45°$ could rupture foam films if the particles are oriented with their diagonal being perpendicular to the film plane. Indeed, Dippenaar[58] showed experimentally that cubic galena particles with contact angle $\theta_{aw} \approx 80°$ rupture the foam films if the particles are properly oriented. Frye and Berg,[55,63] and Garrett[57] showed that hydrophobic glass particles and poly(tetrafluorethylene) particles with irregular shape have a significant antifoam effect even when the contact angle $\theta_{aw} \approx 40°$. The *in situ* formation of similar, sharp-edged soap precipitates is the most probable reason for the observed significant antifoam effect of calcium soaps in the recent study by Zhang *et al.*[68]

It is worthwhile noting that, if the particle is too hydrophilic to be dewetted by the liquid, the two contact lines on the particle surface acquire equilibrium positions which depend on the contact angle θ_{aw} (see Figure 10.11(f)). The distance between the two equilibrium contact lines can be estimated for spherical particles by the expression

$$h_p = 2R_p \cos \theta_{aw} \tag{10.15}$$

Once such a stable bridge is formed, the evolution of the system depends on the ratio of the average film thickness, $h(t)$, and the value of h_p. When $h(t) > h_p$, the particle remains in the foam film and causes local film thinning. When the average film thickness, h, becomes smaller than h_p, the particle is expelled from the film into the neighboring meniscus regions, which have local thickness equal to h_p. The reason for this migration of the particle outside the film is that the surface energy of the system (film + particle + meniscus) is minimal when the particle does not deform the film surfaces, so that no extra surface energy is created by the particle presence. In foams, such insufficiently hydrophobic particles can obstruct the PBs and stabilize the foam as a result of the reduced rate of water drainage. Moreover, if the particles are of sufficiently high concentration to cover the entire bubble surface, the robust shell formed resists bubble shrinking and could arrest almost completely Ostwald ripening.

More detailed analysis of the role of shape, size and contact angle of solid particles for their antifoam activity can be found in the papers by Garrett,[3,57] Frye and Berg,[55,63] and Aveyard *et al.*[59-62]

10.4 Antifoam Effect of Oil Drops

The possible mechanisms of foam destruction by oil drops compared to solid particles are more versatile, due to the possibilities for oil bridge deformation and oil spreading. Several scenarios of foam destruction by oils were proposed in the literature and are discussed in this section.

In most cases, the oil drops destroy the foams through an initial accumulation in the PBs and nodes, *i.e.* the drops behave as slow antifoams.[5–8,14–18,54] This is due to the relatively high entry barrier when the drops are dispersed in surfactant solutions typical for detergency. The oil drops are able to act as fast antifoams and to break the foam films by the bridging mechanisms, described in Section 10.4.1, if the entry barrier is low (*e.g.* the surfactant adsorption layers are incomplete due to low concentration and/or slow kinetics of adsorption). One efficient way to reduce the entry barrier and to transform the oil into a fast antifoam is to add hydrophobic solid particles, and thus to form oil–solid compounds which are considered in Section 10.5. The spreading of the oil makes possible other, non-bridging mechanisms of foam film rupture, which are discussed in Section 10.4.2. Due to the important role of the entry barrier for the antifoam activity and for the specific mode of foam destruction, it is considered separately in Section 10.4.3. In many studies, the antifoam activity is correlated with the so-called entry, E, spreading, S, and bridging, B, coefficients; hence these coefficients are discussed throughout the section, in relation to the various mechanisms of antifoam action.

10.4.1 Bridging–stretching and bridging–dewetting mechanisms

In this sub-section, we first describe two bridging mechanisms of foam film rupture by oil-based antifoams. Then we discuss the formation and stability of the oil bridges and the related entry and bridging coefficients.

10.4.1.1 Bridging–dewetting mechanism

This mechanism is often discussed in the literature in relation to oil-based antifoams,[3–6,9,28,32,59–63] by analogy with the foam film rupture by hydrophobic solid particles. The mechanism implies that, once formed, the oil bridges are dewetted by the aqueous phase due to the hydrophobic nature of the oily surface (see Figure 10.12). One should note, however, that in the surfactant solutions with concentration ⩾ CMC, the three-phase contact angle θ_{aw} is usually below 90° even for very hydrophobic surfaces.[22,26–28,59–62] Therefore, dewetting of a spherical oil drop is improbable in such solutions, unless the surfactant concentration is low and/or the adsorption is very slow.

In contrast, when the antifoam globule is deformable (oil drop or oil–solid mixture with a large excess of oil), it acquires an equilibrium, non-spherical shape after

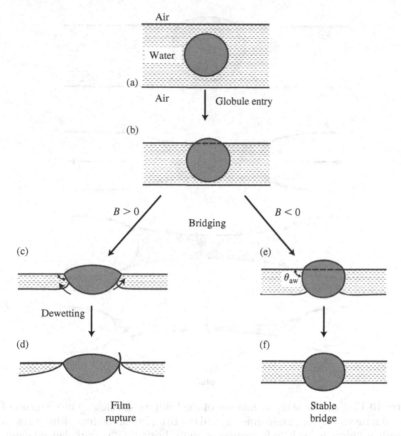

Figure 10.12 Schematic presentation of the bridging of foam film surfaces by an oily globule. (c) and (d) If $B > 0$, the foam film can be ruptured by the bridging–dewetting mechanism or by the bridging–stretching mechanism shown in Figure 10.13. (e) and (f) If the bridging coefficient $B < 0$, the bridge is stable and no film rupture is effected.

the first entry into one of the foam film surfaces.[3,8,11,61,69,70] If the three-phase contact angle θ_{aw} (now air–water–oil measured through water) is larger than 90°, the drop acquires the shape of a bi-convex lens, as shown in Figure 10.12(c). Simple geometrical consideration[3] shows that such a lens can be dewetted by the opposite foam film surface at the moment of oil bridge formation, if no significant change of the lens shape occurs during dewetting (Figures 10.12(c) and (d)). At the present time, there is no unambiguous evidence that the bridging–dewetting mechanism is operative for oil-based antifoams. However, there are no arguments to discard this possibility (*e.g.* for viscous, non-spreading oils) so that future experiments are expected to clarify this issue.

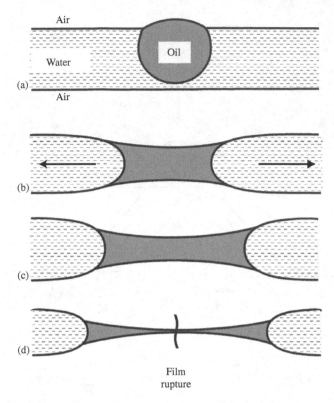

Figure 10.13 Schematic representation of the bridging–stretching mechanism of foam film rupture by fast antifoams. (a) and (b) Bridging of the foam film surfaces by antifoam globule leads to formation of an oil bridge with non-balanced capillary pressures at the oil–water and oil–air interfaces. (c) and (d) The bridge stretches with time until a thin, unstable oil film is formed in the bridge center. The rupture of this oil film leads to destruction of the entire foam lamella.

10.4.1.2 Bridging–stretching mechanism

Optical observations[8,10] with surfactant solutions containing silicone oil-based antifoams showed that, once an oil bridge was formed in the foam lamella, it acquired a biconcave shape with the thinnest region being in the bridge center (see Figure 10.13). Such a bridge is unstable due to uncompensated capillary pressures at the oil–water and oil–air interfaces. As a result, the oil bridge spontaneously stretches with time in a radial direction, so that eventually a thin unstable oil film is formed in the bridge center. The rupture of this oil film results in the perforation of the entire foam lamella. An important requirement for realization of this bridging–stretching mechanism is the possibility for deformation of the antifoam globule. Therefore, such a mechanism cannot be realized with oil drops which are gelled by polymerization or in the presence of solid particles at high concentration or of inappropriate hydrophobicity.[23] A detailed description of the bridging–stretching mechanism can be found in the original papers[10,11] and in the recent review.[8]

10.4.1.3 Entry coefficient

Whatever the mechanism is of foam destruction by antifoam globules, it requires the globules first to enter the solution surface for oil spreading and/or bridge formation to occur. Two different types of factors, thermodynamic and kinetic ones, determine the possibility for realization of drop entry. The thermodynamic aspect is usually discussed in terms of the oil entry coefficient, E, whereas the kinetic aspect is discussed in terms of the drop entry barrier.

The oil entry coefficient[3,61,70] can be calculated from the interfacial tensions of the air–water, γ_{aw}, oil–water, γ_{ow} and oil–air, γ_{oa} interfaces (see Figure 10.14)

$$E = \gamma_{aw} + \gamma_{ow} - \gamma_{oa} \qquad (10.16)$$

The value of E depends not only on the oil used but also on the type and concentration of surfactant, electrolyte and co-surfactant, as well as on various other factors which affect the interfacial tensions.

The thermodynamic analysis[3,61,70] shows that negative values of E correspond to complete wetting of the oil drop by the aqueous phase. This means that, even if an oil drop has appeared on the solution surface (*e.g.* as a result of oil deposition from the air phase), this drop would spontaneously immerse into the aqueous phase because this is the thermodynamically favored configuration (see Figures 10.14(a) and (b)). Pre-emulsified oil drops with $E < 0$ remain immersed inside the aqueous phase and cannot form oil bridges between the surfaces of the foam films or PBs. As a result, oils with negative E are inactive as antifoams[3,61,70] (note that $E < 0$ implies that the other two coefficients are also negative, $S < 0$ and $B < 0$). In contrast, positive values of E correspond to a defined equilibrium position of the oil drop/lens at the air–water surface. Hence, when the oil has positive E and the entry barrier is not too high, stable or unstable oil bridges can be formed in the foam films.[3,7,8,61,69]

To illustrate the relation between the entry coefficient, E, and the entry barrier, let us draw an analogy with the concepts used in chemical kinetics. A positive value of E is the thermodynamic condition for the existence of an equilibrium position of the oil drops at the air–water surface and for formation of oil bridges in the foam films, whereas the entry barrier plays the role of a kinetic barrier which can preclude the realization of these thermodynamically favored configurations, *viz.* the oil drop can remain arrested in the aqueous phase for kinetic reasons.

10.4.1.4 Stability of oil bridges in foam films: bridging coefficient

The first theoretical study of the stability of oil bridges in foam films in relation to antifoaming was made by Garrett.[69] He analyzed the conditions for mechanical

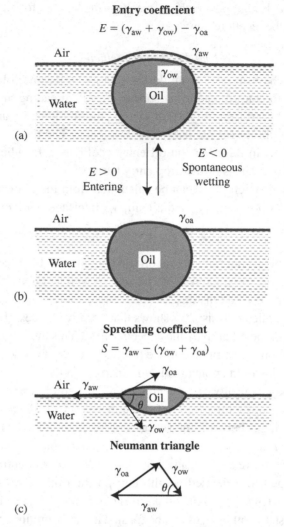

Figure 10.14 Schematic presentation of the meaning of the entry, (a) and (b), and spreading (c) coefficients.

equilibrium of an oil bridge, which is placed in a foam film with perfectly planar surfaces (see Figure 10.15). The mechanical equilibrium of such a capillary system requires the balance of (i) the capillary pressures across the various interfaces and (ii) the interfacial tensions acting on the three-phase contact lines. The second balance can be expressed by the Neumann triangle, illustrated in Figure 10.14(c).

To assess the oil bridge stability, Garrett checked whether the capillary pressures across the oil–air and oil–water interfaces ($P_{oa} \equiv P_o - P_a$ and $P_{ow} \equiv P_o - P_w$) and the Neumann triangle at the bridge periphery can be simultaneously balanced (Figure 10.15). The analysis showed that if the contact angle $\theta_{aw} > \pi/2$, the capillary

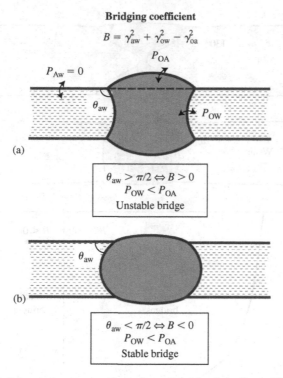

Bridging coefficient

$$B = \gamma_{aw}^2 + \gamma_{ow}^2 - \gamma_{oa}^2$$

(a)

$$\theta_{aw} > \pi/2 \Leftrightarrow B > 0$$
$$P_{OW} < P_{OA}$$
Unstable bridge

(b)

$$\theta_{aw} < \pi/2 \Leftrightarrow B < 0$$
$$P_{OW} < P_{OA}$$
Stable bridge

Figure 10.15 Schematic presentation of an oil bridge in a foam film with planar surfaces according to Garrett's model.[3,69] (a) Unstable bridge with positive bridging coefficient, $B > 0$ ($\theta_{aw} > 90°$) and (b) stable bridge with negative bridging coefficient, $B < 0$ ($\theta_{aw} < 90°$).

pressure P_{ow} is always smaller than P_{oa}. In other words, it is impossible to achieve mechanical equilibrium of an oil bridge with $\theta_{aw} > \pi/2$. Such bridges are considered unstable and they rupture the foam films by either of the bridging mechanisms discussed above (Figure 10.15(a)). In contrast, when $\theta_{aw} < \pi/2$, both the Neumann triangle and the pressure balance can be satisfied, so that the respective oil bridges are mechanically stable and no antifoam effect of the oil is expected (Figure 10.15(b)).

By applying the cosine theorem to the Neumann triangle, Garrett[69] proved that the requirement $\theta_{aw} > \pi/2$ is equivalent to the condition

$$B \equiv \gamma_{aw}^2 + \gamma_{ow}^2 - \gamma_{oa}^2 > 0 \tag{10.17}$$

where B is the bridging coefficient. It can be shown theoretically that positive values of B necessarily mean positive entry coefficient, E, while the reverse statement is not always true.[37,70] In conclusion, Garrett's analysis predicts that oils with $B > 0$ would form unstable bridges and *vice versa*.

An important assumption in Garrett's model is that the surfaces of the foam film are perfectly planar. A more complex model, which accounts for the possible deformation

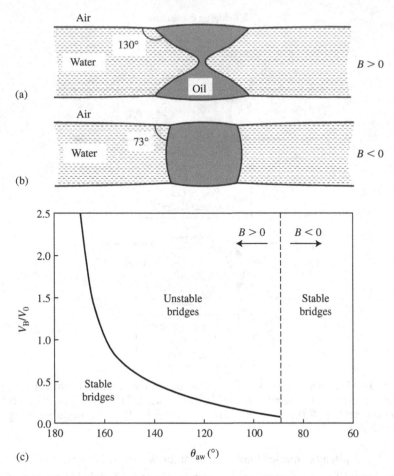

Figure 10.16 Equilibrium shape of oil bridges with different air–water–oil contact angles being (a) 130° and (b) 73°. (c) The dimensionless bridge volume, V_B/V_0, plotted as a function of the contact angle θ_{aw}, where $V_0 = (\pi h^3/6)$. The remaining parameters in the calculations are $\gamma_{oa} = 20.6 \, \text{mN} \, \text{m}^{-1}$, $\gamma_{ow} = 4.7 \, \text{mN} \, \text{m}^{-1}$ and $h = 2 \, \mu\text{m}$. Adapted from Ref. [11]; with permission of the American Chemical Society.

of the foam film surfaces by the oil bridge, was developed in Ref. [11]. This model revealed that the bridge stability and shape depend not only on the contact angle θ_{aw}, but also on the bridge volume, V_B. Another important conclusion of this model was that the oil bridges could acquire equilibrium shapes (satisfying Neumann triangle and balanced capillary pressures) for both positive and negative values of B. These equilibrium shapes usually include a certain deformation of the foam film surfaces not accounted for in Garrett's model. As an illustration, Figures 10.16(a) and (b) shows theoretically calculated equilibrium shapes of two oil bridges with $\theta_{aw} = 130°$ and 73°, respectively. Note that the equilibrium bridge shown in Figure 10.16(a) is with $\theta_{aw} > 90°$, which corresponds to $B > 0$.

The theoretical analysis in Ref. [11] revealed that some of the bridge shapes describe stable equilibrium configurations, whereas other shapes describe unstable equilibrium configurations, corresponding to a local maximum of the system energy. Therefore, the complete theoretical analysis of foam film stability in the presence of oil bridges requires one to clarify the domains of stable and unstable equilibrium bridges. The numerical calculations showed[11] that the bridge stability depends mainly on two factors: the value of θ_{aw} and the ratio V_B/V_0, where $V_0 = (\pi h^3/6)$ is the volume of an imaginary oil drop with diameter equal to the film thickness. As seen from Figure 10.16(c), only stable bridges exist at $B < 0$, in accordance with Garrett's prediction.[69] However, both stable and unstable equilibrium bridges can be formed at $B > 0$. The large bridges are always unstable, whereas the small bridges could be stable or unstable, depending on the value of θ_{aw}.[11]

10.4.2 Role of oil spreading in antifoam activity

10.4.2.1 Spreading coefficient

The spreading affinity of oils is usually discussed in terms of their spreading coefficients (see Figure 10.14(c))[3,60–62,70,71]

$$S = \gamma_{aw} - \gamma_{ow} - \gamma_{oa} \qquad (10.18)$$

It is important to distinguish between the initial spreading coefficient, S_{IN} (defined by using γ_{aw} in the absence of spread oil on the solution surface) and the equilibrium spreading coefficient, S_{EQ} (γ_{aw} in the presence of spread oil).[3,61,70] Rigorous thermodynamic analysis shows that $S_{EQ} \leqslant 0$, while S_{IN} might have arbitrary sign (see Refs. [3,61,70] for more explanation). Note that $S_{EQ} \leqslant S_{IN}$ because γ_{aw} decreases upon oil spreading.

The signs of S_{IN} and S_{EQ} bring direct information about the spreading behavior of the oils. The initial spreading affinity, after the oil is first deposited on the solution surface, is characterized by S_{IN}. Negative value of S_{IN} means that the oil does not spread on the surface. Positive S_{IN} means that the oil would spread as a thin or thick layer. Likewise, S_{EQ} brings information about the thickness of the equilibrium spread layer: if $\gamma_{aw} = \gamma_{ow} + \gamma_{oa}$ (*i.e.* $S_{EQ} = 0$), the Neumann triangle predicts $\theta_{aw} = 0$ and the oil spreads as a thick layer (so-called "duplex film"), whereas negative S_{EQ} and positive S_{IN} imply a thin equilibrium layer, possibly co-existing with oil lenses. The comparison of equations (10.16) and (10.18) shows that if $S > 0$, the entry coefficient is also positive, $E > 0$, because $(E - S) = 2\gamma_{ow} > 0$.

10.4.2.2 Relation between oil spreading and antifoaming

The first study relating the antifoam effect of oils with their spreading behavior was published by Ross.[72] He found that most of the oils with noticeable antifoam activity had positive spreading coefficients. Since that time, there has been an ongoing debate in the literature about the role of oil spreading in the antifoam mechanisms, *e.g.* see Refs. [3–17, 22, 28, 32, 54, 59–65, 70, 73–78]. The discussions are usually made in the context of the assumed mode of antifoam action. Ross[72] speculated that the oil should first connect the two foam film surfaces (*i.e.* an oil bridge is formed) and then spread as a thick layer, in order to replace a portion of the stable aqueous foam film by an unstable oil bridge so that the film rupture can occur. Later experimental studies[3,32] showed that foam film rupture through oil bridge formation is possible at negative values of S_{EQ} and S_{IN}, which means that there is no specific requirement for spreading of the oil as a thick layer (note that the theoretical analysis of the oil bridge stability requires $B > 0$ without imposing any requirement for the spreading behavior of the oil). In other words, oil spreading is not a necessary condition to have antifoam activity of the oils and oil-based compounds.[3,8,10,32] Nevertheless, the correlation between the spreading ability of the oils and their antifoam activity, which is observed in many systems, suggests that the spreading could either facilitate the foam destruction process by some of the bridging mechanisms discussed in Section 10.4.3 or induce another non-bridging mechanism of foam destruction.

Several non-bridging mechanisms were proposed in the literature. The "spreading-fluid entrainment" mechanism[3,28] implies that once an oil drop with positive S_{IN} enters either of the foam film surfaces, the oil spreads in a radial direction from the formed oil lens. This spreading is assumed to drag water in the foam film away from the oil lens, inducing in this way local film thinning and subsequent rupture. Several theoretical models based on this idea were published in the literature, but this mechanism has not found unequivocal confirmation in experimental studies.

In other studies, oil spreading on the surfaces of foam films was directly observed and related to the film rupture.[8,14,15] These observations showed that the spreading oil induces capillary waves of large amplitude on the surface of the foam films ($\Delta h \sim$ hundreds of nanometers). These waves covered almost the entire foam film and often led to film rupture within several seconds, even at relatively large average film thickness $h \approx 1\ \mu\text{m}$. As discussed in Ref. [8], the spreading oil probably "sweeps" some of the surfactant adsorbed on the foam film surface, which results in film destabilization for two main reasons.[8] First, the induced capillary waves (related to the decreased surface elasticity and viscosity of the diluted surfactant adsorption layers) lead to the formation of locally thin regions in the film, which allows film rupture at relatively large average thickness. Second, the diluted adsorption layers cannot stabilize efficiently the local thin spots against rupture, due to the reduced surface charge density on the film surfaces (hence suppressed electrostatic repulsion) and/or to the appearance of attractive hydrophobic forces.[35,37] This mode of foam film rupture was termed the "spreading-wave generation" mechanism of antifoam action.[8]

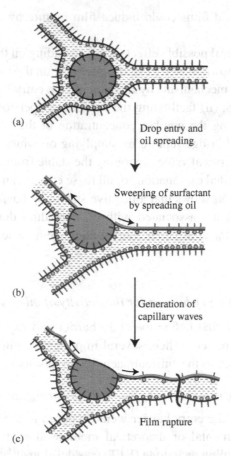

(a)

Drop entry and
oil spreading

Sweeping of surfactant
by spreading oil

(b)

Generation of
capillary waves

Film rupture

(c)

Figure 10.17 Schematic presentation of the "spreading-wave generation" mechanism. (a) and (b) Entry of an oil drop in the region of the Plateau channel leads to oil spreading on the surfaces of the neighboring foam films. (b) and (c) The spreading oil partially sweeps the surfactant from one of the film surfaces leading to the appearance of capillary waves. The local thinning of the foam film and the loose adsorption layer lead to foam film rupture.

An additional factor for capillary wave generation and foam film destabilization could be the asymmetric surfactant distribution which appears after the oil spreads on only one of the foam film surfaces (see Figure 10.17 for schematic presentation and Ref. [59] for experimental results), which is the typical case in defoamer applications. As shown by Binks *et al.*,[78] foam destabilization can be induced by oil vapor which adsorbs first on only one of the foam film surfaces. Then the oil molecules diffuse across the foam films to establish an equilibrium distribution of the oil and surfactant molecules on both surfaces of the foam films. Ivanov, Danov and co-workers[79–81] showed experimentally and analyzed theoretically that the diffusion of surface-active substances across liquid films could induce film rupture by a Marangoni type of instability.

Let us mention several possible effects of oil spreading on the bridging modes of antifoam action. As shown in Ref. [22], oil spreading on the foam film surfaces can facilitate the bridging mechanisms by (i) reducing the entry barriers of the emulsified antifoam globules, (ii) facilitating the antifoam dispersion inside the foaming solution, thus increasing the number concentration of the antifoam globules and (iii) facilitating the oil bridge rupture by supplying oil which increases the bridge volume V_B above the critical value separating the stable from unstable bridges (see Figure 10.16(c)). Detailed explanations of all these effects can be found in Ref. [8].

The above discussion shows that a positive spreading coefficient, S_{IN}, and high spreading rate, which are associated with oil spreading during foaming, could enhance significantly the antifoam activity without being a necessary pre-requisite for antifoam action.

10.4.3 Role of entry barrier for the activity of oil-based antifoams

In this sub-section we first define the entry barrier and explain the experimental method for its measurement. Then, several important results relating the magnitude of the entry barrier to the antifoam activity are discussed.

10.4.3.1 Film trapping technique for measuring the entry barrier

Several definitions of the entry barrier were proposed in the literature, which are related to the experimental or theoretical methods used for its determination. Recently, the film trapping technique (FTT) was developed[16,18] for quantifying the entry barrier of oil drops and mixed oil–solid antifoam globules. In this technique, the capillary pressure of the air–water surface is measured at the moment of drop/globule entry, $P_C^{CR} = (P_a - P_w)$, see Figure 10.18.

The use of P_C^{CR} as a characteristic of the entry barrier provides several important advantages in comparison with the other quantities used in the literature for this purpose. First, P_C^{CR} has a clear physical interpretation with respect to the antifoam action – it corresponds to the capillary pressure which compresses the oil drops in the actual foam (by the surfaces of the thinning foam film or by the walls of shrinking PBs) at the moment of drop entry. Thus the value of P_C^{CR} can be related to foam properties, such as foam height, bubble size, rate of water drainage, which affect the capillary pressure in real foams. Second, P_C^{CR} can be measured by the FTT for oil drops of micrometer size, possibly containing solid particles, like those encountered in practical systems. Therefore, no additional hypotheses are needed to transfer the conclusions from the model FTT experiments to real foams. Third, it allows one to study the effect on the entry barrier of various important factors, such as globule size, oil

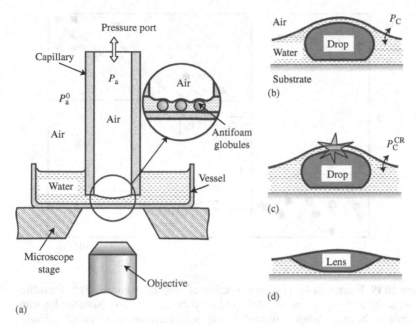

Figure 10.18 Scheme of the experimental setup and the principle of operation of the FTT.[18] (a) and (b) Vertical capillary partially immersed in surfactant solution containing oil drops is held close above the bottom of the experimental vessel. The air pressure inside the capillary, P_a, is increased and the air–water meniscus in the capillary is pressed against the glass substrate. Some of the oil drops remain trapped in the wetting glass–water–air film and are compressed by the meniscus. (c) At a certain critical capillary pressure, $P_C^{CR} = (P_a - P_w)$, the asymmetric film formed between the oil drop and the solution surface ruptures and the drop enters the air–water surface forming a lens (d).

spreading and hydrophobicity and concentration of solid particles in the antifoam compounds. Last but not least, the FTT requires inexpensive equipment and an experienced operator can obtain a large set of data in a relatively short period of time.

10.4.3.2 Role of entry barrier for the general mode of antifoam action

The application of the FTT to various antifoam-surfactant systems has shown that the entry barrier, P_C^{CR}, plays a key role in the antifoam activity and in determining the specific mode of antifoam action.[8,13,16,18–24] In Figure 10.19, we show summarized results for the foam half-life as a function of P_C^{CR} for various surfactant-antifoam pairs. One sees that the data fall into two distinct regions: (i) Systems in which the foam was destroyed in less than 10 s and the entry barrier was always below 15 Pa, (ii) systems in which the foam half-life was longer than 5 min and the entry barrier was above 20 Pa. These results show that there is a threshold value of the entry barrier, $P_{TR} \approx 15$ Pa, which separates the two distinct domains of foam half-life.

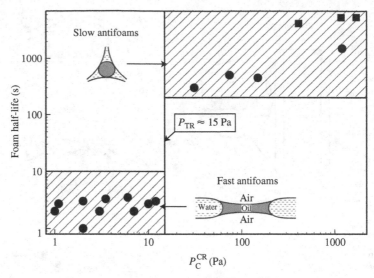

Figure 10.19 Foam half-life, $t_{1/2}$, as a function of drop entry barrier, P_C^{CR}, for different surfactant-antifoam pairs (see Ref. [16] for their description). Note the logarithmic scale on the axes. Adapted from Ref. [16]; with permission of Marcel Dekker.

Microscopic observations showed that, if $P_C^{CR} < P_{TR}$, the antifoam globules are able to easily enter the solution surface and to break the foam films in the early stages of their thinning, *i.e.* the antifoam acts as a fast one.[8,10,19] In contrast, if $P_C^{CR} > P_{TR}$, the antifoam globules are expelled from the films into the neighboring PBs, *i.e.* the antifoam acts as a slow one.[8,14–18]

10.4.3.3 *Role of entry barrier and drop size for the activity of slow antifoams*

Here, as another example of the important role of the entry barrier for the antifoam activity, we explain theoretically the height of the residual foam, H_{RES}, which is observed experimentally in the presence of slow antifoams, see Section 10.2.3 and stage IV in Figure 10.9(a).[14,16] On this basis, we discuss the role of co-surfactants, which are used as foam boosters in the presence of oily antifoams.

Let us consider a foam column with height $H_F(t)$ which contains oil drops with mean radius R_D and entry barrier, P_C^{CR}. From equations (10.2–10.4) one can calculate the equilibrium cross-sectional radius of the PBs, R_{CS}, and the respective capillary pressure, P_C^{PB}, at the top of the foam column (*i.e.* at $Z = H_F$) where the PBs are narrowest and the pressure is highest

$$R_{CS}(H_F) \approx 0.155 \frac{\gamma_{aw}}{\rho g H_F} \tag{10.19}$$

$$P_C^{PB}(H_F) \approx \rho g H_F \tag{10.20}$$

Equations (10.19) and (10.20) predict that at equilibrium, $P_C^{PB} \approx 10^3 \, \text{Pa}$ and $R_{CS} \approx 5 \, \mu\text{m}$ for a foam column with height $H_F = 10 \, \text{cm}$, whereas $P_C^{PB} \approx 100 \, \text{Pa}$ and $R_{CS} \approx 50 \, \mu\text{m}$ for $H_F = 1 \, \text{cm}$ ($\gamma_{aw} = 30 \, \text{mN m}^{-1}$ and $\rho \approx 10^3 \, \text{kg m}^{-3}$ are used in these estimates). Therefore, P_C^{PB} decreases and R_{CS} increases when the foam column decreases its height, *e.g.* as a result of antifoam induced decay.

If the oil drops trapped in the PBs of the initially formed foam with height H_{F0} have entry barrier $P_C^{CR} < P_C^{PB}(H_{F0})$ and radius $R_D > R_{CS}(H_{F0})$, then foam destruction would begin after a certain period of water drainage because the asymmetric oil–water–air films formed between the trapped oil drops and the walls of the Plateau channels (see Figure 10.9(c)) would be unable to resist the compressing capillary pressure. The foam destruction would continue until the foam height becomes so small that $P_C^{PB}(H_F) \approx P_C^{CR}$ (*i.e.* the asymmetric films become stable) or the cross-section of the Plateau channels becomes approximately equal to the drop size, $R_{CS}(H_F) \approx R_D$ (*i.e.* the oil drops are not compressed anymore by the PB walls, Figure 10.9(d)). Hence, the height of the residual foam can be evaluated from the relation $H_{RES} \approx \max\{H_P, H_R\}$, where

$$H_P = \frac{P_C^{CR}}{\rho g} \tag{10.21}$$

$$H_R = 0.155 \frac{\gamma_{aw}}{\rho g} \frac{1}{R_D} \tag{10.22}$$

One can use the dimensionless ratio

$$\frac{H_P}{H_R} = \frac{P_C^{CR} R_D}{0.155 \, \gamma_{aw}} \tag{10.23}$$

to determine whether H_{RES} is governed by the entry barrier of the oil drops or by their size. If $(H_P/H_R) > 1$, which corresponds to large drops and/or high entry barrier, H_{RES} is determined by the entry barrier. In this case, the oil drops are compressed but the asymmetric films are stable. In contrast, if $(H_P/H_R) < 1$ (*i.e.* for small drops and/or low barrier), H_{RES} is determined by the drop size while the entry barrier is not important because the oil drops are too small to be compressed at the end of the foam destruction process.

The relevance of the above estimates to real foams was verified by comparing the predictions of equations (10.21–10.23) to experimental results obtained with a series of surfactant-antifoam pairs.[16] The entry barrier, P_C^{CR}, was measured by the FTT, H_{RES} was measured by foam test and the oil drop size was determined by optical microscopy. As expected, at high entry barriers, $P_C^{CR} \geqslant 400 \, \text{Pa}$, corresponding to taller foam columns, $H_{RES} \approx H_P$; see the continuous line in Figure 10.20 which is

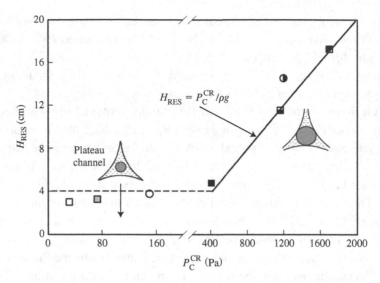

Figure 10.20 Experimental results (symbols) for the height of the residual foam H_{RES} (from Ross–Miles test) *versus* the entry barrier P_C^{CR} measured by the FTT with different surfactant + co-surfactant mixtures containing 0.1 wt.% silicone oil as a slow antifoam. The continuous line is a theoretical estimate from equation (10.21). Adapted from Ref. [16]; with permission of Marcel Dekker.

drawn according to equation (10.21) without any adjustable parameter. At lower entry barriers, $P_C^{CR} < 400$ Pa, the final foam height $H_{RES} \approx H_R$ was independent of P_C^{CR}; see the experimental points below the horizontal dashed line in Figure 10.20, because the Plateau channels were too wide to compress the emulsified oil drops in the respective short foam columns.

Equations (10.21) and (10.22) predict that one can vary the entry barrier and/or the oil drop size to control the final foam height in the presence of slow antifoams. Indeed, FTT measurements[14–16] showed that the addition to the main surfactant of different co-surfactants, such as dodecanol, betaines and aminoxides, led to a significant increase in the oil drop entry barrier at fixed total surfactant concentration. In agreement with equation (10.21), enhanced foam stability was found in the foam tests[14,15] (see Figure 10.21(a)). In complementary experiments, the foam stability was found to be higher when the oil was dispersed into smaller drops (at fixed composition of the surfactant solution), as predicted by equation (10.22); see Figure 10.21(b) for illustrative results. Therefore, one can use appropriate co-surfactants as foam boosters which improve the foam stability by increasing the entry barrier and/ or by facilitating the emulsification of the antifoam into smaller drops. For enhancing the foaminess of the respective solutions, faster kinetics of surfactant adsorption is also essential.

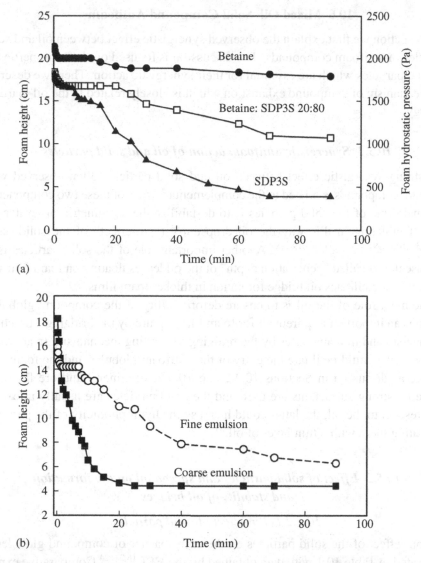

(a)

(b)

Figure 10.21 (a) Effect of betaine (co-surfactant used as foam booster) on foam stability in the presence of 0.1 wt.% silicone oil as a slow antifoam. The primary surfactant is the anionic SDP3S and the total surfactant concentration is 0.1 M. The left-hand axis shows the foam height and the right-hand axis shows the respective hydrostatic pressure at the top of the foam column. As verified by FTT, the main role of betaine is to increase the drop entry barrier, which was measured to be very similar to the hydrostatic pressure of the residual foam (after 70 min): $P_C^{CR} \approx 400$ Pa for SDP3S, ≈ 1100 Pa for the surfactant mixture and nearly 2000 Pa for betaine. (b) Effect of oil drop size on foam stability for a 0.1 M SDP3S solution. Adapted from Ref. [14]; with permission of the American Chemical Society.

10.5 Mixed Oil–Solid Compound Antifoams

In this section we first explain the observed synergistic effect between oil and solid particles in antifoam compounds. The discussion is focused on those properties of oils and particles which are essential for their synergistic action. Then, we describe the mechanism of compound exhaustion which is closely related to the oil–particle synergy.

10.5.1 Synergistic antifoam action of oil and solid particles

The strong synergistic effect between oil and solid particles often observed with antifoam compounds is related to the complementary roles of these two components. The main role of the solid particles is to destabilize the asymmetric oil–water–air films, facilitating in this way the oil drop entry ("pin-effect" of the solid particles).[3,5,10,16,19,23,28,32,54,59–61,82–86] Another important role of the solid particles is to increase the so-called "penetration depth" of the oil lenses floating on foam film surfaces, which facilitates oil bridge formation in thicker foam films.[6,11,54,63]

The main role of the oil is to ensure deformability of the compound globules, which is an important requirement for foam film rupture by the bridging–stretching mechanism and in many cases by the bridging–dewetting mechanism. In addition, oil spreading could facilitate the entry of the antifoam globules and the foam film rupture as discussed in Sections 10.4.2 and 10.5.2. In mineral and ore flotation, where no strong surfactants are used and the solid particles are usually in excess with respect to the oil, the latter could increase the hydrophobicity of the particles by coating them with a thin layer of oil.[63]

10.5.2 Effect of solid particles and spread oil on the formation and stability of oil bridges

10.5.2.1 Pin-effect of solid particles

The pin-effect of the solid particles on the entry barrier of compound globules is illustrated in Table 10.1 with data obtained by the FTT.[16,21–24] Comparative experiments were performed with drops of silicone oil (without silica) and with globules of silicone oil + silica in solutions of two different surfactants. To clarify the effect of oil spreading on the entry barrier, two types of experiments were performed for each system – with and without a pre-spread layer of silicone oil.

The data in Table 10.1 show that the entry barrier is strongly reduced by the presence of hydrophobic silica in the compound globules (see also Figure 10.22). The mechanistic explanation of the "pin-effect" (both in the presence and in the absence of spread oil) can be given by using the concepts from Section 10.3.2, where we discussed the antifoam effect of solid particles with sharp edges. In accordance with the Derjaguin approximation, equation (10.13), small solid particles of nanometer

Table 10.1 *Entry barriers, P$_C^{CR}$, of different antifoams in 10 mM AOT and 1 mM Triton X100 (nonionic) solutions in the presence and in the absence of a pre-spread layer of silicone oil.*

Antifoam	Spread layer	P_C^{CR} (Pa)	
		AOT	Triton X100
Silicone oil	No	28 ± 1	>200
	Yes	19 ± 2	>200
(Silicone oil + silica) – 1	No	8 ± 1	30 ± 1
	Yes	3 ± 2	5 ± 2
(Silicone oil + silica) – 2	No	20 ± 5	22 ± 1
	Yes	4 ± 1	7 ± 1

In foam tests, both compounds behave as a fast antifoam, the silicone oil acts as a slow antifoam whereas the reference surfactant solutions (without antifoam) are stable. The difference between the silicone oil + silica samples 1 and 2 is explained in Ref. [22], where the data are taken from.

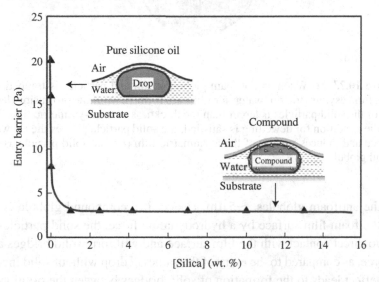

Figure 10.22 Dependence of the entry barrier of silicone oil + silica compound globules on the silica particle concentration in the compound. The entry barrier is measured by the FTT in 10 mM AOT solution. Adapted from Ref. [24]; with permission of the American Chemical Society.

size and/or having sharp edges adsorbed on the surface of the oil drop (Figure 10.23) can come into direct contact with the foam film surface much easier than the drop surface itself, because a repulsive force of lower magnitude has to be overcome.[3] For typical silica agglomerates of fractal shape used in antifoam compounds, one can approximate R_P in equation (10.13) by the radius of the primary silica particles, ~5 nm, which is three orders of magnitude smaller than the typical

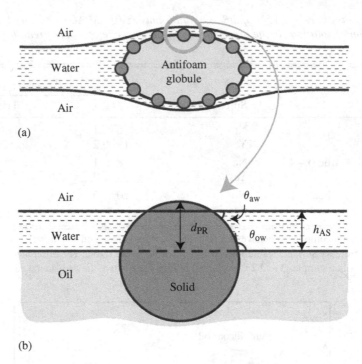

(a)

(b)

Figure 10.23 (a) When an antifoam globule of oil and particles is trapped in a foam film, asymmetric oil–water–air films are formed. (b) If the protrusion depth, d_{PR}, of the solid particles is larger than the thickness of the asymmetric film, h_{AS}, and the condition for dewetting is satisfied, the solid particle pierces the air-water surface and induces rupture of the asymmetric film (*i.e.* the solid particle assists the oil globule entry).

size of the antifoam globules, ~5 μm. Hence, if a compound globule is pushed against the foam film surface by a hydrodynamic force, the solid particles would come into direct contact with the film surface and will form solid bridges at much lower force, as compared to the entry of the same oil drop without solid inclusions. The pin-effect leads to the formation of solid bridges between the oil globule and the foam film surface (see Figure 10.23(b)). If the solid particles are sufficiently hydrophobic, these solid bridges lead to oil emergence on the film surface, *viz.* particle-aided entry of the oil globule.

10.5.2.2 Effect of particle hydrophobicity

To test how the antifoam performance and the entry barrier of mixed globules of silicone oil + silica particles depend on the hydrophobicity of the particles, the following procedure for compound preparation was used.[21,23] Hydrophilic parti-cles were mixed with silicone oil (PDMS) at room temperature and this mixture was stored for months under mild stirring. During this period, PDMS molecules slowly adsorbed on the silica surface rendering it more hydrophobic with time. In parallel experiments, spherical glass beads of millimeter size were hydrophobized by the same procedure and their hydrophobicity was assessed by measuring the particle contact angle at the PDMS-surfactant solution interface.[22,26] As seen from Figure 10.24, the hydrophobization process is very slow at room temperature which allows one to study the effect of silica hydrophobicity on the antifoam activity of the compounds.[21,23]

The FTT experiments showed a pronounced minimum of the entry barrier, P_C^{CR}, as a function of the silica particle hydrophobicity,[21,23] which corresponded to a maximum in the compound durability (Figure 10.24(b)). These results point to the presence of an optimal silica hydrophobicity for best antifoam performance. The observed minimum in P_C^{CR} was explained by a combination of two opposite requirements for the silica particles.[8,23] The first requirement is that the solid particles should protrude sufficiently deep into the aqueous phase in order to bridge the surfaces of the asymmetric oil–water–air film formed between the oil globule and the foam film surface. The protrusion depth of solid spheres can be estimated from the expression $d_{PR} = R_P(1 + \cos\theta_{ow})$, where R_P is the particle radius and θ_{ow} is the contact angle oil–water–solid measured into the aqueous phase (Figure 10.23(b)). One sees that deeper protrusion is ensured by more hydrophilic particles.

The second requirement[3,61] is that the particles should be sufficiently hydrophobic to be dewetted by the oil–water and air–water interfaces in order to induce globule entry. For complete dewetting of a spherical particle to occur, the three-phase contact angles θ_{ow} and θ_{aw} formed at the two contact lines (Figure 10.23(b)) should satisfy the condition

$$\theta_{ow} + \theta_{aw} > 180°, \quad \text{unstable asymmetric film} \qquad (10.24)$$

Therefore, if a solid bridge is formed and condition (10.24) is satisfied, the oil from the drop "uses" the bridge to come into direct contact with the foam film surface. In contrast, when $(\theta_{ow} + \theta_{aw}) < 180°$, there are well defined equilibrium positions of the three-phase contact lines on the particle surface such that the particles can even stabilize the oil–water–air film against rupture.[61]

The two opposite requirements described above explain why an optimal hydrophobicity of the solid particles in compounds would ensure lowest entry barrier. For spherical particles, the optimal contact angle can be estimated by assuming that the particle protrusion depth, d_{PR}, should be equal to the equilibrium thickness of the asymmetric oil–water–air film, h_{AS} (note that the films between the antifoam globules and the foam film surface are usually of micrometer size and thin very rapidly so that the kinetic aspects are not important here)[21]

$$\cos\theta_{ow} \approx h_{AS}/R_P - 1 \qquad (10.25)$$

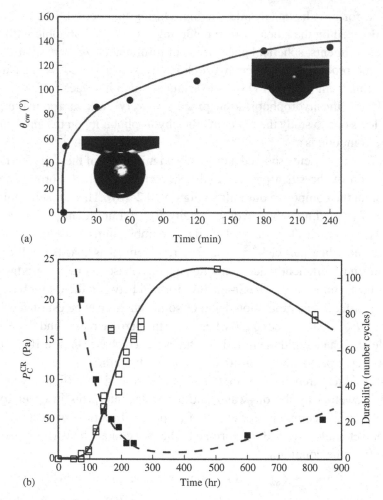

Figure 10.24 (a) Three-phase contact angle measured through water of spherical glass particles attached to the interface between a 10 mM AOT solution and silicone oil as a function of the time of contact of the particles with oil (*viz.* the time for hydrophobization of the particle surface by adsorption of silicone oil molecules). The images show photographs of the respective particles. (b) Critical pressure for globule entry, P_C^{CR} (full squares, left ordinate) and durability (empty squares, right ordinate) of silicone oil + silica compound in 10 mM AOT solution as a function of the time of silica hydrophobization. Adapted from Ref. [23]; with permission of the American Chemical Society.

which predicts that the optimal contact angle decreases with an increase in h_{AS} and with a decrease in particle size. The above mechanistic approach for analyzing the effect of solid particles on the entry barrier of compound globules was further developed to include the case with a spread oil layer on the foam film surfaces.

10.5.2.3 Effect of the spread oil on the entry barrier of compound globules

The experimental data shown in Table 10.1 show that all entry barriers of the studied oil + silica compounds were reduced in the presence of spread oil and were well below the threshold value separating the fast from slow antifoams, $P_{TR} \approx 15\,Pa$, which is in agreement with their high antifoam activity observed in the foam tests.[22,23] In Triton X100 nonionic surfactant solutions, the barriers were even higher than the value of P_{TR} in the absence of spread oil. The most important and non-trivial conclusion from all these data is that the fast antifoam action observed with the studied compounds in Triton X100 solutions is due to the combined action of the solid particles in the globules and the spread oil layer on the solution surface. Without spread oil, the entry barrier would be too high to allow fast antifoam action. In the following, we give a mechanistic explanation of the effect of spread oil on the entry barrier of compound globules.

Since the repulsive barrier between the small solid particles and the solution surface is always expected to be low due to the particle pin-effect, one can expect that the spread oil does not significantly affect the conditions for formation of the solid bridges between the compound globules and the foam film surfaces. However, the conditions for dewetting of the solid bridges formed change in the presence of spread oil, as shown by optical observations and theoretical analysis in Ref. [22]. Once a hydrophobic solid particle comes into direct contact with the air–water surface covered by a spread oil layer, the oil starts to accumulate in the area of the contact line forming an oil collar (see Figure 10.25). This process is driven by the particle hydrophobicity and is energetically favored to displace the aqueous phase contacting the particle surface by oil. The lower end of the oil collar slides along the particle surface with the accumulation of oil around the contact zone.[22] The penetration depth of the collar below the level of the air–water surface, d_{CL}, increases with the value of the contact angle, θ_{ow}, and with the collar volume. When d_{CL} becomes sufficiently large, the two oil phases (in the antifoam globule and on the solution surface) coalesce with each other and globule entry is effected.[22] The necessary condition for realization of this process is

$$\theta_{ow} > 90°, \quad \text{unstable asymmetric film} \tag{10.26}$$

Therefore, the condition for oil entry mediated by solid particles is condition (10.26) in the presence of spread oil, instead of condition (10.24) in the absence of spread oil. Experiments with hydrophobized glass particles/surfaces in the presence of strong surfactants[22,26,28,55] showed that typically $\theta_{ow} \approx 130$–$150° > 90°$, whereas $\theta_{aw} \approx 30$–$70° < 90°$ above the CMC. This means that condition (10.26) is always satisfied with hydrophobic particles, whereas condition (10.24) might not be satisfied for typical surfactant solutions. In the latter systems, the presence of spread oil on the foam film surfaces is an important factor for having fast antifoam action (*e.g.* the results for Triton X100 in Table 10.1).[22]

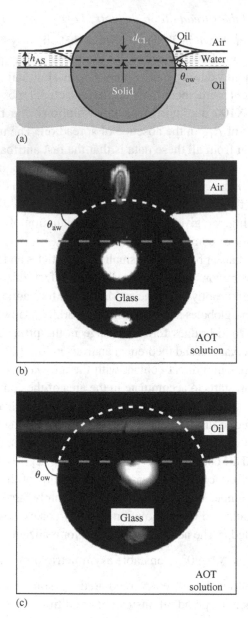

(a)

(b)

(c)

Figure 10.25 (a) Schematic representation of the formation of an oil collar after a hydrophobic solid particle pierces the air–water surface which is covered by a layer of spread oil. (b) Photograph of a hydrophobic glass sphere attached to the air–water surface in the absence of spread oil. (c) Photograph of the same particle after spreading silicone oil on the solution surface. Note the formation of the oil collar and the subsequent change of the three-phase contact angle on the particle surface. The horizontal dashed lines in (b) and (c) indicate the position of the flat oil–water interface if the solid particle bridged the oil–water and air–water interfaces as shown in Figure 10.23(b). Taken from Ref. [22]; with permission of the American Chemical Society.

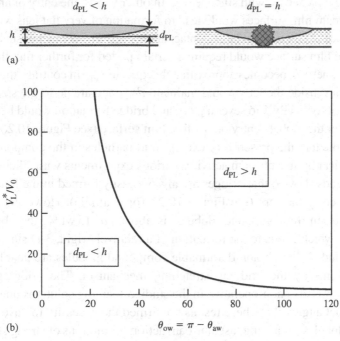

Figure 10.26 (a) Schematic presentation of the increase of the penetration depth, d_{PL}, of an oil lens due to the presence of a solid particle. (b) Plot of the calculated dimensionless volume V_L^*/V_0 of a lens with $d_{PL} = h$ in the absence of solid particles as a function of the contact angle $\theta_{ow} = (\pi - \theta_{aw})$. The scaling volume is $V_0 = \pi h^3/6$. The interfacial tensions in the calculations are typical for silicone oil: $\gamma_{oa} = 20.6 \, \mathrm{mN \, m^{-1}}$ and $\gamma_{ow} = 4.7 \, \mathrm{mN \, m^{-1}}$.

10.5.2.4 Effect of solid particles on lens penetration depth

Another aspect of the oil–solid particle synergy in compounds is related to the fact that the solid particles can facilitate the formation of unstable oil bridges in foam films by increasing the penetration depth, d_{PL}, of the oil lenses floating on the film surfaces. Indeed, material contact between the bottom of the lens and the opposite film surface (needed for oil bridge formation) is possible only after the film thickness becomes equal to d_{PL} (see Figure 10.26(a)).

In the absence of solid particles, d_{PL} can be very small.[10,11] This effect is illustrated in Figure 10.26(b), where we show the calculated volume, V_L^*, of the oil lenses for which d_{PL} is equal to the film thickness h (for convenience V_L^* is scaled with $V_0 = \pi h^3/6$). At given contact angles, lenses with volume $V_L > V_L^*$ "touch" the opposite surface of the foam film so that an oil bridge can be formed. Lenses with $V_L < V_L^*$ could not make a bridge because d_{PL} is too small. As seen from Figure 10.26(b), excessively large lenses are needed to form a bridge if $\theta_{aw} \to 180°$ (Figure 10.12), which is the typical case with silicone oils. In other words, the entry of an oil drop on one of the foam film surfaces would lead to formation of very flat lens with small d_{PL} in the absence of silica and at large contact angles θ_{aw}. The contact of such a lens with the opposite film surface would require a certain period for further film thinning until the film thickness, h, becomes approximately equal to d_{PL}. In contrast, the presence of solid particles inside the lens would maintain d_{PL} comparable to the size of particle agglomerates (typically 1 to several μm), and bridge formation would become possible soon after the globule entry on the first film surface (see Figure 10.26(a)).

Let us note that the presence of excess solid particles in the compound can suppress significantly its antifoam activity. Various experiments with silicone oil + silica compounds showed that bridges are always easily formed in the foam films due to the low entry barriers (see Figure 10.22 for example). However, if the silica concentration in the compound globules is above *ca.* 15 wt.%, they become non-deformable which is due to the formation of a relatively rigid, 3-D silica network in the compound.[23,24] Such non-deformable compound globules are unable to rupture the foam films by the bridging–stretching mechanism. The bridging–dewetting mechanism was also non-operative in the studied foaming solutions due to inappropriate contact angles.[21,23] These results confirmed the necessity to have deformable oil drops/globules for having fast antifoam action in solutions of strong surfactants at a concentration above the CMC. The effect of silica concentration on the activity of the antifoam globules is related also to the process of compound exhaustion, considered in the following sub-section.

10.5.3 Mechanisms of exhaustion of antifoam compounds

In this section we discuss briefly the exhaustion (deactivation) of antifoam compounds because (i) this process is a very illustrative example of the important effect on foam stability of the detailed structure/composition of the dispersed antifoam entities and (ii) the rapid compound exhaustion is a serious problem in practical applications.

10.5.3.1 Exhaustion and reactivation of oil–solid compounds

The process of antifoam exhaustion was illustrated in Figure 10.2 with results from the foam rise method. In Figure 10.27 we show results obtained by another method, the automatic shake test (AST), with a foaming solution of anionic surfactant containing 0.005 wt.% of silicone oil + silica compound. In this test, the foam is generated in a series of shake cycles and the defoaming time, τ_D, is measured after each cycle.[8,12] As seen from Figure 10.27, the initial high activity of the antifoam (short τ_D) is almost constant within the first 20 cycles. Afterwards, a

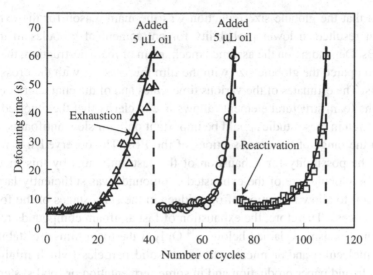

Figure 10.27 Consecutive periods of exhaustion/reactivation of 0.005 wt.% silicone oil + silica compound in an 11 mM AOT solution. An initially active antifoam (defoaming time $\tau_D > 5$ s) gradually loses its activity with the number of shaking cycles and the antifoam is considered as exhausted. The introduction of 5 μl silicone oil (0.005 wt.%) results in complete restoration of antifoam activity and τ_D falls to 5 s again. Three consecutive exhaustion profiles and two reactivations are shown. Adapted from Ref. [12]; with permission of the American Chemical Society.

relatively rapid increase of τ_D is observed and the compound gets exhausted. Note that the compound exhaustion occurs only in the process of foam destruction; if the compound is kept in the same foaming solution without agitation, its antifoam activity remains virtually constant for many hours.

The addition of a new portion of silicone oil (deprived of silica particles) into the foaming solution containing exhausted compound leads to complete restoration of the antifoam activity (see Figure 10.27). This phenomenon is called antifoam reactivation.[12] Note, that the oil used for reactivation does not contain silica and has no antifoam activity in the timescale of interest. Hence, the reactivation process certainly involves the silica particles introduced in the original compound. As seen from Figure 10.27, the consecutive periods of exhaustion–reactivation can be repeated several times without noticeable change of the compound exhaustion profile.

10.5.3.2 Mechanisms of exhaustion

Several possible mechanisms were proposed in the literature to explain the compound exhaustion. In several studies,[6,28,54] a significant reduction of the size of the antifoam globules was observed upon compound exhaustion, from 5–50 μm for fresh antifoam emulsions down to 2–8 μm in exhausted ones. Hence, the authors suggested that the globule size reduction was the main reason for the exhaustion because it resulted in lower probability for entrapment of globules in the films and/or PBs. Depending on the assumed mechanism of foam destruction, the various authors compared the globule size with the film thickness or with the cross-section of the PBs. The estimates of the various timescales and of the characteristic dimensions of the foam structural elements allows one to clarify that the reported globule size reduction in these studies could be important only for slow antifoams, because it falls in the range of the cross-sections of the PBs. The observed size reduction excludes the possibility for exhaustion of the fast antifoams by this mechanism because the globule size of the exhausted compounds was sufficiently large (well above 1 μm) to allow formation of oil bridges in the early stages of the foam film thinning process. Therefore, the exhaustion of fast antifoam compounds requires a different mechanism explained below.[12,24] Only if the foam films are stabilized by polymer molecules and/or micrometer-sized solid particles (which might be the case in pulp and paper production and in some fermentation or food systems), can film thinning be much slower and the equilibrium film thickness larger than the typical cases illustrated in Figure 10.5. In such systems, the size reduction of the film-breaking antifoam globules could have a strong impact on their activity.

Alternatively, Racz *et al.*[86] suggested that the foam films are destroyed mainly by the spread oil layers, possibly containing solid particles. Hence, these authors suggested that the emulsification of spread oil, at the moment of foam film rupture, is the main reason for the antifoam exhaustion. They found by surface tension measurements that the antifoam exhaustion correlated well with the moment when the layer of spread silicone oil disappeared from the solution surface. Note, however, that most of the commercial antifoams are produced as emulsions, which means that this mechanism is incomplete and needs further development to explain the exhaustion of such pre-emulsified compounds.

Pouchelon and Araud[27] observed the formation of macroscopic white agglomerates in surfactant solutions containing over-exhausted oil + silica compounds. Infrared analysis of these agglomerates revealed that they contain silica at very high concentration (up to 17 wt.% in comparison with 2.5 wt.% in the original compound). Based on this observation, the authors suggested that the accumulation of silica into dense oil + silica agglomerates, which are inactive as antifoam entities, is the reason for the compound exhaustion.

Recent studies[12,24] showed that the exhaustion of silicone oil + silica compounds is actually a combination of two inter-related processes which occur in parallel during foam destruction. (i) The oil and silica gradually segregate into two distinct, inactive populations of antifoam globules – silica-free (deformable) and silica-enriched (non-deformable). (ii) The layer of spread oil disappears from the solution surface (see Figure 10.28). The silica-free drops are unable to enter the foam film surfaces

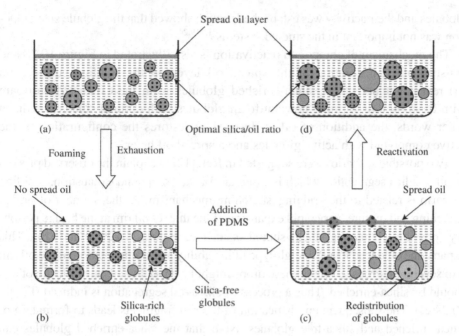

Figure 10.28 Schematic representation of the processes of exhaustion and reactivation of silicone oil + silica antifoam. (a) Initially active antifoam contains globules of optimal oil/silica ratio and layer of spread oil is formed on the solution surface. (b) The foam destruction leads to gradual segregation of oil and silica into two inactive populations of globules (silica-free and silica-enriched); the spread oil disappears and the antifoam becomes exhausted. (c) The introduction of a new portion of oil leads to restoration of the spread oil layer and to redistribution of the silica so that active oil + silica globules are formed again. (d) The antifoam is reactivated. Adapted from Ref. [12]; with permission of the American Chemical Society.

because their entry barrier is too high (*cf.* Figure 10.22). The silica-enriched globules are able to enter the foam film surfaces and to make bridges. However these globules also cannot break the foam films in the absence of spread oil because neither the bridging–stretching mechanism (which requires deformability of the globules) nor the bridging–dewetting mechanism (which requires appropriate contact angles, unrealized in the studied systems[24]) is operative. Since the silica-free globules do not enter the solution surface and the silica-enriched globules do not supply sufficient oil for spreading,[24] the layer of spread oil is eventually emulsified. Ultimately, the spread oil entirely disappears from the solution surface and both types of globules, silica-enriched and silica-free, become unable to destroy the foam films. The compound thus transforms into an exhausted state. These inter-related processes are schematically shown in Figure 10.28. No correlation between the size of the compound globules and their activity was established, which showed that the globule size reduction was not important in the studied systems.[12,19,24]

The mechanism of compound reactivation is also illustrated in Figure 10.28 and consists of (i) restoration of the spread oil layer on the solution surface and (ii) rearrangement of the silica-enriched globules from the exhausted antifoam with fresh oil thus forming new antifoam globules with optimal silica content. In other words, the addition of oil in the system restores the configuration of the active compound with active globules and a spread oil layer.

Two possible scenarios were suggested in Ref. [12] to explain the observed process of oil + silica segregation which is essential for the compound exhaustion. The first scenario is related to the bridging–stretching mechanism. At the moment of bridge stretching and rupture, very rapid expansion of the thicker oil rim at the bridge periphery (possibly containing silica) should occur (see Figures 10.29(a) and (b)). This expansion could lead to a Rayleigh-type of instability and fragmentation of the oil rim into several oil drops. Some of these drops might be devoid of silica, while the others should be silica-enriched. Thus a process of silica–oil segregation is induced (Figure 10.29(c)). The subsequent emulsification of the rim fragments leads to formation of silica-enriched and silica-free globules. Note that the silica-enriched globules can again enter the solution surface and recombine with other globules and with oil lenses. Therefore, the silica-enriched globules in the exhausted samples could be even larger than the initial antifoam globules (*cf.* the white agglomerates observed in Ref. [27]).

The second possible scenario is illustrated in Figures 10.29(d)–(f). The foam film rupture leads to ultra-rapid contraction of the film surfaces. The oil spread on the contracting surfaces is forced to form lenses, some of them devoid of silica (Figure 10.29(e)), which are dragged by the expanding perimeter of the hole in the broken foam film. Thus, the lenses are projected with high velocity towards the PBs where they can be emulsified (Figure 10.29(f)). Subsequent cycles of globule entry → oil spreading → film rupture → emulsification of the spread oil could lead to oil + silica segregation, because the silica particles are not included in the spreading thin layers of silicone oil.

10.5.4 Optimal oil viscosity and globule size in antifoam compounds

The exhaustion mechanisms discussed in Section 10.5.3 explain the fact that there is an optimum viscosity of the oils used for compound preparation.[23,25] If oil with low viscosity is used, the antifoam compound often exhibits high initial activity but exhausts rapidly due to the fast oil spreading and oil + silica segregation.[5,6,25,54] On the other hand, too viscous oils make compounds of low antifoam activity which can be explained by several factors: (i) the dispersion of the viscous compounds into numerous active globules is difficult,[25,28,54] (ii) the rate of oil spreading is low[25,28] and (iii) the deformation of the antifoam globules which is necessary for realization

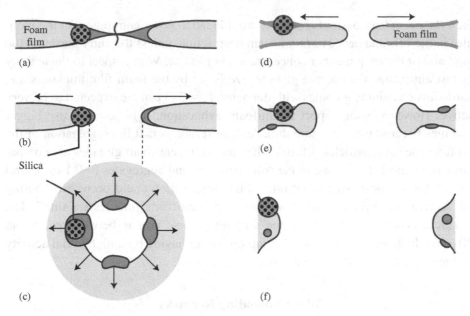

Figure 10.29 Schematic representation of two possible mechanisms of oil + silica segregation in the process of foam film rupture. (a) and (b) After an oil bridge ruptures, the hole formed in the film rapidly expands. (c) The oil rim remaining from the bridge is stretched and fragments into several smaller oil droplets. Some of these droplets contain silica particles while others are deprived of silica. These droplets hit the adjacent PBs with high velocity and are emulsified there. (d) Part of the spread oil layer can be emulsified at the moment of foam film rupture. (e) The expansion of the hole in the film leads to rapid contraction of film surfaces and the excess of spread oil forms oil lenses which are dragged towards the PBs by the perimeter of the expanding hole. (f) The impact of these lenses with the PBs could lead to oil emulsification. Adapted from Refs. [10, 12]; with permission of the American Chemical Society.

of the bridging–stretching mechanism becomes too slow for efficient film rupture. Thus, an optimal oil viscosity ensuring both high initial activity and maximum durability is required.

The mechanisms discussed of foam destruction and antifoam exhaustion allow one to estimate what could be the expected optimal globule size in fast and slow antifoams. The main difficulty in making this estimate is that virtually all mechanisms imply that larger antifoam globules would be more active if the globules were considered individually. However, at fixed total weight concentration of the antifoam, the number concentration of the globules rapidly decreases with their size. Therefore, an optimal size of the globules appears, at which the globules are still very active and of sufficiently high number concentration.

The estimates of the residual foam height made in Section 10.4.3 show that the globule diameter of the slow antifoams should be larger than the cross-section of the PBs, *i.e.* at least 5 μm. Larger size would lead to faster entrapment of the globules in the PBs and hence to faster foam destruction, unless the entry barrier is too high and/or the drop number concentration is too low. With respect to the activity of fast antifoams, the optimal globule size is set by the foam film thickness, *i.e.* antifoams containing globules with diameter $d_A \approx 2$–3 μm are expected to be very active. However, with respect to antifoam exhaustion, it is better to use bigger globules because they are more durable. The reason is that the segregation of the oil from the solid particles is faster when the initial antifoam globules are comparable in diameter to the size of the solid particles and aggregates (~0.1 to several μm).[24] Furthermore, significant oil + silica segregation could occur even during the fabrication of the antifoam emulsion if the antifoam globules are small. The practical experience shows that the optimal globule size is between *ca.* 5 and 30 μm, which seems to be a good compromise for having both high initial activity and reasonable durability of the antifoam emulsions.

10.6 Concluding Remarks

Successful control of the foaming and foam stability of aqueous foams can be achieved by using appropriate hydrophobic solid particles, oil drops and oil–solid compounds. The particular mode of foam destruction depends on the type of antifoam used. It is convenient to classify the antifoams with respect to the location where the antifoam entities enter the air–water surface and begin the foam destruction process. From this viewpoint one can distinguish:

- Fast antifoams which destroy the foam films in the early stages of the film thinning process. The fast antifoams significantly reduce the foaminess of the surfactant solutions and destroy completely the quiescent foams in less than 1 min after stopping the foam generation.
- Slow antifoams which destroy the foam only after the antifoam globules are entrapped and compressed by the shrinking walls of the PBs and nodes in the processes of foam drainage. Several stages in the foam evolution are observed under the action of slow antifoams, and residual long-standing foam remains in the last stage of the foam decay process.

The key factor determining whether a given active antifoam would act as fast or slow is the entry barrier, which characterizes how difficult it is for pre-dispersed antifoam globules to enter the foam film surfaces. The entry barrier was quantified by the FTT, and a threshold value of ≈ 15 Pa was established (in terms of critical capillary pressure leading to drop entry) which separates the fast from slow antifoams.

The characteristic features of the different types of antifoam entities can be summarized as follows.

Solid particles

- The solid particles destroy the foam films by the bridging–dewetting mechanism, which consists of two stages: (i) the solid particle comes into contact with the two opposite surfaces of the foam film making a solid bridge between them and (ii) the liquid dewets the particle and the foam film gets perforated at the particle surface.
- The antifoam efficiency of solid particles depends mainly on their hydrophobicity, shape and size. For complete dewetting of solid particles with smooth convex surface (spheres, ellipsoids), the three-phase contact angle air–water–solid should be higher than 90°. Particles with sharp edges (cubes, prisms, cones, needles, star-shaped and irregularly shaped particles) can destroy the foam films even when their contact angles are as low as 30–40° if the particles are properly oriented in the foam film. In addition, the presence of sharp edges strongly facilitates particle entry and bridge formation. The size of the solid particles becomes an important issue for their antifoam action only if they are too small (with radius below *ca*.1 μm).
- If the solid particles are too hydrophilic to act as antifoams, they can strongly enhance the foam stability by several mechanisms; stabilizing very thick equilibrium foam films, decelerating the water drainage from the foam or arresting the bubble coarsening through gas diffusion across the films.

Oil drops

- Oil drops can destroy foams by various mechanisms; bridging–stretching, bridging–dewetting and several mechanisms related to oil spreading.
- The necessary requirements for having active antifoams depend on the particular mechanism of foam destruction. For the bridging mechanisms, the bridging coefficient, B, should be positive to ensure unstable bridges. For the mechanisms related to oil spreading, the initial spreading coefficient S_{IN} should be positive to ensure oil spreading at least as a thin layer.
- The antifoam efficiency of oil drops correlates well with their entry barrier. If the entry barrier is below the threshold value of 15 Pa, the oil drops behave as fast antifoams and break the foam films by the bridging mechanisms (if $B > 0$). If the entry barrier is higher, the oil drops destroy the foam as slow antifoams by bridging or spreading mechanisms after drop entry in the Plateau borders. In typical surfactant solutions at a concentration above the CMC, the oil drops usually behave as slow antifoams because their entry barrier is above 15 Pa.
- The drop entry barrier depends on various factors, such as the presence of co-surfactant and electrolyte, presence of solid particles, oil drop size and chemical nature of the oil which can be used for efficient control of the antifoam effect to achieve a desired result (fast foam destruction or foam boosting).

Oil–solid compounds

- The oil–solid compounds with large excess of oil can destroy the foam by the same mechanisms as the oil drops. The main difference between compounds and oil drops is that the compound globules usually have a much lower entry barrier (due to the pin-effect of the solid particles), which allows them to act as fast antifoams even in surfactant solutions of high concentration.

- The strong synergistic effect between oil and solid particles in the antifoam compounds is due to the complementary roles of the two components. The main role of the solid particles is to destabilize the asymmetric oil–water–air films, facilitating in this way the oil drop entry (pin-effect) and in some systems to increase the penetration depth of the oil lenses. The main role of the oil is to ensure deformability of the compound globules and to spread on the solution surface. The globule deformability is an important pre-requisite for foam film rupture by the bridging–stretching mechanism and in many cases by the bridging–dewetting mechanism. Oil spreading facilitates the entry of the antifoam globules and the foam film rupture. In some systems related to mineral flotation (when the solid particles are in excess and no strong surfactants are used), the oil can coat the particle surface rendering it more hydrophobic.

- The exhaustion (deactivation) of the fast oil–solid compounds is mainly due to the segregation of the oil and solid particles in the course of foam destruction into two inactive populations of globules: particle-free and particle-enriched. The particle-free globules are unable to enter the foam film surfaces due to their high entry barrier, whereas the particle-enriched globules are non-deformable and, hence, cannot break the foam films.

At the end, let us note that the short review presented cannot cover all aspects of the mechanisms of antifoam action. The systems and processes involved are so complex that new experimental results can always surprise us and require further upgrade of the various mechanisms outlined above. One class of systems which is still not very well understood and deserves more systematic investigation is the antifoams used for non-aqueous foams. The recent advance in the application of new experimental methods for studying the modes of antifoam action and the new ways for synthesis of various solid particles with desired properties suggest that the antifoam research area will continue to develop rapidly.

Acknowledgments

The authors are grateful to Dr. S. Tcholakova from Sofia University for the numerous useful discussions and for a critical reading and help in the preparation of the manuscript. Our colleagues from the Laboratory in Sofia, who contributed in many different ways to the original studies (see Refs. [10–26]), are gratefully acknowledged for their quality efforts.

References

1. R.K. Prud'homme and S.A. Khan (eds.), *Foams: Theory, Measurements and Applications*, Surfactant Science Series, Vol. 57, Marcel Dekker, New York, 1996.

2. A. Prins, in *Food Emulsions and Foams*, ed. E. Dickinson, Royal Society of Chemistry Special Publication, Vol. 58, Cambridge, 1986, p. 30.

3. P.R. Garrett, in *Defoaming: Theory and Industrial Applications*, ed. P.R. Garrett, Surfactant Science Series, Vol. 45, Marcel Dekker, New York, 1993, Chapter 1.

4. P.R. Garrett, (ed.), *Defoaming: Theory and Industrial Applications*, Surfactant Science Series, Vol. 45, Marcel Dekker, New York, 1993.

5. D.T. Wasan and S.P. Christiano, in *Handbook of Surface and Colloid Chemistry*, ed. K. Birdi, CRC Press, New York, 1997, Chapter 6.

6. D.T. Wasan, K. Koczo and A.D. Nikolov, in *Foams: Fundamentals and Applications in Petroleum Industry*, ed. L.L. Schramm, ACS Symposium Series No. 242, ACS, New York, 1994, Chapter 2.

7. D. Exerowa and P.M. Kruglyakov, *Foams and Foam Films: Theory, Experiment, Application*, Elsevier, Amsterdam, 1998.

8. N.D. Denkov, *Langmuir*, **20** (2004), 9463.

9. S. Ross and G. Nishioka, in *Emulsion, Latices and Dispersions*, eds. P. Becher and M.N. Yudenfrend, Marcel Dekker, New York, 1978, p. 237.

10. N.D. Denkov, P. Cooper and J-Y. Martin, *Langmuir*, **15** (1999), 8514.

11. N.D. Denkov, *Langmuir*, **15** (1999), 8530.

12. N.D. Denkov, K. Marinova, H. Hristova, A. Hadjiiski and P. Cooper, *Langmuir*, **16** (2000), 2515.

13. N.D. Denkov and K.G. Marinova, in *Proceedings of 3rd Eurofoam Conference*, Verlag MIT Publishing, Bremen, 2000, p. 199.

14. E.S. Basheva, D. Ganchev, N.D. Denkov, K. Kasuga, N. Satoh and K. Tsujii, *Langmuir*, **16** (2000), 1000.

15. E.S. Basheva, S. Stoyanov, N.D. Denkov, K. Kasuga, N. Satoh and K. Tsujii, *Langmuir*, **17** (2001), 969.

16. A. Hadjiiski, N.D. Denkov, S. Tcholakova and I.B. Ivanov, in *Adsorption and Aggregation of Surfactants in Solution*, eds. K. Mittal and D. Shah, Marcel Dekker, New York, 2002, p. 465.

17. L. Arnaudov, N.D. Denkov, I. Surcheva, P. Durbut, G. Broze and A. Mehreteab, *Langmuir*, **17** (2001), 6999.

18. A. Hadjiiski, S. Tcholakova, N.D. Denkov, P. Durbut, G. Broze and A. Mehreteab, *Langmuir*, **17** (2001), 7011.

19. K.G. Marinova and N.D. Denkov, *Langmuir*, **17** (2001), 2426.

20. N.D. Denkov, K. Marinova, S. Tcholakova and M. Deruelle, in *Proceedings 3rd World Congress on Emulsions*, CME, September 2002, Lyon, Paper 1-D-199.

21. K.G. Marinova, N.D. Denkov, P. Branlard, Y. Giraud and M. Deruelle, *Langmuir*, **18** (2002), 3399.

22. N.D. Denkov, S. Tcholakova, K.G. Marinova and A. Hadjiiski, *Langmuir*, **18** (2002), 5810.

23. K.G. Marinova, N.D. Denkov, S. Tcholakova and M. Deruelle, *Langmuir*, **18** (2002), 8761.

24. K.G. Marinova, S. Tcholakova, N.D. Denkov, S. Roussev and M. Deruelle, *Langmuir*, **19** (2003), 3084.

25. N.D. Denkov, S. Tcholakova, K.G. Marinova, N. Christov, N. Vankova and M. Deruelle, in preparation.

26. K.G. Marinova, D. Christova, S. Tcholakova, E. Efremov and N.D. Denkov, *Langmuir*, **21** (2005), 11729.

27. A. Pouchelon and C. Araud, *J. Disp. Sci. Technol.*, **14** (1993), 447.

28. V. Bergeron, P. Cooper, C. Fischer, J. Giermanska-Kahn, D. Langevin and A. Pouchelon, *Colloids Surf. A*, **122** (1997), 103.

29. T.A. Koretskaya, *Koll. Zeit.*, **39** (1977), 71.
30. A. Bonfillon-Colin and D. Langevin, *Langmuir*, **13** (1997), 599.
31. R. Chaisalee, S. Soontravanich, N. Yanumet and J.F. Scamehorn, *J. Surf. Det.*, **6** (2003), 345.
32. P.R. Garrett, J. Davis and H.M. Rendall, *Colloids Surf. A*, **85** (1994), 159.
33. D. Weaire and S. Hutzler, *The Physics of Foams*, Clarendon Press, Oxford, 1999.
34. G. Narsimhan and E. Ruckenstein, in *Foams: Theory, Measurements, and Applications*, eds. R.K. Prud'homme and S.A. Khan, Surfactant Science Series, Vol. 57, Marcel Dekker, New York, 1996, Chapter 2.
35. P.A. Kralchevsky, K.D. Danov and N.D. Denkov, in *Handbook of Surface and Colloid Chemistry*, ed. K.S. Birdi, 2nd Expanded and Updated Edition, CRC Press, New York, 2002, Chapter 5.
36. I.B. Ivanov, (ed.), *Thin Liquid Films: Fundamentals and Applications*, Marcel Dekker, New York, 1988.
37. P.A. Kralchevsky and K. Nagayama, *Particles at Fluid Interfaces and Membranes*, Elsevier, Amsterdam, 2001.
38. N.D. Denkov, H. Yoshimura and K. Nagayama, *Ultramicroscopy*, **65** (1996), 147.
39. K.P. Velikov, F. Durst and O.D. Velev, *Langmuir*, **14** (1998), 1148.
40. R.G. Alargova, D.S. Warhadpande, V.N. Paunov and O.D. Velev, *Langmuir*, **20** (2004), 10371.
41. A.D. Nikolov, D.T. Wasan, P.A. Kralchevsky and I.B. Ivanov, *J. Colloid Interf. Sci.*, **133** (1989), 1, 13.
42. V. Bergeron and C.J. Radke, *Langmuir*, **8** (1992), 3020.
43. I.B. Ivanov and D.S. Dimitrov, in *Thin Liquid Films: Fundamentals and Applications*, ed. I.B. Ivanov, Marcel Dekker, New York, 1988, Chapter 7.
44. T.T. Traykov, E.D. Manev and I.B. Ivanov, *Int. J. Multiphase Flow*, **3** (1977), 485.
45. E. Manev, R. Tsekov and B. Radoev, *J. Disp. Sci. Technol.*, **18** (1997), 769.
46. J.E. Coons, P.J. Halley, S.A. McGlashan and T. Tran-Cong, *Adv. Colloid Interf. Sci.*, **105** (2003), 3.
47. D. Weaire, S. Hutzler, G. Verbist and E. Peters, *Adv. Chem. Phys.*, **102** (1997), 315.
48. S.J. Cox, D. Weaire, S. Hutzler, J. Murphy, R. Phelan and G. Verbist, *Proc. Roy. Soc. London A*, **456** (2000), 2441.
49. S.A. Koehler, S. Hilgenfeldt and H.A. Stone, *Langmuir*, **16** (2000), 6327.
50. A. Saint-Jalmes, M.U. Vera and D.J. Durian, *Europhys. Lett.*, **50** (2000), 695.
51. A. Saint-Jalmes and D. Langevin, *J. Phys.: Condens. Matter*, **14** (2002), 9397.
52. W.W. Mullins, *J. Appl. Phys.*, **59** (1986), 1341.
53. S. Hilgenfeldt, S. Koehler and H.A. Stone, *Phys. Rev. Lett.*, **86** (2001), 4704.
54. K. Koczo, J.K. Koczone and D. Wasan, *J. Colloid Interf. Sci.*, **166** (1994), 225.
55. G.C. Frye and J.C. Berg, *J. Colloid Interf. Sci.*, **127** (1989), 222.
56. K. Roberts, C. Axberg and R. Österlund, *J. Colloid Interf. Sci.*, **62** (1977), 264.
57. P.R. Garrett, *J. Colloid Interf. Sci.*, **69** (1979), 107.
58. A. Dippenaar, *Int. J. Min. Proc.*, **9** (1982), 1.
59. R. Aveyard, P. Cooper, P.D.I. Fletcher and C.E. Rutherford, *Langmuir*, **9** (1993), 604.
60. R. Aveyard, B.P. Binks, P.D.I. Fletcher, T.G. Peck and C.E. Rutherford, *Adv. Colloid Interf. Sci.*, **48** (1994), 93.
61. R. Aveyard and J.H. Clint, *J. Chem. Soc. Faraday Trans.*, **91** (1995), 2681.
62. R. Aveyard, B.D. Beake and J.H. Clint, *J. Chem. Soc. Faraday Trans.*, **92** (1996), 4271.
63. G.C. Frye and J.C. Berg, *J. Colloid Interf. Sci.*, **130** (1989), 54.
64. R.D. Kulkarni, E.D. Goddard and B. Kanner, *Ind. Eng. Chem. Fund.*, **16** (1977), 472.
65. R.D. Kulkarni, E.D. Goddard and B. Kanner, *J. Colloid Interf. Sci.*, **59** (1977), 468.

66. K. Golemanov, S. Tcholakova, P.A. Kralchevsky, K.P. Ananthapadmanabhan and A. Lips, *Langmuir*, submitted.
67. B.V. Derjaguin, *Theory of Stability of Colloids and Thin Liquid Films*, Plenum, New York, 1989.
68. H. Zhang, C.A. Miller, P.R. Garrett, and K.H. Raney, *J. Colloid Interf. Sci.*, **279** (2004), 539.
69. P.R. Garrett, *J. Colloid Interf. Sci.*, **76** (1980), 587.
70. R. Aveyard, B.P. Binks, P.D.I. Fletcher, T-G. Peck and P.R. Garrett, *J. Chem. Soc. Faraday Trans.*, **89** (1993), 4313.
71. W.D. Harkins, *J. Chem. Phys.*, **9** (1941), 552.
72. S. Ross, *J. Phys. Colloid Chem.*, **54** (1950), 429.
73. P.M. Kruglyakov and T.A. Koretskaya, *Koll. Zeit.*, **36** (1974), 627.
74. K. Koczo, L. Lobo and D.T. Wasan, *J. Colloid Interf. Sci.*, **150** (1992), 492.
75. B.K. Jha, S.P. Christiano and D.O. Shah, *Langmuir*, **16** (2000), 9947.
76. V. Bergeron, M.E. Fagan and C.J. Radke, *Langmuir*, **9** (1993), 1704.
77. P.M. Kruglyakov and N.G. Vilkova, *Colloids Surf. A*, **156** (1999), 475.
78. B.P Binks, P.D.I. Fletcher and M.D. Haynes, *Colloids Surf. A*, **216** (2003), 1.
79. B. Dimitrova, I.B. Ivanov and E. Nakache, *J. Disp. Sci. Technol.*, **9** (1988), 321.
80. K. Danov, I.B. Ivanov, Z.Z. Zapryanov, E. Nakache and S. Raharimalala, in *Synergetics, Order and Chaos*, ed. M.G. Velarde, World Scientific, London, 1988, p. 178.
81. D.S. Valkovska, P.A. Kralchevsky, K.D. Danov, G. Broze and A. Mehreteab, *Langmuir*, **16** (2000), 8892.
82. M. Aronson, *Langmuir*, **2** (1986), 653.
83. P.R. Garrett and P.R. Moore, *J. Colloid Interf. Sci.*, **159** (1993), 214.
84. H. Zhang, C.A. Miller, P.R. Garrett and K.H. Raney, *J. Colloid Interf. Sci.*, **263** (2003), 633.
85. R.E. Patterson, *Colloids Surf. A*, **74** (1993), 115.
86. G. Racz, K. Koczo and D.T. Wasan, *J. Colloid Interf. Sci.*, **181** (1996), 124.

Nikolai Denkov (left) received an M.S. degree in 1987 and his Ph.D. degree in Physical Chemistry in 1993, at the Laboratory of Chemical Physics and Engineering, Sofia University. As senior researcher he spent 1 year in each of the Nagayama Protein Array Project, Japan (1994/1995), Rhodia Silicones Europe, Lyon (1997/98) and Unilever R&D, Edgewater, USA (2003/04). His main research interests are in the areas of foam and emulsion stability and rheology, emulsification and foaming, particle interactions and formation of 2D-colloid crystals, and capillary and surface forces in thin liquid films. He has published 91 research papers. In 2004 he was co-organizer of the symposium "Physics and Design of Foams" in Edgewater. Currently Dr. Denkov is Associate Professor of Theoretical Chemistry in the Laboratory of Chemical Physics and Engineering, Faculty of Chemistry, Sofia University, Bulgaria.

Krastanka Marinova (right) received an M.S. degree in 1992 in the Faculty of Physics at Sofia University. She defended her Ph.D. thesis in the area of antifoams in 1999 at the Laboratory of Chemical Physics and Engineering in Sofia University. As guest researcher she spent 6 months in Rhodia Silicones Europe, Lyon, studying the mechanisms of foam destruction by silicone-based antifoams, and 5 months in the Department of Chemical Engineering, University of Patras, Greece. Her main research interests are in the areas of antifoams, kinetics of surfactant adsorption and interfacial rheology (including development of new experimental methods in collaboration with Krüss GmbH, Germany), surface forces, and stability of foam and emulsion films. She has published 21 papers in these areas. Currently Dr. Marinova is Assistant Professor in the Laboratory of Chemical Physics and Engineering, Faculty of Chemistry, Sofia University, Bulgaria.

Tel.: + 359-2-962531; *Fax:* + 359-2-962643; *E-mail:* nd@lcpe.uni-sofia.bg

11

Metal Foams: Towards High-Temperature Colloid Chemistry

Norbert Babcsán and John Banhart

Department of Materials Science, Hahn-Meitner-Institute,
Berlin D-14109, Germany
Institute of Materials Science and Technology, Technical
University of Berlin, Berlin D-10632, Germany

11.1 Introduction

Liquid foams are collections of gas bubbles uniformly dispersed in fluids and separated from each other by self-standing thin films. If the distance between bubbles is comparable to the bubble size one prefers to speak of bubble dispersions. In foams, bubble arrangements are usually disordered and gas volume fractions are high. If a liquid foam is solidified, a solid foam is obtained. Solid foams show many interesting properties which is the reason for their wide use, *e.g.* in civil engineering, chemistry or the food industry.[1]

Any liquid matter should be foamable and so is liquid metal. The prospect of being able to make light durable metallic foams already triggered research more than half a century ago. In 1943 Benjamin Sosnick[2] attempted to foam aluminium with mercury. He first melted a mix of Al and Hg in a closed chamber under high pressure. The pressure was released, leading to vaporisation of the mercury at the melting temperature of aluminium and to the formation of a foam. Less hazardous processes were developed in the mid-1950s when it was realised that liquid metals could be more easily foamed if they were pre-treated to modify their properties. This could be done by oxidising the melt or by adding solid particles. Elliott[3] at Bjorksten Research Laboratories (BRL) developed an aluminium foaming process in the 1950s. BRL subsequently entered into an agreement with the LOR Corporation to develop commercial uses for foamed aluminium. A pilot plant was constructed at BRL to produce large wall panels. Other potential uses, such as crash bumpers for cars, were also investigated. In the late 1960s, the entire operation was sold to the Ethyl Corporation and the pilot plant was moved to Baton Rouge.

BRL continued for several years to investigate methods for foaming other metals such as lead and zinc.

Two methods for foaming metals were used in those days, and they are still used today.[4] In the first of these, gas is injected continuously to create foam. The foam accumulates at the surface of the melt and the result looks somewhat like a glass of draught beer. In the second method, gas-releasing propellants are added to the percursor, akin to the yeast of the baker (see Figure 11.1). Aluminium was found to be particularly amenable to foam production. The Ethyl Corporation produced material of remarkably high quality which was given to the Ford Motor Company for evaluation. Why was this initial development not successful? It was not the time for lightweight materials in the era of seemingly unlimited energy supply, and issues of safety and recycling were not so important as now. Whatever the reason, the excitement and the level of research and development activities both declined after 1975.

By the end of the 1980s, there was a resurgence in metal foam research through-out the world. Japanese engineers at Shinko Wire Company developed what is now known as the Alporas process.[5] Norsk Hydro in Norway[6] and Alcan Corporation in Canada[7] independently developed a foaming process for particle-stabilised melts. In 1990 an old powder compact foaming route developed in the late 1950s by Allen *et al.*[8] was re-discovered and brought to a considerable level of sophistication at the Fraunhofer Laboratory in Bremen, Germany.[9] These and other variants have been continually refined up to the present day.[10,11]

Today, a small number of companies produce aluminium foams. To our know-ledge, there are three in Germany, two in Austria and one each in Japan, Korea and Canada.[12] Here we do not count manufacture of cellular metals by sintering, elec-troplating or casting. The corresponding structures are frequently called foams but actually belong to a quite different class of material. There are some applications for aluminium foams now including stiffening parts for cars, crash bumpers for light railways, a lifting arm for a lorry and stiff beams for working machines.[13] The market is still very small but slowly expanding.

The scientific challenges now are to improve foam properties and to make the production process more reliable. For this, some knowledge of foam stability is indispensable. Surprisingly, research on the physics of metal foaming is quite restricted. Of about 300 journal papers listed on a dedicated web site,[12] only about 20 are concerned with investigations of liquid metal foams, the remainder con-centrating on processing, properties and applications of solid foams. Only very recently was the issue of metal foam stabilisation addressed and traced back to the presence of micro- or even nanometer-sized solid particles in the liquid metal.[14] The time has come to understand liquid metal foams as an independent field of research and to look at these systems in the framework of colloid chemistry.

(a)

(b)

Figure 11.1 (a) Aluminium foam blown with air from a particle-stabilised melt (left) and a glass of beer (right). (b) Zinc foam (left) and a bread roll (right) both foamed by internal gas creation. Photographs courtesy of the Hahn-Meitner-Institute, Berlin.

11.2 The Making of Metal Foams

11.2.1 Processing routes

There are many ways to make metal foams and the various methods can be classi-fied in different ways. For the purpose of this treatise the classification given in Table 11.1 is particularly useful. We shall concentrate on aluminium alloys in the following although the phenomena which will be discussed occur for other metals too. As already mentioned in Section 11.1, we distinguish the way the gas comes into the melt, *i.e.* the gas source. Bubble creation can be internal or external. In the former case, gas bubbles are created by gas evolution from within the melt. Nucleation of dissolved gas triggered by changes in pressure or temperature is a possible mechanism; the decomposition of a chemical blowing agent is another. Often hydrides or carbonates are used as blowing agents in which case hydrogen or carbon dioxide evolve from the blowing agent. Alternatively, chemical reactions in the melt can create, *e.g.* water vapour, which then drives foam expansion. In con-trast, external bubble creation is caused by injecting gas into the melt continuously from outside, *e.g.* through a capillary or a porous frit. Clearly, the two methods imply different rheological phenomena during foaming as the bubbles are created at many locations in the melt in the former case while they have to travel a certain distance in the latter, usually from the injection point to the surface.

Another feature which can be used for classification is the type of melt which is foamed. In the simplest case we deal with a pure molten metal. It is known from water that it is hard to make a foam from pure liquids and this is also true for metals. If bubbles are created in a pure melt they quickly rise to the surface and vanish

Table 11.1 *Classification of metal foam making processes. For each of the eight possible categories either a company, trade or process name or a reference is given. See text for details.*

| Type of gas source | Type of melt | | | |
	Pure melt	Particles added	Particles created *in-situ*	Molten powder compact
Internal	Pötschke *et al.*[16] AMF[18] Gasar[19]/Lotus[20]	Formgrip[24] Foamcast[25]	LOR, Ethyl[26,27] Alporas[5] VFT[28] DCP[29,30]	Alulight[23]/ Foaminal[8,9,15] Thixo-foam[31]
External	Trial described in Ref. [17]	Alcan[7] Hydro[6] Metcomb[22]	Trial by the authors	Trial by the authors

there. Approaches to overcome this problem include creating bubbles in a melt by chemical reactions under weightlessness.[16] However, even in the absence of any buoyancy force, one observes that bubbles tend to coalesce quickly in such systems and no foam with a significant and uniform porosity is obtained. Another approach involved bubbling argon gas through a highly pure magnesium melt while keeping the temperature near the melting point, and viscosity as high as possible, and solidifying while bubbling continued.[17] A certain level of porosity could be preserved but the resulting material could not really be called a foam. A related processing route has been investigated recently. In the so-called amorphous metal foaming (AMF) process, highly viscous bulk amorphous glass was foamed by internal gas formation and a quite high porosity level was obtained.[18] A further approach is the Gasar[19] or Lotus[20] process. Gas is dissolved in a melt under high pressures and is then solidified directionally. One observes gas nucleation at the solidification front. Gas bubbles are pinned to the already existing pores so that they cannot float to the surface but form large and elongated pores. Again, temperature control prevents the bubbles from escaping but no real foam is obtained. Another strategy known from the literature is "foaming" a solid directly by creep expansion[21] which we shall not consider further here.

Scientists quickly began to realise that gas bubbles in pure melts are too volatile and too prone to coalescence for making stable metallic foams. It is common knowledge in metal processing that porosity in castings or welding can be quite large and unwanted porosity is especially notorious when the metal to be cast or welded contains a high degree of non-metallic impurities. One therefore suspected that the presence of solid particles in the melt can help tackle problems of foam stability. Such solid particles can be brought into the melt either by simply admixing them to the melt (Table 11.1, 3rd column), by *in-situ* creation in the melt (4th column) or by making use of the oxidised surface layer of metal powders which are used during processing (5th column). We shall now describe some of the various possibilities, not in a historical but more in a logical order.

If metal matrix composites (MMC) containing a high-volume fraction (10–20%) of oxide or carbide particles are melted one obtains a suspension of these particles in the melt. Gas injection through a nozzle, frit, impeller, *etc.* then leads to bubbles that rise to the surface and form a stable layer of foam there. The bubbles are still buoyant but coalescence is reduced to a minimum by the addition of particles. This foaming route was developed simultaneously by Alcan and Hydro Aluminium in the late 1980s who used aluminium alloys and silicon carbide particles. A technology called Metcomb[22] uses a similar approach and allows for the manufacture of very uniform pore morphologies as shown in Figure 11.2(a).

The same MMC can be used for foam making using an internal gas source instead of injecting gases. In the so-called Formgrip process,[24] a blowing agent (titanium

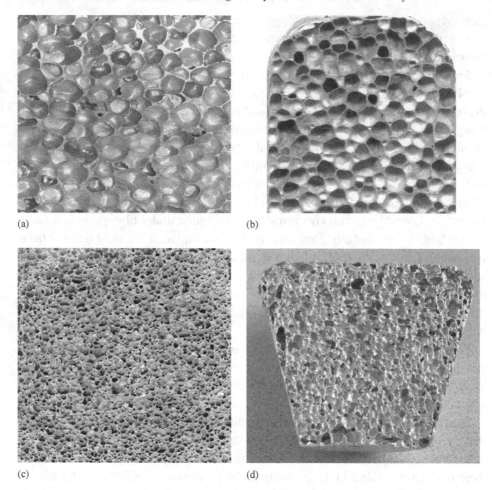

Figure 11.2 Macrostructures of various metal foams. (a) Metcomb Al-foam (courtesy D. Leitlmeier), (b) Formgrip Al-foam, (c) Alporas Al-foam, (d) VFT Mg-foam, (e) Alulight Al-foam,[23] (f) melted powder compact foamed by gas injection, (g) Alporas alloy foamed by gas injection, (h) Amorphous metal foam (Pd-based). Width of picture is (a) 20 mm, (b) 30 mm, (c) and (d) 80 mm, (e) 60 mm, (f) and (g) 30 mm, (h) 2 mm.

hydride, TiH_2) is added to a molten MMC. After sufficient stirring the melt is solidified. Owing to a prior pre-treatment of the hydride, gas evolution during stirring is not too pronounced and the solidified precursor contains less than 14% pores. The actual foam making step comprises re-melting of the precursor and holding at a baking temperature at which the blowing agent decomposes according to $TiH_2 \rightarrow Ti + H_2$. Unlike blowing gas through a nozzle, gas bubbles are now created simultaneously in the entire volume by the decomposing hydride particles.

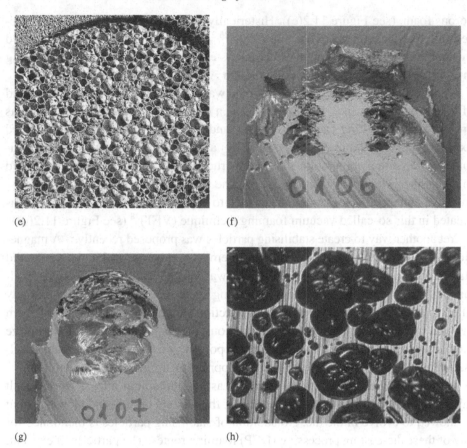

(e)   (f)

(g)   (h)

Figure 11.2 (*continued*).

The bubbles grow to a stable and uniform foam (see Figure 11.2(b)). An alternative technique mixes particle-free aluminium alloys and titanium hydride in a die-casting machine.[25] Due to rapid solidification there is no need to use pre-treated TiH_2 in this Foamcast method. To improve foamability of the precursor, some alumina or oxidised aluminium powder is admixed to the blowing agent prior to injection.

Instead of preparing suspensions of particles in liquid metals one can condition melts in a different way to form stable foams. One idea was to create particles in the melt deliberately by *in-situ* reactions, *e.g.* oxidation. In the Japanese Alporas process, calcium metal is added to an aluminium melt after which the melt is stirred in the presence of air for some minutes during which oxide particles are formed *in-situ* in the melt.[5] After this, TiH_2 is added to the conditioned melt and dispersed by stirring. The hydride sets hydrogen free and turns the molten metal into a highly porous foam, (see Figure 11.2(c)). Historically, this was the way foams were successfully produced in the 1960s and 1970s.[26,27] At that time particles were created by bubbling, *e.g.* pure oxygen gas or water steam, through the melt using rotating injectors or by admixing dry ice into molten Al 7–22 wt.% Mg alloy.

Quite recently a modification of this route was successfully demonstrated. Instead of creating particles in a well-defined way, a contaminated magnesium melt was prepared by re-melting casting overflows and machining chips which contained oxides, hydroxides, entrapped gas, *etc.* TiH_2 as an internal gas source was substituted by pressure manipulation. The entire crucible containing the melt was placed in a chamber which was then evacuated. The pressure drop caused the gases dissolved or entrapped in the contaminated melt to expand and a quite regular foam was created in this so-called vacuum foaming technique (VFT)[28] (see Figure 11.2(d)).

Yet another way to create stabilising particles was proposed recently.[29] A magnesium alloy containing about 10% alloying elements was processed in the semi-solid state in a thixo-moulding machine. The melt was injected into a closed mould simultaneously with some MgH_2 serving as blowing agent. The mould was only partially filled to allow for foam evolution after injection. It was found that a metal foam evolved in the mould which could be taken out after rapid solidification. From the bubble size distribution as a function of local porosity, the existence of a foam stabilisation mechanism was deduced. A similar approach was proposed for aluminium.[30] A normal die-casting machine is used for casting aluminium alloys. During melt injection, MgH_2 powder is added. As for Mg, the mixture of melt and blowing agent is allowed to evolve in the die. No addition of stabilising particles is mentioned for any of these die-casting processing (DCP) foaming routes. The particles, if existent, must then have been created *in-situ* during processing.

Another foaming method starts from metal powders. These are mixed with a blowing agent powder (usually TiH_2 for Al alloys) after which they are consolidated to a dense material by hot pressing or extrusion in the solid[15] or casting in the semi-solid state.[31] Re-melting this material then triggers foam expansion as the blowing agent releases hydrogen. This processing route, called Alulight or Foaminal, is analogous to foaming by the Formgrip method already mentioned. Unlike the Formgrip method, no particles have been added. Particles present in the melt are the previously oxidised surfaces of the individual metal particles and they are obviously sufficient for stabilisation. Figure 11.2(e) shows an Alulight sample made by this technique.

Table 11.1 shows two entries which have not yet been discussed. The question is whether *in-situ* oxidised melts such as the ones used for making Alporas foams or re-melted metal powder compacts without blowing agent can also be foamed by external bubble formation by gas injection through a nozzle. We carried out trials and obtained the "foams" shown in Figures 11.2(f) and (g). Obviously foaming is more difficult here which we shall discuss later.

11.2.2 Stages of metal foam evolution

The term foamability quantifies how much good foam can be made from a liquid. The change of a foam from formation until collapse is called foam evolution. The evolution stages during foaming vary with the strategy chosen, *e.g.* external or internal gas injection. In most cases the temperature course during foaming includes a heating stage, an isothermal holding stage and a cooling step. The principal stages are shown in Figure 11.3.

Foaming with external gas bubbling (see Figure 11.3(a)) includes melting the alloy to be foamed, injecting gas (usually at a constant temperature) and then some further isothermal holding, either deliberately to study stability or to allow for processing the foam into shaped parts, and then cooling to room temperature. During gas injection the bubbles rise from the injection point to the surface. After detachment from the injector the volume of the bubble changes in a well-defined way; heating of the gas contained in the bubble to the temperature of the liquid metal and the decreasing hydrostatic pressure during rise may lead to a certain expansion. We shall see later that during travelling the bubbles collect particles which adhere to the bubble surfaces. Provided that all parameters are appropriate, the bubbles are then stable and build up a foam layer. Decay phenomena can lead to changes in the foam layer during isothermal holding. Solidification can also influence foam structure.

Making foams utilising a blowing agent, *i.e.* internal gas bubbling, can be more complicated. In Figure 11.3(b), the foaming stages are shown for the case in which a precursor containing a blowing agent is melted. This applies to both the Formgrip-type and the Alulight-type foaming routes. In these foams the bubbles continuously grow during foaming due to the gas supplied by the blowing agent and due to thermal expansion. Expansion is governed by the continuous decomposition of the blowing agent (mostly a hydride or carbonate). Complicated relationships exist between the decomposition kinetics of the blowing agent and the melting of the alloys, and the kinetics depends on the particle size of the individual blowing agent particle. Therefore, bubbles are not expected to grow uniformly. A further complication is oxide layers which can be grown on the surface of blowing agent particles and which act as diffusion barriers to modify the decomposition kinetics of the blowing agent.[32]

If foaming is triggered by an external gas source the various stages of foaming can be well separated from each other. The actual foam formation takes only seconds after which the foam is usually held at a constant temperature. This facilitates studies of liquid foam and the interpretation of results. In contrast, during foaming of precursors containing a blowing agent, foaming starts already during heating of the precursor and the gas supply depends in a complicated way on the decomposition kinetics of the blowing agent. Development of a fully expanded foam can require several minutes. The study of such foams is therefore more difficult since bubble nucleation and growth, drainage and coalescence overlap. Other foaming

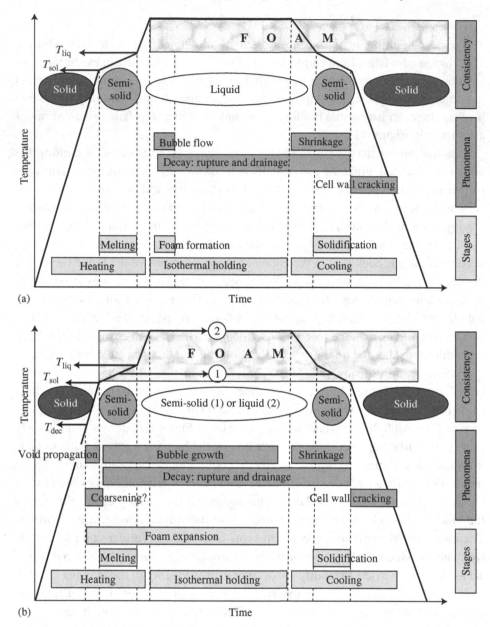

Figure 11.3 Schematic representation of foam evolution as observed in (a) foams made by external gas injection and (b) foams made by foaming precursors containing blowing agents, *i.e.* by internal gas generation.

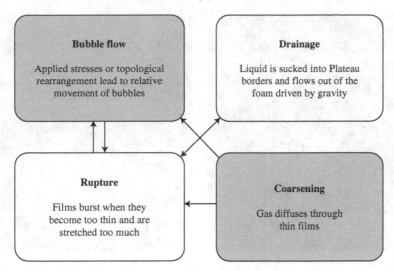

processes such as the VFT or the Alporas process have foaming stages slightly different to the two considered. These processes are usually isothermal which facilitates their study slightly, but gas generation still follows a complicated law.

The range of phenomena in liquid foams is wide and they all influence one another. The main effects are drainage, rupture (coalescence), coarsening and flow. Their interdependence is indicated by arrows in Figure 11.4. Drainage can lead to rupture since it gradually reduces the thickness of films and therefore increases the probability for an instability which eventually destroys the films. In metallic foams there is little coarsening due to the inability of gas to diffuse through the relatively thick films in the short times provided. One strategy in foam research is to eliminate some of these effects to be able to study the remaining ones. This is a strong motivation for using micro-gravity which avoids drainage. Low gravity experiments on metal foams were performed in TEXUS (Technologische Experimente unter Schwerelosigkeit) sounding rockets[33] or parabolic flights.[34] Flow of bubbles can be avoided by choosing stationary conditions in which foam expansion has come to an end. Under these circumstances rupture can be studied without any disturbing interference.

11.3 Stabilisation of Metal Foams

11.3.1 Phenomenology of stabilisation

The term foam stability informs about the lifetime of a foam under given conditions and is related to the absence of cell wall rupture and to a limitation to drainage effects which eventually destroy foam structure. While phenomena related to

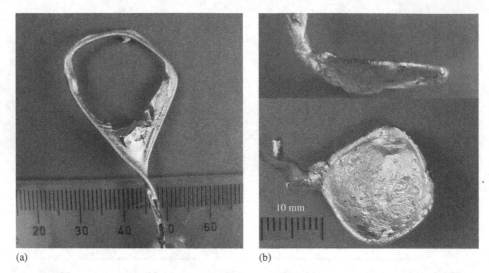

(a) (b)

Figure 11.5 Pull-out tests with a wire frame. (a) Aluminium melt without particles, (b) aluminium melt containing particles. Taken from Ref. [46]; with permission of Wiley–VCH.

aqueous foam stability are discussed in the literature in large monographs[36–44] and in Chapter 10 of this book, work concerned with metal foams is still quite restricted.[14,34,45–50]

In Section 11.2.1 we saw that the manufacture of stable metal foams requires the presence of solid particles in the melt. This can also be shown by experiments in which a single metal film is produced by dipping a wire frame into a liquid and pulling it out. The attempt to produce a film from a particle-free aluminium melt with a wire frame fails (Figure 11.5(a)). On pulling out the film it ruptures immediately. The edge of the film remnant is only 2 μm thick. When the melt contains particles, however, stable liquid films can be pulled out by both vertically and horizontally arranged wire frames (Figure 11.5(b)). The thickness of these films is roughly the same as the average thickness of the cell walls in foams made of the same material. This simple experiment is convincing empirical evidence of the necessity of solid particles in stabilising thin liquid metal films. The large variety of metal foaming techniques demonstrates that the origin of these particles can be quite different ranging from added ceramic particles, oxides created *in-situ* in the melt, oxides originating from powder particles used in processing to solid particles which are a natural component of any semi-solid melt. On the phenomenological level their necessity is evident although the way they act needs clarification.

While traditional aqueous foam can be considered a colloidal system because the diameter of the stabilising elements, *e.g.* surfactant molecules, is well below 1 μm, liquid metal foams are stabilised by particles which cover a wide range of

Table 11.2 *Solid particles used for metal foam stabilisation.*

Particle class	Colloid class	Description	Shape of particles	Size range
Added particles	Liquid–metal suspensions → can sediment → no true colloids	MMC melt	Smooth or angular polyhedral	0.1–50 μm
In-situ particles generated	Liquid–metal sols → no sedimentation → true colloids	Bi-films in melt	Complex shaped	0.05–1 μm thick 10–50 μm wide
Oxide remnants	Liquid–metal gels → no sedimentation → true colloids	Melted powder compacts	Irregular filaments	20 nm thick 50 μm wide

size. Mixing particles with liquid metals and achieving a uniform mixture is more difficult for small particles. Therefore, such systems only exist above 100 nm particle size.[51] *In-situ* oxidation techniques are not that restricted and allow for achieving smaller particles as, *e.g.* for Alporas-type foams. Oxide remnants of powders used for processing are even smaller at least in one direction. The range of particles which has been used for foam stabilisation is listed in Table 11.2.

Figure 11.6 shows the size range which starts with 20 nm for the thickness of oxide filaments stabilising Alulight/Foaminal foams and ends with about 50 μm for the coarsest particles in Metcomb-type foams containing added SiC particles. In the former case we can therefore speak of high-temperature colloids, while in the latter the term suspension seems more appropriate and we draw a somewhat artificial boundary at a value of 1 μm. It is interesting to note that the volume fraction of particles needed for foam stabilisation is related to their size. The larger the particles, the more one needs for effective stabilisation.[47]

11.3.2 Open questions and theoretical approaches

Often analogies between aqueous and liquid metal foams are drawn. Pure single-component liquids cannot be foamed. Bubbles both in pure water and pure aluminium immediately rupture when they arrive at the surface. By adding surface active agents, however, water becomes foamable and the same applies to metal foams when they are treated by one of the ways outlined in Section 11.2.1. However, there are also obvious differences. Stable aqueous films can be made as thin as 6 nm

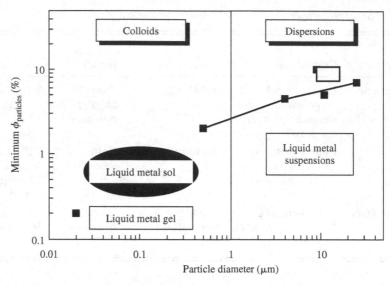

Figure 11.6 Size range of stabilising particles for metal foams of various types. For each size the volume fraction of particles needed for stabilisation is shown. Taken from Ref. [47]; with permission of Elsevier.

(black film),[38] whereas metal foam films are rarely thinner than some tens of micrometers.[31,52] The question is why?

To prevent an aqueous film from thinning up to rupture, a local force is necessary which acts against local destabilisation caused by perturbations such as thermal fluctuations or surface waves. Surfactants reduce surface tension but their main effect is to increase surface elasticity. In many types of foams, the origin of stability is the Gibbs–Marangoni effect. This occurs whenever a thin film is stretched and the film holds surfactant molecules in solution which lower surface tension. As stretching increases the surface area of the film, the density of surfactant molecules per unit area of the film falls and surface tension increases. The Gibbs–Marangoni effect dictates that a stretched film will contract like an elastic skin. Hence, a force in the plane of the film occurs. The effect is quantified by a Gibbs elasticity E_G defined as

$$E_G = -A \frac{d\gamma_{la}}{dA} \qquad (11.1)$$

where A is the surface area and γ_{la} is the surface tension of the liquid.[53] E_G ranges from 0.01 to $0.04\,\mathrm{N\,m^{-1}}$ for water.[54] Values for liquid alloys or liquid metal colloids are not yet known but should be determined.

There are also forces acting perpendicular to the films. The balance of such forces is quite complicated. Besides attractive van der Waals forces there are repulsive forces, called disjoining forces, opposing film contraction. Electrical double layers on the film surfaces can generate such forces. The balance of these two forces describes foam stability in the so-called Derjaguin–Landau–Verwey–Overbeek (DLVO) theory.[55] However, other forces such as hydration forces, steric forces, structural forces, peristaltic or hydrodynamic forces can also be important.[56]

Aqueous foams can also be stabilised by solid particles through different mechanisms. Totally wettable particles present in Plateau borders slow down foam drainage. Partially wettable particles form layers on the surfaces of the liquid films similar to surfactants.[57] Colloidal particles form long-range ordered microstructures in the liquid films which are stabilised by a non-DLVO surface force called the structural force.[58] The structural force appears during the thinning of liquid films containing colloidal particles, *e.g.* surfactant micelles, macromolecules or solid particles. It arises from long-range interactions in concentrated colloidal dispersions. The question still remains as to how particles, their action described by their concentration, their contact angle with the liquid–vapour system, their size and shape, affect the stability of liquid foams.

Foam stability of a model system of water and 3.88 μm surface modified polystyrene particles was characterised by Wilson[59] as a function of the contact angle which the particles made with the air–water surface. The contact angle was controlled by either salt or surfactant addition. Between 0° and 33° no foaming was found, between 33° and 67° slight foaming occurred, between 67° and 85° good foaming was seen and between 85° and 95° very good foaming was observed. In the presence of surfactants, Johansson and Pugh[60] examined the foam stabilising effect of micrometer-sized quartz particles. A stabilising effect was found in the range of 60–90° with an optimum at 75°. Below 44 μm particle size, foam stability was enhanced with respect to larger particles. Sun and Gao examined the effect of 1–75 μm poly(tetrafluoroethylene) (PTFE), polyethylene (PE) and polyvinyl chloride (PVC) particles in surfactant-free ethanol–water dispersions.[61] The wetting angle of the particles was tuned by varying the ethanol concentration. An optimum wetting angle range of 75–85° for obtaining stable foams was found. Smaller particles produced more stable foams. Sethumadhavan *et al.*[62] stabilised aqueous liquid films with silica particles ranging in size from 3 to 39 nm. It has to be noted that non-wetting particles are used even as anti-foaming agents where the defoaming ability depends on both the contact angle (>90° leading to good defoaming) and the surface roughness of the particles.[39] An increase of particle concentration decreases the apparent surface tension of a water suspension.[63] Recently, Binks and Horozov[64] showed how to prepare stable aqueous foams in the absence of any other surface active agent. Depending on the hydrophobicity of the silica nanoparticles, foaming was most effective for particles possessing 32% SiOH groups.

In metal foams, the electrostatic forces are shielded in the bulk because the surfaces are charged with electrons. The stabilisation mechanism must therefore be different. As metal foams contain solid particles it is sensible to look for an analogy between the particle stabilisation mechanism of aqueous foams and that of metallic foams. While there is agreement that particles are needed for stabilisation, there is still some dispute about how particles act in metallic foams. Table 11.3 gives an overview of how foams based on different fluids are stabilised and what the most important properties of the fluid are.

A dynamic picture of foam stability is proposed by Gergely *et al.*[49,66] and states that the vertical motion of a liquid can be damped by increasing the viscosity of the liquid by adding particles or by reducing the temperature to the range of the semi-solid state. Kumagai *et al.*[67] suggested that solid particles in aqueous foams lead to flatter curvatures around the Plateau borders which reduces the suction of metal from the cell wall into the border. Viscosity was also held responsible for the observed stability of metal foams made by foaming powder compacts.[68] The films present in these foams contain up to 1 wt.% oxygen[69] which is thought to be present as thin oxide pockets around volumes of molten aluminium within the films. The idea is that although the liquid aluminium in the cell walls itself has a low viscosity, the entire system behaves like an extremely viscous fluid because the molten aluminium is contained in the oxide pockets by capillary forces or by the mechanical barrier effect of oxides. The viscosity could even approach infinity if the oxides immobilised the metallic component in this way.

Körner *et al.*[70] used a cellular automata model to show the effect of bulk viscosity and surface tension on foam evolution. They found that neither increased viscosity nor decreased surface tension resulted in a stable foam. However, introduction of a disjoining force stabilises the system. The dynamic model was criticised by Körner *et al.*[50] who pointed out that it could not explain the observed long-term

Table 11.3 *Summary of the properties of different foams.*

	Water[38]	Polymer[86]	Glass[65]	Metal[31]
Surface tension of liquid (mN m^{-1})	72	4–30	500	1000
Viscosity of liquid (mPa s)	1	$100-2.5 \times 10^7$	100–1000	3
Cell wall thickness before rupture	6 nm	400–850 nm	100–500 nm	20–30 μm
Stabiliser	Surface active molecules and/or particles	Silicone molecules or block co-polymers	Glass forming and surface active elements (P, Na, Si, B, *etc.*)	Particles

stability of some foams and took this as a proof that a different explanation must apply. Some modelling of metal foam stability started based on the concept of a metal foam containing isolated added particles. The stability map by Jin *et al.*[7] was interpreted by Kaptay[71] on the basis of theoretical considerations. He pointed out that the wetting angle between particles and liquid has to be in a certain range for particles to be able to stabilise the gas–liquid bubble surface. Using a static model of a 3-D network of solid spherical particles he attempted to explain the force transfer between two interfaces of a foam and in this way stability *via* a disjoining pressure.[72] Various configurations of particles were considered in these models, of which three are shown in Figure 11.7. The basic idea is that a partially wetted particle is

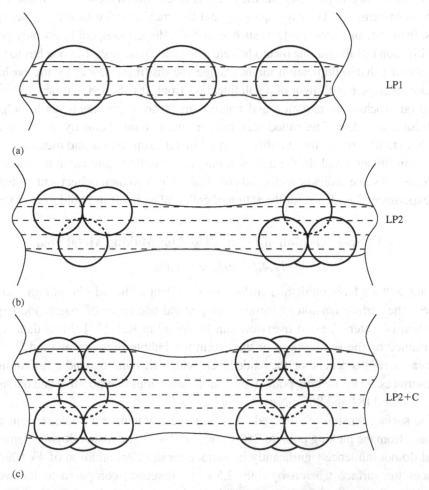

Figure 11.7 Possible particle configurations which could create a disjoining pressure in a liquid film: loosely packed single layer (LP1), loosely packed double layer (LP2), loosely packed double layer plus clusters (LP2+C). Taken from Ref. [72]; with permission of Elsevier.

pinned to a surface since it has its lowest energy when immersed into the liquid at a given depth. A single layer of loosely packed particles (LP1), also discussed by Ip *et al.*,[48] stabilises a film if the contact angle is below 90°. The two surfaces of the liquid are then pinned symmetrically to the particles. Two layers of particles (LP2) do an even better job. Whenever the contact angle is below 145° they keep the two surfaces well separated. The attractive forces are compensated by the stiffness of the particles which are in contact. A third version of the model was proposed since the experimental evidence in many foams showed that the two layers do not touch in reality. In the LP2+C model, the two surface layers are separated by mechanical bridges which are additional clusters (C) of particles keeping the two layers apart.

The distribution of particles and the forces acting between them are well known in aqueous suspensions. Their transparency and low melting point facilitate investigations. In high-temperature systems such as metallic dispersions, until now only post-solidification metallographic methods were available. In such studies one has to take into account that solidification might change the distribution, *e.g.* by the pushing effect or particle engulfment of a solidification front. It is hoped that new methods based on synchrotron radiation and microfocus X-ray experiments could help to establish a new discipline called high-temperature colloid chemistry which would then help to understand the stability of liquid metal foams and liquid metal colloids (LMC) in further detail. In the next sections, we shall first summarise the physical properties of some ordinary melts and colloidal melts, and then collect and systematise experimental evidence for the action of colloidal particles in liquid metal foams.

11.4 Characterisation of Liquids Used for Making Metal Foams

11.4.1 Surface tension

Various authors have published surface tension data of liquid aluminium and its alloys. The surface tension of metals is in general one order of magnitude larger than that of water. A good overview can be found in Ref. [73]. Most data were determined by the sessile drop or the maximum bubble pressure method.[74–77] In both cases, results depend on the oxide layers which are hard to avoid on aluminium melt surfaces. At the melting point, typical surface tensions of oxide-free and oxidised Al melts are 1184 and 865 mN m^{-1} respectively.[73]

The surface tension of oxidised Al is only weakly temperature dependent and changes from the melting point to 750°C only by 3%.[75] The main alloying elements of Al do not influence significantly its surface tension. An addition of 1 wt.% Mg reduces the surface tension by only 2.5%.[78] Therefore, compared to the weak effect of metallic alloying elements, oxidation has a strong effect by reducing surface tension by 27%. Kaptay[73] analysed surface tension data of liquid Al_2O_3 and extrapolated this data to low temperatures. It was found that the surface tension of

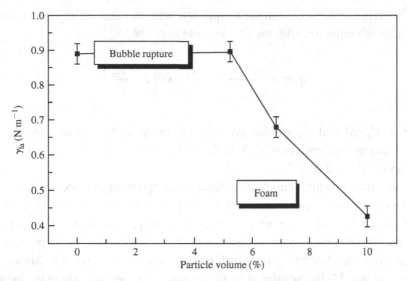

Figure 11.8 Apparent surface tension of Al melt (composition Si: 0.4–0.8, Fe: 0.7, Cu: 0.3, Mn: 0.004, Mg: 0.74 in wt.%) as a function of Al_2O_3 particle content at 700°C. Taken from Ref. [45]; with permission of Wiley–VCH.

oxidised molten aluminium agreed with the extrapolated data of Al_2O_3 which indicates that oxygen covering the surface of molten aluminium resembles a supercooled state of Al_2O_3.

The kinetics of oxidation depends on the oxygen partial pressure. Ricci *et al.*[79] estimated the time required to form an adsorbed oxygen monolayer on various metallic melts exposed to atmospheres containing different oxygen contents. An oxygen monolayer on molten tin exposed to an atmosphere containing 1000 ppm, 1% and 10% of oxygen, forms in 44, 4.4 and 0.4 ms, respectively. Measurements of the surface tension of molten Sn/Pb solder as recently published by Howell *et al.*[80] using oscillating jets show the importance of progressive oxidation of melt surfaces. Heterogeneous systems can be characterised by an apparent surface tension. Melt suitable for foaming, *i.e.* a suspension of ceramic particles in a liquid aluminium alloy, was investigated using a high-temperature maximum bubble pressure tensiometer.[45] In this case, Al containing a certain amount of alumina particles of 11 μm diameter was studied. It was found that increasing the Al_2O_3 particle content reduces the surface tension which is an analogous effect to the oxidation effect mentioned above (see Figure 11.8).

11.4.2 Viscosity

The bulk viscosity of homogeneous liquid metals is comparable to that of water. Kaptay[81] recently derived a unified equation for the dynamic viscosity of pure liquid metals combining Andrade's equation with the activation energy concept and Andrade's equation with the free volume concept

$$\eta = A \cdot \frac{M^{1/2}}{V_m^{3/2}} \cdot T_m^{1/2} \cdot \exp\left(B \cdot \frac{T_m}{T} \right) \tag{11.2}$$

where η, V_m, M and T_m are the dynamic viscosity, molar volume, atomic mass and melting point, respectively. T is the temperature, $A = (1.8 \pm 0.39) \times 10^{-8}$ $(\mathrm{J\,K^{-1}mol^{1/3}})^{1/2}$ and $B = 2.34 \pm 0.20$.

Heterogeneous liquids can be described by an apparent or effective viscosity in many cases. One example is semi-solid alloys. The presence of solidified primary α-particles in the melt is known to increase viscosity.[82] For high solid fractions in the range of 50%, such systems can exhibit non-Newtonian flow behaviour with a shear-rate-dependent viscosity which has a practical relevance for thixo-casting or rheo-casting. Melts containing solid ceramic particles also show an increased bulk viscosity. This was measured *e.g.* for lead.[83] Melts produced by re-melting compacted lead powders containing oxidised surfaces have a viscosity which is more than twice as high as low-oxide lead melt (see Figure 11.9).

The viscosity of the melts used for making Alporas foams, obtained by adding Ca to Al and allowing for some oxidation during stirring, can reach 10 times the

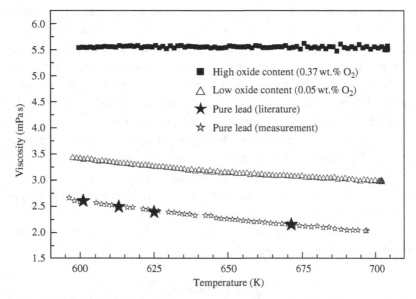

Figure 11.9 Bulk viscosity *versus* temperature measured by a rotational viscometer for pure Pb and two different Pb foam precursors. Taken from Ref. [83]; with permission of the University of Bremen.

value of the oxide-free alloy depending on stirring time.[84] Certain alloys can be transferred into a glassy state by continuous cooling from the liquid state. In some alloys this transition is obtained at very low cooling rates, *e.g.* less than $1 \, \mathrm{K \, s^{-1}}$. The viscosity of such alloys can reach very high values which has encouraged researchers to create foams from them.[18] Some typical viscosities of metallic and non-metallic liquids are listed in Table 11.4.

Viscosities of heterogeneous systems measured in a viscometer show an averaged behaviour of the liquid. In foams, the composition of the liquid varies, with higher particle concentrations near the surface of a film. Therefore, a gradient in viscosity can be expected with a surface viscosity above the average. The situation is even more complex in the compacted and re-melted powders which are used in the Alulight/Foaminal process. Here the meaning of the term viscosity is unclear. Films in the foams contain a network of oxide filaments which was created during powder atomisation. These filaments retain the liquid metal and lead to a very viscous appearance of the material which could possibly approach infinity if

Table 11.4 *Viscosities of various liquids.*

Liquid metals	η (mPa s)	Reference	Non-metallic liquids	η (mPa s)	Reference
Pure Al at T_m	2 ± 0.8	[85]	Pure water at 20°C	1.025	[38]
Pure Pb at T_m	2.61	[85]	Polymer melts	100–1013	[86]
Pure Fe at T_m	6.92	[85]	Slag melt at 1500°C	200 300	[87]
Melt of Pb powder, 0.37 wt.% O_2 (0.05 wt.% O_2) at T_m	5.5 (3.2)	[83]	Glycerine at 20°C	1508	[88]
Alporas melt at 720°C	1.5–14	[84]	Glass melt $T = T_1$	10,000	[88]
AlSi6.5/SiC/10p at 700°C	30	[89]	Glass melt $T > T_{\mathrm{softening}}$	5×10^9	[88]
AlSi6.5/SiC/20p at 700°C	50	[89]	Glass melt $T > T_g$	5×10^{16}	[88]
AlSi6.5/SiC/30p at 700°C	300	[89]	–	–	–
Bulk met glass at T_1	1000	[18]	–	–	–

T_m: melting point, T_1: liquidus temperature, T_g: glass forming temperature.
Notes: MMC designations: alloy/particle/vol.%

liquid flow is blocked completely. Upon stirring or shearing, however, the oxides agglomerate which changes conditions irreversibly.[50] Therefore, viscosity depends on history in this case and one should be very careful in using the term.

11.4.3 Reactions between particles and melt

Aluminium MMC are the raw material of Cymat and Metcomb foams. Aluminium has a very high chemical reactivity in the liquid state. Al melts do not only react with most metals but also with some ceramics. Only if MMC are made by a solid-state process are interfacial reactions insignificant. If, however, liquid phase infiltration of ceramics or admixture of ceramic particles to a melt are applied to produce MMC,[90] then reactions are inevitable. The reaction temperatures and times during MMC production and aluminium composite melt foaming are comparable which allows us to use the experience in MMC production for assessing the conditions during foaming. The reaction products often modify the surface of the particles in the melt resulting in a kind of coating. These coatings not only impair mechanical properties but also could influence foamability of such melts. Some common reaction data are summarised in Table 11.5.

Among the materials listed, the SiC and Al_2O_3 systems are most frequently used in practice. Some reactions can be both eliminated and promoted by alloying.[98,99]

Table 11.5 *Chemical reactions of particles with Al melt below 1000°C.*

Ceramics	Reaction with pure Al	Reaction products	Reference
AlB_2	No	–	[91]
TiB_2	No	–	[91]
Carbon	Depends on surface structure	Al_4C_3	[98,99]
Al_4C_3	No	Al_4C_3 reacts with water forming CH_4	[98,99]
B_4C	Yes	Al_3BC, AlB_2	[99]
SiC	Yes	Al_4C_3	[98,99]
TiC	Yes	Al_4C_3, Al_3Ti	[99]
AlN	No	–	[92]
BN	Yes	AlN	[92]
Si_3N_4	Yes	AlN	[93]
Al_2O_3	No	–	[98,99]
CaO	No	$CaAl_2O_4$ in the presence of O_2	[94,95]
MgO	Yes	$MgAl_2O_4$	[96]
SiO_2	Yes	Al_2O_3, Al_2SiO_5, $Al_6Si_2O_{13}$	[98,99]
TiO_2	Yes	Al_2O_3, AlTi, Al_3Ti, TiO	[97]

SiC reacts with liquid Al.[100] The associated formation of aluminium carbide is detrimental to composite properties because of its brittleness and reactivity with water. It is also undesirable in the melt stirring process because it increases melt viscosity. Aluminium carbide formation can be prevented by adjustment of the matrix composition, by coating SiC particles with a physical barrier (Al_2O_3, TiO_2, TiB_2, SiO_2, TiC, TiN, *etc.*) or by applying a sacrificial layer (Ni, Cu, *etc.*).[99] To prevent the formation of Al_4C_3, pure Al has to be alloyed with Si. The minimum Si content required for suppressing carbide formation was found to be 8.5 wt.% at 610°C and 13 wt.% at 825°C with a linear interpolation rule in between.[101] The presence of as little as 1 wt.% Mg in pure Al leads to the formation of Mg_2Si and facilitates the formation of aluminium carbide.[102] Contradictory to this, in AlSi9Mg/SiC/10–20p, the maximum overheating temperature was found to be 750°C which is 60°C higher than in the same alloy without Mg.[101] In the Al–SiC system above 1400°C, Viala *et al.*[103] found other complex carbides such as Al_4SiC_4 and Al_8SiC_7. If SiC is oxidised, the resulting SiO_2 surface layer can react with the liquid Al forming Al_2O_3, Al_2SiO_5 or $Al_6Si_2O_{13}$ (mullite). In the presence of Mg the reaction product is Al_2MgO_4.[104]

Alumina, Al_2O_3, is stable in liquid aluminium but reacts in the presence of magnesium either to Al_2MgO_4 (spinel) for $1 < c_{Mg}$ (wt.%) < 4 or to MgO for $c_{Mg} > 4$ wt.%.[105] The thin passivation layer of MgO prohibits further reactions. Detrimental spinel formation is continuous and can be limited by either (i) using a mixed oxide such as $Al_6Si_2O_{13}$ (mullite), (ii) forming a MgO barrier layer exploiting the reaction itself, (iii) using an aluminium matrix with a low magnesium content ($c_{Mg} < 1$ wt.%) or (iv) further alloying the Al matrix with an element like Sr which selectively segregates at the interface and inhibits spinel growth.[106] Al_2MgO_4 is observed for $c_{Mg} > 0.5$ wt.% using infiltration casting but no reaction occurs during squeeze casting. Schuster and Skibo[107] reported interface spinel formation of AA6061/Al_2O_3/10–20p (industrial alloy) as well.

The thermodynamic stability range of reactions of liquid Al and ceramics as a function of alloying elements have been calculated and summarised in the literature, Table 11.6.

11.4.4 Wetting of particles by melt

Wetting properties can be estimated by measuring the contact angle θ, measured into the liquid metal phase, in sessile drop experiments with accuracies around 1–5°. Contact angle and the interfacial energies are connected through Young's equation

$$\gamma_{pl} = \gamma_{pa} - \gamma_{la} \cos \theta \qquad (11.3)$$

Table 11.6 *Calculated thermodynamic stability range of reactions between liquid Al and ceramics.*

Ceramic	Reaction products	Reaction interval	Remarks
SiC	Al_4C_3	Below 7 wt.% Si content at the melting point[108]	Increasing temperature and adding extra Mg, the reaction interval predicted by the model increases; good fit with experiments.
Al_2O_3	$MgAl_2O_4$	$0.02 < c_{Mg}$ (wt.%) < 1 at 700°C[109]	Calculation not explained in detail; the author's experimental findings agree with his calculation.

Table 11.7 *Contact angle of pure Al melt/particle/gas interface measured into the melt phase. Taken from Ref. [77]; with permission of Trans Tech Publications.*

	TiB_2	WC	TiC	BN	SiO_2	SiC	Graphite	Al_2O_3	TiN
θ (°)	0	5	10	15	23	27	52	63	86

where γ_{pa} is the ceramic particle–air surface energy, γ_{pl} is the ceramic particle–liquid aluminium surface energy and γ_{la} is the surface tension of the liquid. The Young–Dupré equation can also be applied

$$W = \gamma_{la}(1 + \cos \theta) \qquad (11.4)$$

where W is the adhesion energy. The ceramic particle is perfectly wetted by the liquid at $\theta = 0°$ or $W \geq 2\gamma_{la}$. The contact angle depends on time, temperature, composition, heat treatment of ceramics (impurity or absorbed gas removal), surface roughness[110] and atmospheric conditions during the experiment. For Al, the oxide layers inevitably influence the measurements. Some examples for the contact angle of pure Al melt/particle/gas systems at 1100°C (no oxide layer on liquid Al, purified Ar atmosphere at 10^{-8} bar residual pressure) are summarised in Table 11.7.[77]

Among all metal/ceramic combinations, Al–SiC has been one of the most widely studied systems. The contact angle between aluminium and SiC can be modified by adding Si, Mg or Cu. The achievable changes can be as high as 40° for some compositions and temperatures. Si in liquid Al protects the SiC particles against interfacial reactions.[111] The change of the contact angle with time was measured in the system AlSi18/SiC at 800°C under high vacuum 4×10^{-5} Pa.[112] The initial value was found to be 120°. SiC was so well protected against any reaction with Al by the high Si content of the melt that even after 3 h the measured contact angle of 60° was more than twice as large as that of non-alloyed Al melt (27°).[112] The enhancement of wetting in the absence of Si is still not clear. Formation of Al_4C_3 is a possible explanation but the contact angle between Al_4C_3 and Al[113] of 55° is even higher than the contact angle of Al–SiC melt (27°).

Laurent *et al.*[112] measured the dependence of the contact angle of the pure Al–Al_2O_3 couple on oxygen partial pressure. Close to the melting point of Al, the contact angle was 160° at 10^{-3} Pa but 100° at 10^{-5} Pa. Above 900°C the oxide layer reacted with liquid aluminium and evaporated in the form of Al_2O, thus the pressure dependence of contact angle disappeared and θ decreased down to 90°. Alloying with Li and Ca improved wetting due to the formation of Li or Ca aluminium oxides on the surface of the alumina particles and decreased the surface tension.[114] The initial and time-dependent contact angles of (0001)-oriented α-Al_2O_3 single crystals under less than 5×10^{-4} Pa oxygen partial pressure were recently investigated.[115] The contact angle was found to have a tendency to decrease with increasing temperature. Between 700°C and 800°C surplus oxidation was found without any essential change of contact angle. Heated above 800°C, the pattern of oxygen on the (0001) plane varies from (1 $\times$ 1) to an oxygen deficient ($\sqrt{31} \times \sqrt{31}$R $\pm$ 9°) Wood index structure while the contact angle decreases. Due to the reaction between Al melt and Al_2O_3 a strong decrease of wetting angle is observed above 1100°C. It was also found that while the contact angle of the (0001) plane changes with time (at low temperature increasing, at high temperature decreasing), other crystallographic planes are not sensitive with respect to time. It can be summarised that 75° (30 min, 1500°C) seems to be the minimum and 130° for the maximum values of the contact angles (30 min, 700°C).[115]

11.5 Investigation of Foaming Process

Phenomena related to metal foam evolution and stability have been studied *ex-situ* after solidification or *in-situ* while the foam is still in its liquid state. Both methods are complementary. *Ex-situ* methods allow us to apply a larger variety of analytical tools but the evolution of foams has to be investigated by considering different samples representing the stages of foaming. In contrast, *in-situ* methods allow us to follow the evolution directly but are only available for a restricted set of analytical methods.

11.5.1 Ex-situ *analysis of metallic foams after solidification*

Ex-situ analysis of metal foams is an important tool for studying the microstructure and architecture of metal foams. By freezing the foam one hopes to preserve its morphological features and the arrangement of solid particles in the cell walls at least to a certain extent. One has to be aware of effects during solidification such as interactions of the solidification front with particles and pressure changes which might lead to a difference between the real situation in the foam and the picture obtained by analysing solid foams. We shall now first analyse the microstructure of foams made by some of the processes listed in Table 11.1. We search for the particles responsible for stabilisation and for oxide films formed during foaming. Then the structure of cell walls will be analysed to reveal the influence of the stabilisation mode on the thickness and the shape of films.

11.5.1.1 Microstructure of foams and importance of particles

11.5.1.1.1 Added particles and external gas blowing (Metcomb-type foams) In this type of metal foam it is most obvious that particles are responsible for liquid metal foam stability. A schematic stability map giving limits for particle content and particle size required for metal foam creation was published by Jin *et al.*[7] Ip *et al.*[48] later found that particle attachment to the gas–liquid interface is necessary for the extended stability of the aluminium foam. This occurs when, after external gas injection, bubbles travel through the liquid and interact with the dispersed particles, an effect analogous to flotation. Leitlmeier *et al.*[22] measured the volume

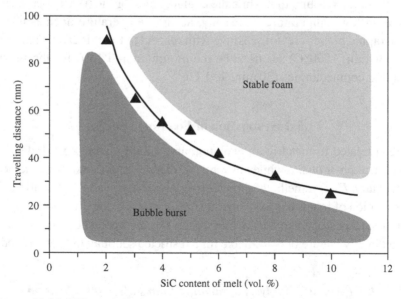

Figure 11.10 Effect of particle concentration on the travelling distance of bubbles through the melt and its effect on foam stability. Taken from Ref. [22]; with permission of Wiley–VCH.

fraction of particles in aluminium-based foams as a function of the travelling distance of bubbles in the melt, given by the distance from the injector to the metal surface, and found the stability criterion shown in Figure 11.10. Accordingly, below a critical travelling distance, no stable foam could be produced since the bubbles did not collect enough particles. The particle concentration in the cell walls increases up to a saturation value[22] because of the long travelling path of the rising bubbles in the MMC melt. After only some centimeters of rising in the MMC melt, the SiC particles collected were only sufficient to cover the surfaces of the bubbles (Figure 11.11(a)). A longer travelling distance then yields a configuration with surface coverage plus some particles inside the cell wall (Figure 11.11(b)). Interestingly, a meniscus of the melt around the particles located on the cell wall surface is not observed (Figure 11.12).

Stable foams were successfully produced from aluminium alloy melts containing SiC and Al_2O_3 particles (MMC melts).[22] It is interesting to note that Al_2O_3 particles were coated with nanometer-sized spinel (Al_2MgO_4) crystals in some successful foaming experiments in which the melt contained magnesium.[47] This surface reaction was found to improve foam stability. In contrast to this situation it was reported that SiC loses its ability to stabilise a melt after overheating. X-ray diffraction analysis showed the formation of Al_4C_3 in the melt at 950°C.[52] The microstructure of the bubble remnants produced under these conditions showed a wavy surface.

There is some evidence that the contact angle of the particles is the determining parameter in liquid metal foam stability.[47] Therefore, selecting stabilising particles by exploiting tabulated data for contact angles[72] could be a successful strategy. In the example given in the previous paragraph, aluminium carbide formation on the surface of stabilising SiC particles probably modifies the contact angle of the particle putting it outside the preferred region.

The foamability of an Al composite containing TiB_2 particles (AlSi10 + 15 wt.% TiB_2 particles of 3–6 μm size) was also examined.[47] The contact angle between TiB_2 and pure liquid Al is reported to be 98° at 900°C,[116] although one has to be aware of differing data giving values down to 0°.[77] It was attempted to create aluminium foam by gas injection into the melt at 700°C and it was found that TiB_2 is not an effective stabiliser for this particular system. Only irregular bubble remnants could be produced. TiB_2 particles fall out of the bubbles leaving a significant amount of powder on the surface of the melt. Particles can be extracted from a melt only if they are not wetted, indicating that TiB_2 is actually not wetted in this particular situation.

An additional important effect on foamability and microstructure is created by the reaction of the liquid metal with an oxidising blowing gas such as air.[45,46] In the presence of oxygen, aluminium melts oxidise almost instantaneously. An oxide film continuously grows and reaches several hundred nanometers in thickness.

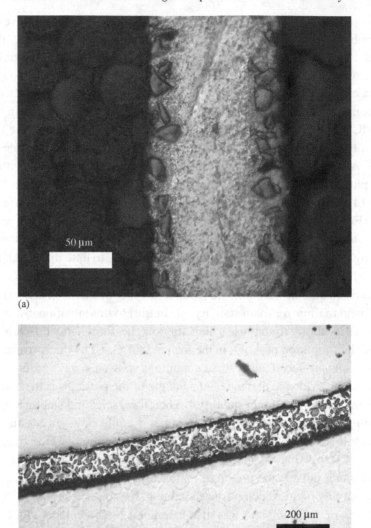

(a)

(b)

Figure 11.11 Cell wall cross-section of aluminium foam stabilised with SiC particles. The bubble rising height in the melt was (a) 4 cm and (b) 15 cm. Courtesy of the Hahn-Meitner-Institute, Berlin.

The oxidation effect on stability will be discussed in Section 11.5.2.2. Here we concentrate on its influence on surface segregation of particles. A typical oxide skin on the inner surface of an air-blown foam is shown in Figure 11.13.

Ip *et al.*[48] suggested that complete coverage of the bubble surface by particles is not required to attain a stable liquid metal foam and assumed a minimum surface

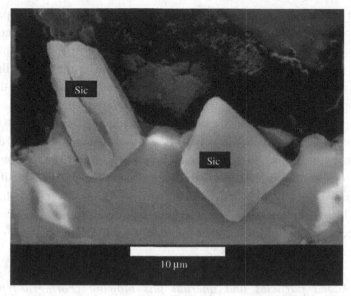

Figure 11.12 SEM cross-section showing SiC particles at the surface of a Metcomb foam. Taken from Ref. [52]; with permission of the Univeristy of Miskolc.

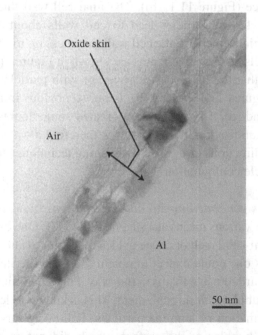

Figure 11.13 Oxide layer on the cell wall of a foam from a MMC (Cymat, Canada). Taken from Ref. [46]; with permission of Wiley–VCH.

coverage of 50%. The maximum surface coverage of Al_2O_3 and SiC particles in Metcomb foams was found to be 45% when nitrogen was used for foaming. The uniform coverage of a cell wall can be seen for Metcomb foams containing Al_2O_3 particles in Figure 11.14(a). The particle concentration in the cell walls is much higher than in the raw material, sometimes approaching 30 vol.%[22] and the surface areas are especially enriched having a local concentration of 2–4.5 times the bulk value. The surface coverage decreases when oxidising foaming gas is used.[46] Wrinkled oxide layers cover the cell wall surfaces whenever foams are blown with air or oxygen (Figure 11.14(b)). Such layers are similar to those observed on the surface of aluminium castings. In order to make the particles visible, a cell wall surface was gently polished until the surface oxide layer had disappeared (Figure 11.14(b)). The surface coverage observed here did not exceed 26%. The local surface concentration was found to be only 1.5 times the particle concentration of the raw material in this case.

11.5.1.1.2 Added particles and internal gas blowing (Formgrip-type foams)
Scanning electron microscopy (SEM) analysis of cell walls in solidified Formgrip foams shows a dense coverage with particles if SiC particles with 13 μm average size are used (Figure 11.15(a)). If stabilisation with much larger particles (70 μm) is attempted, the situation changes and only a small amount of the particles can be found on the surface (Figure 11.15(b)). The final cell wall thickness also varies with particle size; 13 μm particles lead to cell walls about 85–100 μm thick, whereas 70 μm particles increase the cell wall thickness up to 300 μm. Cross sections of the cell walls of these two foams are shown in Figures 11.15(c) and (d).

It is interesting that the high surface coverage with particles as shown by the face-on images, Figures 11.15(a) and (b), is not so obvious in the cross-sections, Figures 11.15(c) and (d). This might have created some doubts about the mechanism of particle stabilisation in the past. It is therefore always necessary to take images from both directions or to use 3-D imaging techniques to obtain a reliable picture of the particle configuration.

11.5.1.1.3 Particles generated in-situ *and external gas blowing* Particles can be generated *in-situ* by various means as listed in Table 11.1. One can rely on the solid component of a semi-solid melt or create oxides by *in-situ* oxidation. After injecting gas into such melts one could expect to obtain foams. Experimental work shows that in reality it is hard to make foams in this way. Pure aluminium melts were pre-treated for 1 h by bubbling air through them. This is known to lead to inner oxidation and to the formation of alumina particles. Following this treatment, it was possible to obtain bubbles at the surface which did not rupture immediately. However, the bubbles were not stable enough to form a foam layer. Coalescence

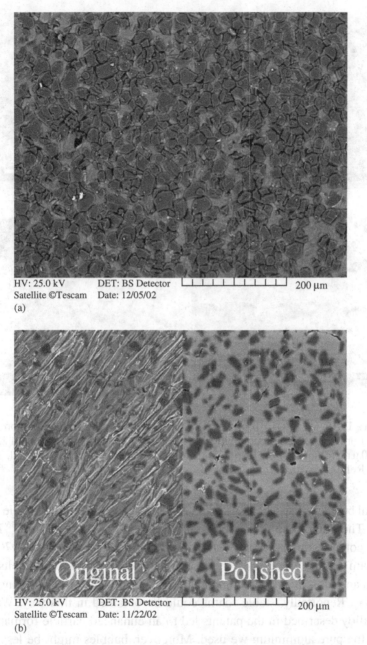

HV: 25.0 kV   DET: BS Detector   200 μm
Satellite ©Tescam   Date: 12/05/02
(a)

HV: 25.0 kV   DET: BS Detector   200 μm
Satellite ©Tescam   Date: 11/22/02
(b)

Figure 11.14 SEM images of the surfaces of a cell wall of an Al_2O_3 particle-containing Duralcan aluminium composite (W6D22A). (a) Untreated and foamed with nitrogen and (b) Left-untreated, and right-polished, with 1 μm sized SiC grains and foamed with air. Taken from Ref. [46]; with permission of Wiley–VCH.

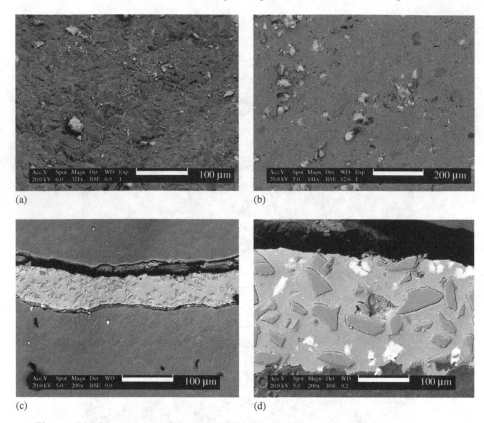

Figure 11.15 (a) and (b) Top view SEM images of a cell wall in a foam prepared from Formgrip material containing 10 vol.% SiC of average diameter (a) 13 and (b) 70 μm. (c) and (d) Corresponding cross-sections of the same material. Taken from Ref. [47]; with permission of Elsevier.

and partial bubble collapse led to an onion-like structure into which bubble remnants piled up. The wall of one such bubble remnant showed a wavy surface.[46] It is interesting to compare this approach with patents from the 1960s[26] and 1970s[27,117] in which foam making processes were described based on pre-treating melts by inner oxidation and subsequently foaming them by addition of the gas releasing blowing agent TiH$_2$. Reportedly, stable foams could be obtained in this way. We suspect that the alloy described in the patents led to an enhanced particle formation compared to the pure aluminium we used. Moreover, bubbles might be less prone to collapse when formed more gently by adding a gas releasing blowing agent as compared to gas injection where the metal films are stretched rather quickly. Moreover, the strong movement of a large single bubble generated within the melt might lead to an agglomeration of oxide bi-films[46,118] as observed for different materials by Körner *et al.*[50]

The Alporas foaming method leads to very homogeneous foams. The method uses an Al melt in which particles are created *in-situ* by adding Ca and allowing for oxidation (see below). The question now is if it is possible to make stable foams from these melts by injecting gas instead of adding titanium hydride. A trial experiment on a AlCa3 alloy stirred for 30 min showed that bubbles of about 10 mm diameter are created in the melt which then rise to the top of the melt and merge there. However, some large bubbles remain stable after some minutes of isothermal holding (see Figure 11.2(g)). A very inhomogeneous cell wall thickness distribution was found ranging from 10 μm to some hundred μm. The surfaces of the cells are wavy (Figure 11.16(a)). The foam microstructure contains many complex-shaped phases in the eutectic phase of the alloy (Figure 11.16(b)).

11.5.1.1.4 Particles generated in-situ *and internal gas blowing (Alporas-type foams)*
It is appropriate to ascribe the stabilisation effect in Alporas foams to the solid particles formed during stirring the aluminium alloy containing Ca under the influence of air which is dragged into the melt through the vortex.[14] Microstructural analysis of Alporas foams reveals the presence of oxides.[46] A commercial Alporas foam sample[119] was polished and etched with a 0.5 vol.% HF solution. SEM and EDX (Energy Dispersive X-ray) examinations show deeply etched valleys with high oxygen content and secondary phases attached to line-shaped pores with higher Ca and/or Ti content than the matrix. These line-shaped pores, most probably oxide bi-films,[118] form networks decorated with secondary phases. We suspect that these decorated bi-films play the crucial role in the stabilisation of Alporas-type metal foams. The size and volume fraction of these inclusions is estimated to be below 1 μm and 1 vol.%, respectively. No obvious segregation of these bi-films to the surface of the cell walls was found.

Recent investigation of the cell wall surface of the above mentioned Alporas foam sample show bi-films (Figure 11.17). Another Alporas-type foam with different composition[120] clearly shows sub-micron sized particles besides the bi-films (Figure 11.18). These particles could be primary crystals formed on the amorphous oxide skin during oxidation.[121] Yang and Nakae[122] were able to produce foam from an AlSi7Mg0.5 alloy without any Ca addition. They stirred an alloy melt in air for 20 min with 600 rpm and added 5 wt.% Al powder of 60 μm particle diameter. After this, foaming was triggered by TiH_2 addition. This technique actually combines oxide formation of the Alporas process with some ideas of the powder compact technique (see Section 11.5.1.1.5). As the aluminium powder added is oxidised, the oxide films entrained in the liquid assist foaming in addition to the oxides created during stirring.

11.5.1.1.5 Molten powder compacts and internal gas blowing (Alulight-type foams)
Foams are produced by melting powder compacts containing a blowing agent. It has been shown that the oxygen content of the precursor material is crucial for the

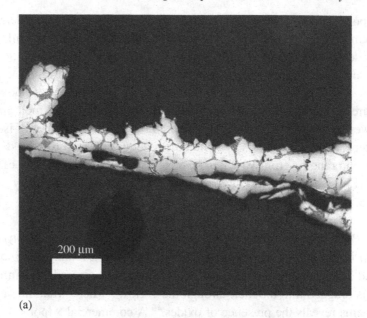

(a)

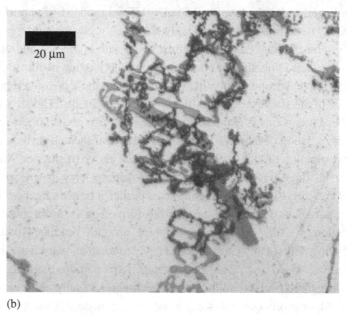

(b)

Figure 11.16 Optical micrographs of "foam" created by blowing a TiH$_2$-free Alporas precursor (same sample as shown in Figure 11.2(g)). (a) Wavy cell wall and (b) cross-section etched with HF containing complex-shaped phases in the eutectic phase. Courtesy of the Hahn-Meitner-Institute, Berlin.

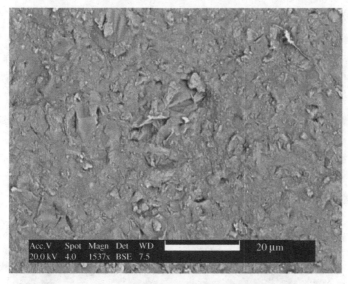

Figure 11.17 SEM image of oxide bi-films on Alporas foam cell wall surface with surface concentration of 3.62 O, 4.51 Ca, 0.62 Ti, 3.62 Fe with remainder Al (in wt.%). Courtesy of the Hahn-Meitner-Institute, Berlin.

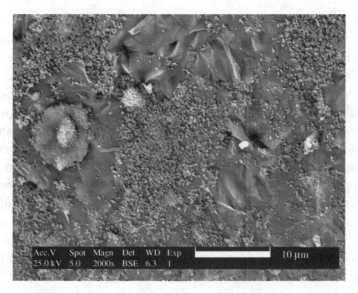

Figure 11.18 SEM image of the surface of an Alporas-type foam cell wall containing wrinkled oxide bi-films and sub-micron particles with surface concentration of 9.06 O, 1.71 Ca, 0.35 Ti with remainder Al (in wt.%). Courtesy of the Hahn-Meitner-Institute, Berlin.

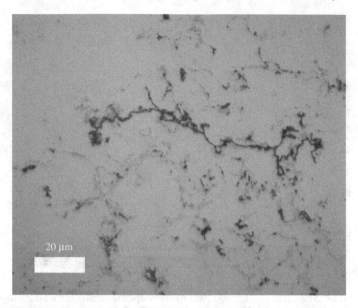

Figure 11.19 Optical micrograph showing the complex phases in re-melted pure Al powder compact. Courtesy of the Hahn-Meitner-Institute, Berlin.

stability of such foams. A study by Weigand showed that aluminium powders with very low oxide contents did not lead to very stable foams[69] and a similar observation was made recently when argon and air-atomised aluminium powders were compared.[50] Air-atomised powders not only exhibit a higher oxide content but also a distribution of oxygen which includes the interior of each powder particle.[50,69] In lead foams,[14,83] a similar dependence of foam stability on oxide content was found. Foams containing 0.06 wt.% oxygen did foam but significant collapse and continuous drainage was observed during the foaming stage. A content of 0.16 wt.%, however, lead to stable foams. Figure 11.19 shows the oxide phases in a melted pure aluminium powder compact (not containing any blowing agent). Wübben *et al.*[34] used micro-gravity experiments to demonstrate that the primary action of the solid stabilising component of powder compact foams is to prevent films from coalescing while their influence on viscosity was seen as less important. Foaming of pure Al powder compacts could be further stabilised with TiB_2, Al_2O_3 and SiC particle addition in this order.[123]

11.5.1.1.6 Molten powder compacts and external gas blowing Our recent trial to foam Al–Si powder compacts using a foam generator was not successful (see Figure 11.2(f)). Although the same material can be foamed by internal gas creation, *e.g.* by embedded TiH_2 particles, this is not the case when the gas enters the liquid from one injection point.

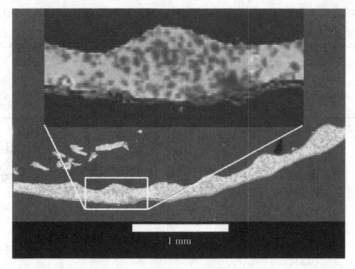

Figure 11.20 SEM image of the wavy surface surrounding agglomerated particles of cell wall made from AlMg1SiCu/Al$_2$O$_3$/22p raw material. Taken from Ref. [52]; with permission of the University of Miskolc.

11.5.1.2 Structure of cell walls

A quantitative characterisation of cell walls and Plateau borders is essential to understand the driving forces for stability. The cell wall thickness can be measured easily by analysing metallographic sections. More complex data such as curvature of Plateau borders requires 3-D tomographic methods. Many metal foams with low densities show a polyhedral cell structure with homogeneous cell wall thickness distributions between two Plateau borders.[52] Foams with higher densities usually have smaller and predominantly spherical cells mostly with diameters below 3 mm.

Although most metal foams show smooth cell walls (roughness in the range of the size of the solid particles present in the melt), some foams deviate from this pattern especially when their cell walls contain agglomerated particles (see Figure 11.20).[45,52] This particularly happens in foams made from powder compacts or in mixed particle systems. Films of metallic melt cannot be stretched to such an extent as aqueous films. Liquid metallic films containing particles usually rupture below a critical thickness which ranges from about 30 to 150 μm[31] for foams made from powder compacts and slightly lower values for foams produced from MMC.[45] In the latter case, it could be shown that the cell wall thickness depends on (i) the particle diameter – larger particles produce thicker cell walls whereas smaller particles give thinner walls, (ii) the composition of the alloy, (iii) the composition of the particle, (iv) the foaming temperature, (v) the cell size and (vi) the concentration of the particles.[45,66]

Table 11.8 *Cell wall thicknesses of aluminium foams. All foams were produced at 1 bar atmospheric pressure.*

Alloy	Particle type and size (μm)	Minimum cell wall thickness (μm)	Average cell wall thickness (μm)	Remark	Reference
Foams blown by internal gas formation					
Al	Powder oxides	100	345	–	[127*]
Al99.9	Powder oxides	–	215	TiH$_2$ (size 5 μm)	[50]
Al99.9	Powder oxides	–	165	TiH$_2$ (size 2 μm)	[50]
AlSi2	Powder oxides	80	250	–	[127*]
AlSi6Cu4	Powder oxides	51	–	Measured in liquid	[31*]
AlSi7	Powder oxides	52	–	Measured in liquid	[31*]
AlSi7	Powder oxides	27–36	88	–	[128]
AlSi10Mg	Powder oxides	60–70	100–170	–	[129*,**]
AlSi10Mg0.6	Powder oxides	30–50	–	Low resolution	[130*,**]
AlSi12	Powder oxides	80	205–320	–	[127*]
AlSi12Mg0.6	Powder oxides	60–80	–	–	[131*,**]
AlCa1.5	*In-situ* oxides	–	~100	–	Own measurement
AlSi9Mg0.5	SiC (13)	–	~40	–	[24*]
AlSi9Mg0.5	SiC (63)	–	~210	–	[24*]
AlSi10Mg	SiC (13)	–	~85	–	[47*]
AlSi10Mg	SiC (70)	–	~260	–	[47*]
Foams blown by external gas formation					
AlSi10Cu3Ni1.5	20% SiC (13)	20–30	55	0 min, 700°C, 10 mm	[52]
AlSi10Cu3Ni1.5	20% SiC (13)	33 (17 ox.)	44	10 min, 700°C, 10 mm	[52]
AlSi10Cu3Ni1.5	20% SiC (13)	33	44	100 min, 700°C, 10 mm	[52]

Table 11.8 (*continued*)

Alloy	Particle type and size (μm)	Minimum cell wall thickness (μm)	Average cell wall thickness (μm)	Remark	Reference
AlSi10Mg	12.5% Al_2O_3 (23)	–	35–50	No spinel, 0 min, 700°C, 10 mm	[52]
AlMg1SiCu	10% Al_2O_3 (11)	30–43	112–86	Spinel, 695– 765°C, 10 min, 10 mm	[52]
AlSi10Cu3Ni1.5	10% Al_2O_3 (11)	21	74	Spinel, 725°C, 10 min, 10 mm	[52]
AlSi10Cu3Ni1.5 Mg3	10% Al_2O_3 (11)	20	61	Spinel, 725°C, 10 min, 10 mm	[52]

*No temperature data, **estimated from a graphical representation.

Detailed analysis of the cell wall thickness of added particle foams has been carried out.[45,46,52] It was found that the surface composition of the particles is strongly related to the cell wall thickness. Alumina particles with spinel-coated surfaces lead to thicker cell walls than uncoated SiC particles of the same diameter.[52] Lognormal distributions of the cell wall thickness were found in all cases. In Metcomb foams, no significant difference has been found between cell wall thicknesses of air- and nitrogen-blown aluminium foams. The cell wall thickness can also depend on the collection efficiency of the particles by the bubble surface during foam evolution. The cell wall thickness increases with the travelling distance of the bubbles, eventually reaching a saturation value.[124] Relationships between cell diameter and cell wall thickness of foams manufactured by Cymat (Canada) were given by Wood.[125] The cell wall thickness can be considered constant, around 50 μm, for large cell sizes above 8 mm diameter. At a diameter of 3 mm, the thickness is 85 μm.

The effect of ceramic particles on the cell wall thickness of Cymat-type foams was examined by Dequing and Ziyuan.[126] It was found that the cell wall thickness increases with particle size and concentration and decreases with foaming temperature. Average and minimum cell wall thicknesses should be determined more systematically in the future, preferably as a function of particle parameters (size, shape, amount, bulk and surface composition, surface roughness), foam cell sizes, alloy characteristics (composition, temperature) and bubble pressure. The data available at present is given in Table 11.8.

From the values in Table 11.8 we can conclude that; (i) in foams blown by external gas formation the cell wall thickness does not decrease significantly with isothermal holding time (4th and 5th columns),[52] (ii) in both foaming routes, addition of Mg and Si leads to a decrease of cell wall thickness, (iii) the minimum cell wall thickness is 20–30 μm in aluminium foams, (iv) spinel-coated Al_2O_3 particles yield thicker cell walls than SiC particles of the same size and volume fraction.

11.5.2 In-situ *investigation of liquid metallic foams*

In order to understand the evolution of metal foams it is useful to obtain data characterising the state of the foam *in-situ* during expansion while the metal foam is fluid. The technique which is probably the easiest to handle is the "expandometer" which is a specially constructed dilatometer for measuring foam volume as a function of time.[132] Other *in-situ* techniques which have been employed include thermoanalysis,[133] thermogravimetry[134] and ultrasound probing.[135] All these techniques do not reveal the details of structural changes inside an evolving metal foam. *In-situ* X-ray monitoring experiments on metal foam yield such information. The first experiments of this kind were carried out in the Alcan laboratories 15 years ago.[136] With improved equipment, such experiments were later carried out with synchrotron X-ray radiation,[68] and recently using a laboratory X-ray source.[137]

11.5.2.1 Foams created by internal gas evolution

Foamable precursors of different kinds were foamed and observed by *in-situ* X-ray radioscopy. Of the many results, some connected to metal foam stability are reviewed here. Foam evolution of both hot-pressed and thixo-cast foamable precursors of AlSi6Cu4 alloy containing TiH_2 as a blowing agent was investigated under a normal atmosphere.[138,139] Applying an appropriate heating profile, stable foams were found even after 10 min of isothermal holding in both the semi-solid and liquid state (below and above liquidus temperature, respectively). The difference between hot-pressed and thixo-cast precursors is shown in Figure 11.21, which shows samples near the maximum of expansion. Significant drainage was observed in the foam made from the thixo-cast sample while the foam made by foaming the pressed powder sample does not show any drainage. An explanation can be that the network of oxides is destroyed during thixo-casting, *e.g.* by agglomeration of the oxides to larger structures which are not effective in stabilisation. Such effects could be provoked by stirring molten powder compacts.[50] If this happens, the delicate gel-like structure of inter-locked filaments is damaged leading to increased drainage.

Cell rupture is often observed when foamable precursors expand. Synchrotron radioscopy revealed that the time for liquid films to disappear lies below 55 ms.[31] Recent investigations on Formgrip foams by X-ray radioscopy show the effect of

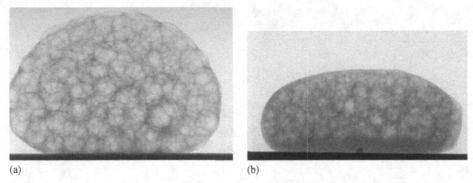

(a)   (b)

Figure 11.21 X-ray radioscopic pictures of a hot-pressed precursor foam. (a) $T = 600°C$ (below liquidus), thixo-cast precursor foam with significant drainage. (b) $T = 625°C$ (above liquidus). Width of samples is 20 mm. Taken from Ref. [139]; with permission of Trans Tech Publications.

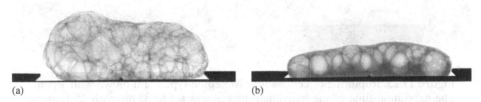

(a)   (b)

Figure 11.22 Real-time X-ray radioscopy pictures obtained using a synchrotron beam of a liquid Al foam produced from Formgrip precursors containing AlSi9Cu3 + 10 vol.% SiC and 0.5 wt.% TiH_2 as blowing agent stabilised by (a) 13 µm and (b) 70 µm SiC particles. Taken from Ref. [140]; with permission of the Institute of Materials and Machine Mechanics.

particle size on stability very clearly. Precursors containing three types of SiC particle were foamed under equal conditions and observed *in-situ*. It was found that bubble coalescence was much more pronounced for the precursor containing the coarse, *i.e.* 70 µm, particles than for the finer particles, *i.e.* 3 and 13 µm. Figure 11.22 shows two foams at a late stage of foaming. Obviously, the foam stabilised with fine particles is still stable while the one containing coarse particles has already collapsed.

11.5.2.2 Foams created by external gas injection

A recent investigation[141] with high-speed microfocus X-ray radioscopy showed that, during foaming of Metcomb-type foams with argon blowing, the rupture time of the cells is below 40 ms (Figure 11.23). Drainage is a common and well-investigated

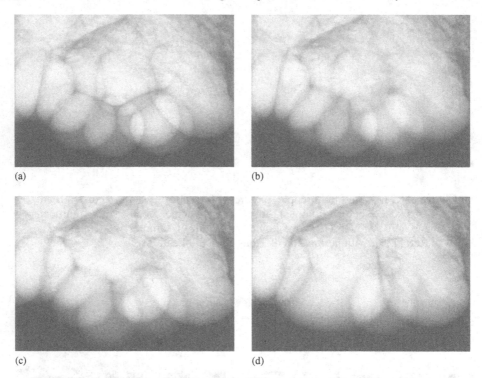

(a)   (b)

(c)   (d)

Figure 11.23 Rupturing of cell walls in Metcomb-type foam blown with argon. The exposition time of the individual images was set to 33 ms with 25 frames being recorded per second. Width of each image is 14 mm. F3S20S Duralcan composite was used. Taken from Ref. [141]; with permission of MetFoam 2005.

phenomenon in transparent aqueous foams. In order to characterise drainage in aluminium foams, real-time X-ray radioscopic experiments were carried out on liquid aluminium foam stabilised with SiC particles and blown with argon and air gases. The exposition time of each image was 110 ms. Particle-stabilised aluminium foams were produced from F3S20S Duralcan MMC precursor at 700°C. The foams were generated using Metcomb technology.[22] Bubbles of approximately 7 mm in diameter were created on the bottom of the melt, controlling the flow rate and the gas pressure with the foam generator of HKB, Austria. Approximately 42 and 25 cm³ of foam were produced applying 4.2 s (argon) and 2.5 s (air) gas pulses, respectively.[142]

The liquid metal foam instantaneously forms a polyhedral structure. In spite of the significant flow of bubbles during foam formation, no foam rupture was detected at this stage of the process in the air-blown foams. Characteristic images of foam evolution, made by an X-ray scanner, are shown in Figure 11.24 and Figure 11.25. The two figures differ by the blowing gas used, corresponding to argon

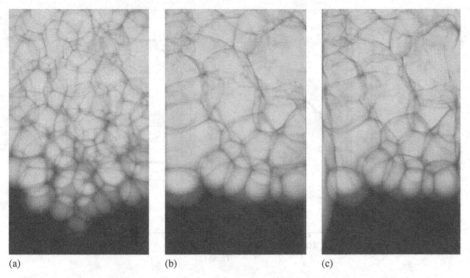

(a)   (b)   (c)

Figure 11.24 X-ray images showing the evolution of argon-blown foam (a) just after foam formation (liquid), (b) at the end of isothermal holding (5 min, liquid), (c) after solidification (solid). Sample widths are 40 mm at the bottom. Taken from Ref. [142]; with permission of Trans Tech Publications.

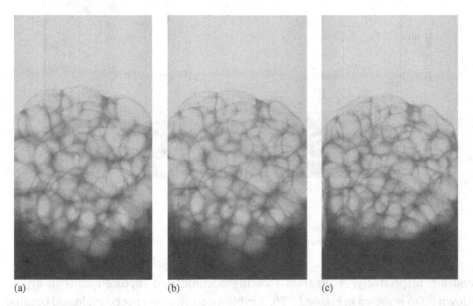

(a)   (b)   (c)

Figure 11.25 X-ray images showing the evolution of air-blown foam (a) just after foam formation (liquid), (b) at the end of isothermal holding (5 min, liquid), (c) after solidification (solid). Sample widths are 40 mm at the bottom. Taken from Ref. [142]; with permission of Trans Tech Publications.

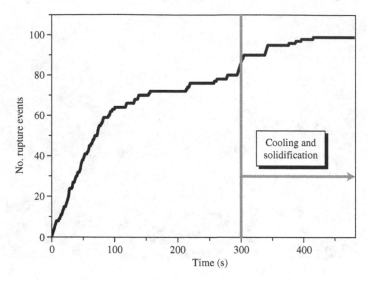

Figure 11.26 Accumulated rupture events in argon-blown foam. Taken from Ref. [142]; with permission of Trans Tech Publications.

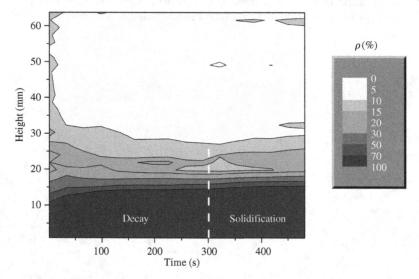

Figure 11.27 Drainage curves of argon-blown foam with different volume fractions of liquid being shown. Taken from Ref. [142]; with permission of Trans Tech Publications.

and air, respectively. A diagram showing accumulated rupture events in argon-blown foam is shown in Figure 11.26. Air-blown foam could not be evaluated because of the lack of rupture events. A diagram showing the vertical density profile of the foam as a function of height, h (drainage diagram) for argon-blown foam is shown in Figure 11.27. In these figures ρ represents the density calculated from integrals over horizontal lines of image intensities.

The uniformity of bubble size is obvious from the first images in both Figure 11.24 and Figure 11.25 reflecting the situation just after foam creation. Isothermal holding leads to foam coarsening and a slight degradation of uniformity in argon-blown foams while air-blown foams remain almost unchanged. Air-blown foams are therefore stable against rupture. It is known from previous *ex-situ* investigations[46] that a thick oxide skin develops on the inner surfaces of the cells whenever the foaming gas contains oxygen. This oxide skin improves the stability of foams. During solidification, foams shrank significantly in both cases.

Argon-blown foams show some rupture and drainage during isothermal holding. According to Figure 11.26 rupture is notable during the first 100 s and occurs at a constant rate. It then levels off and is nearly constant for another 200 s. Cooling triggers some film rupture events which were already observed for Alulight-type foams previously.[31] The drainage diagram in Figure 11.27 shows significant drainage only in the first 20 s. In air-blown foam, drainage was hardly detectable. This difference can be explained using experimental results on single Plateau borders of Koehler *et al.*[143] in aqueous foams. There it was shown that the mobility of film surfaces determines the drainage rate. Immobile surfaces slow down drainage. Using the analogy with aqueous foams, the oxide can be considered as an immobiliser of surfaces, thus leading to less drainage.

11.5.2.3 Synchrotron tomography

From the metallographic investigation of various metallic foams and an associated analysis of particle distributions in the cell walls, the mechanism of stabilisation cannot be described with certainty. Depending on the type of foam or the location of the images, the picture varies. In some images the surface coverage of the metal films with particles is very clear (Figure 11.11(a)), while on other images portions of the surface devoid of particles can be detected (Figure 11.15(b)). It is not evident from the images how mechanical forces can be transmitted from one side of a film to the other *via* particle networks as postulated, *e.g.* by Kaptay,[72] which would be required to create a disjoining force. There are two explanations for these difficulties. Firstly, the 2-D character of metallographic analysis could obscure the structure of the postulated 3-D networks. Secondly, it is not guaranteed that the arrangement of particles in solid foams is identical to that in liquid foams. During cooling, the interaction of the solidification front with the particles, either engulfing or rejecting them, could change their arrangement to such an extent that the postulated networks cannot be detected in solid foams.

One way to obtain 3-D information is using synchrotron tomography. Solid metal foams of the Formgrip-type (see Section 11.2.1), based on an AlSi10Mg matrix and stabilised with SiC particles of different sizes, were investigated by high-resolution tomography with monochromatic radiation allowing for a separation of the two components.[144] The images obtained showed that the individual particles are indeed

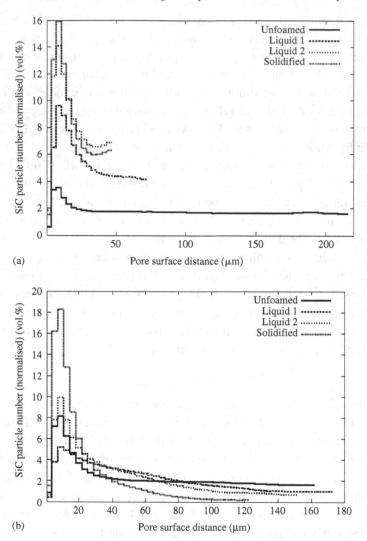

Figure 11.28 Correlation between individual SiC particles and pores in liquid Al foam as obtained by synchrotron tomography. Foams created by melting Formgrip precursor containing SiC particles of diameter (a) 70 and (b) 13 μm. For each material four foaming stages are given: unfoamed solid precursor material, early stage of foaming (liquid 1), late stage of foaming (liquid 2) and the solidified foam. The particle density given is a measure of the likelihood of finding a particle at a certain distance to a pore. Taken from Ref. [145]; with permission of the Hahn–Meitner–Institute, Berlin.

not all connected with each other and do not form networks which would allow for a transmission of forces across the cell walls. Therefore, the evidence from the metallographic studies does not seem to be an artefact of 2-D imaging. In order to assess possible effects during solidification, *in-situ* tomography was carried out on liquid metal foams of the same composition.[145] The foams were created in a small glass cylinder which gave them some mechanical support at the end of the foaming process and prevented the foams from moving during the 20 min of exposure. After acquiring a tomogram of the precursor prior to foaming, two tomograms were obtained in the liquid state – one soon after maximum expansion, the other after a holding time – after which the foams were solidified. Then one more tomogram was taken. The data obtained was analysed by 3-D image analysis yielding the correlation between pores and SiC particles as shown in Figure 11.28. These correlation curves yield a measure of the likelihood of finding particles at a given distance to the pores. Figure 11.28(a) shows data corresponding to foams containing very coarse SiC powder which does not lead to stable foams. Quite clearly in the unfoamed state (which contains only few small pores) the correlation between bubbles and particles is weak, as the SiC density is almost constant. In the liquid state a pronounced correlation exists which becomes stronger with time. This correlation is expressed by a peak for small distances to a pore and a lower density for larger distances. Solidification, however, does not change the picture very much. In foams stabilised with smaller SiC particles (Figure 11.28(b)), the picture looks different. From the unfoamed to the liquid foamed state the correlation increases to some extent. Solidification, however, leads to an increase in correlation. Apparently the particle arrangement is changed by the advancing solidification front and particles are pushed to the surface. This could be an indication that networks of solid particles might exist in the liquid but cannot be detected after solidification. However, a direct proof for this is still lacking.

11.6 Conclusions

The experimental work found in the literature demonstrates that probably all metal foams contain at least one solid phase which is responsible for stabilisation. The liquid in metal foams therefore forms a high-temperature suspension or colloid. The origin, nature and spatial arrangement of the solid component in the fluid vary in different metal foam types. The way solid particles stabilise foams is only partially understood. Coming from aqueous or glass foams, explanations in terms of viscosity, surface tension/elasticity and surface activity have been given. It is tempting to look for a unifying scheme according to which particles act and there have been claims that such a universal mechanism exists.[50] The present authors are sceptical that this is true and suspect that various distinct mechanisms apply. Foams of the Metcomb- or Formgrip-type are stabilised by individual, fairly large particles which can move around in the liquid. In solidified foams they seem to segregate preferentially at the liquid–gas surfaces of films with only a minor fraction of particles being in the interior of the films (Figure 11.29(a)).

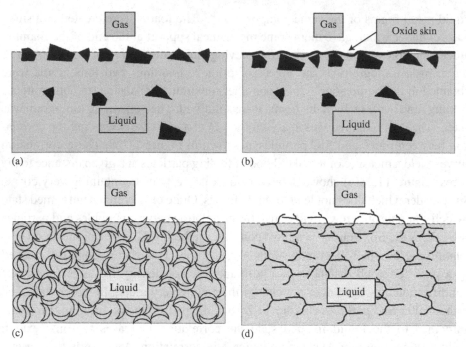

Figure 11.29 Some of the particle–liquid configurations found in real metal foams (one liquid–gas surface is shown). (a) Partially segregated particles on Metcomb-type foams blown with argon, (b) the same system blown with oxidising blowing gas, (c) interlocked oxide filaments in Alulight-type foams, and (d) partially connected bi-films in Alporas-type foam.

It is timely to give an interpretation in terms of models such as the LP2+C model by Kaptay[72] (see Figure 11.7(c)). Accordingly, the particles are pinned to each of the interfaces of a film by means of interfacial forces. This would explain micrographs such as Figure 11.11(a). However, it is not clear how the attractive interactions between the two interfaces are balanced. Kaptay postulates that mechanical forces are responsible for this counter pressure, the disjoining pressure, which in his model are transferred through locally densely packed layers of particles. Experimental evidence, however, does not support the existence of such bridges between the two surface layers. The particles which are completely immersed in the liquid phase are not interconnected as shown clearly by high-resolution tomography.[144] However, there is some evidence that surface segregation of particles increases during solidification (Figure 11.28) so that the particle content in the film could be higher in the liquid state than actually observed in the solid. This could be a loophole for saving the model since the existence of bridges connecting the two surface particle layers cannot be completely excluded. Another experimental fact is that there is no meniscus around individual particles sitting at each interface (see

Figure 11.12) as one would expect from the picture of partially wetted particles. Instead, the interface is rather smooth around particles. Furthermore, the oxide skin created during foaming with air plays an important role in stabilisation (Figure 11.29(b)). Such skins are not included in the current models of foam stability. In conclusion, we believe that the stabilisation mechanism of Metcomb- or Formgrip-type foams is still waiting for a full explanation.

Foams of the Alulight-type show a completely different stabilisation pattern. The existence of oxide filaments within aluminium powder particles, within pressed powder mixtures and within foams has been shown by various authors either by direct observation[50,69] or indirectly by proving that low oxide systems have a lower foam stability.[50,83] Wübben[83] could even show that the oxide content is proportional to the inner surface area of foams. These filaments do not segregate at the surface but rather span the entire volume of a film or Plateau border. The system of filaments is thought to hold the liquid aluminium alloy by capillary action or by merely locally enclosing small volumes of melt and preventing them to flow into the Plateau borders. The individual filaments seem to be interlocked so that they could act like a rigid sponge holding the liquid (Figure 11.29(c)). The system is quite fragile and seems to resemble a gel in terms of rheology. Poking into an aluminium foam with a steel needle can result in an outflow of melt. Körner *et al.*[50] claim that individual oxide filaments form larger networks which form a kind of meta-particles with a huge internal porosity. They assume that these meta-particles stabilise the cell walls by forming an LP1 layer in Kaptay's notation (see Figure 11.7(a)). The idea sounds convincing but is based on very few and mostly indirect observations, *e.g.* the observed waviness of the cell walls. The traditional explanation that the foam is stabilised by the largely increased apparent viscosity of the liquid films by simply preventing fluid flow seems also possible with the current state of knowledge. To some extent, the two viewpoints might not differ too much and the distinction between dynamic (viscosity-based) and static (surface activity-based) action might be just semantics in these gel-like fluids.

In addition to these foams, Alporas-type foams exist in which oxide bi-films are created *in-situ* by reaction of the Al alloy with Ca or Mg in the presence of air. These films could act in a similar way to the oxide filaments in Alulight-type foams. The oxide bi-films are possibly less connected so that these foams are more fluid and can reach lower relative densities after full expansion (see Figure 11.29(d)).

Another point worth discussing is the possibility that a solid phase formed from the liquid during cooling into the semi-solid range of the phase diagram could stabilise foams. It was suspected that primary aluminium grains in foaming aluminium–silicon melts could enhance stability in addition to the oxide filament stabilisation mechanism.[68] As foams were observed also to be stable, although less, in the fully melted state this mechanism could not be responsible for stability alone and the literature is not specific on a mechanism of stabilisation (viscosity enhancement or surface activity). Körner *et al.*[29] later baptised this effect endogeneous stabilisation and made it responsible for the stability of Mg foams created in DCP foams. They further suggest that the primary Mg particles are surface active and form layers analogous to the ceramic particles used for Metcomb-type foams. Although this idea is convincing there is no proof for it at present. Foams made by DCP were obtained by injecting MgH_2 into pure Mg or Al.[29,30] As such, processing can hardly be carried out in an oxygen-free environment. Spontaneous oxidation of the melt might also have added some oxides which then could have stabilised the foam. It remains to prove that alloys in the semi-solid state can be foamed in the absence of any reactive gas.

Finally, the foams made from extremely viscous metallic glasses present a challenge. Some of these foams show polyhedral cells which are obviously stabilised by interfacial forces, probably by remnants of the blowing agent B_2O_3 used.[18] Others, especially the foams in the early stages which are more like bubble dispersions, show bubbles which have merged to one volume while still maintaining their original shape indicating the absence of interfacial stabilisation (see Figure 11.2(h)). This indicates the absence of stabilisation and the dominance of viscosity in foam formation. In conclusion, there are many open questions associated with metallic foams. More experimental work is required to clarify the role of solid particles in liquid metal colloid systems. Naturally, the main effort will be directed towards improving the materials. The aims are to produce aluminium foams with better stability, more uniform cell structures and *via* a cheaper manufacture. Improved liquid metal colloids could open the road to such foams.

Acknowledgements

Various people have supported us by providing samples of their foams or raw materials for foam making which we could use for further analysis. These are Vlado Gergely of Cambridge University, Bo-Young Hur of Chinju, Korea, Dietmar Leitlmeier of HKB, Austria and Tetsuo Miyoshi, Japan. Astrid Haibel and Francisco García-Moreno have provided some graphs from their papers. Support by the European Space Agency (ESA) *via* an International Trainee Grant for one of us (NB) is gratefully acknowledged.

References

1. L. Gibson and M.F. Ashby, *Cellular Solids*, Cambridge University Press, Cambridge, 1997.
2. B. Sosnick, *US Patent*, 2 434 775 (1948).
3. J.C. Elliot, *US Patent*, 2 751 289 (1956).

4. J. Banhart and D. Weaire, *Phys. Today*, **55** (2002), 37.
5. S. Akiyama, K. Imagawa, A. Kitahara, S. Nagata, K. Morimoto, T. Nishikawa and M. Itoh, *European Patent Application*, 0 210 803 A1 (1986).
6. W.W. Ruch and B. Kirkevag, *WO Patent*, 9 101 387 (1991).
7. I. Jin, L.D. Kenny and H. Sang, *US Patent*, 5 112 697 (1992).
8. B.C. Allen, M.W. Mote and A.M. Sabroff, *US Patent*, 3 087 807 (1963).
9. J. Baumeister, *German Patent*, 4 018 360 (1991).
10. J. Banhart, *Progr. Mat. Sci.*, **46** (2001), 559.
11. H.P. Degischer and B. Kriszt, *Handbook of Cellular Metals*, Wiley-VCH, New York, 2002.
12. http://www.metalfoam.net, 2006.
13. J. Banhart, *Int. J. Vehicle Des.*, **37** (2005), 114.
14. J. Banhart, *J. Metals*, **52** (2000), 22.
15. F. Baumgärtner, I. Duarte and J. Banhart, *Adv. Eng. Mater.*, **2** (2000), 168.
16. J. Pötschke, T. Hoster and P. Pant, *Report to German Ministry of Science and Technology (BMFT)*, 1984.
17. M. Niemeyer, H. Haferkamp and D. Bormann, *Mat.-Wiss. u. Werkstofftech.*, **31** (2000), 419.
18. J. Schroers, C. Veazey, M.D. Demetriou and W.L. Johnson, *J. Appl. Phys.*, **96** (2004), 7723.
19. V. Shapovalov and L. Boyko, *Adv. Eng. Mater.*, **6** (2004), 407.
20. S.K. Hyun and H. Nakajima, *Adv. Eng. Mater.*, **4** (2002), 741.
21. M.W. Kearns, P.A. Blenkinshop, A.C. Barber and T.W. Farthing, *Int. J. Powder Met.*, **24** (1988), 59.
22. D. Leitlmeier, H.P. Degischer and H.J. Flankl, *Adv. Eng. Mater.*, **4** (2002), 735.
23. http://www.alulight.com, 2006.
24. V. Gergely and T.W. Clyne, *Adv. Eng. Mater.*, **2** (2000), 175.
25. J. Banhart, J. Baumeister, A. Melzer and M. Weber, *German Patent*, DE 198 13 176 (1998).
26. W.S. Fiedler, *US Patent*, 3 214 265 (1965).
27. C.B. Berry Jr., *US Patent*, 3 669 654 (1972).
28. K. Renger and H. Kaufmann, *Adv. Eng. Mater.*, **7** (2005), 117.
29. C. Körner, M. Hirschmann, V. Bräutigam and R.F. Singer, *Adv. Eng. Mater.*, **6** (2004), 385.
30. W. Knott, B. Biedermann, M. Recksik and A. Weier, *German Patent Application*, DE 101 27 716 (2001).
31. H. Stanzick, M. Wichmann, J. Weise, J. Banhart, L. Helfen and T. Baumbach, *Adv. Eng. Mater.*, **4** (2000), 814.
32. B. Matijasevic-Lux, J. Banhart, S. Fiechter, O. Görke and N. Wanderka, *Acta Materialia*, **54**, (2006), 495.
33. European Space Agency, Erasmus Experiment Archive, http://spaceflight.esa.int/eea, 1982.
34. T. Wübben, H. Stanzick, J. Banhart and S. Odenbach, *J. Phys. Condens. Matter*, **15** (2003), 427.
35. J. Banhart, in *Foams and Films*, eds. D. Weaire and J. Banhart, MIT-Verlag, Bremen, 1999, p. 73.
36. D. Weaire and S. Hutzler, *The Physics of Foams*, Oxford University Press, Oxford, 1999.
37. J.J. Bikerman, *Foams*, Springer-Verlag, New York, 1973.
38. D. Exerowa and P.M. Kruglyakov, *Foam and Foam Films*, Elsevier, Amsterdam, 1998.

39. P.R. Garrett, *Defoaming: Theory and Industrial Application*, Marcel Dekker, New York, 1993.
40. A.J. Wilson, *Foams: Physics, Chemistry and Structure*, Springer Series in Applied Biology, New York, Vol. 1, 1989.
41. E. Dickinson and G. Stainsby, *Advances in Food Emulsions and Foams*, Elsevier Applied Science, Amsterdam, 1988.
42. K.L. Mittal and P. Kumar, *Emulsions, Foams and Thin Films*, Marcel Dekker, New York, 2000.
43. I.D. Morrison and S. Ross, *Colloidal Dispersions: Suspensions, Emulsions, and Foams*, Springer, New York, 1989.
44. L.L. Schramm, *Emulsions, Foams, and Suspensions-Fundamentals and Applications*, VCH, Weinheim, 2005.
45. N. Babcsán, D. Leitlmeier and H.P. Degischer, *Mat.-Wiss. u. Werkstofftech.*, **34** (2003), 22.
46. N. Babcsán, D. Leitlmeier, H.-P. Degischer and J. Banhart, *Adv. Eng. Mater.*, **6** (2004), 421.
47. N. Babcsán, D. Leitlmeier and J. Banhart, *Colloids Surf. A*, **261** (2005), 123.
48. S.W. Ip, Y. Wang and J.M. Toguri, *Can. Met. Quart.*, **38** (1999), 81.
49. V. Gergely and T.W. Clyne, *Acta Mater.*, **52** (2004), 3047.
50. C. Körner, M. Arnold and R.F. Singer, *Mat. Sci. Eng. A*, **396** (2005), 28.
51. D. Kenny, Alcan, Canada, Private communication, 2004.
52. N. Babcsán, *Ph.D. Thesis*, University of Miskolc, Hungary, 2003.
53. A.I. Rusanov and V.V. Krotov, *Prog. Surf. Membrane Sci.*, **13** (1979), 415.
54. A. Prins and M. van den Tempel, *J. Phys. Chem.*, **73** (1969), 2828.
55. P.M. Claesson and M.W. Rutland, in *Handbook of Applied Surface and Colloid Chemistry*, ed. K. Holmberg, Wiley, New York, 2001, p. 933.
56. P.M. Claesson and M.W. Rutland, in *Handbook of Applied Surface and Colloid Chemistry*, ed. K. Holmberg, Wiley, New York, 2001, pp. 934–964.
57. B.P. Binks, *Curr. Opin. Colloid Interf. Sci.*, **7** (2002), 21.
58. A.D. Nikolov and D.T. Wasan, *Langmuir*, **8** (1992), 2985.
59. J.C. Wilson, *Ph.D. Thesis*, University of Bristol, UK, 1980.
60. G. Johansson and R.J. Pugh, *Int. J. Min. Proc.*, **34** (1992), 1.
61. Y.Q. Sun and T. Gao, *Metall. Trans.*, **33A** (2002), 3285.
62. G.N. Sethumadhavan, A.D. Nikolov and D.T. Wasan, *J. Colloid Interf. Sci.*, **240** (2001), 105.
63. T. Okubo, *J. Colloid Interf. Sci.*, **171** (1995), 55.
64. B.P. Binks and T.S. Horozov, *Angew. Chem. Int. Ed.*, **44** (2005), 3722.
65. C. Nexhip, S. Sun and S. Jahanshahi, *Metall. Mater. Trans. B.*, **31B** (2000),1105.
66. V. Gergely, L. Jones and T.W. Clyne, *Trans. JWRI*, **30** (2001), 371.
67. H. Kumagai, Y. Torikata, H. Yoshimura, M. Kato and Y. Yano, *Agric. Biol. Chem.*, **55** (1991), 1823.
68. J. Banhart, H. Stanzick, L. Helfen and T. Baumbach, *Appl. Phys. Lett.*, **78** (2001), 1152.
69. P. Weigand, *Untersuchung der Einflußfaktoren auf die pulvermetallurgische Herstellung von Aluminiumschäumen*, MIT-Verlag, Bremen, 1999.
70. C. Körner, M. Thies and R.F. Singer, *Adv. Eng. Mater.*, **4** (2002), 765.
71. G. Kaptay, in *Cellular Metals and Metal Foaming Technology*, eds. J. Banhart, M.F. Ashby and N.A. Fleck, MIT-Verlag, Bremen, 2001, p. 117.
72. G. Kaptay, *Colloids Surf. A*, **230** (2004), 67.
73. G. Kaptay, *Mater. Sci. Forum*, **77** (1991), 315.

74. A. Pamies, C. Garcia Cordovilla and E. Louis, *Scripta Metall.*, **18** (1984), 869.
75. R.A. Saravanan, J.M. Molina, J. Narciso, C. Garcia-Cordovilla and E. Louis, *Scripta Mater.*, **44** (2001), 965.
76. R.A. Saravanan, J.M. Molina, J. Narciso, C. Garcia-Cordovilla and E. Louis, *J. Mater. Sci. Lett.*, **21** (2002), 309.
77. G. Kaptay, E. Bader and L. Bolyan, *Mater. Sci. Forum*, **329–330** (2000), 151.
78. C.Garcia-Cordovilla, E. Louis and A. Pamies, *J. Mater. Sci.*, **21** (1986), 2787.
79. E. Ricci, A. Passerone, P. Castello and P. Costa, *J. Mater. Sci.*, **29** (1994), 1833.
80. E.A. Howell, C.M. Megaridis and M. McNallan, *Int. J. Heat Fluid Flow*, **25** (2004), 91.
81. G. Kaptay, *Z. Metallkd.*, **96** (2005), 24.
82. D. Brabazon, D.J. Brown and A.J. Carr, *Mater. Sci. Eng.*, **A356** (2003), 69.
83. T. Wübben, *Ph.D. Thesis*, University of Bremen, Germany, 2003.
84. L. Ma and Z. Song, *Scripta Mater.*, **39** (1998), 1523.
85. T. Iida and R.I.L. Guthrie, *The Physical Properties of Liquid Metals*, Clarendon Press, Oxford, 1988.
86. D. Klempner and K.C. Frisch, *Handbook of Polymeric Foams and Foam Technology*, Hanser Publishers, Munich, 1991.
87. S.M. Jung and R.J. Fruehan, *ISIJ Int.*, **40** (2000), 348.
88. D.S. Viswanath and G. Natarajan, *Data Book on The Viscosity of Liquids*, Hemisphere Pub. Corp., New York, 1989.
89. H.K. Moon, J.K. Cornie and M.C. Flemings, *Mater. Sci. Eng.*, **A144** (1991), 253.
90. H.P. Degischer, in *Encyclopedia of Materials: Science and Technology*, eds. K.H.J. Buschow, R.W. Cahn, M.C. Flemings, B. Ilschner, E.J. Kraner and S. Mahajan, Elsevier, Amsterdam, 2001, pp. 5452–5455.
91. J. Fjellstedt, A.E.W. Jarfors and L. Svendsen, *J. Alloy Comp.*, **283** (1999), 192.
92. K.B. Lee, J.P. Ahn and H. Kwon, *Metall. Trans. A*, **32A** (2001), 1007.
93. K.B. Lee and H. Kwon, *Metall. Trans. A*, **30A** (1999), 2999.
94. S. Akiyama, H. Ueno, K. Imagawa, A. Kitahara, K. Morimoto, T. Nishikawa and M. Itoh, *US Patent*, 4 713 277 (1987).
95. V. Gergely, H.P. Degischer and T.W. Clyne, *Compr. Compos. Mater.*, **3** (2000), 797.
96. M. Thomas, D. Kenny and H. Sang, *US Patent*, 5 622 542 (1997).
97. E.Y. Gutmanas and I. Gotman, *J. Eur. Ceramic Soc.*, **19** (1999), 2381.
98. T.P.D. Rajan, R.M. Pillai and B.C. Pai, *J. Mater. Sci.*, **33** (1998), 3491.
99. S. Vaucher and O. Beffort, *MMC-Assess Thematic Network*, EMPA Thun, **9** (2001); http://mmc-assess.tuwien.ac.at
100. V.M. Bermudez, *Appl. Phys. Lett.*, **42** (1983), 70.
101. D.J. Lloyd, *Compos. Sci. Tech.*, **35** (1989), 159.
102. C. Cayron, *EMPA Report No. 250*, Thun (2001).
103. J.C. Viala, P. Foretier and J. Bouix, *J. Mater. Sci.*, **25** (1990), 1842.
104. A. Bardal, *Mater. Sci. Eng.*, **A159** (1992), 119.
105. I. Hansson and D.J. Lloyd, *World Patent*, WO9317139 (1993).
106. D.J. Lloyd, I. Jin, A.D. McLeod and C.M. Gabryel, *World Patent*, WO9308311 (1993).
107. D.M. Schuster and M.D. Skibo, *J. Met.*, **45** (1993), 26.
108. M.S. Yaghmaee and G. Kaptay, *Mater. Sci. Forum*, **473–474** (2005), 415.
109. A.D. McLeod, and C.M. Gabryel, *Metall. Trans.*, **23A** (1992), 1279.
110. T. Onda, S. Shibuichi, N. Satoh and K. Tsujii, *Langmuir*, **12** (1996), 2125.
111. D.S. Han, H. Jones and H.G. Atkinson, *J. Mater. Sci.*, **28** (1993), 2654.
112. V. Laurent, D. Chatain, D. Chatillon and N. Eustathopoulos, *Acta Metall.*, **36** (1988), 1797.

113. V. Laurent, C. Rado and N. Eustathopoulos, *Mater. Sci. Eng.*, **A205** (1996), 1.
114. S. Suresh, A. Mortensen and A. Needlman, *Fundamentals of Metal-Matrix Composites*, Butterworth-Heinemann, Stoneham, 1993, p. 51.
115. P. Shen, H. Fujii, T. Matsumoto and K. Nogi, *Scripta Mater.*, **48** (2003), 779.
116. A.R. Kennedy, in *Proc. Int. Conf. High Temperature Capillarity*, eds. N. Eustathopoulos and N. Sobczak, Cracow, 1997, p. 395.
117. L.M. Niebylski, C.P. Jarema and T.E. Lee, *US Patent*, 3 816 952 (1974).
118. J. Campbell, *Castings: The New Metallurgy of Cast Metals*, 2nd edn., Butterworth-Heinemann, Oxford, 2003.
119. Supplied by Dr. Miyoshi, Shinko Wire Co., Japan.
120. Supplied by Prof. Hur, Gyeong Shang National University, South Korea.
121. G.M. Scamans and E.P. Butler, *Metall. Trans. A.*, **6A** (1975), 2055.
122. C.C. Yang and H. Nakae, *J. Mater. Process. Technol.*, **141** (2003), 202.
123. A.R. Kennedy and S. Asavavisitchai, in *Cellular Metals and Metal Foaming Technology*, eds. J. Banhart, N.A. Fleck and A. Mortensen, MIT-Verlag, Berlin, 2003, p. 147.
124. N. Babcsán and D. Leitlmeier, Unpublished work.
125. J.T. Wood, in *Metal Foams*, eds. J. Banhart and H. Eifert, MIT-Verlag, Bremen, 1998, pp. 31–35.
126. W. Dequing and S. Ziyuan, *Mater. Sci. Eng.*, **A361** (2003), 45.
127. B. Kriszt, P. Cekan and K. Faure, in *Cellular Metals and Metal Foaming Technology*, eds. J. Banhart, M.F. Ashby and N.A. Fleck, MIT-Verlag, Bremen, 2001, pp. 77–82.
128. O. Brunke, S. Odenbach and F. Beckmann, *Eur. Phys. J. Appl. Phys.*, **29** (2005), 73.
129. C. Körner, F. Berger, M. Arnold, C. Stadelmann and R.F. Singer, *Mat. Sci. Tech.*, **16** (2000), 781.
130. B. Olurin, M. Arnold, C. Körner and R.F. Singer, *Mat. Sci. Eng.*, **A328** (2002), 334.
131. A.E. Markaki and T.W. Clyne, *Mat. Sci. Tech.*, **16** (2000), 785.
132. J. Banhart, J. Baumeister and M. Weber, *Proc. Eur. Conf. on Advanced PM Materials*, 1995, pp. 201–208.
133. D. Lehmhus and G. Rausch, *Adv. Eng. Mater.*, **6** (2004), 313.
134. F. von Zeppelin, M. Hirscher, H. Stanzick and J. Banhart, *Compos. Sci. Technol.*, **63** (2003), 2293.
135. U. Laun and J. Banhart, in *Cellular Metals and Metal Foaming Technology*, eds. J. Banhart, M.F. Ashby and N.A. Fleck, MIT-Verlag, Bremen, 2001, p. 255.
136. I. Jin, Private communication, 2004.
137. F. Garcia-Moreno, M. Fromme and J. Banhart, *Adv. Eng. Mater.*, **6** (2004), 416.
138. F. García-Moreno, N. Babcsán and J. Banhart, *Colloids Surf. A*, **263** (2005), 290.
139. F. Garcia-Moreno, N. Babcsán, J. Banhart, M. Haesche and J. Weise, *Cellular Metals and Polymers*, eds. R.F. Singer, C. Körner and V. Altstädt, Trans Tech Publications, Uetikon-Zuerich, 2005, p. 31.
140. N. Babcsán, J. Banhart and D. Leitlmeier, in *Advanced Metallic Materials*, eds. J. Jerz, P. Sebo and M. Zemankova, Institute of Materials & Machine Mechanics, Bratislava, 2004, pp. 5–15.
141. N. Babcsán, F. García-Moreno and J. Banhart, presented at *MetFoam 2005*, Kyoto, September 2005.
142. N. Babcsán, F. García-Moreno, D. Leitlmeier and J. Banhart, *Mater. Sci. Forum*, **508** (2006), 275.
143. S.A. Koehler, S. Hilgenfeldt, E.R. Weeks and H.A. Stone, *Phys. Rev. E*, **66** (2002), 040601.

144. A. Haibel, Unpublished work.
145. A. Haibel, A. Bütow, A. Rack and J. Banhart, presented at *16th World Conference on Non-Destructive Testing*, Montreal, September 2004.

Norbert Babcsán (left) obtained his diploma in 1996 at the University of Miskolc, Hungary in the field of Engineering Physics with work related to single crystal growth. He received his Ph.D. at the University of Miskolc in 2003 in the field of Materials Science and Technologies and spent 1 year at NASA Marshall Spaceflight Center. Between 1996 and 2001, he was an assistant lecturer at the Department of Non-metallic Materials at University of Miskolc, Hungary. He was scientific advisor for Leichtmetall-Kompetenzzentrum Ranshofen, Austria, where he carried out aluminium foam research between 2001 and 2002. He has been a post-doc. in the metal foam group at the Technical University of Berlin and Hahn-Meitner-Institute in Berlin since 2003. He is one of the founders of ADMATIS Ltd., a Hungarian Space R&D company.

John Banhart (right) is a Professor in the Institute of Materials Science and Technology at the Technical University of Berlin and head of the Department of Materials Science at the Hahn-Meitner-Institute in Berlin. Current working fields and research interests are light-weight materials including aluminium alloys, bulk metallic glasses, nanocrystalline alloy composites and metal foams. His department runs facilities for small-angle X-ray and neu-tron scattering and operates various methods for tomography. He is a physicist and earned his Ph.D. in physical chemistry at the University of Munich in 1989. After working in theor-etical alloy physics he changed to application oriented work at the Fraunhofer-Institute in Bremen where a process for foaming metals was developed in close cooperation with indus-try. He obtained his habilitation degree in 1998 in solid state physics at the University of Bremen.

Tel.: +49-30-80622710; *Fax*: +49-30-80623059; *E-mail*: banhart@hmi.de

Index